JOSEPH J. BARBORIAK, Sc.D.
Research Service (151B)
Veterans Administration Center
Wood, Wisconsin 53193

NUTRITION AND DIET THERAPY
IN GASTROINTESTINAL DISEASE

TOPICS IN GASTROENTEROLOGY

Series Editor: **Howard M. Spiro, M.D.**
Yale University School of Medicine

PANCREATITIS
Peter A. Banks, M.D.

MEDICAL ASPECTS OF DIETARY FIBER
Edited by Gene A. Spiller and Ruth McPherson Kay

NUTRITION AND DIET THERAPY IN GASTROINTESTINAL DISEASE
Martin H. Floch, M.S., M.D., F.A.C.P.

A Continuation Order Plan is available for this series. A continuation order will bring delivery of each new volume immediately upon publication. Volumes are billed only upon actual shipment. For further information please contact the publisher.

NUTRITION AND DIET THERAPY
IN GASTROINTESTINAL DISEASE

Martin H. Floch, M.S., M.D., F.A.C.P.

Clinical Professor of Medicine
Yale University School of Medicine and
Chairman, Department of Medicine and
Chief, Section of Gastroenterology and Nutrition,
Norwalk Hospital

PLENUM MEDICAL BOOK COMPANY

NEW YORK AND LONDON

Library of Congress Cataloging in Publication Data

Floch, Martin H
  Nutrition and diet therapy in gastrointestinal disease.

  (Topics in gastroenterology)
  Includes index.
  1. Digestive organs—Diseases—Diet therapy. 2. Digestive organs—Diseases—
Nutritional aspects. 3. Nutrition. I. Title. II. Series.
RC802.F46   616.3'30654   80-20721
ISBN 0-306-40508-3

© 1981 Plenum Publishing Corporation
233 Spring Street, New York, N. Y. 10013

Plenum Medical Book Company is an imprint of Plenum Publishing Corporation

Printed in the United States of America

To my wife, Gladys,
and our children,
Jeffrey, Craig, Lisa, and Neil

Foreword

A physician with a broad consultative practice, Dr. Floch combines his clinical experience with a zeal for exploring what has been written by others. Chief of Medicine at the Norwalk Hospital for the past decade and still an active consulting gastroenterologist, Dr. Floch has given us a volume which every clinician dealing with digestive disorders will want to have at his or her desk.

Not everyone will agree with all that Dr. Floch has prescribed in the way of detailed dietary help for the common afflictions of mankind's gut, but in this book the reader can get at the background of the controversy. All clinicians have had problems in assessing when to use elemental diets, how to apply advances in peripheral and intravenous alimentation, and in many other matters which are discussed in detail in this fine volume. Dr. Floch displays what is available in dietary therapy, evaluates the nutritional inadequacies surrounding most digestive disturbances, and calmly evaluates competing claims. He gives a brief overview of gastrointestinal physiology pertaining to an understanding of nutritional complications as well as the genesis of the major gastrointestinal disorders. In this sense his book can be read as a mini-physiological text.

I am delighted to have this book in our gastrointestinal series and I hope that the reader will profit from it as much as I have. It should be of interest to medical students, internists, and family practitioners, as well as to nutritionists, dietitians, and nurses for whom the vivid summaries of medical knowledge should provide fine background information.

Howard M. Spiro, M.D.

Preface

My involvement in the development of the field of gastroenterology during the last 25 years has led me to the writing of this book. In the 1950s the invention of peroral small-bowel biopsy techniques plunged gastroenterologists into intensive study of digestion and absorption of nutrients as they relate to small-bowel structure and physiology. Diseases of malnutrition and malabsorption, e.g., tropical sprue and gluten enteropathy, were redefined.

In the 1960s the field turned to research on bile acids and the microflora. The pathophysiology of gallstones was clarified and a possible medical therapy developed. Quantitative and qualitative relationships of organisms within the gut were more clearly defined and information came forth to explain some important disease processes, e.g., hypersecretion in cholera.

In the 1970s the field of nutrition received a great thrust of interest with the realization that lipids have a role in arteriosclerotic disease, and that the so-called underdeveloped societies of Asia and Africa have little of the degenerative diseases plaguing Western societies. While nutritionists were gaining strength for their theories from these epidemiologic observations, most gastroenterologists were ecstatically developing their new diagnostic flexible fiberoptic endoscopes so that the depths of the gut could be probed from without. However, some continued to study nutrient metabolism as it related to the intestine, while others opened new vistas by discovering a whole host of gut proteins. These polypeptides were definitely classified as hormones and found to be produced and involved in intestinal physiology, and hence related to nutritional factors. It is at this point in history that the fields of gastroenterology and nutrition cross, and where I hope this book can help bring them together.

The first section discusses basic gastrointestinal physiology as it relates to nutrition. It includes chapters on eating; normal nutrient requirements; observations on relationships of fiber to luminal metabolism; and how weight loss occurs and is diagnosed.

The second section on pathophysiology relates disease processes to diet therapy, an aspect of medicine that is still in its infancy. Diet therapy should become the most important approach if the observations indicating that the gut is the source of early control of metabolic disease are correct. The advent of enteral and parenteral feedings changed the outlook in the early treatment of

anorexia nervosa. An understanding of the effects of foods on the lower esophageal sphincter gives us a new approach to the diet therapy of inflammatory
esophageal disease. The quandary that exists in dietary therapy of peptic disease of the stomach and duodenum persists, but a realization that frequent
feeding of foods is helpful is now accepted as clinical fact. The treatment of
chronic liver disease with specific amino acids (branched chain) holds promise,
and, certainly, the manipulations of fluid and salt intake in conjunction with
vitamin and mineral replacement can be effective in maintaining cirrhotic subjects. The vast amount of information concerning physiologic factors in small-
bowel disease is reviewed. The chapter on colon disease discusses the recent
advances in fiber and carbohydrate metabolism. The relationship of the irritable bowel syndrome and food allergies to the gut is reviewed in detail.

A disease long considered primarily a hormonal metabolic disturbance,
diabetes mellitus, now appears to be modified by a high-carbohydrate diet. This
revolution in diet therapy in diabetes epitomized the role of the gut in systemic
disease.

The final section of the book reviews methods of nutrition therapy including enteral formulas, total parenteral nutrition, and 32 diet plans. Review of the
literature and diet manuals was a challenging event. The author would like to
point out the many different diet plan forms included in the book. Those selected have all been proven therapeutically successful, either in the literature
or in the author's own experience. Tables are included that I hope will be
useful. Each diet is cross-referenced to the text and recommended for specific
disease entities.

I would like to stress that diet therapy never causes a dramatic, rapid
response and both the patient and the therapist must have patience with it.
When the gastrointestinal evaluation is correct, there must be a valid trial
period that should not be rushed. A true test of the diet takes at least a fortnight
to a month. Only after that period of time can dietary failure be accepted.
Finally, success is dependent on patient education, an area that has been
greatly neglected. Often the diet is handed to a patient and that is the end of the
role of the nutritionist, dietitian, or physician. If diet therapy is to succeed, it
must include careful patient education, a process that includes appropriate
instruction, reevaluation, and testing of patient compliance.

I have had a strong commitment to help popularize the field of nutrition as
it relates to the gastrointestinal tract and hope that this book helps to accomplish that goal.

Martin H. Floch

Norwalk, Connecticut

Acknowledgments

I would first like to thank my mentors. As a Master's student, I learned how to approach a research project under the tutelage of Dr. Wilbur Bullock and the late Dr. George Moore at the University of New Hampshire. It was there I learned the effectiveness of a single cell in the digestive and excretory process while studying the snail digestive gland. At New York Medical College, the late Dr. Lois Lillick taught me the role of microorganisms in the human body. I then learned the gross and functional anatomy of the GI tract from Dr. Abraham Geffen at Beth Israel Hospital. I shall be eternally grateful for the opportunity offered me by Drs. Milton Rubini and William Meroney to study the small bowel at the United States Army Tropical Research Laboratory in Puerto Rico. It was there I learned about tropical and small-bowel disease from many Puerto Rican physicians, and Dr. Thomas Sheehy to whom I am grateful for teaching me to enjoy thinking out a protocol and researching a clinical project. Later, Dr. Rubini again helped further my interests by offering me the opportunity to be one of the assistant editors of the American Journal of Clinical Nutrition.

At Yale, Dr. Howard Spiro has unselfishly given me (as well as hundreds of others) the opportunity to work, the encouragement to fulfill my ambition, and the guidance necessary to be successful.

Finally, when discussing mentors, I am presently, and hope to be in coming years, taught by the fellows in our training program. It is the youth of our training programs who continually teach us the most.

My next acknowledgment must go to the colleagues—members of the Norwalk Medical Group—who were instrumental in helping me develop the initial clinical experience in gastroenterology that I hope I have captured in parts of this book. At the Norwalk Hospital, my close associate, Dr. John Sacco, has been an invaluable confidant in both clinical and philosophic matters in education and the practice of medicine. My good friends in pathology, Dr. Roy Barnett and Dr. Irwin Weisbrot, have added ongoing information in their observations in the field of gastrointestinal pathology. Dr. Richard Sallick has been a constant friend and colleague in the field of education. I am also grateful for the support received from the Broad of Trustees at Norwalk Hospital, its president, Mr. Norman Brady, and the Department of Medicine at Yale and its chairman, Dr. Samuel Thier.

I shall always be deeply grateful to a loyal administrative associate, Mrs. Aurelia Gould, who helped formulate the Department of Medicine at the Norwalk Hospital, always loyal, creative, and an advisory confidante. It has been my good fortune to have several secretaries who assisted me during the creation of this book, most recently Mrs. Gail Krebs and Ms. Anne Dhunjishaw. The major burden of the typing of the manuscript has been borne by Miss Nancy Dunham and Mrs. Sandra Kaplan. To both I owe great thanks for their technical assistance. My research associate, David Vargo, has been of immeasurable assistance. Paul Ostreicher assisted in the early literature search, Neil Floch in creation of tables, and Gladys Floch in the arduous task of editing. I also had the good fortune to train many fellows during the writing of this book. Drs. William Pintauro, Marc Wolfman, and Richard Doyle, as fellows in GI, have had to bear with many frustrations and the inability to spend necessary time together due to the demands of this preparation. I am also grateful to the members of the Gastroenterology and Nutrition section who have been supportive of our general efforts at Norwalk Hospital, notably Dr. James Tracey and former fellows who stayed on, Drs. H. Martin Fuchs, Arthur D'Souza, and Paul Schwartz. Also, thanks to the nurses and assistants who helped run our section, where much of our clinical impressions are developed, Aileen True, Peggy Marron, and Sydney Stevens, and to the woman who helped build our educational program, Mrs. Margaret Covington.

I have been blessed with a fantastic wife and four great children who have always given me joy. In my early years, I was supported and encouraged by my late father, Samuel Floch, and to this day by a wonderful mother, Jean Floch; also supportive, a great sister and brother-in-law, Gloria and Bunny Lobel. It is somewhat unusual to acknowledge a mother-in-law, but I have been blessed with one who has always been supportive and inspirational by her enthusiasm, Mrs. Bess Wisser Eigen.

To all these people I give thanks and hope that this book will give them some pride for whatever contribution it makes to further science and good to humans.

MHF

Contents

Part III: Methods of Nutritional Treatment

1

Basic Gastrointestinal Physiology of Nutrition

The Physiology of Eating

Eating is a complex event that involves psychologic, sociologic, economic, environmental, cultural, and internal physiologic components. All of these influences are expressed in the final food that is chosen, handled, and eaten by the individual. The relationship of foods to the environment has been studied by cultural scientists, but these studies have not been applied to any great extent to the actual physiology of eating. In this chapter, I will briefly describe these factors as they relate to the physiology of eating in man.

1.1. THE PSYCHOLOGY AND NEUROPHYSIOLOGY OF EATING

1.1.1. The Brain Center

The hypothalamus had been considered the main neurocontrol center for eating,[1-5] but neural tracts are now considered just as important.[6-10] Damage to the ventromedial hypothalamus produces severe overeating and obesity[2,3] in animals so that body weight increases quickly but stabilizes at a new weight level that is then maintained. Thus, this system appeared to demonstrate some regulatory inhibitory control over eating in a so-called "satiety center." Lesions produced in the animals in the lateral hypothalamus cause both aphagia and adipsia, resulting in great weight loss. Consequently, the lateral area probably contains a so-called "feeding center."[4] The feeding center theory is enforced by the fact that electrical stimulation of the lateral hypothalamus elicits feeding or drinking in animals.[5] Although the hypothalamus had been considered the central and almost exclusive area in the brain for regulating eating and drinking, there is now experimental evidence that many neural tracts that pass through the hypothalamus may have major influences on eating. This has been demonstrated in the excellent microsurgery techniques of Grossman and his associates,[6-8] who were able to demonstrate similar effects by cutting neural tracts, and by the chemical lesions produced by other investigators.[9,10] It is apparent from these studies and the literature that there is a limited understanding of the neuroanatomy involved in eating. Whether it plays a major role in man or not is not readily apparent at this time because most previous work has been in animal models. More applicable information in man may not become available in this century.

1.1.2. Regulation of Food Intake

Indeed, this is a very complex mechanism that is poorly understood. The stimulation to eat revolves about psychologic and visceral components. Stimulants such as taste and smell that affect the integrated psychologic components are obviously important but not readily evaluated.

Tastebuds in the tongue are present at the base of circumvallate and fungiform papillae and have pores that permit substances to enter the bud, which contains many taste cells. Taste fibers from the facial and seventh cranial nerves travel via the chorda tympani to the facial sensory area and the opercular surface of the Sylvan fissure of the cortex. Sweet, sour, salty, and bitter sensations can be determined by this pathway. The number of tastebuds varies tremendously from person to person and also decreases with age.[10a] Taste sensations can become highly developed in certain individuals. Although the influence of taste on food regulation is little understood, it is obvious that it must have a major regulatory factor for those of us who love sweets and others who love salty foods. Why and how these variations develop are poorly understood.

The tongue also has temperature- and pain-sensation capability. How much this is involved in eating habits is also poorly understood. Nevertheless, the tongue does attempt to regulate intake by controlling temperature extremes. Taste thresholds can be measured with some accuracy, although they are difficult to measure and involve a great deal of subjective evaluation. Techniques that appear to be reproducible and simple have been described in which test solutions of various dilutions are used to assess the ability to discern the four basic taste sensations.[11,12] Taste threshold norms and deficiencies have been published and the literature reviewed by Kelty and Maier.[12] The recent work of Stinebaugh and associates[13] clearly reflects the uncertainty in this field as they were able to carefully assess taste sensitivity for salt but were not able to associate the taste threshold for salt with renal sodium excretion and sodium balance, indicating that an individual sensitivity to salt is not related to body salt physiology. The entire field of taste and its relationship to food regulation needs a vast amount of research. Interestingly, the mineral zinc has clearly been established as important in taste metabolism. Deficiencies in zinc result in decreased taste sensations[14] that are correctable by zinc replacement therapy.

Smell and its effect on human eating habits is even less well understood than taste. Olfactory receptors in the nose transmit impulses through the second cranial nerve to olfactory receptive areas in the uncus and adjacent portions of the hippocampal gyrus of the temporal lobe. The odors that are emitted from foods are unending in their combinations. The obvious stimulatory and inhibitory odors are known to all of us and the choices and differences between societies are also obvious. However, the effects the less obvious odors have in regulating our eating habits are not understood. Destruction of the olfactory

Table 1-1. Mechanisms of Stimuli Activating or Inactivating Eating[a]

Stimulus	Satiety system	Feeding system	Eating
1. Normal satiety signals and electrical.	Activated →	Inhibited →	None
2. No satiety signals. Focal lesions— chemical or physical. Norepinephrine injection (a-antagonist).	Inactivated →	Uninhibited →	Active
3. Hypoglycemia. Electrical or chemical.		Activated →	Active
4. Focal lesions. Hyperglycemia. Isoproteranol injection (β antagonist).		Inactivated →	None

[a]Modified from traditional model of neural control of feeding by Van Itallie *et al.*[15]

pathways or cortex results in anosmia. Lesions that irritate the olfactory centers in the brain may cause olfactory hallucinations that can result in sensations of peculiar odors and tastes.

Regulation of eating resolves about activation and inactivation of both the *feeding and satiety centers.* Satiety centers are the strongest in controlling feeding. When activated, they inhibit the feeding centers, and when inactivated, the release of that inhibition will permit feeding.[15] Furthermore, there appear to be stimuli that will directly activate the feeding system. These are schematized in Table 1-1 and include electrical stimulation, chemostimulation, and hypoglycemia.

Table 1-2 lists satiety stimuli and methods of conduction that occur when food enters the stomach and subsequently is released into the small intestine for absorption. These biochemical satiety signals are instituted almost immediately after feeding, the activation of satiety occurring with gastric distension, increased tone of luminal contents, and nutrient substances that enter the duodenum and subsequently are absorbed and enter the portal circulation.

Intestinal hormones such as secretin, enterogastrone, glucagon, and cholecystokinin (CCK) are all capable of initiating satiety. The work of Antin, Gibbs, and associates[16,17] has clearly demonstrated the satiety effect of CCK in rats, but this work needs more elaboration in man.

In summary, there are definite feeding centers in the brain that can be stimulated to increase food intake. They are activated by electrical stimulation, hypoglycemia, and chemical stimuli. These centers can be inactivated directly by hyperglycemia, focal lesions, and all of the stimuli that create satiety such as described above.

In addition to the mechanisms for short-term regulation of eating, long-term regulation also exists.[15] The long-term component appears to be necessary in order for humans to maintain a relatively constant weight. Although there is wide variation in caloric intake, most healthy young men and women

Table 1-2. Possible Short-Term Satiety Signals[a]

Phase[b]	Food-derived stimulus	Peripheral transducers	Satiety signals[c]
Ingestive (preabsorptive)	Gastric distension	Intragastric stretch receptors	Neural (vagus)
	Tonicity of gastric–duodenal contents	Duodenal osmoreceptors	Neural (vagus) (inhibition of gastric motility and emptying)
	Substrates entering duodenum	Gut chemoreceptors	Gut secretion into circulation of cholecystokinin and other "satiety hormones"
Postingestive (absorptive)	Portal circulation		
	Glucose, amino acids	β-Cells of pancreas	Insulinemia
	Glucose	Hepatic-vagal glucoreceptors	Neural (vagus)
	Glucose	Rate of hepatic metabolism of food-derived substrate	Neural or "liver-derived humoral signals"
	Amino acids	Hepatic thermoreceptors	
	Systemic circulation		Tonicity of blood
			Temperature of blood (thermic response to food)
			Glucosemia
			Amino acidemia
			Chylomicronemia
			Insulinemia
			"Other humoral satiety signals"
			Hormone ratios (e.g., insulin/glucagon)

[a]Modified from Van Itallie et al.[15]
[b]The ingestive and postingestive phases are sequential but may overlap temporarily.
[c]Some satiety signals (i.e., the thermic response to food and the concentrations in plasma of various hormones) may also serve as "long-term" signals reflecting nutritional status.

maintain their weight at stable levels even in the face of deliberate food variations.[18,19] The mechanism of long-term control of body weight is not clear in humans. However, all sorts of theories are postulated, including control by humoral mechanisms, release of fatty acids from adipose tissue, and central nervous system regulatory centers.

1.1.3. Eating Habits, Food Faddism, and Cults

No monograph on nutrition is complete without some mention of eating habits and faddism. Humans are concerned with eating for preservation, but as intelligent beings they attempt to eat not only to satisfy themselves but to eat what is good for their health. How humans gain satisfaction and knowledge of what is good eating is a fascinating and complex sociologic phenomenon. One must investigate environment, culture, regional and historical factors, functional aspects, and quantitative availability as well as clinical factors.[20] Our experiences in this century with nutritional habits in so-called highly developed countries as compared with the underdeveloped countries of Africa and Asia exemplify the social and economic influences on eating habits. A specific example is the highly refined diet developed in the highly developed countries. During the first few decades of this century, sophisticated machinery developed in the industrial revolution enabled us to finely process foods. Consequently, refined sugar developed. Human ingenuity in developing high-protein foods also made them readily available for consumption. During the middle of this century, the development of communications media such as radio, television, and newspapers enabled sophisticated advertising techniques to flourish. Furthermore, the highly developed society lost concern with the previous century's mode of cooking and detailed food preparation and moved to fast-food processing. As a consequence, the diet changed to include refined simple carbohydrates, high protein, and high animal fats and proteins as staple food items. This sociologic phenomenon can easily be contrasted with that of the underdeveloped countries, where unrefined carbohydrates are still consumed. Because scientific information indicates that the diet of highly developed societies may be deleterious, a growing awareness of the benefits of good foods is causing a change in the diet. Less animal fat is being consumed. Recent trends indicate that we will probably also move away from refined sugar to more complex polysaccharide ingestion. These examples of trends in this century within our own experience clearly demonstrate the tremendous effect society has on eating habits. This entire subject is reviewed by both Todhunter[21] and Robson.[22]

Although this may seem remote to the practicing physician, it is of utmost importance. Physicians often prescribe a diet, but yet the cultural and sociologic factors do not make the prescription realistic. In order to obtain an effective change in diet, the physician must realize that the patient is in an environmental situation that may have to be modified in order to obtain the desired clinical result. The pediatrician who prescribes spinach daily is not realistic.

Cults and faddism are often reflected in clinical problems. The definition of a fad is the pursuit of an interest with exaggerated effort for a brief time. If it is only for a brief time, some of the consequences of a fad diet are usually not serious. An example of this in recent times is the high-animal-fat diets to lose weight. Although it is possible to lose a considerable amount of weight with these diets, the fad dieter cannot follow it for a long period of time, and consequently the hyperlipemia that is associated is short-lived. However, fads may be carried to extremes and prove harmful or even fatal as evidenced by the arrythmias associated with long-term liquid-protein diets.[23,24] Most often, fads are short-lived and as a consequence not harmful. A careful diet history by the physician will reveal a fad and any of its consequences can be readily corrected. The fad dieter may harbor a serious underlying personality disorder.

Consequently, the discovery of faddism in a patient may lead the physician to treat any personality deficiency that is causing the patient to resort to fad eating.

There are many cults that have taught us the benefits of certain food habits. Vegetarianism is an example that demonstrates very specific intake with elimination of all animal products. (The details of physiologic variations caused by vegetarian diets are discussed in Chapters 4 and 12.) Faddists often can harm themselves by following highly restricted diets for extended periods of time. A vegetarian must supplement his/her diet with B_{12} to avoid the consequences of vitamin B_{12} deficiency.

Diets in cults are often very carefully designed and any deficiencies that occur in these groups are often corrected by design, such as the vegetarian who supplements the diet with vitamin B_{12}. Nevertheless, it behooves the clinician to take a careful diet history and ascertain whether there is any deficiency due to either fad or cult diets.

1.1.4. Behavior Modification

In 1927, Pavlov[25] described his findings concerning conditioned reflexes of animals. Since that time there has been a steady but very slow progress in our understanding of the behavior of a subject as it relates to eating.[26,27] Modification of behavior has gained popularity because of recent limited successes with this form of treatment for obesity. In 1959 Stunkard[28] published the poor results obtained with traditional diet management of obesity. He and others demonstrated that more significant short-term weight loss can be obtained with behavior modification techniques,[29-31] but several workers have not obtained similar results and others yet caution on the long-term benefits of this type of management.[32,33]

Behavior modification has also shown promise in the fields of alcoholism, mental retardation, chronic pain, depression, hysteria, phobias, anorexia nervosa, childhood autism, and in the common social problem of cigarette smoking.[34] Although this is presently a most acceptable method of treating behavioral disorders, much more research and understanding in this field are needed before it can be accepted as the primary method of treatment in these disorders.

Pomerleau and associates[34] describe two basic concepts involved in behavior modification. First is "contingency management," which is based on the observation that consequences of behavior (reinforcing stimuli) determine the pattern of subsequent behavior. A reward that acts as a reinforcer will modify behavior. A simple example is money given by a mother to her child for eating a particular food. The second concept is that of "stimulus control." A particular stimulus can affect behavior when it is associated with a reinforcer, or when it signals a situation in which the behavior is associated with a reinforcer. This has been employed most effectively in obesity by analysis of the environmental situations involved in overeating. Patients are advised to keep detailed records of all physical, emotional, geographic, and social situations and successful weight loss is associated with an active role by the patient. Self-control methods take the patient from a passive therapeutic role to an active role in their own treatment.

Bruch[33] has categorized overeaters into four groups. The most common group is composed of those who overeat but are able to successfully contain the problem and maintain their body weight. This group includes the vast majority of people. The second group is composed of those who are unable to maintain weight by themselves. Of these there are some who are able to lose weight with the assistance of some group or physician. In the third group are those who fail to lose weight despite advice and assistance. The fourth group is composed of those who do not want to succeed and will neither apply self-control themselves nor seek advice on how to do it. It is well to remember that although the developed countries' populations usually have a plethora of food, the majority of people are able to use self-control to maintain their own body weight.

The treatment of obesity is well outlined in many manuscripts and texts.[29-36] The purpose of this book is not to repeat those works. Suffice it to say that the present methods usually require a baseline understanding of the patient, a treatment protocol employing behavior modification and diet advice, and then a follow-up of the phases of treatment.

Modification of eating habits is important not only in obesity but in a great number of digestive diseases. Modification of diet appears to be important in the prevention of coronary artery disease. Epidemiologists also indicate that dietary factors are of great importance in a wide variety of diseases such as colon cancer and arteriosclerosis. In order for preventive medicine to effectively decrease the incidence of the diseases caused by diet deficiencies, the ability to modify diets to include, or exclude, certain components is important.[37] The psychology involved in modifying long-term eating habits is complex. The development of behavior modification techniques of greater sophistication will probably become necessary if the need of altering American dietary patterns to prevent long-range consequences is real. Our own experience in a simple experimental attempt to induce more regular bowel habits among individuals indicated that forced dietary recommendations were not readily acceptable.[38] Mass media education, however, can cause shifts in food consumption in populations that are subject to those media. Recent evidence indicates a shift from ingestion of primarily animal fats to vegetable fats,[39] and that with this shift there is a trend toward less coronary artery disease.[40] The experience of the Norwegian recommendations of decreased animal fat and increased polysaccharides on dietary control is fascinating. It has shown that the diet of an entire population can be modified by a forceful recommendation to an intelligent population that is concerned with long-range health benefits.[41] It is hoped that our own population can adopt these recommendations. Although governments are making such recommendations, there are still doubters in the scientific nutrition field. The doubts are raised on sound scientific reasoning, but the statistical epidemiologic trends cannot be denied. Our own concerns with a deficiency in fiber in the ordinary diet of highly industrialized countries reveal that there is definitely inadequate fiber intake, and it will be necessary for mass public education to correct this.

Behavior modification holds great promise in both the fields of therapy and preventive medicine.

1.2. GASTROINTESTINAL PHYSIOLOGY OF EATING

The details of gastrointestinal physiology that occurs between the time a morsel of food first crosses the lips to when the remains leave the anus encompass all of the physiology of man. Perhaps it is scientifically grandiose to make that statement, yet the muscular, neurologic, and circulatory systems are involved either directly in actual consumption, digestion, and absorption, or indirectly by the effects of consumption. It is not my intent to describe all of these aspects, but rather the immediate and direct effects of foods as they pass through the digestive system. As we review each area, appropriate references will be made to more detailed monographs and texts for the interested reader.

1.2.1. Mouth and Pharynx

The anatomical drawings and descriptions of the mouth and pharynx of Netter and Oppenheimer[42] have visually captured the role of the mouth including the mandible, temporomandibular joint, organelles of the floor and roof of the mouth, the muscles involved in mastication, and the roles of the tongue, teeth, and salivary glands and secretion during eating. Their work also vividly depicts the role of the pharynx and all of the blood and nervous supply in this area.

After food enters the mouth, it is chewed, the extent of chewing varying from subject to subject. Variables such as personality and the condition of the teeth will affect the degree and amount of chewing. Subjects capable of chewing coarser foods may seek those out, and obviously those who cannot chew because of limitation of the muscular skeletal structure or teeth will seek softer and more liquid foods.

Once food enters the mouth and the chewing process begins, the salivary glands pour saliva into the oral cavity. Small salivary glands are distributed throughout the lining of the oral cavity, and the bilateral parotid, submandibular, and sublingual glands are distributed along the jaw from the base of the ear to the chin. The parotid is primarily serous, the submandibular mixed, and the sublingual primarily mucoid secreting. Basket cells are situated in most of the glands and account for the contractile action that results in the gush of saliva. The serous cells secrete primarily amylase, which acts optimally at pH 6.5 and helps convert long-chain dextrose into simple disaccharides. However, salivary amylase is inactivated at a pH below 4.5 and consequently is destroyed in the acidic stomach. The more mucoid secretions of the sublingual gland have a protective and lubricating effect. The total amount of saliva secreted daily is approximately 1000 to 1500 ml. The pH varies from 6.2 to 7.6. Organic substances such as proteins, including immunoglobulins, urea, and uric acid, may be in the saliva as well as chloride, phosphate, and bicarbonate anions, and calcium, sodium, and potassium cations. Although food directly acts as the best stimulant for salivary secretion, calcitonin, secretin, and pancreozymin can all stimulate and affect salivary amylase secretion.[43,44]

Once the food is chewed, it is then ready for swallowing. The jaw shuts and the soft palate becomes elevated. It forms a partition closing off the nasal cavity that ordinarily communicates with the pharynx. As the bolus of food moves backward over the tongue, the epiglottis is tilted forward, closing off the laryngeal airway. The tongue then moves backward and forces the food through the pharynx and over the epiglottis into the first part of the esophagus. Recovery then ensues with the tongue moving back, the epiglottis opening so that respiration continues, and the soft palate simultaneously dropping, tilting the nasopharynx. This entire process ordinarily takes a little over 1 sec.[40,45] For more detail on this subject, the interested reader is referred to Davenport's text.[46]

1.2.2. Esophagus

As the bolus of food leaves the pharynx, it enters the esophagus. The esophagus is a muscular tubelike structure, approximately 25 cm long from the upper esophageal sphincter to its lower esophageal sphincter. It is innervated by the vagus and sympathetic nerves, but the complexities of nervous control still remain unclear. The esophagus can be divided into thirds. The upper third contains striated muscle and the lower two-thirds smooth muscle. The upper third is innervated by the glossopharyngeal and vagus nerves, which are not autonomic, whereas the smooth muscle is supplied by the vagus that has an autonomic distribution. It is the innervation and control of the lower esophageal sphincter that remain somewhat unclear. In any event, during swallowing as the epiglottis closes off the trachea and the larynx rises up, the upper esophageal sphincter (UES) opens and the bolus of food is forced through that area into the lower esophagus. The lower esophageal sphincter (LES) usually opens as the bolus enters. The area of the UES is about 3 cm long and represents the junction of the pharyngeal–esophageal area. It is ordinarily contracted at rest, and pressures in the area may measure as high as 60 cm of water. The cricopharyngeus muscle is at the level of this segment, and it is presumed to be the main source of the sphincteral action.[46-48] As the UES relaxes, the bolus of food passes into the lower two-thirds of the esophagus that is controlled primarily by smooth muscle involuntary action. The *primary* wave of peristalsis occurs moving down the esophagus as a continuation of oral and pharyngeal movements in a sequential fashion. *Secondary* peristalsis also occurs, but these contractions originate below the UES and are not initiated by any previous oral or pharyngeal contractions. The peristaltic wave that originates just below the sphincter generates pressures from 30 to 120 mm Hg. The pressure reaches a peak at approximately 1 sec that may last as long as 0.5 sec and subside in approximately 1 sec. The entire duration of the contraction at any given point in the esophagus may last from 3 to 7 sec. Motility studies can trace the orderly sequential contractions through the esophagus. These push the bolus of food down the esophagus until it reaches the area of the LES. The LES is usually open at the time food arrives, having opened at the time the UES pushes the bolus through.

The LES is an extremely important structure because it maintains a barrier against reflux. Anatomically, it is difficult to define as a sphincter, although manometric and radiographic studies define the segment ranging from 2 to 3 cm in the contraction phase that represents a high-pressure zone as acting as a sphincter. Although there are discrepancies between manometric studies, the mean pressure in normal subjects ranges around 25 mm Hg. And patients with reflux usually have pressures below 15 mm Hg, with a range usually between 0 and 10 mm Hg.[46] The neurogenic control is not clear, and there appears to be some gastrointestinal hormone influence on the sphincter. Although hormones can be shown to have an effect, how significant their physiologic role is not clear. It is readily apparent that exogenous doses of gastrin increase the tone of the sphincter, whereas secretin, cholecystokinin, glucagon, and prostaglandins all can decrease LES pressure. How these hormones actually work in a synergistic *in vivo* effect in man is not yet completely understood.[50]

It is also clear that different types of food have different effects on the LES sphincter. This has definite significance in clinical situations as will be discussed in the chapter on diseases of the esophagus. It is apparent that protein will raise the LES significantly, carbohydrate will raise it less so, and strictly fat foods will significantly lower the LES. In all probability these effects of foods are mediated through hormonal action.

1.2.3. Stomach

The bolus of swallowed food enters the stomach through the LES. The stomach connects the esophagus with the small intestine. Its shape varies, but anatomists divide it according to its function and histology. It is divided into the cardis, fundus, body, antrum, and pylorus. The lesser curvature is shorter than the greater curvature. The fundus and body contain most of the acid- and pepsin-secreting cells, whereas the antrum contains cells that secrete gastrin. Intrinsic factor is produced in the neck cells throughout the stomach. The terminal 2 to 3 cm of the stomach is often referred to as the pylorus. The blood supply arrives from the celiac trunk and the major branches of the celiac, the left gastric, common hepatic, and splenic arteries, splitting off into numerous branches that provide the rich supply to the organ. The autonomic innervation of the stomach is primarily parasympathetic through the vagus nerve and sympathetic through fibers that arise from T6 to T10. As with all of the viscera of the gastrointestinal tract, it is composed of four layers: mucosa, submucosa, muscularis, and serosa. The mucosa is distinctive in that it contains gastric glands. The glands of the cardia and pyloric areas are similar, and much thinner than the gastric or fundic glands. The pyloric glands contain cells that secrete gastrin, whereas all of the glands contain cells that richly secrete mucus. The fundic and gastric glands are distinctive in that they contain chief cells that secrete pepsinogen and parietal cells that secrete hydrochloric acid.

The stomach may begin its role in digestion before the bolus of food enters by actively secreting from impulses that were induced by taste, smell, chewing, or emotional factors.[51] These factors are mediated through neurogenic control.[51] Stimulation of secretions may be selective depending upon the type of food.

Pepsinogens can be classified into two groups. Group I are secreted in the proximal stomach and group II throughout the stomach and the duodenum. At rest there is a small amount of pepsinogen. The greatest stimulant to pepsinogen secretion is vagal; however, hormonal stimulation may occur primarily through secretin and less so through gastrin. The secretion of pepsinogen via gastrin may occur through gastrin's stimulation of acid, which acts as a direct stimulant on the mucosa to increase pepsinogen secretion. The major amount of pepsinogen secretion occurs immediately after the bolus hits the stomach. Approximately 3 hr after feeding, the chief cells are virtually devoid of zymogen granules, which contain pepsinogen, but they gradually reaccumulate. Pepsinogen is converted to pepsin, which is maximally active between pH 1 and 3. Conversion occurs in the presence of acid and pepsin itself. The enzyme is active in splitting large proteins into polypeptides.

Gastric juice also contains two other enzymes, lipase and gelatinase. Lipase is stable at pH 2 and active at pH 4 to 7. It is most active in the hydrolysis of short-chain and medium-chain triglycerides that are found in milk.

Gelatinase is found in crude preparations of pepsin and is much more effective in liquifying gelatin than pepsin. Urease is also present in the stomach. Most ureases in the body are of bacterial origin; however, gastric urease is not and may play an important role in the splitting of urea and formation of ammonia when large amounts of urea are ingested, or in uremic patients.

One of the most important functions of the stomach is secretion of hydrochloric acid. Some gastrointestinal physiologists have spent their research lives trying to understand this phenomenon. Although we have only been able to demonstrate that hydrochloric acid functions to convert pepsinogen to pepsin, it seems that it must play a more significant role in man's physiology. Ordinarily the gastric mucosa secretes sodium, potassium, and calcium cations and chloride and bicarbonate anions. At rest or during basal secretion, the hydrogen ion concentration is approximately 20 mN and rises sharply to approximately 125 mN during maximal secretion, whereas sodium conversely falls from approximately 85 to 10 mN. There is a concomitant rise of chloride ion from 120 to 155 mN. The potassium ion during secretion rises slightly but is relatively stable between 10 and 20 mN. In normal states, the greatest stimulus for secretion of hydrochloric acid is the eating process. The strongest mechanism of stimulation is the hormone gastrin. The exact mediation of secretion is still in question, but it is undoubtedly related to vagal innervation and direct stimulation of gastrin with release of histamine at the end organ. Parietal cells secrete maximally and the amount of acid secreted is related to the number of cells. In healthy men and women, basal secretion averages between 0 and 5 meq H^+/liter, and maximal stimulation with acid histamine or pentagastrin ranges between 20 and 35 meq H^+/liter. The composition of gastric juice is ordinarily a function of the rate of secretion, but it may be independent of the rate. By techniques of intragastric titration, the amount of acid present during the eating of a meal can be measured. Those measurements correlate well with techniques of maximal histamine stimulation.[50] The energy for acid secretion is derived from ATP. Intracellular reactions are probably controlled by cyclic AMP. In the process of acid secretion, the hydrogen ion concentration of the plasma is raised from 0.00005 mN to 150 to 170 mN in the gastric juice. This requires approximately 1500 cal per liter of juice secreted. Even though tremendous amounts of research have been done involving the area of actual acid secretion, two theories of redox scheme and ATPase are unproven. The interested reader is referred to Davenport's[46] text, which describes these theories. What is best understood today is that within the parietal cell, carbon dioxide is converted to H_2CO_3, which is then converted into H^+ and HCO_3^-. The H^+ reacts with Cl^- that is actively transported through the cell and forms free secreted Hcl. The Cl^- is present in serum at approximately 105 mN, and by an active transport mechanism against an electrochemical gradient, it is concentrated to approximately 170 mN in the gastric juice.

Of some importance is the fact that increased blood flow and normal blood flow are essential to the gastric mucosa in order to maintain prolonged normal gastric secretion. The gastric blood flow is increased by cholinergic stimulation, epinephrine, histamine, and gastrin; it is decreased by an infusion of vasopressin or norepinephrine, and secretin. As will be discussed under specific organ diseases, different foods stimulate enzyme and acid secretion in the stomach at different rates. However, one can generalize and state that maximal acid and pepsinogen secretion are parallel: they reach maximum at 45 min to 1 hr after eating a meal and gradually decline to normal levels by 3 hr after a meal.

Another important factor in normal nutrition is the so-called gastric mucosal barrier that prevents penetration of hydrochloric ions back through the gastric mucosa. The barrier is believed to be composed of a glycoprotein matrix that can be broken by many substances. The break results in disease to the mucosa. Ethanol, natural and synthetic detergents, bile acids and lysolecithin, and aerophytic acids such as acetic, propionic, and butyric all break the mucosal barrier by being fat soluble. Many drugs including acetylsalicylic acid, salicylic acid, indocin, and butazoladine also break the barrier.

When the bolus of food is within the stomach and active secretion begins, chyme is created. An important function of the stomach is to delay the passage of chyme into the intestine so that the intestine is not overwhelmed, and so that the food can be properly mixed for initial digestion to begin. Consequently, the motility of the stomach becomes one of its most important functions. Liquid foods leave the stomach much more rapidly than solid foods. The mechanisms by which gastric motility is controlled are reviewed carefully by Cooke.[52] The empty stomach exhibits three types of contractions: simple low-amplitude type I waves that last 5 to 20 sec, type II waves of greater amplitude that last 12 to 16 sec, and type III waves that are seen in the antrum and that occur 10 to 30% of the time, are of high amplitude, and last approximately 1 min. In the resting stomach, type I waves are seen about 25% of the time, type II waves about 50%.

The movements of the full stomach are complex. Foods tend to mix selectively with liquids. When the stomach is full, the peristaltic wave follows the normal basic rhythm. It appears to begin high on the greater curvature and moves over longitudinal muscle through the body, antrum, and pylorus. In man the frequency of peristalsis is approximately 3 waves/min. Two or three waves might actually be present at any given time. As the pylorus opens and a portion of the chyme moves through it into the intestine, the remaining contents of the chyme are retropulsed back into the stomach and this results in a mixing motion. The stomach empties at the rate of approximately 20 ml/min for a 750-ml liquid meal.

The electrical events that occur within the stomach are modified by both neurogenic and hormonal influences. Liquid contents empty much faster than solid contents. It is suggested that the body and fundus are more influential in emptying liquids and that the antrum has control of emptying solids. There is also some evidence to indicate that the pyloric sphincter may have a selected influence on emptying solids but little influence on emptying liquids.

Hyper- and hypoosmotic solutions as well as fats appear to delay emptying of the stomach. This occurs by a reflux mechanism from receptors situated in the duodenum.

Although cholecystokinin, secretin, and glucagon have been indicated as playing an experimental role in gastric emptying, their physiologic role is not yet clearly understood.[52] The role of gastrin in feeding will be discussed in more detail below. However, in the feeding process it is important to remember serum gastrin reaches its peak elevation at 30 min after feeding and gradually returns to normal over a 2- to 3-hr period. One can correlate the rise in acid secretion with gastrin secretion. Figure 1-1 demonstrates the acid–gastrin feed-

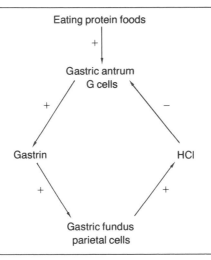

Figure 1-1. Gastric secretion in man. (+) Stimulates, (−) inhibits.

back mechanism that is related to feeding. As gastrin rises, gastric acidity rises and as that reaches peak levels it acts as an inhibitory influence on the gastrin-producing cells to cut off further production of gastrin and consequently the stimulus for acid secretion. This is a true negative feedback mechanism.

1.2.4. Duodenum, Pancreas, Liver, and Gallbladder

As the bolus of food is selectively delivered through the pylorus into the duodenal bulb and sweep, a sequence of electrolyte, enzyme, and hormone secretions evolves. This is probably one of the most dynamic areas of biochemical activity in the human body. The duodenal mucosa produces some forms of gastrin, all of the secretin, pancreozymin–cholecystokinin, and probably enterogastrone and enterocrinin. These hormones then stimulate and selectively inhibit the secretion and function of organs feeding into the duodenum. The exact details of the hormones are described in the narrative below and in Tables 1-3 and 1-4. Duodenal anatomy is unique in that the secretions from the liver and gallbladder pass through the common duct where most often they are joined by the ducts from the pancreas to enter the second part of the duodenum through the papillae of Vater. The histology of the duodenum is similar to that of the jejunum and ileum, which are described in more detail in Chapter 13. The duodenum is unique because of the presence of Brunner's glands. These glands occupy primarily the upper half of the duodenum in the mucosa and

Table 1-3. Gastrointestinal Hormones[a]

Hormone	Location produced	Cell type
	Stomach and Intestine	
Substance P	Gastrointestinal tract	EC_1
Motilin	Small intestine	EC_2
	Stomach fundus	ECL
Gastrin	Antrum, duodenum	G
Glucagon	Stomach fundus	A
Somatostatin	Antrum, upper intestine	D
	Stomach, intestine	D_1
	Stomach, duodenum	X
Secretin	Small intestine	S
CCK	Duodenum, jejunum	I
GIP	Small intestine	K
Enteroglucagon	Intestine	EG
VIP	Intestine	H
	Pancreas	
Insulin	Islets	B
Glucagon	Islets	A
Somatostatin	Mainly islets	D
VIP	Islet and nonislet	D_1
Pancreatic polypeptide	Mainly nonislet	D_2

[a]Modified from Pearse *et al.*[72]

Table 1-4. Gastrointestinal Hormones

Hormone status	Main stimulus	Main action
Gastrin	Protein foods	Stimulate gastric acid
Secretin	Acid in duodenum	Stimulate pancreatic and liver H_2O and HCO_3
CCK	Fat in duodenum	Stimulate pancreatic enzyme gallbladder contraction
Glucagon		Hepatic glycogenolysis Inhibit motility
Probable and suggested hormones		
Enteroglucogen	CHO and fat in duodenum	Inhibit motility
GIP	CHO and fat	Decrease of gastric motility Stimulate insulin
VIP	Uncertain	Vasodilator ?Regulate gastric emptying
Somatostatin	Uncertain	Uncertain in gut
Protein polypeptide	Uncertain	Uncertain
Motilin	Alkalinization of duodenum	Stimulate or regulate motility
Chymodenin	Uncertain	Stimulate pancreatic chymotrypsim
Enterooxyntin	Protein in small intestine	Stimulate gastric secretion
Bulbogastrone	Acid in duodenal bulb	Inhibit acid secretion
Urogastrone	Uncertain	Inhibit gastric secretion
Gastrone	Uncertain	Inhibit gastric secretion
Enterocrinin	Uncertain in duodenum	Stimulate intestinal secretion
Incretin	Uncertain in duodenum	Release insulin
Villikinin	Uncertain in duodenum	Stimulate villi contraction
?Enterogastrone		

submucosa and secrete a very viscous alkaline material of approximately pH 8.2 to 9.3 whose main function is to neutralize acid gastric secretions. The epithelial cells of the duodenum are rich in enzymes as are those of the jejunum that are active in absorptive and digestive processes as described in Chapter 13. The crypt cells of the duodenal glands contain most of the hormone-producing cells that stimulate the pancreas, liver, and gallbladder as well as exert an inhibitory influence on gastric function.

As the bolus of chyme enters the duodenum, the very sensitive receptors begin the process of slowing the gastric emptying as well as stimulating intestinal digestion. The motility and rhythm of the duodenal bulb are different than both the antrum and the postbulbar area. It appears to be slightly more active in rate than in the duodenum, which has segmental contractions occurring at approximately 12/min.[53] In a relatively healthy human, the bulb projects the bolus of food forward into the postbulbar area, where segmental contractions begin the process of mixing food and enzymes. Gastric emptying may be affected also by the shape of the duodenum as an abnormal loop is associated with delayed emptying and increased gastrin levels.[54]

The acidity of the chyme is rapidly neutralized by duodenal secretions. The duodenum receives chyme at varying pH and osmolality but delivers chyme at constant pH and osmolality.[55] However, the acid content is the greatest stimulus for secretin production, whereas the fat and carbohydrate content of the chyme is the greatest stimulus for CCK–pancreozymin secretion. Both of these hormones actively stimulate pancreas, gallbladder, and liver secretions. The release of secretin has been shown to be uneven and hence very sensitive to pH changes.[56]

The pancreas is the major site of enzyme production for use in intraluminal digestion of food.[57] All one has to do to appreciate this fact is see the malnutrition that follows removal of the pancreas, or the severe steatorrhea that is present in patients with pancreatic insufficiency. The pancreas is situated retroperitoneally and its head is surrounded by the sweep of the duodenum. It is composed of acini, which contain cells that contain zymogen granules. The acini are spherical in shape and secrete into intercalary ducts that merge into excretory ducts and finally into intralobular ducts. The cells lining the ducts become progressively more cuboidal as they move out into the periphery. Cells of the ducts do not contain zymogen granules, but they do stain for mucin, and the walls of the largest ducts contain elastic fibers and muscle cells that permit them to contract. The large intralobular ducts communicate with the main or accessory duct of the pancreas which then combines in the head of the gland with the common bile duct to exit into the duodenum through the ampulla of Vater. There are numerous anomalous arrangements for the exit of both the main pancreatic duct and the occult accessory duct of Santorini. The pancreas also contains islets of alpha and beta cells that secrete glucagon and insulin into the circulation. They have a rich blood supply in order to receive these secretions. The vagus nerve supplies the pancreas and the sympathetic nervous supply is predominantly through the celiac ganglion and splanchnic nerves. The gland itself is divided into head, body, and tail.

The function of the pancreas can be divided into its exocrine and endocrine secretions.[57] The very rich exocrine secretion can be divided into its enzyme, fluid, and electrolyte secretions. The enzyme secretions occur through the zymogen granules that are uniquely formed within the cell complex, modified within the Golgi bodies, neatly wrapped so that it contains all of the necessary enzymes, and are secreted as a package (Fig. 1-2).

The approximately 1500 ml of fluid solution secreted by the pancreas arises from a continuous secretion in the intralobular ducts that is hormonally stimulated in the extralobular ducts. The sodium content is stable at approximately 157 mN, the potassium content at 7 mN. However, the chloride and bicarbonate contents appear to be inversely proportional to hormonal stimulation. At base levels the secretion of chloride ion is approximately 100 mN but

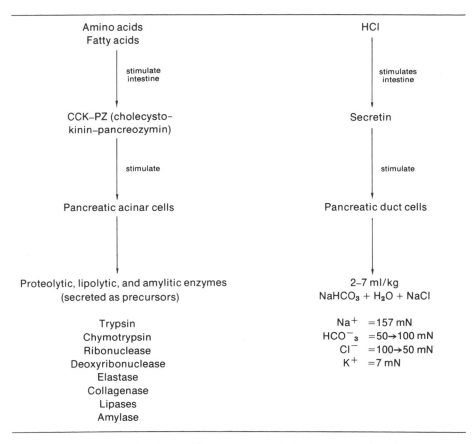

Figure 1-2. Pancreatic secretory stimulants and products.[46,53]

falls to 50 mN, and the bicarbonate is at a base of 50 mN and rises to about 100 mN after secretin stimulation.

The enzymes contained within the zymogen granules are as follows.

Trypsinogen is rapidly converted to its active form of trypsin in the presence of trypsin itself, calcium, magnesium, and enterokinase. There is a trypsin inhibitor within the pancreatic gland that apparently prevents the conversion within the gland. Trypsin has a molecular weight of approximately 25,000 and is the enzyme responsible for breaking proteins into proteoses and peptides.

Chymotrypsinogen is converted to chymotrypsin in the presence of trypsin. It has a molecular weight of approximately 25,700 and its actions are very similar to that of trypsin. However, it shows a preference to cleave peptide bonds that have a carbonyl group containing the amino acids tyrosine and phenylalanine. Both trypsin and chymotrypsin are endopeptidases that split internal protein bonds.

The pancreas also secretes exopeptidases carboxypeptidase A and carboxypeptidase B, which cleave end amino acids of chains. Similar enzymes are also readily present in the intestinal epithelium.

Other proteolytic enzymes secreted by the pancreas are elastase, collagen-

ase, ribonuclease, and deoxyribonuclease. All of these have actions on proteins similar to their names.

Amylase of pancreatic origin is essentially for splitting the 1,4-α-glycoside
linkage in starch and glycogen in order to release disaccharides. It is referred to
as an alpha amylase and has a molecular weight of 5000.

Lipase is essential for normal fat absorption; its absence causes massive
steatorrhea. In conjunction with bile salts it hydrolyzes a wide variety of insoluble esters of glycerol. Other lipases are also present and perform specific
functions, e.g., *phospholipase A,* which splits off a fatty acid from lecithin to
form lysolecithin.

Animal experiments definitely reveal enzyme adaptation to diets. However, evidence in man is less readily available. The hormones insulin and
glucagon will be discussed below.

As protein and fatty materials enter the duodenum, the intestinal mucosa
is stimulated to produce CCK–PZ (cholecystokinin–pancreozymin) and secretin. These in turn stimulate the acinar cells of the pancreas (Fig. 1-2) to secrete
enzymes and stimulate duct cells to increase their production of bicarbonate
and the flow of fluid. In healthy humans, the pancreas is capable of excreting
enough enzymes to digest all of the food eaten, as well as enough bicarbonate
and fluid to neutralize the acid presented in the chyme so that the pH in the
duodenum and jejunum remains between 7 and 9 for optimal digestion and
absorption of all nutrient factors.

There appears to be some nervous control of pancreatic secretion as well
as the hormonal influence. This is not as well understood, although vagus nerve
stimulation does increase pancreatic secretion and sympathetic stimulation
may increase or decrease selectively.

The liver (Chapter 10) has a pivotal role in nutrition. It is a chemical
factory for production of proteins and modification and homeostasis of chemical balance. It also has an important role in secreting fluid and bile acids
important for digestion and absorption. The two components of bile important
in nutrition are the *bile-acid-independent fraction* and the *bile-acid-dependent
fraction.* When the chyme enters the intestine, the same hormones that stimulate pancreatic secretion have their effect on the liver and the gallbladder. The
bile-acid-independent fraction is composed primarily of water and electrolytes
(Fig. 1-3) and is secreted primarily by duct cells. The bile-acid-dependent fraction contains bile acids returned through the enterohepatic circulation, newly
synthesized bile acids, and lecithin and cholesterol that are all secreted in a
micellar solution.

The total output of biliary secretion in man ranges from 250 to 1100 ml
per day. Although secretion is stimulated by hormones, it is an active process
and not a simple ultrafiltration. It is dependent upon back pressures that can
limit flow. Concentrations of the various components also vary. When the ratio
of bile-acid-dependent components is disturbed from normal, cholesterol may
then precipitate and form gallstones. This occurs in so-called lithogenic bile.
The ordinary baseline electrolyte concentrations of bile are noted in Fig. 1-3.
When secretin stimulation occurs, as in pancreatic juice, there is a reciprocal

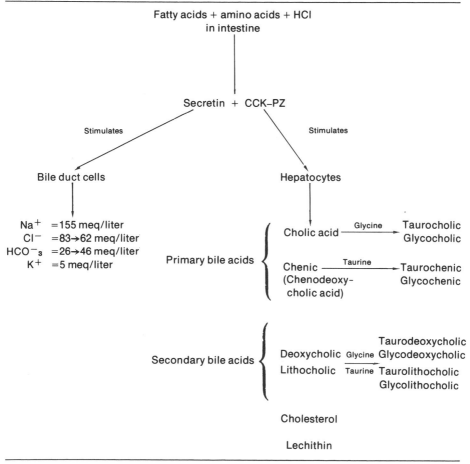

Figure 1-3. Biliary secretion.[46,54,55]

relationship between chloride and bicarbonate, bicarbonate increasing from approximately 25 to 80 meq/liter and chloride falling from 90 to 65 meq/liter.[58]

The liver produces two primary bile acids, cholic and chenic (chenodeoxycholic) (Fig. 1-3). These are conjugated with both glycine and taurine to respectively produce glycocholic, taurocholic, glycochenic, taurochenic, etc. These primary bile acids are secreted in their conjugated form into the duodenum. Bacterial 7 α-dehydrogenase, deconjugates the protein fraction leaving free cholic or chenic acid, which is either further dehydroxylated in the small bowel to respectively deoxycholic or lithocholic bile acids, or absorbed back into the portal system. The dehydroxylated and converted bile acids are similarly absorbed. This entire process results in the so-called enterohepatic circulation (Fig. 1-4). The concentration of bile salts and basal secretion ranges from 20 to 55 mM, whereas during peak flow stimulated by secretin it falls to approximately 10 mM. Ursodeoxycholic acid, the 7β epimer of chenic as well as other epimers, appears important in maintaining bile function.[78,79]

Bile acids are powerful choleretics, and they stimulate active secretion of themselves by hepatocytes, which is accompanied by passive increased flow of water and other electrolytes. The secretion of the bile-acid-dependent fraction is governed by the rate of return of bile acids to the liver through the enterohepatic circulation. The rate of return is controlled in its cycle by stimulation for bile secretion and gallbladder contraction.[46] Consequently, if one markedly increases feeding, bile acid secretion will also increase. However, this may merely increase the number of times that bile acids cycle through the enterohepatic circulation. The total normal bile acid pool in man ranges from 2 to 3.25 g, and from 3 to 12 cycles/day depending on the number of feedings. In the enterohepatic circulation bile acids are primarily reabsorbed in the distal ileum but may be reabsorbed in the colon. Man synthesizes enough bile acids to make up what he usually loses, in the range of approximately 500 mg/day. The interested reader is referred for more detail on the subject to the pioneering works by Hofmann, Small, and their colleagues.[59-61]

The enterohepatic circulation consists of the following components in a closed pathway (Fig. 1-4). Primary bile acids, chenic and cholic, are produced in the liver, conjugated with glycine and taurine, and then secreted in the bile into the duodenum. Taurine conjugates are anions at pH 6 to 7.7 whereas glycine conjugates are less ionized at intestinal pH. These are extremely active

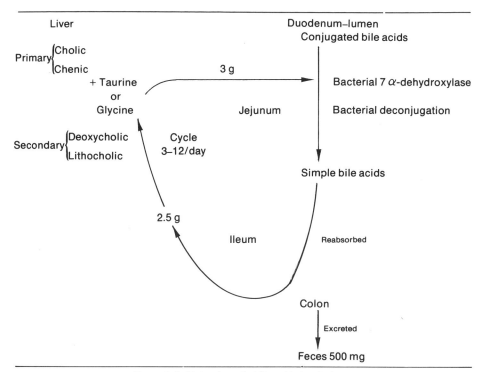

Figure 1-4. Enterohepatic bile acid circulation.

in fat absorption and in the formation of micelles (see Chapter 13). In the intestine bacterial enzymes are available for deconjugation so that cholic and chenic acids are set free. Bacterial 7 α-dehydroxylase converts cholic acid and chenodeoxycholic acid to the secondary bile acids deoxycholic acid and lithocholic acid, respectively. All of the simple bile acids, primary and secondary, are very actively absorbed in the distal ileum but may be passively absorbed in the jejunum and colon. All of the simple bile acids are transmitted through the portal circulation to the liver where all four of these may be reconjugated with taurine or glycine and once again passed into the duodenum along with the newly synthesized bile acids that replace the quantity lost in the feces.

The gallbladder in healthy humans concentrates bile. When choleretics are secreted, it contracts and forces bile through the ampulla of Vater into the duodenum. Normal bile contains sodium at approximate concentrations of 140 to 150 meq/liter. The columnar surface epithelial cells of the gallbladder contain a very powerful sodium pump mechanism capable of concentrating the sodium to approximately 340 meq/liter. This is done by rapid removal of water, which also results in increased concentrations of bile acids, cholesterol, and pigments. In healthy humans the critical ratio of bile acids plus phospholipids to cholesterol is maintained so that cholesterol remains in solution. When cholesterol precipitates it is because of either an increase in its concentration or a decrease in bile acid or phospholipid, resulting in so-called "lithogenic bile."

The capacity of the gallbladder ranges from 40 to 60 ml. As is well known, humans can live normally without a gallbladder. When the gallbladder is not present, compensatory mechanisms then develop so that adequate bile is secreted at the time food enters the duodenum. Cholecystectomy does not result in malabsorption or in increased steatorrhea.

Within 30 min of eating a meal, the gallbladder begins to contract and empties at its maximum in anywhere from 20 to 105 min. It never empties completely, as evidenced by oral cholecystogram studies. When the gallbladder contracts, bile fluid forced into the duct raises the pressure so that the sphincter at the ampulla opens and bile enters the duodenum. The gallbladder contraction is affected by both humoral CCK and vagal nervous pathways and begins as early as the cephalic phase of digestion. The greatest stimulus for contraction is the presence of fat in the duodenal sweep because it causes the prompt release of CCK from the duodenal mucosa. Other hormones such as caerulein and gastrin have similar terminal amino acid structures and are also capable of causing contraction of the gallbladder. They are referred to as cholecystogues. The role of foods in relation to gallbladder function and secretion will be discussed in more detail in Chapters 11 and 12.

Active and passive processes of absorption begin once pancreatic and biliary secretions mix with the chyme passed from the stomach into the duodenum. Solids leave the stomach at similar rates, but liquids are emptied more rapidly.[62] Digestion is rapid so that within 15 to 30 min after chyme passes into the intestine, the major carbohydrate, protein, and fat absorptions are completed. (The digestive and absorptive processes are described in Chapter 13.)

When these are accomplished, the jejunum and ileum are then involved in electrolyte and fluid balance. Multilumen balloon-tube small-bowel studies reveal that nutrient mixtures in the duodenum elicit jejunal secretion at sites distal from the food.[63] The ultimate balance of fluid and electrolytes evolves after nutrient absorption and is described in sections on the small intestine and diarrhea. Once fluid and electrolyte exchange is stabilized, the material is passed into the right side of the colon through the ileocecal valve. The functions of the colon are to absorb water and stabilize waste excretion. This is described in more detail in Chapter 14.

An important integrating function of the small and large intestines is that the major absorption of nutrient materials occurs in the duodenum and proximal jejunum and rapidly tapers off so that minimal absorption of materials occurs in the ileum. Of all nutrient materials, fats may be absorbed the greatest in the distal part of the small bowel.

Contractile activity increases in the small intestine after a meal. There is an increased frequency of so-called "slow waves." This same increased activity occurs in the sigmoid colon. The actual integrated motility of the gut has been studied little in man and most studies have been performed in animals. Although primary secretion, digestion, and absorption occur in the upper digestive tract, there is also evidence to indicate that the more distal parts of the small intestine stimulate gastric secretory responses to meals.[63] In normal volunteers chyme diverted at the ligament of Treitz resulted in decreased gastric secretion, supporting theories that the small intestine contributes appreciably to the magnitude of gastric secretion during meals.[70]

1.2.5. Intestinal Microflora

The role of bacteria in the gastrointestinal tract has been poorly understood. Their role in the upper and midgut is not yet fully apparent. The enterohepatic circulation as noted above is a very specific function for intestinal bacteria. They produce the enzyme 7 α-dehydroxylase, which is necessary for deconjugation and conversion of bile acids in order to maintain the enterohepatic circulation. Other such specific functions are not known for bacteria.

The stomach is usually relatively free of bacteria. The acid secretion cleanses the flora that is swallowed. Fungi, streptococci, and lactobacilli do pass regularly into the upper small bowel, but colony counts rarely exceed 10^4/ml intestinal fluid. As one progresses down the jejunum to the area of the distal ileum, the intestine rapidly becomes colonized by a wide variety of organisms. The anaerobic flora reaches average colony counts of 10^9 and the aerobic flora 10^8. As one progresses into the colon, the anaerobic counts rise to 10^{11}. Predominant anaerobic species are usually *Eubacteria* and *Bacteroides,* and aerobic species *Escherichia* and *Streptococci*. There are huge variations in species hosted *among* individuals, but in an individual the flora usually remains relatively stable. The interested reader is referred to several monographs and texts on the subject.[64-68]

1.2.6. Gastrointestinal Hormones

During the past two decades the greatest contributions in endocrinology have come forth in the understanding of gastrointestinal hormones. Many of them have been biochemically isolated, purified, and accepted as true hormones. Others are in the process of being completely understood and evaluated. Grossman[71] defines a gastrointestinal hormone in the following way: (1) stimulation to one part of the tract produces a response in a distant target; (2) the effect persists after cutting all nerves connecting the site of stimulation and the target; (3) the effect is produced by an extract of the part to which the stimulus was applied; and (4) the effect is produced by infusing exogenous hormone in amounts in molecular forms similar to the increase in blood concentration seen after the stimulus for endogenous release. Secretin, cholecystokinin, gastrin, and gastric inhibitory peptides have all reached hormonal status by virtue of these criteria. Hormones originating in the pancreas have long been accepted as true hormones. In Table 1-3 the organs and their cell types and hormones are listed according to histochemical and radioimmunoassay determinations as organized and classified by Pearse and his colleagues.[72]

Table 1-4 lists all true hormones and those considered hormones, their major actions, and the foods and stimuli to produce them.

1.2.6.1. Gastrin[46,73,74]

The first gastrin described was a 17-amino-acid straight-chain peptide. This has been lovingly referred to as "little gastrin." Following that a 34-chain peptide, "big gastrin," was described, and a 14-chain peptide has since been isolated, "mini gastrin." Other variants have been isolated from the pig, cat, and dog. All of the physiologic activity is exhibited by the last four amino acids of the C-terminus of the gastrin. The gastrins may be sulfated or nonsulfated, but sulfation has no effect on potency. Pentagastrin is the synthetic C-terminal tetrapeptide to which β-alanine has been added. It is a stable compound and commercially available. Gastrins are produced in the antrum and the duodenum. The half-life of big gastrin is approximately 42 min and that of little gastrin 7 min. They are equally potent and both are primarily inactivated in the kidney. The normal fasting plasma concentration is between 50 and 100 pg/ml with the upper limit of normal at approximately 200 pg/ml.

The major actions of gastrin can be divided into those considered physiologic or pharmacologic. The major physiologic action is the stimulation of acid. However, it also stimulates pepsinogen secretion, gastric blood flow, and contraction of the circular muscle of the stomach. It has also been demonstrated that growth of the stomach, small intestine, and pancreas is dependent on gastrin.

Of its pharmacologic actions, gastrin stimulates all water, bicarbonate, and electrolyte secretions of the small intestine, pancreas, and liver; enzyme secretions by the pancreas and small intestine; release of insulin, glucagon, and calcitonin; contraction of the lower esophageal sphincter and the gallbladder, small intestine, and colon smooth muscle. Pharmacologically gastrin has certain inhibitory effects, such as inhibiting contraction of the pyloric, ileocecal, and ampulla of Vater sphincters and potent inhibition of gastric emptying. Some evidence exists that it decreases absorption of glucose and electrolytes in the small intestine.

The main stimulus for release of gastrin from G cells is a protein-containing meal, which is much more potent in stimulating gastrin release than either fat or carbohydrate meals. It is released into the serum usually within a few minutes and peak serum levels occur at approximately 30 to 40 min. Both distension of the antrum of the stomach as well as cholinergic agents will also stimulate its release, whereas acidification of the stomach inhibits release (Fig. 1-1). Although gastrin release is promoted by vagal stimulation, denervated stomachs nonetheless secrete approximately half the normal gastrin level.

Extremely high gastrin levels are noted in the Zollinger–Ellison syndrome. These patients have a non-β-islet-cell tumor of the pancreas that secretes gastrin and is known as a gastrinoma. The range of serum levels in these patients has been reported anywhere from 300 to 350,000 pg/ml. In duodenal ulcer subjects there is controversy as to whether serum gastrin is increased or decreased. Some investigations indicate that increased serum gastrin levels do occur in association with hypersecretion of duodenal ulcer.[75]

1.2.6.2. Secretin[44,46,73,76,80]

Secretin is a polypeptide consisting of 27 amino acids, all of which are required to obtain a response. It is secreted by the S cells present in the duodenum.

The main action of secretin is to stimulate the duct cells of the pancreas and liver to produce bicarbonate and water. Lesser actions attributed to the hormone in the stomach are stimulation of pepsinogen secretion, stimulation of pyloric sphincter contraction, and inhibition of gastrin-stimulated acid secretion and gastric motility. There also appears to be a synergistic effect in which it increases some enzyme secretion in the pancreas as well as insulin release and inhibition of glucagon release. It is also reported to inhibit contraction of the lower esophageal sphincter. Secretin production is stimulated by the presence of acid in the duodenum from the stomach. In the range of pH 0–3 secretin release is proportional to the amount of acid. However, when the pH rises above 3, there is a marked decline in the amount of secretin released. It has been reported that above pH 4.5 insignificant amounts of secretin are released. As the pancreas is stimulated to produce bicarbonate, the duodenal contents are neutralized so that the production of secretin is cut off. Good studies are not yet available to correlate the intake of various foods with this hormone's activity. But logic makes one reason that foods producing high acid will cause the greatest secretin release, i.e., proteins.

1.2.6.3. Cholecystokinin[44,46,73,76]

Cholecystokinin is the term now used to include both its own action and what was formerly referred to as pancreozymin. It is referred to in the literature in the abbreviated form CCK–PZ, and it is a polypeptide containing 33 amino acids. The last five amino acids in its C-terminal sequence are exactly those found in gastrin. Both hormones have the same physiologic actions, but they differ greatly in the qualitative responses they elicit. The C-terminal octapeptide of cholecystokinin has all of the hormone's actions and is used often in experimental work. It is produced in the duodenum and the jejunum.

The actions of the hormone can be divided into physiologic and pharmacologic. Cholecystokinin's main physiologic function is to stimulate contraction of the gallbladder, relaxation of the ampulla of Vater, or sphincter, and stimulation of enzyme secretions of the pancreas. It weakly stimulates bicarbonate and water secretion of the pancreas and liver; as such its role is synergistic with secretin. It has an effect of slowing gastric emptying and is a definite trophic hormone of the pancreas.

Pharmacologic doses weakly stimulate acid secretion of the stomach, completely inhibit gastrin-stimulated acid secretion, weakly stimulate pepsinogen secretion, inhibit the lower esophageal sphincter, stimulate motility of the colon, and stimulate secretion of duodenal glands. It also has been shown to stimulate release of insulin and glucagon. As noted in Section 1.1.2, it can be a mediator of feeding behavior by producing satiety.

Release of the hormone is stimulated by L-isomers of amino acids, fatty acids, hydrochloric acid, and foods. The greatest stimulus for release of digestive enzymes is L-tryptophan; in decreasing order of potency are the levorotatory forms of phenylalanine, lysine, methionine, arginine, valine, isoleucine, histidine, leucine, threonine, and finally totally ineffective are the dextrorotatory forms and nonessential amino acids. Fatty acids stimulate pancreatic secretion, and those of 18 carbon lengths are greater stimulants than shorter chain length acids. Oleic acid is the most potent, then stearic, palmitic, lauric, and decanoic in decreasing potency, and octonoic extremely weak.

Acid appears to be a weak releaser of cholecystokinin, although the gallbladder will definitely contract after the presence of acid in the duodenal sweep. Secretin is released at anywhere from 5 to 24 times the amount of cholecystokinin from acid stimulation. One would expect that fats are the strongest stimuli for CCK–PZ, but the amino acid L-tryptophan is the most potent and this probably relates to its pancreatic enzyme stimulatory effect. This double role of CCK–PZ must relate it closely to all foods.

Caerulein is a decapeptide found in the skin of the Australian frog *Hyla caerulea*. It has the same C-terminal tetrapeptide as gastrin and cholecystokinin and has a sulfated tyrosyl residue, in a position corresponding to cholecystokinin. Therefore, it has some of the properties of both of the hormones and is a particularly strong cholecystogue.

1.2.6.4. Glucagon[72,76] and Enteroglucagon

Pancreatic glucagon is a single-chain polypeptide composed of 29 amino acids with a molecular weight of approximately 3500. A substance physiologically and pharmacologically similar to glucagon has been isolated from the duodenum and jejunum. Its exact chemical structure is uncertain and its molecular weight is approximately 2900. However, it is thought to be glucagonlike and consequently is referred to as enteroglucagon. In the gut there are two types of immunoreactive glucagon cells.

Glucagon stimulates glycogenolysis so that glucose is made available from stored carbohydrate. It activates the adenyl cyclase of the liver cell membrane. It has definite interactions with insulin. Insulin inhibits the hepatic production of glucose, ketone bodies, and urea when its production range is high. Increased glucagon levels in a state of relative insulin lack, such as diabetes, intensify the insulin deficiency. Glucagon has been suspect in the production of diabetes. Enteroglucagon can also effect glycogenolysis. In pharmacologic doses both glucagon and enteroglucagon have similar activity. These include inhibition of pancreatic enzyme in volume secretion and a most potent effect of inhibition of motility of the entire gut. It is thought that glucagon inhibits the effects of secretin and gastrin selectively.

After eating there is a sharp rise of enteroglucagon, but not of glucagon. The stimulation of production of the hormone by food is one of the major differences between the two. Stimulation of enteroglucagon by food comes primarily from carbohydrates and long-chain triglycerides. In subjects with the dumping syndrome, it increases greatly during rapid transit through the intestine. Although these differences are noted between the pancreatic and gut glucagon, the fact that the intestinal form has not been produced synthetically prevents further differentiation between the effects of the two.

Insulin is the best known and best studied of hormones. Its actions will be discussed in the chapter on pancreatic disease.

The following compounds have hormonal-type activity but have not as yet been completely accepted as hormones.

1.2.6.5. Gastric Inhibitory Polypeptide[72,76]

Gastric inhibitory polypeptide (GIP) is produced in the small intestine by so-called K cells. It consists of 43 amino acid radicals and has a molecular weight of approximately 5100. The 17 C-terminal amino acids are unique to this hormone.

Its physiologic action is twofold. First, it inhibits gastric acid secretion that is stimulated by pentagastrin, gastrin, insulin, and histamine as well as inhibiting pepsin secretion. Second, it appears to be a very potent stimulator of insulin secretion. It is suggested that GIP is responsible for the greater release of insulin noted after oral feeding of glucose than intravenous administration of glucose.

GIP release from the intestine is biphasic. There is an initial peak stimulated by glucose and a much later peak stimulated by fat. The release of GIP by glucose is exaggerated both after vagotomy and pyloroplasty.

1.2.6.6. Vasoactive Intestinal Polypeptide[72,76,77]

Vasoactive intestinal polypeptide (VIP) contains 20 amino acids and is related to the secretin–glucagon–GIP group. Because it is related to the other hormones, its actions are diverse and similar. It is a potent stimulator of pancreatic bicarbonate and water excretion, is capable of increasing the liver glycogenolysis, stimulates insulin and glucagon release, stimulates small intestinal juice production, and inhibits gastric acid output. However, its greatest action is as a potent vasodilator with hypotensive effect and consequently is named vasoactive intestinal polypeptide.

Although VIP is produced in the intestine, its mechanism of release is not understood. There is only one significant report of an increase noted after ingestion of fat and ethanol and infusion of HCI but not after infusion of amino acids, glucose, or a mixed meal.[77] The exact physiologic role is still not clear. In the so-called pancreatic–cholera disease syndrome consisting of watery diarrhea, hypocalemia, and achlorhydria, increased levels of VIP in plasma have been recorded. Unfortunately, once this syndrome is noted, the patients manifesting the symptom complex often have metastasis from pancreatic tumor producing the VIP.

1.2.6.7. Motilin[72],[76]

Motilin is a polypeptide containing 22 amino acids with a molecular weight of 2700. It has been synthesized and through immunoelectron cytochemical studies has been shown to be produced by EC cells with the highest concentration in the jejunum and significant amounts occurring in the duodenum and upper ileum.

Its physiologic action causes powerful contractions in the antrum and duodenum of the dog, but when infused into human subjects, there appears to be decreased gastric emptying. The present understanding indicates that it might be a regulator of gut motility, but its exact role is not yet understood. It has also been shown to enhance pepsin secretion in man. The exact stimulus for motilin production is also not well understood at this time.

1.2.6.8. Somatostatin[72],[76]

Somatostatin is a tetradecapeptide that is produced primarily in the hypothalamus. It has been shown to be present and produced in the antrum and upper intestine so-called D cells.

Its main action is to suppress the secretion of growth hormone and thyroid-stimulating hormone. It has been shown also to totally block the release of insulin and glucagon. As a hormone, it demonstrates the interaction between the brain and gut axis. It also is capable of inhibiting gastrin release that occurs both during fasting and after meals, pepsin output, motilin, and both gallbladder contraction and pancreatic enzyme production. It is felt that its role is primarily peripheral and that significant circulating levels do not occur as relatively large amounts are needed to produce these effects.

1.2.6.9. Pancreatic Polypeptide[72],[76]

Pancreatic polypeptide is produced in the pancreas and is a 36-amino-acid straight-chain polypeptide.

Although plasma concentrations have been reported to rise rapidly after ingestion of a meal, in amounts similar to that of insulin, the exact role in man is not reported. The effect of bovine pancreatic polypeptide has been studied in dogs with conflicting results. The clearest effects are those of inhibition of gallbladder contraction and of pancreatic enzyme output. However, there appears to be varying effect with varying dosage. Future studies should reveal its relationship to insulin and glycagon and carbohydrate metabolism.

1.2.6.10. Other Digestive Hormones[46,65,69]

Chymodenin, bulbogastrone, urogastrone, gastrone, enterooxyntin, enterocrinin, villikinin and incretin have all been suggested as hormones, but too little is known about their role in man to be accepted as having a physiologic meaningful effect. The interested reader is referred to other writings on those hormones for more detail.[76]

Bombesin is an active peptide that is found in amphibian skin and duodenum. Like caerulein, mentioned above, it has significant gastrointestinal and smooth muscle activity when injected into man. It is capable of stimulating gastric acid secretion and actually is the only agent that produces a pH-independent gastrin release. It is also capable of stimulating contraction of the gallbladder and relaxation of the ampulla sphincter as well as stimulation of protein-rich pancreatic secretion. Although it may be a useful research tool, as yet it has reached no clinical significance in man.

1.3. SUMMARY

A series of complex physiologic events occur that begin with mastication and swallowing followed by the detailed secretory phenomena of the stomach and upper duodenum. The bolus of food that enters the stomach is modified by acid and pepsinogen secretion, which is stimulated by gastrin. The chyme then enters the duodenal sweep and, through its acid and protein content, further stimulates cholecystokinin secretion that further stimulates pancreatic, liver, and intestinal secretions to enhance digestion and absorption. Gastrointestinal hormones have regulatory mechanisms in eating. Cholecystokinin acts in some form as a satiety agent, and hormones such as glucagon and insulin act in some regulation of motility of the gut. This fantastic integrated brain, gut, and hormone system is being elaborated upon as further research studies ensue. However, there is no question from our present state of knowledge that the central nervous system is integrated with the gut and that the gut produces hormones that have marked influence on the central nervous system. The hormones of the gut also have an effect on other metabolic functions within the body.

What is so desperately missing is the information on which foods and what combinations of foods turn on, or off, certain hormones or metabolic pathways. Recent advances in nutrition indicate that selective foods may have marked selective effects on selected body systems, specifically the evidence that red pepper increases gastric acid secretion, that certain fiber substances such as lignin combine with bile salts and hence have a potential effect on cholesterol metabolism, and that other fiber substances such as pectin have hypocholesterolemic effects. The mechanisms by which selective food substances have their selective metabolic effects are first becoming elaborated via research studies. Where this information is available, it will be reviewed in the following chapters under selective digestive disorders.

REFERENCES

1. Novin D, Wyrwicka W, Bray GA: *Hunger: Basic Mechanisms and Clinical Implications.* New York: Raven Press, 1976.
2. Hetherington AW, Ranson SW: Experimental hypothalamic-hypophysial obesity in the rat. *Proc Soc Exp Biol Med* 41:465, 1939.
3. Brobeck JR, Tepperman J, Long CNH: Experimental hypothalamic hyperphagia in the albino rat. *Yale J Biol Med* 15:831, 1943.
4. Anand BK, Brobeck JR: Hypothalamic control of food intake in rats and cats. *Yale J Biol Med* 24:123, 1951.
5. Delgado JDR, Anand BK: Increase of food intake induced by electrical stimulation of the lateral hypothalamus. *Am J Physiol* 172:162, 1953.
6. Grossman SP: Role of the hypothalamus in the regulation of food and water intake. *Psychol Rev* 82:200, 1975.
7. Hennessy JW, Grossman SP: Overeating and obesity produced by interruption of the caudal connections of the hypothalamus: Evidence of hormonal and metabolic disruption. *Physiol Behav* 17:103, 1976.
8. Grossman SP, Hennessy JW: Differential effects of cuts through the posterior hypothalamus on food intake and body weight in male and female rats. *Physiol Behav* 17:89, 1976.
9. Ungerstadt V: Adipsia and aphagia after 6-hydroxydopamine induced degeneration of the negio-striatal dopamine system. *Acta Physiol Scand Suppl* 367:95, 1971.
10a. Schiffman SS, Hornack K, Reilly D: Increased taste thresholds of amino acids with age. *Am J Clin Nutr* 32:1622, 1979.
10. Stricker EM, Zigmond MJ: Recovery of function after damage to central catecholamine-containing neurons: A neurochemical model for the lateral hypothalamic syndrome, in Epstein N, Sprague J (eds): *Progress in Psychobiology and Physiological Psychology,* vol VI. New York: Academic Press, 1976.
11. Henkin RI, Gill J, Bartter F: Studies on taste thresholds in normal man and in patients with adrenal cortical insufficiency: The role of adrenal cortical steroids and of serum sodium concentration. *J Clin Invest* 42:727, 1963.
12. Kelty MF, Maier J: Rapid determination of taste thresholds: A group procedure. *Am J Clin Nutr* 24:177, 1971.
13. Stinebaugh BJ, Velasquez MI, Schloeder FX: Taste thresholds for salt in fasting patients. *Am J Clin Nutr* 28:14, 1975.
14. Henkin RH, *et al:* Idiopathic hypoqeusia with dysqeusia, kyposmia, and dysosmia. *JAMA* 217:434, 1971.
15. Van Itallie TB, Smith NS, Quartermain D: Short-term and long-term components in the regulation of food intake: Evidence for a modulatory role carbohydrate status. *Am J Clin Nutr* 30:742, 1977.
16. Antin J, Gibbs J, Holt J, *et al:* Cholecystokinin elicits the complete behavioral sequence of satiety in rats. *J Comp Physiol Psychol* 89:784, 1975.
17. Gibbs J, Smith GP: Cholecystokinin and satiety in rats and rhesus monkeys. *Am J Clin Nutr* 30:758, 1977.
18. Robinson MF, Watson PE: Day-to-day variations in body weight of young women. *Br J Nutr* 19:225, 1965.
19. Campbell RG, Hashim SA, Van Itallie TB: Studies of food-intake regulations in man. Responses to variations in nutritive density in lean and obese subjects. *N Engl J Med* 285:1402, 1971.
20. Grivetti LE, Pangborn RM: Food habit research: A review of approaches and methods. *J Nutr Ed* 5:204, 1973.

21. Todhunter EH: Food habits, food fadism and nutrition, in Recheigl M (ed): *Food, Nutrition and Health. World Review of Nutrition and Dietetics,* vol 16. Karger, Basel, 1973.

22. Robson JRK: Food habits: Cultural determinants and methodology of change, in Robson JRK (ed): *Malnutrition—Its Causation and Control,* vol 2. New York: Gordon & Breach, 1972.

23. Singh BN, Gaarder TD, Kanegae T, *et al:* Liquid protein diets and torsade de pointes. *JAMA* 240 (2):115, 1978.

24. Van Itallie T: Liquid protein mayhem. *JAMA* 240:144, 1978.

25. Pavlov IP: *Conditioned Reflexes.* New York: Dover Publications, 1927.

26. Skinner BF: *The Behavior of Organisms.* New York: Appleton–Century–Crofts, 1938.

27. Rachlin H: *Introduction to Modern Behaviorism.* San Francisco: W H Freeman & Company, 1970.

28. Stunkard A, McLaulin-Hume M: The results of the treatment for obesity. *Arch Intern Med* 103:79, 1959.

29. Stunkard A: From explanation to action in psychosomatic medicine: The care for obesity. *Psychosom Med* 37:195, 1975.

30. Stuart RB: Behavior control of overeating. *Behav Res Ther* 5:357, 1967.

31. Abramson EE: A review of behavioral approaches for weight control. *Behav Res Ther* 11:347, 1973.

32. Curry H, Malcolm R, Riddle E, et al: Behavioral treatment of obesity. *JAMA* 237:2829, 1977.

33. Bruch H: The treatment of eating disorders. *Mayo Clin Proc* 51:266, 1976.

34. Pomerleau O, Bass F, Crown V: Role of behavior modification in preventive medicine. *N Engl J Med* 292:1277, 1973.

35. Bray GA, Bethune JE: *Treatment and Management of Obesity.* Hagerstown, Md: Harper & Row, 1974.

36. Craddock D: *Obesity and Its Management,* ed 3. Churchill Livingstone, 1978.

37. Christakis G: The case for balanced moderation, or how to design a new American nutritional pattern without really trying. *Prev Med* 2:329, 1973.

38. Kahaner N, Fuchs H-M, Floch MH: The effect of dietary fiber supplementation in man. I. Modification of eating habits. *Am J Clin Nutr* 29: 1437, 1976.

39. Friend B: Nutritional review. National Food Situation. USDA Economic Research Service, November, 1973 (available as CFE 299-8 from Consumer and Food Economics Institute, ARS, USDA, Hyattsville, Md 20782).

40. Levy RI: Stroke decline: Implications and prospects. *N Engl J Med* 300:490, 1979.

41. Dorfmann S, Ali M, Floch MH: Low fiber content of Connecticut diets. *Am J Clin Nutr* 29:87, 1976.

42. Netter FH, Oppenheimer E: *The Ciba Collection of Medical Illustrations,* vol 3, *Digestive System.* Part 1: Upper Digestive Tract, pp 3–31. Ciba, 1959.

43. Drack GT, Koelz HR, Blum AL: Human calcitonin stimulates salivary amylase output in man. *Gut* 17:620, 1976.

44. Mulcahy, Fitzgerald HO, McGeeney KF: Secretin and pancreozymin effect on salivary amylase concentration in man. *Gut* 13:850, 1970.

45. Rushmer RF, Hendron JA: *J Appl Physiol* 3:622, 1951.

46. Davenport HW: *Physiology of the Digestive Tract,* ed 4. Chicago: Year Book Medical Publishers, 1976.

47. Ingelfinger FJ: How to swallow and belch and cope with heartburn. *Nutr Today* 8:12, 1973.

48. Christensen T: The controls of oesophageal movement. *Clin Gastroenterol* 5:15, 1976.

49. Cohen S, Harris LD: Anatomy and normal physiology of the esophagus and pharynx, in Dietschy JM (ed): *Disorders of the Gastrointestinal Tract.* New York: Grune & Stratton, 1976.

50. Fisher RS, Cohen S: The influence of gastrointestinal hormones and prostaglandins on the lower esophageal sphincter. *Clin Gastroenterol* 5:29, 1972.

51. Moore JG, Motoki D: Gastric secretory and hormonal responses to anticipated feeding in five men. *Gastroenterology* 76:71, 1979.

52. Cooke AR: Control of gastric emptying and motility. *Gastroenterology* 68:804, 1975.

53. Misiewicz J: Motility of the gastrointestinal tract, in Dietschy JM (ed): *Disorders of the Gastrointestinal Tract.* New York: Grune & Stratton, 1976.

54. Thommesen P, Fisker P, Lovegreen NA, et al: The influence of an abnormal duodenal loop on basal and food-stimulated serum gastrin concentrations. Scand J Gastroenterol 13:979, 1978.
55. Miller LJ, Malagelada J-R, Go VLW: Postprandial duodenal function in man. Gut 19:699, 1978.
56. Pelletier MJ, Chayuialle JAP, Minaire Y: Uneven and transient secretin release after a liquid meal. Gastroenterology 75:1124, 1978.
57. Janowitz HD, Banks PA: Normal functional physiology of the pancreas, in Dietschy JM (ed): Disorders of the Gastrointestinal Tract. New York: Grune & Stratton, 1976.
58. Waitman AM, Dyck WP, Janowitz H: Effect of secretin and acetozolamide on the volume and electrolyte composition of hepatic bile in man. Gastroenterology 56:286, 1969.
59. Hoffmann NE, Hofmann AF: Metabolism of steroid and amino acid moieties of conjugated bile acids in man. Gastroenterology 67:887, 1974.
60. Small D, Dowling R, Redinger R: The enterohepatic circulation of bile salts. Arch Intern Med 130:552, 1972.
61. Hofmann AF: The enterohepatic circulation of bile acids in man. Clin Gastroenterol 6:3, 1977.
62. Grimes DS, Goddard J: Gastric emptying of wholemeal and white breads. Gut 18:725, 1977.
63. Wright JP, Barbezat GO, Clain JE: Jejunal secretion in response to a duodenal mixed nutrient perfusion. Gastroenterology 76:94, 1979.
64. Drasar BS, Hill MT: Human Intestinal Flora. New York: Academic Press, 1974.
65. Moore WEC, Holdeman LV: Human fecal flora: The normal flora of 20 Japanese-Hawaiians. Appl Microbiol 27(5):961, 1974.
66. Floch MH, Gorbach SL, Luckey TD: Intestinal microflora. Am J Clin Nutr 23:1425, 1970.
67. Luckey TD, Floch MH: Intestinal microecology. Am J Clin Nutr 25:1291, 1972.
68. Floch MH, Hentges DJ: Intestinal microecology. Am J Clin Nutr 27 (symposium issue):1211, 1974.
69. Luckey TD: Intestinal microecology. Am J Clin Nutr 32:105, 1978.
70. Clain JE, Malagelada JR, Go VLW, et al: Participation of the jejunum and ileum in postprandial gastric secretion in man. Gastroenterology 73:211, 1977.
71. Grossman MD: Physiologic effects of gastrointestinal hormones. Fed Proc Fed Am Soc Exp Biol 36:1930, 1977.
72. Pearse AGE, Polak JM, Bloom SR: The newer gut hormones. Gastroenterology 72:746, 1977.
73. Chey WY, Brooks FP: Endocrinology of the Gut. Thorofare, NJ: Charles B Slack, Inc, 1974.
74. McGulgan JE: Serum gastrin in health and disease. Am J Dig Dis 22:712, 1977.
75. Gedde-Dahl D: Serum gastrin response to food stimulation and gastric acid secretion in male patients with duodenal ulcer. Scand J Gastroenterol 10:187, 1975.
76. Rayford PL, Miller TA, Thompson JC: Secretin, cholecystokinin and newer gastrointestinal hormones. N Engl J Med 294:1093, 1157, 1976.
77. Schaffalitsky M de, Fahrenkrug J, Holst JT, et al: Release of vasoactive intestinal polypeptide (VIP) by intraduodenal stimuli. Scand J Gastroenterol 12:793, 1977.
78. Federowski T, Salen G, Tint S: Transformation of chenodeoxycholic acid and ursodeoxcholic acid by human intestinal bacteria. Gastroenterology 77:1068, 1979.
79. Ponz De Leon M, Carulli N, Loria P: Cholesterol absorption during bile acid feeding. Gastroenterology 78:214, 1980.
80. Wormsley KG: The physiological implications of secretin. Scand J Gastroenterol 15:513, 1980.

Normal Nutrient Requirements

In order to maintain normal body function, humans must maintain a balance between energy requirements and caloric intake. Chemical factors not produced by the body such as minerals and vitamins must be obtained in order to utilize ingested calories for energy. In this chapter the details of nutrition requirements are summarized so that one can have a basic understanding of the art and science of nutrition. Monographs on each nutrient factor are referenced so that the interested reader can extend his scope of understanding as desired.

All nutrition intake is subject to tremendous social and economic influences. One only has to compare malnourished children suffering from kwashiorkor to calorie-conscious members of suburban societies to understand the influence of social and economic factors. Because the scope of this book is nutrition in digestive disorders, the social and economic factors will not be stressed, but their importance should never be minimized. It may be possible to diagnose and prescribe for chronic pancreatic insufficiency, but it may be impossible to have the financial resources or social comprehension to pay for the enzymes to correct the disease.

The human goal is to maintain optimal body weight. At our present state of knowledge it is well accepted that malnutrition and obesity are undesirable conditions. It is assumed by the scientific world that it is best to maintain a stable weight, although it is not absolutely clear as to whether fluctuations in weight are significant. The accepted standard optimal weight for our society is that recorded by the Metropolitan Life Insurance Company (see Table 2-1). These recommendations are similar to those listed with recommended daily allowances for nutrient factors in order to maintain the recommended weight.[1]

Body weight is maintained by a balance between caloric intake minus energy expended plus energy required for the basal state. Stable body weight = caloric intake − (calories expended + basal caloric need).

Table 2-1. *Desirable Weights of Adults (Indoor Clothing), Ages 25 and Over*[a]

Height (in shoes)			Small frame		Medium frame		Large frame	
ft	in.	cm	lb	kg	lb	kg	lb	kg
				Men				
5	2	157.5	112–120	50.8–54.4	118–129	53.5–58.5	126–141	57.2–64
5	3	160	115–123	52.2–55.8	121–133	54.9–60.3	129–144	58.5–65.3
5	4	162.6	118–126	53.5–57.2	124–136	56.2–61.7	132–148	59.9–67.1
5	5	165.1	121–129	54.9–58.5	127–139	57.6–63	135–152	61.2–68.9
5	6	167.6	124–133	56.2–60.3	130–143	59.0–64.9	138–156	62.6–70.8
5	7	170.2	128–137	58.1–62.1	134–147	60.8–66.7	142–161	64.4–73
5	8	172.7	132–141	59.9–64	138–152	62.6–68.9	147–166	66.7–75.3
5	9	175.3	136–145	61.7–65.8	142–156	64.4–70.8	151–170	68.5–77.1
5	10	177.8	140–150	63.5–68	146–160	66.2–72.6	155–174	70.3–78.9
5	11	180.3	144–154	65.3–69.9	150–165	68.0–74.8	159–179	72.1–81.2
6	0	182.9	148–158	67.1–71.7	154–170	69.9–77.1	164–184	74.4–83.5
6	1	185.4	152–162	68.9–73.5	158–175	71.7–79.4	168–189	76.2–85.7
6	2	188	156–167	70.8–75.7	162–180	73.5–81.6	173–194	78.5–88
6	3	190.5	160–171	72.6–77.6	167–185	75.7–83.5	178–199	80.7–90.3
6	4	193	164–175	74.4–79.4	172–190	78.1–86.2	182–204	82.7–92.5
				Women				
4	10	147.3	92–98	41.7–44.5	96–107	43.5–48.5	104–119	47.2–54
4	11	149.9	94–101	42.6–45.8	98–110	44.5–49.9	106–122	48.1–55.3
5	0	152.4	96–104	43.5–47.2	101–113	45.8–51.3	109–125	49.4–56.7
5	1	154.9	99–107	44.9–48.5	104–116	47.2–52.6	112–128	50.8–58.1
5	2	157.5	102–110	46.3–49.9	107–119	48.5–54	115–131	52.2–59.4
5	3	160	105–113	47.6–51.3	110–122	49.9–55.3	118–134	53.5–60.8
5	4	162.6	108–116	49.0–52.6	113–126	51.3–57.2	121–138	54.9–62.6
5	5	165.1	111–119	50.3–54	116–130	49.0–59	125–142	49.4–64.4
5	6	167.6	114–123	51.7–55.8	120–135	54.4–61.2	129–146	58.5–66.2
5	7	170.2	118–127	53.5–57.6	124–139	56.2–63	133–150	60.3–68
5	8	172.7	122–131	55.3–59.4	128–143	58.1–64.9	137–154	62.1–69.9
5	9	175.3	126–135	57.2–61.2	132–147	59.9–66.7	141–158	64.0–71.7
5	10	177.8	130–140	59.0–63.5	136–151	61.7–68.5	145–163	65.8–73.9
5	11	180.3	134–144	60.8–65.3	140–155	63.5–70.3	149–168	67.6–76.2
6	0	182.9	138–148	62.6–67.1	144–159	65.3–72.1	153–173	69.4–78.3

[a]Weights of insured persons in the United States associated with lowest mortality. From *Metropolitan Life Insurance Company Statistical Bulletin* 40, Nov.–Dec. 1959.

2.1. ENERGY AND ENERGY REQUIREMENTS

2.1.1. Energy

The term energy is difficult to define for it is not entirely explicit. Standard dictionaries define energy as the power necessary to move from inertia. In the field of nutrition, energy is defined in chemical terms. As such, the unit of energy most commonly referred to is the calorie and most recently by world health organizations the joule.

A calorie is the amount of heat required to raise the temperature of 1 gram of water 1° centigrade (from 15 to 16° C). A kilocalorie is the amount of heat required to raise the temperature of 1 liter of water 1° centigrade. The kilocalorie or calorie is the term most commonly used by nutritionists.

A joule is defined as the amount of work done for heat generated by an electric current of 1 ampere acting for 1 second against the resistance of 1 ohm, all by force of 1 newton acting through a distance of 1 meter.

One calorie equals 4.184 joules and one kilocalorie is equal to 4.184 kilojoules (kJ). Most experts accept a rounded value of 4.2 kJ. Therefore, a daily allowance of 2000 kcal would equal a daily allowance of 8400 kJ. Because the use of joules requires many digits, another unit that is being used is the megajoule, and 1000 kcal is equal to 4.184 megajoules, 1000 kJ equaling 1 MJ.

Man's energy is obtained from food sources. However, there is poor understanding and poor agreement between estimates of energy intake and energy output. There appears to be tremendous daily variation in human energy requirements, as well as variation due to the type of food supply.[2]

Regardless of the total energy intake, what is most important is the *available energy* that is derived from the intake. Only energy that is converted to adenosine triphosphate (ATP) is available as stored energy for use in the body. The maximum yield from glucose and fat is approximately 38–40%, whereas the maximum yield from protein is between 32 and 48% of that ingested. The rest of the derived energy is wasted, or better put, used up in the process of creating the ATP energy store. The picture is further complicated by the fact that more energy is used up if a compound such as glucose is converted to fats. The details of these intricate interrelationships of energy use have been discussed in detail by Hegsted.[3]

In summary, it is important to remember that all energy is captured as ATP in energy-rich phosphate bonds. When energy is required ATP is broken down to ADP, releasing energy.

2.1.2. The Basal State

Basal metabolism can be defined as the energy required to carry on essential body processes at rest. The essential body work that occurs at rest in order to maintain life can be placed into three categories:

1. *Mechanical functions:* energy needed to maintain vital functions such as respiration movement, circulation, movement of body secretions, and maintenance of normal muscle tone.
2. *Excretory functions:* energy to continually get rid of body waste products such as in urine formation, transport of carbon dioxide, and maintenance of colonic stability.
3. *Synthesis functions:* energy needed to build all body products such as red blood cells, hormones, and enzymes.[4,5]

Table 2-2. Basal Metabolism Standard (male)

Age	kcal/m^2/hr
3	60
5	56
10	48
15	43
20	40
25	39
30	38
40	36
50	36
60	35
70	33
75	32

Basal metabolism will vary greatly with many factors. Table 2-2 displays the gradual decrease in calories required from age 3 through 75. A great deal of energy is required during the rapid-growth phases from ages 3 to 15; from age 20 to 75 growth has usually ceased, but yet there is a slow decline in the amount of basal energy that is required. There is still not a complete understanding, and one recent study suggests that active older men may not decrease their basal energy need.[5a]

Weight and height as well as age have a marked effect on basal energy requirements. Table 2-3 details the relationship of height and weight to the basal metabolic rate for a 22-year-old male or female according to the recommended dietary allowances. Note that there is a direct linear increase in caloric requirements according to optimal weight and height.

Table 2-3. The Basal Metabolic Rates of Adults According to Weights and Heights[a]

Height		Men			Women		
		Median weight		BMR	Median weight		BMR
in.	cm	lb	kg	kcal/day	lb	kg	kcal/day
60	152				109±9	50±4	1399
62	158				115±9	52±4	1429
64	163	133±11	60±5	1630	122±10	56±5	1487
66	168	142±12	64±5	1690	129±10	59±5	1530
68	173	151±14	69±6	1775	136±10	62±5	1572
70	178	159±14	72±6	1815	144±11	66±5	1626
72	183	167±15	76±7	1870	152±12	69±5	1666
74	188	175±15	80±7	1933			
76	193	182±16	83±7	1983			

[a]From Food and Nutrition Board: *Recommended Daily Allowances,* 8th rev. Washington, DC: National Academy of Sciences–National Research Council, 1974.

JOSEPH J. BARBORIAK, Sc.D.
Research Service (151B)
Veterans Administration Center
Wood, Wisconsin 53193

Table 2-4. Energy Cost of Activity for 70-kg Man/hr[a]

Activity	kcal
Rowing	1120
Swimming	553
Running	490
Dancing	266
Bicycling	175
Ping-Pong	308
Dishwashing or typing	70
Sitting	28
Lying still	7
Walking 3 mph	140
Walking 4 mph	238

[a]Exclusive of BMR and SDA.

2.1.3. Energy Expended

Once one arises from the basal state there is an increased requirement for energy in one of three ways: (1) voluntary activity, (2) external (environmental) stress, and (3) internal (disease) stress. All three of these increase various cellular levels of activity so that more energy is required. The exact increase in energy requirements caused by stress and infectious, metabolic, and degenerative diseases is poorly understood and has been inadequately studied for specific diseases. The best understanding is in surgical disorders.[5]

The increased expenditure of energy during various physical activities has been studied. Table 2-4 lists the calories needed to compensate for the energy expended by various common activities (computed from Ref. 1). It is obvious from the table that one who spends an hour running or swimming per day will be able to consume 500 cal more than one who does not. Although most exercise does not result in extreme use of energy, it does have a significant effect. For instance, comparing two given individuals, the one who spent the hour exercising could eat an extra piece of apple pie per day and gain no weight, whereas the other who did not exercise would gain a significant amount of weight from eating an extra dessert.

2.1.4. Energy Value of Foods

The total energy value of a given food is determined by measuring the amount of heat produced in a calorimeter. The bomb calorimeter is a completely insulated apparatus that records the heat produced from complete oxidation of matter and is measured by a change of the temperature in the water surrounding the apparatus.

Accepted biologic availability of energy for food categories is listed in Table 2-5. The heat of the combustion of protein is 5.6 kcal/g, carbohydrate 4.1 kcal/g, fat 9.4 kcal/g, and alcohol 7.1 kcal/g. However, when one is considering

Table 2-5. Biologic Available Energy (kcal/g)

	Heat of combustion	Loss in urine	Percentage absorbed	Atwater factor
Carbohydrate	4.1	0	99	4
Protein	5.6	1.25	92	4
Alcohol	7.1	Trace	100	7
Fat	9.4	0	95	9

the available energy in the body, one must calculate for loss in urine (protein) and the correct percentage of absorption, which is only 92% for proteins, 95% for fats, and 99% for carbohydrates. After the appropriate calculations as outlined by Atwater, the final kcal/g values are protein −4, carbohydrate −4, fat −9, and alcohol −7.

The ultimate role of ingested food in the body can be summarized as follows:

1. It is used directly for body structure (amino acids that enter directly into cellular structure).
2. It participates directly in specific actions (such as vitamins and minerals).
3. It is used directly for energy (such as glucose).
4. It is converted to energy-rich ATP for storage.
5. The nitrogen excess is excreted, not stored.
6. It passes through the alimentary canal unabsorbed for the purposes of (a) use by the microflora, (b) physiochemical reactions, and (c) excretion.

2.1.5. Recommended Dietary Allowances

In 1943 the Food and Nutrition Board, National Research Council–National Academy of Sciences published recommended dietary allowances. Periodically these have been revised. The most recent revision was in 1980.

"The Recommended Dietary Allowances are the levels of intake of essential nutrients considered, in the judgement of the Food and Nutrition Board on the basis of available scientific knowledge, to be adequate to meet the known nutritional needs of practically all healthy persons."[1]

These recommended dietary allowances are not to be confused with the United States RDA that is published by the Food and Drug Administration for the purpose of food labeling. The setting of two standards is confusing to most observers. Nevertheless, it exists and probably reflects the confusion in the entire field as to what is adequate intake. The United States RDA attempts to define it on the basis of minimal intake, whereas the Food and Nutrition Board readily admits that their RDA recommendations are probably above the minimum requirements.

Table 2–6. Food and Nutrition Board, National Academy of Sciences–National Research Council Recommended Daily Dietary Allowances (Revised 1980)[a]

	Weight		Height		Protein	Fat-soluble vitamins			Water-soluble vitamins							Minerals					
Age (yr)	(kg)	(lb)	(cm)	(in.)	(g)	Vita-min A (μg RE)[b]	Vita-min D (μg)[c]	Vita-min E (mg α-TE)[d]	Vita-min C (mg)	Thia-mine (mg)	Ribo-flavin (mg)	Niacin (mg NE)[e]	Vita-min B_6 (mg)	Fola-cin[f] (μg)	Vitamin B_{12} (μg)	Cal-cium (mg)	Phos-phorus (mg)	Mag-nesium (mg)	Iron (mg)	Zinc (mg)	Iodine (μg)
Infants 0.0–0.5	6	13	60	24	kg×2.2	420	10	3	35	0.3	0.4	6	0.3	30	0.5[g]	360	240	50	10	3	40
0.5–1.0	9	20	71	28	kg×2.0	400	10	4	35	0.5	0.6	8	0.6	45	1.5	540	360	70	15	5	50
Children 1–3	13	29	90	35	23	400	10	5	45	0.7	0.8	9	0.9	100	2.0	800	800	150	15	10	70
4–6	20	44	112	44	30	500	10	6	45	0.9	1.0	11	1.3	200	2.5	800	800	200	10	10	90
7–10	28	62	132	52	34	700	10	7	45	1.2	1.4	16	1.6	300	3.0	800	800	250	10	10	120
Males 11–14	45	99	157	62	45	1000	10	8	50	1.4	1.6	18	1.8	400	3.0	1200	1200	350	18	15	150
15–18	66	145	176	69	56	1000	10	10	60	1.4	1.7	18	2.0	400	3.0	1200	1200	400	18	15	150
19–22	70	154	177	70	56	1000	7.5	10	60	1.5	1.7	19	2.2	400	3.0	800	800	350	10	15	150
23–50	70	154	178	70	56	1000	5	10	60	1.4	1.6	18	2.2	400	3.0	800	800	350	10	15	150
51+	70	154	178	70	56	1000	5	10	60	1.2	1.4	16	2.2	400	3.0	800	800	350	10	15	150
Females 11–14	46	101	157	62	46	800	10	8	50	1.1	1.3	15	1.8	400	3.0	1200	1200	300	18	15	150
15–18	55	120	163	64	46	800	10	8	60	1.1	1.3	14	2.0	400	3.0	1200	1200	300	18	15	150
19–22	55	120	163	64	44	800	7.5	8	60	1.1	1.3	14	2.0	400	3.0	800	800	300	18	15	150
23–50	55	120	163	64	44	800	5	8	60	1.0	1.2	13	2.0	400	3.0	800	800	300	18	15	150
51+	55	120	163	64	44	800	5	8	60	1.0	1.2	13	2.0	400	3.0	800	800	300	10	15	150
Pregnant					+30	+200	+5	+2	+20	+0.4	+0.3	+2	+0.6	+400	+1.0	+400	+400	+150	h	+5	+25
Lactating					+20	+400	+5	+3	+40	+0.5	+0.5	+5	+0.5	+100	+1.0	+400	+400	+150	h	+10	+50

[a] The allowances are intended to provide for individual variations among most normal persons as they live in the United States under usual environmental stresses. Diets should be based on a variety of common foods in order to provide other nutrients for which human requirements have been less well defined.

[b] Retinol equivalents. 1 retinol equivalent = 1 μg retinol or 6 μg carotene.

[c] As cholecalciferol. 10 μg cholecalciferol = 400 IU of vitamin D.

[d] α-Tocopherol equivalents. 1 mg d-α tocopherol = 1 α-TE.

[e] 1 NE (niacin equivalent) is equal to 1 mg of niacin or 60 mg of dietary tryptophan.

[f] The folacin allowances refer to dietary sources as determined by Lactobacillus casei assay after treatment with enzymes (conjugases) to make polyglutamyl forms of the vitamin available to the test organism.

[g] The recommended dietary allowance for vitamin B_{12} in infants is based on average concentration of the vitamin in human milk. The allowances after weaning are based on energy intake (as recommended by the American Academy of Pediatrics) and consideration of other factors such as intestinal absorption.

[h] The increased requirement during pregnancy cannot be met by the iron content of habitual American diets nor by the existing iron stores of many women: therefore, the use of 30–60 mg of supplemental iron is recommended. Iron needs during lactation are not substantially different from those of nonpregnant women, but continued supplementation of the mother for 2–3 months after parturition is advisable in order to replenish stores depleted by pregnancy.

The RDA for 1980 is listed in Table 2-6 and includes recommendations for energy, proteins, vitamins, and minerals. In the past the RDA was established for a 70-kg man or 58-kg woman at age 22 in a 20°C environment with no disease, no stress, and growth phase ended. The 1980 RDA does list allowances according to age, weight, and height, and subsidies for both pregnant and lactating females. It is of importance to note that the 1980 RDA does not list any recommendations for carbohydrates, fats, or fiber; however, the United States Senate Select Committee hearings on nutrition have resulted in recommendations concerning the intake of carbohydrates, fats, and fiber. This is new in the United States, but such recommendations exist in Scandinavia. The committee's recommendations have not been accepted nationally and controversy prevails. Nevertheless, the impetus of recent epidemiologic studies has resulted in a strong commitment by the United States government so that one can be assured that more concrete scientifically accurate recommendations will be forthcoming in the future.[6]

2.1.6. Energy Requirements

As noted above and as listed in Table 2-6 on RDA the energy requirements in kilocalories vary for weight, height, and age and are increased with activity. The RDA does not account for activity, or environmental or disease stresses. Keeping those in mind, the RDA requirements in kilocalories reach a peak at approximately age 22 when the growth phase has totally ended, and then gradually decrease with aging. Table 2-2 lists the gradual decrease in caloric need per square meter in the basal state with aging. Note that the decrease in requirement is slow and gradual from the end of growth phase compared with the rapid decrease in requirement during the first few decades of life.

In summary, Fig. 2-1 is a simplistic outline of energy production and storage. All of human energy comes from foods. The simplest food is glucose, which can be combusted in the presence of oxygen to produce carbon dioxide and water and the release of energy. This energy is then used, or stored, by the conversion of adenosine diphosphate to adenosine triphosphate, and storage of these energy-rich phosphate bonds in man occurs in fat. Other foods can enter

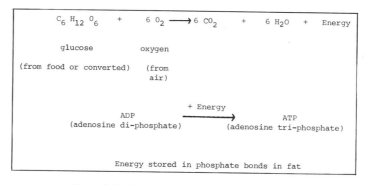

Figure 2-1. Energy production and storage.

Table 2-7. Amino Acids

Essential	Nonessential
Histidine[a]	Alanine
Isoleucine	Arginine
Leucine	Asparagine
Lysine	Aspartic acid
Methionine	Cysteine
Phenylalanine	Cystine
Threonine	Glutamic acid
Trytophan	Glutamine
Valine	Glycine
	Hydroxyproline
	Proline
	Serine
	Tyrosine

[a]Histidine is required for infants, but it is not clear how essential it is for adults.

into this process by being converted to glucose. Carbohydrates and fats yield the maximum conversion to ATP, whereas protein foods are slightly lower energy sources.

2.2. PROTEINS

Proteins reflect the complexity of humans. They are composed of carbon, hydrogen, oxygen, and nitrogen. Other elements such as sulfur, phosphorus, iodine, and iron may be incorporated to form important protein complexes. Nitrogen is the element that makes proteins different from carbohydrates and fats. Humans are capable of producing most proteins, although there are some essential proteins that only plants produce.

2.2.1. Chemistry of Proteins

Protein chemistry is a complex science and the purpose of this book is not to review all those complexities, but to summarize information necessary for a basic understanding of nutrition. Proteins, composed of approximately 16% nitrogen, are polymers of amino acids. There are 22 amino acids in man (see Table 2-7). Eight of the amino acids are considered essential. Histidine appears to be required for infants; its requirement in adult humans is still under debate. All of the amino acids contain the same terminal group as shown in Fig. 2-2. The simplest amino acid, glycine, merely has a hydrogen atom attached to the terminal group. The next simplest amino acid is alanine, which has a carbon attached to the basic group. The other 20 amino acids have a variety of carbon atoms and molecular structures attached to the terminal group. As shown in Fig. 2-2, if two of the amino acids group together, then one has a dipeptide. Glycine plus alanine will give glycylalanine. Because proteins are polymers,

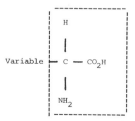

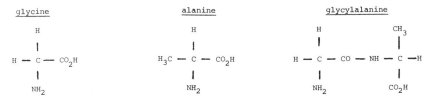

Figure 2-2. Basic amino acid structure.

the molecules are attached in a three-dimensional arrangement. Terms for the different spatial arrangements are alpha, beta, etc. When many dipeptides are bonded together, then one has peptides and so on to form complex proteins. When one considers the great variety of combinations of these 22 proteins, in a great variety of spatial arrangements, then the complexity of protein structure can be appreciated.

2.2.2. Digestion and Absorption

Amino acids, dipeptides, and large peptide bonds may all be absorbed directly through the intestinal mucosa.[7] Proteins from foods are not usually ingested as long peptide chains. Pepsinogen produced in the stomach is activated by acid in the stomach to the enzyme pepsin, which attacks long peptide chains, breaking them into smaller peptide and dipeptide chains. These then pass into the small intestine where they may be further digested into simple dipeptides and amino acids.

The enzymes that assist in performing this action are secreted by the pancreas in the form of proenzymes that become activated once secreted into the gut lumen. The pancreatic enzymes can be divided into *endopeptidases,* trypsin, chymotrypsin, and elastase, which attack the interior peptide bonds; and *exopeptidases,* which attack the end carboxyl bonds and are appropriately called carboxypeptidases.[8] In addition to these enzymes there are numerous peptidases in the brush border of the intestinal epithelial cells that can break peptide bonds and assist in absorption. This is a very complex mechanism and can occur at the cell surface. When one considers the myriad of types of amino acid combinations, one can appreciate the complexity of the enzyme activity at the epithelial cell surface. It now also appears that the secretion of the protein enzymes is governed by a feedback mechanism, the pancreatic cells being stimulated to produce more enzyme when trypsin becomes bound to protein in the gut lumen.[9]

Amino acids, as the final product of protein digestion, were thought to be the major final pathway of absorption into the intestine. Evidence indicates there is great uncertainty concerning this, and in all probability dipeptides might be the important protein substance absorbed. Although amino acids are absorbed by an active competitive carrier mechanism through the small intestine, in which the different amino acids compete with each other for the carrier, it is also readily apparent that dipeptides can be absorbed intact.[10,11] Evidence also indicates that dipeptide and tripeptide proteins are probably more easily absorbed by the same carrier system and hence more beneficial to man than simple amino acids.[7,12] The initial stage of digestion yields approximately 30% neutral and basic amino acids and 70% small peptides. These are transported by specific carrier mechanisms across the brush border. Hydrolysis can occur on either side of the brush border. Approximately 10% of the peptides containing predominantly glycine or proline–hydroxyproline enter and exit through the intestinal cell unhydrolyzed.[8,8a]

The bulk of amino acids pass into the portal vein system to be utilized by the body. However, a significant amount is utilized by intestinal cells for synthesis of important body proteins. A small amount is absorbed as tetrapeptides and appears to be hydrolyzed by the brush border and cell enzymes, and some pass through to the urine when not used by the body.[7] The jejunum is the most active site of protein absorption.

2.2.3. Regulation of Protein Metabolism

The amount of protein ingested in Western societies varies greatly. On average there is an approximate dietary intake of 100 g of protein per day, but the small bowel secretes another 70 g/day in the form of shed intestinal cells. From this mass of ingested plus gut-secreted protein material, less than 10 g is excreted. Consequently, the small intestine not only has an absorptive function, but a regulative function as well in that the shedding of cells and reabsorption of protein material are important aspects of body protein metabolism.[13]

Once proteins pass through the intestinal cells and into the portal blood flow, the liver takes over as a pivotal site in the production of important body proteins. As noted in the chapter on liver diseases (Chapter 10), the mechanisms for production of albumin can easily be stimulated or suppressed by the blood levels of protein that present to the liver. The body muscle mass requires great amounts of amino acids. Brain metabolism is greatly affected by certain amino acids circulating in the blood such as tryptophan. For example, insulin will have a marked effect on the metabolism that occurs both in muscle and in brain by increasing the uptake in muscle as well as increasing the level of serotonin formation.[14]

2.2.4. Requirement

The 1980 RDA dropped the requirement of protein from 0.9 to 0.8 g/kg body weight.[1] Specifically, this means that the adult male between the ages of 23 and 50 now requires only 56 g of protein, whereas previously 65 g was suggested. Similarly, the adult female requirement has dropped from 55 to 46 g.

Some controversy still exists concerning the exact safe requirement. Short-term studies revealed that 0.59 g egg protein/kg body/day was a safe level, as reported in the 1973 FAO/WHO expert committee report on energy and protein requirements. Later studies have revealed that this is not a safe level for long-term use because both young and old men will develop a negative nitrogen balance at this level of protein intake.[15,16] There is much disagreement concerning exact quantitative and qualitative protein requirements. Energy intake has a definite effect on protein requirement.[17] At marginal energy intake the body's need for nitrogen rises.[17] Long-term studies have revealed that at high protein intake, the body does not appear to adapt and holds onto larger amounts of nitrogen.[18]

There are eight and probably nine essential amino acids that must be included in the protein intake. Table 2-8 lists the average grams per day of various amino acids that should maintain normal growth and cell turnover as can be determined from the present state of knowledge.[3] Note the great variability between males and females and among individual proteins—tryptophan need at 0.157 to 0.25 g/day and leucine at 0.62 to 1.1. g/day. It is also important to note that the average American diet often far exceeds the basic recommended daily intake.

2.2.5. Food Protein Requirements

The latest volume of Church and Church lists the food values of portions commonly used and their protein and amino acid contents. The interested

Table 2-8. Estimated Minimal Requirements of Essential Amino Acids for Normal Adults[a]

Essential amino acids[b]	Female (g/day)	Male (g/day)	Recommended daily intake (g/day)	Average American diet (g/day)
Tryptophan	0.157	0.25	0.5	0.9
Threonine	0.350	0.50	1.0	2.8
Isoleucine	0.450	0.70	1.4	4.2
Leucine	0.620	1.10	2.2	6.5
Lysine	0.500	0.80	1.6	4.0
Methionine	0.350	1.10	2.2	3.0
Phenylalanine	0.220	1.10	2.2	4.1
Valine	0.650	0.80	1.6	4.2
Histadine	—	—	—	—

[a]Modified from Hegsted.[3]
[b]Cystine may replace part of the methionine allowance (about one-sixth). Tyrosine may replace part of the phenylalanine allowance (about one-half).

Table 2-9. Amounts of Essential Amino Acids in Various Proteins (mg/g 20N)

	Isoleucine	Leucine	Lysine	Phenyl-alanine	Methio-nine	Threo-nine	Trypto-phan	Valine	Protein score as measure of excellence
Amino acid combination estimated as ideal for man	270	306	270	180	144	180	90	270	100
Egg protein	428	565	396	368	196	310	106	460	100
Beef	332	515	540	256	154	275	75[a]	345	83
Milk protein	402	628	497	334	190	272	85[a]	448	80
Fish	317	474	549	231	178	283	62[a]	327	70
Oat protein	302	436	212[a]	309	84[a]	192	74[a]	348	79
Rice protein	322	535	236[a]	307	142[a]	241	65[a]	415	72
Flour protein	262[a]	442	126[a]	322	78[a]	174	69[a]	262	47
Maize protein	293	827	179[a]	284	117[a]	249	38[a]	327	42
Soya protein	333	484	395	309	86[a]	247	86[a]	328	73
Pea protein	336	504	438	290	77[a]	230	74[a]	317	58
Potato protein	260	304	326	285	87[a]	237	72[a]	339	56
Cassava	118[a]	184[a]	310	133[a]	22[a]	136[a]	131	144	22

[a]Less than the estimated ideal portion.

reader can scrutinize that volume for more detailed analysis of average foods.[19] Table 2-9 lists the amino acid content of some common protein foods and rates the content against that estimated to be ideal for man.[20] Note that egg protein meets all of the requirements.

It is evident from the above that complex scientific data exists on the protein available in foods and its absorptive mechanism. What comes through clearly to the interested clinical nutritionist is that there must be an adequate supply of essential amino acids to the gut and that it can be supplied by ordinary foods in our society. As eggs and beef may not be the most desirable foods, and should not be the only source of protein, protein therefore must come from a balance of foods.[21]

Nitrogen is the essential component of proteins. However, there are other sources of nitrogen in the body, e.g., ammonia, uric acid, urea, purines, creatinine, and creatine. These substances make up a part of the nonspecific nitrogen, which can be defined as "nitrogen that is metabolically available, and leads to minimal toxicity in quantities used."[22] It may include nitrogen that is furnished by nonessential amino acids, nitrogen from excess amino acids, and from nonprotein sources such as listed above. At the present time, the role of nonspecific nitrogen in the nutrition of human beings is not certain. It seems to affect the human protein and specific essential amino acid requirements, but it is not always in the same direction or in the same degree in different individuals. A much better understanding of the importance of nonspecific nitrogen in human nutrition is needed for it is a potential source of protein dietary supplementation.[22]

2.2.6. Carbohydrates

Carbohydrates are the principle source of energy for humans. As noted in the section on energy, it supplies approximately 4 kcal/g. Plants store energy in the form of carbohydrates by their ability to convert carbon dioxide and water in the presence of chlorophyll into carbohydrate. (Remember humans store energy as fat.) Throughout the world carbohydrates vary in their importance as a source of food, but they represent the major source of energy supply to the majority of peoples. Cereals and roots have been the major source of carbohydrates for thousands of years. Only in Western societies have so-called "refined carbohydrates," sucrose, glucose, and fructose, become a major source of carbohydrates. This latter change is presently undergoing intensive scrutiny as a possible cause of a scourge of degenerative diseases.

Table 2-10 classifies the common carbohydrates available to man. Sugar alcohols such as sorbitol and mannitol are rarely used as energy sources; more recently they have been used in drug products. Ethanol is discussed in the chapter on liver diseases (Chapter 10) and surprisingly has a very high percentage intake in Western societies. It may make up as much as 6% of the average caloric intake of some societies. Although ethanol yields 7 kcal/g, it nonethe-

Table 2-10. Carbohydrates

---

1. Monosaccharides
   Hexoses (glucose, fructose, galactose, etc.)
   Pentoses (ribose, deoxyribose, etc.)

2. Sugar alcohols (sorbitol, mannitol) 2 kcal/g

3. Ethanol 7 kcal/g

4. Disaccharides
   Maltose (glucose + glucose)
   Sucrose (glucose + fructose)
   Lactose (galactose + glucose)

5. Polysaccharides
   Amylose (starch)
   Amylopectin (starch)
   Dextrins
   Inulin (starch, fructose)
   Pectin
   Gums
   Celluloses (fiber)
   Hemicelluloses (fiber)
   Lignin (fiber)

---

less must be classified as a drug and has deleterious side effects, as opposed to nutritional carbohydrates. It is essentially a fermentation product of glucose.

Major sources of energy are monosaccharides, disaccharides, and polysaccharides.

Glucose, fructose, and lactose all contain six carbon atoms and therefore are referred to as hexoses. The structures of glucose and fructose are shown below.

$$
\begin{array}{cc}
\text{C–OH} & \text{C–H}_2\text{OH} \\
| & | \\
\text{H–C–OH} & \text{C–O} \\
| & | \\
\text{HO–C–H} & \text{HO–C–H} \\
| & | \\
\text{H–C–OH} & \text{H–C–OH} \\
| & | \\
\text{H–C–OH} & \text{H–C–OH} \\
| & | \\
\text{C–H}_2\text{OH} & \text{C–H}_2\text{OH} \\
\text{Glucose} & \text{Fructose}
\end{array}
$$

Glucose is sweet, soluble in water, found naturally in grapes, sweet fruits, and onions, and constitutes 35% of honey.

Fructose, structurally similar to glucose, is much sweeter and has some markedly different properties. Specifically, it is absorbed much more rapidly in a totally passive process and appears to have no effect on glucose tolerance in diabetic subjects.

Sucrose is a polymer of glucose and fructose and is the major carbohydrate used by Western societies. It is refined cane and beet sugar and is added to most processed foods.

Lactose is the disaccharide sugar found in milk. It is a structural combination of glucose and galactose.

Maltose is the disaccharide sugar that results from enzymatic breakdown of barley. Germinated barley may be heated and dried to give malt. Malt extract contains maltose, which gives it its sweetness.

Trisaccharides and tetrasaccharides, such as raffinose and stachyose, play a very minor role as food substances.

Polysaccharides, in a historical sense, have been the major carbohydrate and hence major energy source of humans since primitive hunters learned to utilize plants. Table 2-10 lists the major polysaccharides used by man. Polysaccharides are long chains of monosaccharides in a variety of spatial arrangements. Molecules such as cellulose may contain up to 3000 or more glucose units. It differs from amylose by the spatial arrangement of glucose units but has extremely different physical and nutritional properties.

Starch is a mixture of amylose and amylopectin. These are broken down exclusively to the disaccharide maltose during digestion. When starch is cooked it is often broken into smaller chains of branched and unbranched amylose and amylopectin called dextrin. Inulin, which is made up of chains of fructose, has very little nutritional value and is found in Jerusalem artichokes.

Pectin, gums, celluloses, hemicelluloses, and lignin are polysaccharides that have low caloric value but a very significant physiologic role in the gut. Until recently, Western societies had processed these substances out of the diet because of their low caloric value. Recent evidence indicates they may be extremely important because of their physical properties, and hence their absence may be related to the cause of degenerative disease (see Chapters 4 and 14). Because of their low nutritional value, they have been poorly studied by recent scientific methods and little information is available.

We do know that pectin is composed of many units of galacturonic acid, which is closely related to galactose. Cellulose is a long chain of glucose units whose glucose molecules are alternately linked in different spatial arrangements. Because of this, it is poorly digested in humans and hence remains relatively intact unless there is bacterial enzymatic breakdown while in the gut. Hemicelluloses and lignins will be discussed in detail in the chapter on fiber (Chapter 4).

2.2.7. Digestion and Absorption of Carbohydrates

When carbohydrates are ingested, they are first digested by the action of amylase from salivary secretions. Most studies indicate amylase has little effect on the breakdown of small chains of saccharides, but attacks the longer chains of amylose and amylopectin present in starch. The major digestion occurs when the chyme enters the duodenum where pancreatic α amylase hydrolyzes the 1,4-α-glucose–glucose linkages in amylose and the 1,6-α-glucose linkages in amylopectin. The ultimate breakdown product of this enzymatic digestion is either maltotriose (three glucose molecules) or α-dextrins. It does not release any free glucose or attack any of the disaccharide chains.

After pancreatic amylase performs its function, short-chain saccharides are presented to the intestinal mucosal epithelium where oligosaccharidases located on the brush border membrane of the superficial and lateral intestinal epithelial cells are active. The oligosaccharidases are specific for disaccharides and dextrins. Lactase is specific for the disaccharide lactose. It is commonly deficient in humans and results in lactase deficiency that may be symptomatic; this is discussed in detail in Chapter 14.

Once the disaccharidases lactase, maltase, and sucrase hydrolyze their respective disaccharide, the resultant hexose is absorbed through the intestinal cells from the brush border. Glucose and galactose are actively transported against a concentration gradient by a sodium-dependent mechanism. This requires an energy processor that is theorized to be a carrier mechanism. Fructose is absorbed by simple diffusion, and although it appears to require carrier mechanisms, it does not require any energy. Once within the intestinal cells, these simple hexoses are transported into the portal blood for utilization in the liver and throughout the body.[23,24]

Carbohydrate metabolism is discussed further under the specific disorders of pancreatic, small-intestine, and liver diseases. It is important to note that glucose is a major source of energy for nervous tissue and must be available at all times. Energy in humans is stored in phosphorus-rich bonds in fat, but a large source of carbohydrate is stored in the form of glycogen in the liver that is readily converted to glucose for rapid sources of the hexose to maintain homeostasis.

2.2.8. Carbohydrate Requirements

Because optimal protein and fat requirements are limited, the major energy source of the total caloric intake should be in the form of carbohydrates. As noted in Section 2.1.5., there is no recommended intake for carbohydrates. Out of necessity it must make up at least 50% of the energy source, and the most recent recommendation of the United States Senate Select Committee studying nutrition was that carbohydrates should make up approximately 60 to 65% of the optimal nutritional caloric intake.[6] In order to perform this task, they should be in the form of cereals and vegetables, as was common prior to the glut of processed foods.[6] This implies that the Western diet that includes a very high intake of sucrose from refined sugar and beet cane is in some way harmful, as opposed to a diet rich in cereal and root carbohydrates that includes starches and high fiber content.[25]

The minimal requirement of carbohydrate is not understood. However, humans require anywhere from 3 to 5 g of carbohydrate or else ketosis will result, regardless of the total caloric intake.[26]

2.2.9. Carbohydrate Foods

Table 2-11 lists the percentage of carbohydrates in a variety of food categories.

Note that cereals are composed of anywhere from 70 to 85% carbohy-

Table 2-11

Food	Approximate % carbohydrate
Cereal grains	
Rice No. 1	79
Wheat No. 2	72–78
Rye	68–78
Corn (maize)	75
Oats	70
Barley	79
Fruits	6–20
Vegetables	3–35
Nuts	10–27
Eggs	0
Meats and fish	0–4

drates, whereas foods such as eggs, cheese, meats, and fish have relatively no carbohydrates. There is little information available on the type of carbohydrates that are present in foods. The difficulties in chemical analysis and the variability among products are first being realized, and only recently studied in more detail. Such studies reveal that there is a great deal of variability in the carbohydrate content of plant foods depending upon the site and growth factors.

In summary, humans use carbohydrates as the major source of food energy. The most recent estimates are that the best carbohydrate sources are cereals, vegetables, and roots. These should make up approximately 60–65% of the energy source. The carbohydrates are digested relatively easily in the small intestine by pancreatic amylase and at the surface of the intestinal epithelial cells by disaccharidases. They are transported through the intestinal cells into the blood with glucose as a ready source of energy for nervous and muscular tissue. The optimal foods to supply this energy for humans are not completely understood, but it is apparent that polysaccharides are the best sources. It is also apparent that poorly digested and absorbed carbohydrates, commonly called fiber substances, are also important in the homeostasis of human gut ecology.

2.2.10. Fats and Lipids

Fats are important to man because of their high energy value (9 kcal/g as opposed to 4 kcal/g for proteins and carbohydrates) and because they are the form in which all energy is stored in the body. The use of fats in foods has varied greatly. Asian and African societies tend to derive as little as 20% of their total calories in the form of fat, whereas Western societies derive as much as 45%. Furthermore, in Western societies there has been a tremendous shift in the type of fats ingested. During the 1940s and 1950s as much as 25% was animal fat, whereas this has fallen during this decade to less than 15% (see Table 2-12).

Table 2-12. Sources of Nutrient Fat in Fats and Oils Consumed per Capita per Day in the United States[a]

| Period | Animal sources (g) | | | | Vegetable sources (g) | | | |
	Butter	Lard	Edible beef fats	Total	Margarine	Shortening	Salad and cooking oils	Total
1947–1949	10.6	16.4	0.5	27.5	5.5	10.6	9.1	25.2
1957–1959	8.2	14.5	1.8	24.5	8.7	9.7	13.5	31.9
1967	5.5	11.1	3.0	19.6	9.7	13.2	18.8	41.6
1971	5.1	9.4	3.0	17.6	10.0	14.9	22.4	47.3
1973	4.7	7.2	2.8	14.7	10.6	15.6	25.5	51.7

[a] Adapted from Mitchell et al.[40]

Table 2-13. Classification of Lipids (Fats and Oils)

1. Simple lipids
   a. Triglycerides—esters of fatty acids with glycerol
   b. Esters of fatty acids with high-molecular-weight alcohols
      Waxes
      Cholesterol esters
      Vitamins A and D esters
2. Compound lipids
   a. Phospholipids—phosphorus-containing lipids
      Lecithins
      Cephalins
      Sphingomyelins
   b. Glycolipids—sugar-containing lipids
      Cerebrosides
      Gangliosides
   c. Sulfolipids—sulfur-containing lipids
   d. Lipoproteins—protein-containing lipids
   e. Lipopolysaccharides—polysaccharide-containing lipids
3. Derived lipids (may be simple or compound lipid)
   a. Fatty acids
   b. Alcohols
      Glycerol
      High-molecular-weight alcohols
   c. Mono- and diglycerides
   d. Sterols
      Cholesterol
      Ergosterol
      Bile acids
      Steroid hormones
      Vitamin D
   e. Hydrocarbons
      Squalene
      Carotenoids
      Aliphatic hydrocarbons
   f. Fat-soluble vitamins
      Vitamin A
      Vitamin E
      Vitamin K

The questions of what type and quantity of fat are correct in the diet remain a problem to Western societies. Due to their great satiating capacity, they continue to be popular. However, as research evolves and as important as fiber and polysaccharides may become, so unimportant or deleterious may fats become. The true value of fats may evolve to be a minimal essential basic requirement.

2.2.11. Chemistry of Fat

Table 2-13 is a classification of all lipids. The term "lipid" is preferred over "fats" in an encompassing classification system because it includes oils and other substances that have physicochemical properties of being insoluble in water, such as sterols.

Fats, solid at room temperature, and oils, liquid at room temperature, make up the major lipid content of foods and are in the form of triglycerides. Triglycerides are esters of glycerol in combination with three fatty acids:

Triglyceride

H
|
H–C–O–fatty acid   (position 1)
|
H–C–O–fatty acid   (position 2) (β)
|
H–C–O–fatty acid   (position 3)
|
H

glycerol + 3 fatty acids

Any of the fatty acids may be attached to the glycerol molecule in a variety of combinations. Foods contain a variety of combinations produced by animals and plants.

An important property of fatty acids is the saturation of carbon atoms with hydrogen. Table 2-14 lists the number of carbon atoms and the number of unsaturated carbon bonds in the most common fatty acids.

Figure 2-3 depicts palmitic acid, a 16-carbon chain that is totally saturated and commonly found in harder animal fats, and linoleic acid, which is an 18-carbon chain, four of which are not fully saturated. Both are commonly found in vegetables and polyunsaturated fats. Foods contain a variety of com-

Table 2-14. Common Fatty Acids

Name	Number of carbon atoms	Number of unsaturated bonds
Butyric	4	0
Caproic	6	0
Caprylic	8	0
Capric	10	0
Lauric	12	0
Myristic	14	0
Palmitic	16	0
Stearic	18	0
Oleic	18	1
Linoleic	18	2
Linolenic	18	3
Arachidonic	20	4

PALMITIC ACID

LINOLEIC ACID

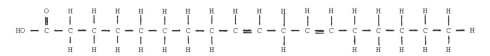

Figure 2-3. Palmitic acid and linoleic acid.

binations of triglycerides and fatty acids. Table 2-15 lists the proportions of saturated, monosaturated, and polyunsaturated fatty acids in common animal and vegetable fats and oils in our diet.

Many compound and derived lipids as listed in Table 2-13 are essential factors in metabolism, e.g., lecithin and lipoproteins. Those derived from the sterol skeleton (see Fig. 2-4), vitamin D, bile acids, cholesterol, and the steroid sex hormones estrogen, progesterone, and testosterone, are all pivotal in homeostasis.

2.2.12. Digestion and Absorption of Fat

There is virtually no digestion of lipid material in the mouth or stomach. Lipids do have a marked effect on gastric emptying, the stomach emptying slower when a rich lipid meal is eaten. Once lipids enter the duodenum, they stimulate it to secrete pancreozymin–cholecystokinin (CCK–PZ), which in turn

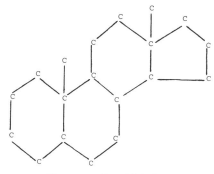

Figure 2-4. Steroid skeleton.

Table 2-15. Analyses of Fatty Acids Typical of Some Fats of Animal and Plant Origin[a,b]

| | Saturated | | | | Monounsaturated | | Polyunsaturated | | | |
	Lauric 12:0	Myristic 14:0	Palmitic 16:0	Stearic 18:0	Palmitic 16:1	Oleic 18:1	Linoleic 18:2	Linolenic 18:3	Arachidonic 18:4	Other polyenoic acids
Animal fats										
Lard	—	1.5	27.0	13.5	3.0	43.5	10.5	0.5	—	—
Chicken	2.0	7.0	25.0	6.0	8.0	36.0	14.0	—	—	—
Egg	—	—	25.0	10.0	—	50.0	10.0	2.0	3.0	—
Beef	—	3.0	29.0	21.0	3.0	41.0	2.0	0.5	0.5	—
Butter	3.5	12.0	28.0	13.0	3.0	28.5	1.0	—	—	—
Human milk	7.0	8.5	21.0	7.0	2.5	36.0	7.0	1.0	0.5	—
Menhaden (fish)	—	9.0	19.0	5.5	16.0	—	—	—	—	48.5
Vegetable oils										
Corn	—	—	12.5	2.5	—	29.0	55.0	0.5	—	—
Peanut	—	—	11.5	3.0	—	53.0	26.0	—	—	—
Cottonseed	—	1.0	26.0	3.0	1.0	17.5	51.5	—	—	—
Soybean	—	—	11.5	4.0	—	24.5	53.0	7.0	—	—
Olive	—	—	13.0	2.5	1.0	74.0	9.0	0.5	—	—
Coconut	49.5	19.5	8.5	2.0	—	6.0	1.5	—	—	—

[a] Adapted from *Dietary Fat and Human Health*, National Research Council Publication 1147, Washington, DC, 1966.[40]
[b] Composition is given in weight percentages of the component fatty acids (rounded to nearest 0.5) as determined by gas chromatography. The number of carbon atoms: number of double bonds are indicated under the common name of the fatty acid. These data were derived from a variety of sources. They are representative determinations, rather than averages, and considerable variation is to be expected in individual samples from other sources.

Luminal phase

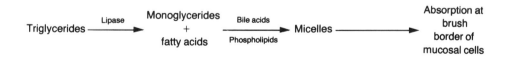

Lipolysis ⟶ Micellar solubilization

Triglycerides →(Lipase) Monoglycerides + fatty acids →(Bile acids, Phospholipids) Micelles ⟶ Absorption at brush border of mucosal cells

Mucosal intracellular phase

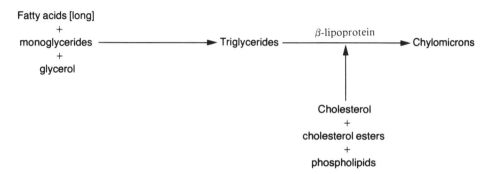

Fatty acids [long] + monoglycerides + glycerol ⟶ Triglycerides →(β-lipoprotein) Chylomicrons

Cholesterol + cholesterol esters + phospholipids

Figure 2-5. Fat digestion and absorption.

stimulates pancreatic and biliary secretions. Intraluminal digestion occurs and rapidly progresses to mucosal cell absorption as outlined in Fig. 2-5.[24] Triglycerides are hydrolyzed by the action of pancreatic lipase, which is secreted rapidly into the duodenum. The triglycerides are broken down into glycerol, fatty acids, and mainly β-monoglycerides with the fatty acid in the number 2 position remaining. Some diglycerides and triglycerides remain, but they are estimated to be less than 2%. At the same time, bile acids are rapidly being secreted into the duodenum. CCK–PZ also stimulates the gallbladder to contract and release bile. The conjugated bile acids are most effective in joining with phospholipids and fatty acids to form micelles that act as a solubilizing agent. The bile acids act as a detergent and the entire micellar complex is now soluble. This complex is presented to the brush border, or microvillous membrane of the mucosal cells. In addition to fatty acids, some cholesterol and steroid substances may be engulfed in the micelles. The micellar complex permits the fatty acids and β-monoglycerides to pass through the water layer of the microvillous membrane by passive diffusion. The mucosal cell also has a lipid cell membrane, but long-chain fatty acids and β-monoglycerides easily pass through this membrane. The conjugated bile acids are not absorbed during this phase and pass on down the jejunum to the ileum. Some bile acid may be absorbed in the jejunum, but the majority is absorbed further down in the ileum after it has been deconjugated by bacterial enzymes.

Once within the cells, fatty acids and β-monoglycerides are reformed to triglycerides in combination with newly formed glycerol. The new triglycerides are then coated with β-lipoprotein to form a chylomicron. Chylomicrons may engulf small amounts of cholesterol, cholesterol esters, and phospholipids into the lipoprotein coating.[28-30]

The absorption of fat-soluble vitamins, A, D, E, and K, is similar to triglyceride absorption. All of the vitamins are hydrolyzed by pancreatic enzymes to free vitamins and long-chain fatty acids, which are solubilized as micelles and transported through the microvillous membrane into the cells. All are absorbed by passive diffusion. Vitamins D, E, and K are not reformed whereas vitamin A is, and all are transported out of the cell as chylomicrons.

2.2.13. Metabolism and Fatty Acids

All stored energy in the body is in the form of fat. As outlined in Table 2-16 fat that is absorbed into the gut through the small intestine circulates into the lymph in the form of chylomicrons and then into the systemic circulation. Short- and medium-chain fatty acids form triglycerides that pass directly through the portal blood to the liver. Once within the liver (see Chapter 10), the hepatocytes actively metabolize triglycerides, phospholipids, and cholesterol and coat them with lipoproteins of either high density, low density, or very low density. This complex is then circulated to tissues in the body. The entire

Table 2-16. Fat Metabolism

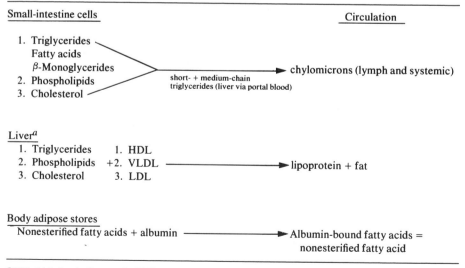

[a]HDL, high-density lipoprotein; VLDL, very-low-density lipoprotein; LDL, low-density lipoprotein.

physiology of lipoprotein combinations with fats is under intense study because of its relationship to atherosclerosis.[30],[31]

Once the fat lipoprotein complex enters the circulation from the liver, it is
then available to most cells in the body. A lipoprotein lipase exists on the
surface of most cells that breaks the triglyceride bond and permits entry of the
triglyceride into the cell, where it is broken down and again resynthesized into
triglycerides for storage. When there is need for fatty acid as a source of
energy, it is then released as a nonesterified fatty acid or as a free fatty acid,
bound to albumin. When they are released into the circulation, they are very
actively metabolized and taken up by the tissues. Consequently, serum levels
are usually low. When there is an excess of free fatty acid, it is rapidly picked
up by the liver, or it is hydrolyzed in tissues to form triglycerides and coated by
the appropriate lipoprotein.[32]

Fatty acids are the long-term source of energy in humans. This can best be
understood by the fact that 90% of all energy is held in fatty acids and 10% in a
glycerol portion of the triglyceride. The fatty acids are first activated by acetyl
coenzyme A. Then a series of β-oxidations results in successive shortening of
the fatty acid chain so that a two-carbon-unit acetyl coenzyme A is finally
produced. Each acetyl coenzyme A is then metabolized so that there are 12
ATPs produced for each acetyl coenzyme A. Therefore, for the cost of one
ATP and coenzyme A, depending on the length of the fatty acid involved, there
will be a multiple of 12 ATPs formed for 2 carbon atoms. Most fatty acids
contain an even number of carbon atoms. However, there are a few that have r-
chain carbon atoms, and those would yield one propionyl coenzyme A that
enters into the metabolism by assisting in gluconeogenesis. Therefore, fatty
acids with an odd number of carbon atoms may be helpful in preventing or
treating ketosis.[33]

The glycerol that is freed when triglycerides are hydrolyzed is converted
to α-glycerol phosphate and may be reused in the formation of new triglycerides or enter into gluconeogenesis. This occurs primarily within the liver, although the process is very active in the intestinal mucosal cells.

2.2.14. Minimum Daily Requirements

The exact requirements of the body for fats are still not clear. Linoleic
acid (Fig. 2-3) is defined as an essential fatty acid. It is the polyunsaturated
fatty acid most prevalent in foods. It is totally effective in relieving the symptoms of fatty acid deficiency.[34] More recently, fatty acid deficiency in infants
has been confirmed in a series of subjects who required prolonged intravenous
alimentation. The fat-free regimen induced deficiencies that were subsequently
corrected by oral feeding of methyl linoleate.[35] Collins et al.[36] described the
essential fatty acid deficiency in a male who had undergone massive intestinal
resection and was maintained on intravenous fluids. Most recently, three patients developed classic dry, scaly skin lesions while on total parenteral nutrition that was corrected when the patients were fed parenteral fat.[37]

The classic syndrome of essential fatty acid deficiency includes the development of a dry, scaly skin lesion that is most typical in body folds but may become generalized. In initial animal experiments, there was also associated loss of hair, emaciation, and impaired reproductive capacity with an associated increased intake of water and food. The syndrome resulted in the death of the rats.[38]

From this available information, it appears that at least 1–2% of the kilocalorie intake of infants must be in the form of polyunsaturated fats in order to prevent any deficiency. Although this may be of importance in malnutrition states and where supportive nutrition is necessary, it does not appear to be a problem in the average American diet and in most disease states. Quite contrary, Western societies appear to consume excess amounts of fat. When one considers that a Western diet only required 1–2 g protein/kg body wt. for maintenance and that carbohydrate intake was limited, the major source of energy had to come from fats. Furthermore, excessive intake of saturated fats has been noted. Most recommendations recognize this and recommend that saturated fat be less than 10% of the total calorie intake and an additional 20% be in the form of polyunsaturated and monosaturated fats.[6] With this average intake, the requirement for essential fatty acids would certainly be easily realized.

2.2.15. Fat in Foods

Table 2-15 lists the saturated, monosaturated, and polyunsaturated fatty acid content of most common foods and vegetable oils. Evidence indicates that polyunsaturated and monosaturated fatty acids should be the major source of fat in the diet. The observer is still uncertain as to what the exact percentage of fat should be of the energy requirement and caloric intake. I believe the evidence will justify that the majority of calories should be from polysaccharides. However, in our present state of the art of nutrition, the recommendation is that energy intake from fats should make up approximately 30% of the calories, 10% from each of the saturated, monosaturated, and polyunsaturated forms. Therefore, this would mean that a majority of fats eaten should be from selected vegetable fat oils as they contain mostly mono- and polyunsaturated fatty acids and can make up 20% of our energy source.

It follows then that animal fats and dairy fats make up less than 10% of our daily energy intake. Consequently, foods that have been common in the Western diet, breakfast meats, pork, and heavy beef dinners, should be eliminated from our diet as a staple and substituted with foods that contain primarily mono- and polyunsaturated fats.

Concomitant with the decrease in saturated fat intake is the recommended decrease in cholesterol intake. Table 2-17 lists the approximate cholesterol content of some common foods. The exact content is difficult to obtain because there is limited biochemical evaluation. Most figures are obtained from extrapolation from a limited number of evaluations.[39] The recommendation is that the cholesterol intake should be less than 300 mg/day. This observer feels that this is much too high and probably should be less than 100 mg/week. When one

Table 2-17. Cholesterol Content of Foods[a]

Food	Measure	Weight (g)	Cholesterol (mg)
Whole milk	8 oz	240	34
Egg	1	50	242
Meat, fish	1 oz	30	21
Chicken (with skin)	1 oz	30	24
Liver	1 oz	30	131
Sweetbreads	1 oz	30	140
Shrimp	1 oz	30	45
Lobster	1 oz	30	25
Crab	1 oz	30	30
Oysters, clams	1 oz	30	15
Cheese, cheddar	1 oz	30	28
Butter	1 tbsp	14	35
Margarine (all veg. oil)	1 tbsp	14	0

[a]Calculated from Feeley et al.[39]

considers that one egg contains roughly 240 mg of cholesterol and that it is common for westernized man to eat half a dozen eggs per week, one can easily say how the dietary habits of westernized man should be changed. (See Chapter 20 for more detail.)

REFERENCES

1. Recommended Dietary Allowances, 1980. Committee on Dietary Allowances, Food and Nutrition Board, Division of Biological Sciences, Assembly of Life Sciences, National Academy of Sciences–National Research Council, ed 9 (revised), Washington, DC.
2. Miller DS, Gluggony PM: 1. An experimental study of overeating low or high protein diets. *Am J Clin Nutr* 20:1212, 1967.
3. Hegsted DM: Energy needs and energy utilization. *Nutr Rev* 32:33, 1974.
4. Krause MV, Mahan LK: *Food, Nutrition, and Diet Therapy*. Philadelphia, W B Saunders Co, 1979.
5. Ballinger WF, Collins JA, Drucker WR, *et al: Manual of Surgical Nutrition*. Philadelphia: W B Saunders Co, 1975.
5a. Calloway DH, Zanni E: Energy requirements and energy expenditure of elderly men. *Am J Clin Nutr* 33:2088, 1980.
6. United States Senate Select Committee Hearings on Nutrition.
7. Adibi SA: Intestinal absorption of amino acids and peptides. *Viewpoints Dig Dis* 10:4, 1978.
8. Gray GM, Cooper HL: Protein digestion and absorption. *Gastroenterology* 61:535, 1971.
8a. Sleisenger MH, Kim YS: Protein digestion and absorption. *N Engl J Med* 300:659, 1979.
9. Greene GM, Olds BA, Matthews G, et al: Protein as a regulator of pancreatic enzyme secretion in the rat. *Proc Soc Exp Biol Med* 142:1162, 1973.
10. Schultz SG, Curran PF: Coupled transport of sodium and organic solutes. *Physiol Rev* 5:637, 1974.
11. Adibi SA: Intestinal transport of dipeptides in man: Relative importance of hydrolysis and intact absorption. *J Clin Invest* 50:266, 1971.
12. Silk DBA, Clarke ML: Intestinal absorption of dipeptides: A major mode of protein absorption in man. *Gastroenterology* 67: 559, 1974.

13. Alpazale PRS, Kinzie JL: Regulation of small intestinal protein metabolism. *Gastroenterology* 64:471, 1973.

14. Crim MC, Monro HN: Protein, in *Present Knowledge in Nutrition,* ed 4. Washington, DC: Nutrition Foundation, Inc, 1976.

15. Gaiza C, Scrimshawns, Young VR: Human protein requirements: A long-term metabolic nitrogen balance study in young men to evaluate the 1973 FAO/WHO safe level of egg protein intake. *Am J Clin Nutr* 107:335, 1977.

16. Zanni E, Calloway DH, Zezulka AY: Protein requirements of elderly men. *J Nutr* 109:513, 1979.

17. Kishi K, Miyatani S, Inoue G: Requirement and utilization of egg protein by Japanese young men with marginal intakes of energy. *J Nutr* 108:658, 1978.

18. Oddoye EA, Margen E: Nitrogen balance study in humans: Long-term effect of high nitrogen intake on nitrogen accretion. *J Nutr* 109:363, 1979.

19. Church CF, Church HN: *Food Values of Portions Commonly Used,* ed 12. Philadelphia: J B Lippincott Co., 1975.

20. Pyke M: *Success in Nutrition.* London: John Murray, Ltd, 1975.

21. Swendseid ME, Harris CL, Tuttle SG: The effect of source of nine essential nitrogen or nitrogen balance in young adults. *J Nutr* 71:105, 1960.

22. Kies C: Non-specific nitrogen in the nutrition of human beings. *Fed Proc Fed Am Soc Exp Biol* 31:1172, 1972.

23. Gray GN: Carbohydrate digestion and absorption. *Gastroenterology* 58:96, 1970.

24. Gray GM: Carbohydrate digestion and absorption. *N Engl J Med* 292:1225, 1975.

25. Connor WE, Connor SL: Sucrose and carbohydrate, in *Present Knowledge in Nutrition,* ed 4. Washington, DC: Nutrition Foundation, Inc, 1976.

26. AMA Council on Foods and Nutrition: A critique of low-carbohydrate ketogenic weight reduction regimes. A review of Dr. Atkin's diet revolution. *JAMA* 224:1415, 1973.

27. Hofmann AF: A physiological approach to the intraluminal phase of fat absorption. *Gastroenterology* 50:56, 1966.

28. Gangyl A, Ockner RK: Intestinal metabolism of lipids and lipoproteins. *Gastroenterology* 68:167, 1975.

29. Ockner RK, Hughes FB, Isselbacher KJ: Very low density lipoproteins in intestinal lymph: Role in triglyceride and cholesterol transport during fat absorption. *J Clin Invest* 48:2867, 1969.

30. Frederickson DS, Levy RI, Lees RS: Fat transport lipoproteins—an integrated approach to mechanisms and disorders. *N Engl J Med* 276:34, 1967.

31. Miller GJ, Miller NE: Plasma-high-density-lipoprotein concentration and development of ischaemic heart-disease. *Lancet* 1:16, 1975.

32. Masoro EJ: Fat metabolism in normal and abnormal state. *Am J Clin Nutr* 30:1311, 1977.

33. Mattson FH: Fat, in *Nutrition Reviews' Present Knowledge in Nutrition,* ed 4. New York: Nutrition Foundation, Inc, 1976.

34. Holman RT, Casterro, Weise HF: The essential fatty acid requirement of infants and the assessment of their dietary intake of linoleate by serum fatty acid analysis. *Am J Clin Nutr* 14:70, 1964.

35. Paulsred JR, Penseler L, Whitten CF, *et al:* Essential fatty acid deficiency in infants induced by fat-free intravenous feeding. *Am J Clin Nutr* 25:897, 1972.

36. Collins FD, Sinclair AJ, Royle JP, et al: Plasma lipids in human linoleic acid deficiency. *Nutr Metab* 13:150, 1971.

37. Riella MC, Broviac JW, Wells M, et al: Essential fatty acid deficiency in human adults during total parenteral nutrition. *Ann Intern Med* 83:786, 1975.

38. Burr GO, Burr MM: A new deficiency disease produced by the rigid exclusion of fat from the diet. *J Biol Chem* 82:345, 1929.

39. Feeley RN, Criner PE, Watt BK: Cholesterol in foods. *J Am Diet Assoc* 61:134, 1972.

40. Mitchell HS, Rynbergen HJ, Anderson L, *et al: Nutrition in Health and Disease,* ed 16. Philadelphia: JB Lippincott, 1976.

Facts about Vitamins and Minerals

3.1. VITAMINS

Ever since James Lynn observed in 1593 that orange juice can prevent scurvy, humans have gradually become more aware that there are noncaloric substances in foods that are necessary for survival. Those substances that have enzymatic activity and are not synthesized by man are generally defined as vitamins. For the purposes of this book, the important facts concerning vitamins are described in a simplified outline form.

In each section the RDA (recommended dietary allowance) as listed in Table 2-6 is noted. Remember that some of these values are arbitrary, the classic example being vitamin C, for which the RDA was lowered from 60 mg in 1968 to 45 mg in 1974 and raised again to 60 mg in 1980, and yet many scientists have stressed the need for much higher doses of vitamin C. Nevertheless, the RDA is the reference point for most of us and the accepted value.

The following 13 substances are considered vitamins and are divided into fat-soluble and water-soluble groups[1]:

Fat-soluble	*Water-soluble*	
A	C (ascorbic acid)	Pantothenic acid
D	Biotin	B_6 (pyridoxine)
E	B_{12} (cyanocobalamin)	B_2 (riboflavin)
K	Folacin	B_1 (thiamine)
	Niacin (nicotinic acid)	

In Sections 3.1.1–3.1.13, which are in outline form, facts concerning each of the 13 vitamins are listed. Appropriate references are given for exact details concerning each vitamin. Characteristics of vitamins in general include the following:

1. They are not synthesized by the body (except D).
2. They are cofactors in reactions.
3. A deficiency causes disease.
4. An overdose may cause disease.

5. Disease states may increase requirement (e.g., hyperthyroid, malabsorption).
6. Multiple deficiency may mask clinical findings.
7. Cofactor substances are similar in action but not essential in diet and are not true vitamins:
    a. Carnitine (vs. Akee fruit poison).
    b. Lipoic acid (vs. mushroom poison).
    c. Inositol.
    d. *para*-Aminobenzoic acid.
    e. Pteridines.
    f. Bioflavonoids.
    g. Ubiquinone.

3.1.1. Vitamin A[2-9]

1. RDA: 1000 μg retinol equivalent (R.E.) in males, 800 μg in females.
2. β-Carotene in yellow vegetables, retinol in animal oils.
3. Common sources: carrots, sweet potatoes, fish oils, liver, milk products, eggs.
4. Major source is absorbed β-carotene; conversion stimulated by thyroxine.
5. Absorption in small intestine aided by bile salts, lecithin, vitamin E.
6. In small intestine one β-carotene may be cleaved to two retinaldehyde molecules.
7. Retinaldehyde converted to retinol by enzyme in small intestine.
8. Preformed A is in diet as retinyl esters and may also be absorbed after conversion to retinol in brush border of small intestine.
9. Retinol (vitamin A) converted to retinal or retinoic acid.
10. Transported in lymph chylomicrons.
11. Retinoic acid is 10% of retinol.
12. Metabolic active form is uncertain.
13. Retinyl esters stored in liver cells in fat and hydrolyzed to retinol before release.
14. Transported from liver by retinal-binding protein.
15. Liver stores depleted in 2 to 3 years.
16. Normal serum retinol = 30–65 mg/100 ml.
17. Deficiency[5-8]
    a. More common in children in rural villages and in chronic parasitosis or infection.
    b. Seen in malabsorption syndromes, excessive mineral oil use, alcoholic cirrhosis.
    c. Associated with hypoalbuminemia.
    d. Causes night blindness, xerophthalmia.
18. Hypervitaminosis, A[9]
    a. Seen in food faddists, eaters of polar bear liver, infants fed fish oils daily.

   b. Transient hydrocephalus in infants.
   c. Fatigue, weakness, anorexia, CNS SX.
   d. Rough skin, cortical thickening, liver damage.
   e. Symptoms disappear when intake cut but organ damage may persist.
   f. Carotene may accumulate in high intake but not converted to cause retinol overdose.

3.1.2. Vitamin D[10-15]

1. RDA: 10 μg.
2. Active compounds found only in dairy products, eggs, and fish.
3. D_2 (calciferol), D_3 (cholecalciferol).
4. Ergosterol in fungi converted to calciferol by ultraviolet rays.
5. Sterols in skin main source and converted to cholecalciferol by sun's ultraviolet rays; therefore, true vitamin status uncertain.
6. Dietary and bile secreted; absorbed from small intestine like vitamin A; bile salts necessary.
7. Liver hydroxylates D_3 to 25-OHD$_3$.
8. Kidney hydroxylates to 1,25-$(OH)_2D_3$ + 24,25-$(OH)_2D_3$.
9. 1,25-$(OH)_2D$ activates Ca-binding protein for Ca absorption in small intestine.
10. Promotes Ca and P absorption and mobilization from bone.
11. Stored in fat.
12. Normal plasma 25-OHD$_3$ = 10–45 ng/ml; plasma 25-$(OH)_2D_3$ = 3–5 ng/ml.
13. Deficiency: rickets in children, osteomalacia in adults, biliary cirrhosis, and small-bowel disease or resection.
14. Hypervitaminosis by overfeeding can cause hypercalcemia and death; symptoms are lethargy, anorexia, and constipation; large stores can cause prolonged hypercalcemia.

3.1.3. Vitamin E (Tocopherol)[16-21]

1. RDA: 8–10 mg α-tocopherol. Expressed as vitamin E activity, 80% by α-tocopherol and 20% by other tocopherols.
2. Common sources: vegetable oils derived from cereals, eggs, margarine, fruits.
3. Antioxidant action protects polyunsaturated fats, vitamins A, C against rancid change.
4. Absorption aided by bile salts and stored in liver.
5. β-Lipoprotein is carrier in circulation.
6. Serum E is greater than 0.5 mg/100 ml.
7. Measured by RBC peroxidase hemolysis.
8. Deficiency is rare and seen in severe malabsorption and in infants.

9. Early depletion gives increase RBC–H_2O_2 hemolysis + creatinuria, ceroid deposition (brown bowel).
10. No known toxicity.

3.1.4. Vitamin K[22,23,23a]

1. No RDA; estimated at 0.03 µg/kg/day.
2. Found in dark green vegetables, seaweed, liver, kidney.
3. Naturally occurring form is fat soluble.
4. K_1 occurs naturally but can be produced synthetically and is water soluble; approximately 60% activity of K.
5. K_2 is produced by intestinal bacteria and available directly by absorption.
6. K_3, K_4, K_5, K_6, K_7, K-S all produced synthetically.
7. Absorption aided by bile salts and stored in liver.
8. Stimulates appearance of coagulation proenzymes produced in liver for factors II (prothrombin), V, VII, and X.
9. Prothrombin (II), V, VII, and X all dependent for maintenance levels.
10. Numerous drugs interfere with its warfarin (coumarin) antagonist action (e.g., ethanol, aspirin, dilantin, etc.).
11. Deficiency
    a. Not seen by diet.
    b. Seen in malabsorption and obstructive jaundice.
    c. Prolonged prothrombin time.
    d. Seen in infants under 15 months.
12. Measure clotting factors to determine effective blood levels.
13. Toxicity not seen clinically.

3.1.5. Vitamin C (Ascorbic Acid)[24-28]

1. RDA: 60 mg for adults.
2. Common sources are fresh fruits, especially black currants, and fresh vegetables, especially tomatoes, fresh peas, potatoes and brussels sprouts.
3. Absorbed by intestine.
4. Destroyed by heat.
5. Main function in hydroxylation reactions.
6. In presence of oxygen it hydroxylates proline to collagen.
7. Active in hydroxylation of other compounds such as norepinephrine, tryptophan, phenylalanine, and tyrosine.
8. Assists in absorption of iron.
9. Protects other vitamins; necessary for conversion of inactive folic acid to active folinic acid.
10. 1.2 mg/100 ml in plasma.
11. Acidifying urine agent; when body stores loaded, excreted in urine.

12. Deficiency
    a. Scurvy.
    b. Impaired collagen formation, loss of connective tissue, and capillary fragility.
    c. Infant (Barlow's disease) cartilage also involved.
    d. Adult depletion takes 4 to 7 months.
    e. Levels less than 0.3 mg are significant but may be zero for 3 months before symptoms.
13. No evidence of toxicity but large doses may give diarrhea, acid urine hemolysis in glucose-6-phosphate deficiency.
14. Large-dose (megavitamin) therapy for treatment or prevention of disease is controversial.

3.1.6. Biotin[29-31]

1. No RDA; estimated at minute amount, 300 μg.
2. Common sources: egg yolk, yeasts, nuts, chocolate, meats, cereal grains, mushrooms, soybeans, fish.
3. Coenzyme for CO_2 fixation (carboxylation).
4. Manufactured by intestinal bacteria as well as readily available in small amounts; therefore, deficiency is rare.
5. Raw egg white contains avidin, which binds biotin; therefore, deficiency seen in raw egg faddists.
6. Whole-blood levels 820 pg/ml.
7. Urine biotin 14–100 μg/day.
8. Deficiency
    a. Raw egg syndrome.
    b. Symptoms of seborrheic dermatitis, lassitude, and anorexia.

3.1.7. Vitamin B_{12} (Cyanocobalamin)[32-38]

1. RDA: 3 μg.
2. In foods of animal origin only; highest in liver, kidney, clams.
3. Gastrin-stimulated gastric intrinsic factor (IF) binds 2 moles of B_{12}.
4. If B_{12} complex adheres to ileal mucosa and in presence of Ca^+, it is released and absorbed optimally at pH 7.5.
5. Transported by carrier protein, transcobalamin.
6. Necessary for RBC maturation.
7. Necessary for maintaining nerve myelination.
8. Interrelated with B_6 and folate metabolism.
9. Possible inhibition by megadoses of ascorbic acid.
10. Serum level of cobalamin (all active forms) less than 100 pg/ml indicates deficiency.
11. Schilling test using radioactive B_{12} with and without IF assists diagnosis.

12. Deficiency may produce megaloblastic anemia, demyelinating disease; seen in
   a. Pernicious anemia.
   b. Malabsorption disorders such as
      1. *D. latum* infestation.
      2. Tropical sprue.
      3. Gluten enteropathy.
      4. Postgastrectomy syndromes.
13. No toxicity known.
14. Injection of 100 μg is curative and monthly injection prevents any deficiency.

3.1.8. Folacin (Folic Acid, Pteroylglutamic Acid)[39-46]

1. RDA: 400 μg.
2. High in green vegetables, beans, nuts, eggs, yeasts.
3. Dietary forms are various polyglutamates.
4. Active absorption of pteroylglutamate in proximal jejunum.
5. Monoglutamate absorbed best.
6. Glucose enhances and starvation decreases absorption.
7. Tetrahydrofolic acid (THF) intermediate form.
8. N^5-Methyl-THF storage in liver.
9. Necessary for RBC maturation (DNA metabolism).
10. Normal RBC folate greater than 140 ng/ml; normal serum folate 4–12 ng/ml.
11. Formiminoglutamic acid is formed from histadine and combines with THF. In folic acid deficiency, formiminoglutamic acid is increased in the urine and histidine load may increase its excretion.
12. Assay with *Lactobacillus casei* or *Streptococcus faecalis*.
13. Deficiency
   a. Due to poor intake.
   b. May occur with increased demand.
   c. Many drugs impair absorption or interfere with metabolism (i.e., oral contraceptives, methotrexate, dilantin, triamterene).
   d. Depletion occurs in 5 to 10 weeks.
   e. Malabsorption occurs in tropical sprue, drug interference, as well as specific absorption defects.
   f. Often associated with B_{12} or Fe deficiency.
   g. Produces progressive megaloblastic anemia; in rare cases small-bowel villi atrophy and malabsorption.
   h. Oral treatment: 250 μg to 15 mg/day.
14. No toxicity known.

3.1.9. Niacin (Nicotinic Acid)[47-50]

1. RDA: 18–20 mg males, 12–14 mg females.
2. Niacin is pyridine 3-carboxylic acid; amide form is nicotinamide.
3. Abundant in wheat bran, brown rice, wheat germ, wholemeal flour, meat, yeast, liver, kidney, fish.
4. Absorbed in small intestine and transported to liver where incorporated into nicotinamide–adenine dinucleotide (NAD) and NAD phosphate.
5. NADP important in ATP formation.
6. Tryptophan is precursor, B_6 necessary to convert to NAD; 60 mg tryptophan equivalent to 1 mg nicotinamide.
7. Many bound forms found in foods, especially cereals.
8. Deficiency: pellagra
   a. Seen with basic corn and low-protein diets.
   b. May occur with increased demand or decreased absorption.
   c. Onset may be insidious.
   d. Pathognomonic sunburn to dark skin lesion.
   e. Neurologic, mental, and functional symptoms such as anorexia may precede skin lesion.
   f. Oral or parenteral replacement therapy.
9. Large doses cause vasodilation, may be hypocholesterolemic, and may cause abnormal liver function tests.
10. Used to treat type II, IV, and V hyperlipidemia in doses of 1.5 to 3 g/day.

3.1.10. Pantothenic Acid[51-53]

1. No RDA but probably between 5 and 10 mg/day.
2. Widely distributed in foods in adequate amounts and stable.
3. Absorbed in small intestine.
4. Converted in liver to coenzyme A and essential in carbohydrate and fat metabolism.
5. Acyl carrier protein is another active form.
6. Deficiency: rare
   a. Lethargy, weakness, peculiar gait, burning feet syndrome, crossed leg position.
   b. Blood level less than 100 μg/100 ml.
   c. Seen in malnutrition; urine less than 1 mg/day (0.8–7 mg normal).
7. No evidence of toxicity.

3.1.11. Vitamin B_6 (Pyridoxine)[54-59]

1. RDA: 2 mg/day.
2. Pyridoxine is most common in plants; pyridoxal, pyridoxol, and pyridoxamine are equivalent forms common in animals.

3. Common sources are meats, bran, vegetables, wholemeal flour, and many foods.
4. Absorbed in upper small bowel.
5. All forms converted to coenzyme PLP (pyridoxal-5-PO_4), important in protein metabolism.
6. Deficiency: rare
   a. Occurs with precipitous increased protein intake, when antagonists such as isoniazid and penicillamine are used, in multiple B deficiencies, in uremia.
   b. Neuritis an early symptom.
   c. Later neurologic symptoms may cause convulsions.
   d. Anemia is microcytic, hypochromic, and responds to high doses and associated iron overload.
   e. Urine below 30–50 mg/day significant.
   f. Tryptophan load increased xanthurenic acid excretion.
   g. Prevented by 50–100 mg/day.
7. Deficiencies also seen as a variety of inborn errors of metabolism.
8. Corrects abnormal tryptophan metabolism induced by estrogens.
9. No toxicity from large doses.

3.1.12. Riboflavin (B_2)[60-62]

1. RDA: 1.4–1.7 mg males, 1.2–1.3 mg females.
2. Good sources: yeast extracts, liver, kidney, meat, wheat germ, cheese, spinach, eggs; less in vegetables.
3. Absorbed in small intestine by site-specific saturable transport mechanism.
4. Converted to flavin mononucleotide (FMN) in intestine and then to flavin adenine dinucleotide (FAD) in the liver.
5. FMN and FAD cofactors of many enzymes important in metabolism of pyridoxine, RBC, xanthine, aldehyde, etc.
6. Deficiency
   a. Three to eight months to deplete body stores.
   b. Common in alcoholics except beer drinkers.
   c. Photophobia, sore throat, angular stomatitis, glossitis, magenta tongue, and many vague symptoms; normocytic, normochromic anemia.
   d. Twenty-four-hour urine excretion less than 100 mg/day when deficient.
7. RBC glutathione reductase test used for borderline deficiencies.
8. No evidence of toxicity from large doses.

3.1.13. Thiamine (B_1) (Aneurine)[63-68]

1. RDA: 1.2–1.5 mg males, 1.0–1.2 mg females.
2. Good sources are brewer's yeast, wheat germ, peanuts, pork, kidney, whole wheat flour; some in meat, fish, eggs.

3. Removed from cereal when milled.
4. Active and passive absorption from small intestine and transported to liver.
5. Phosphorylation to thiamine pyrophosphate (TPP) occurs in intestine, liver, kidney.
6. TPP acts as coenzyme in oxidative decarboxylation and many oxidation reactions.
7. Important in nerve physiology.
8. Thiaminase, an antivitamin, may be found in raw fish such as clams and carp.
9. Deficiency: Beriberi
   a. Acute form can cause acute congestive heart failure and is seen in the Orient.
   b. Dry type gives neurologic symptoms with slow (1- to 2-week) onset.
   c. Wet type gives edema and effusions.
   d. High output cardiac failure may be lethal.
   e. In alcoholics, associated with Wernicke's encephalopathy.
   f. Seen in combination with riboflavin and niacin deficiencies.
   g. Most common form in children 3–10 months old is a cardiac type.
10. Large doses are not toxic, but a rare form of "thiamine shock" is reported from even small doses.

3.2. MINERALS

The body consists of approximately 4% mineral elements. Table 3-1 lists the relative percentages of the more common minerals. Calcium and phosphorus are the most abundant because of their integral roles in bone structure. Although some of the minerals are found in minute amounts in the body and consequently are referred to as trace elements, they nevertheless are essential to body function. Specific functions of the various minerals are demonstrated by the role of cobalt in vitamin B_{12} metabolism, iodine in thyroid function, and zinc in insulin metabolism. In addition to those minerals listed, zinc, iodine,

Table 3-1 Relative Mineral Composition of Adult Body[a]

Element	Percentage of ash
Calcium	39
Phosphorus	22
Potassium	5
Sulfur	4
Chlorine	3
Sodium	2
Magnesium	0.7
Iron	0.15

[a]Modified from Mitchell.[69]

fluorine, copper, chromium, selenium, cobalt, manganese, molybdenum, vanadium, tin, silicon, and nickel are all thought to be essential components of the human body. No essential role has been yet described for cadmium, lead, mercury, arsenic, boron, lithium, or aluminum.

A brief description of the important facts concerning each mineral follows. Wherever one of the minerals is a key dietary factor in a specific disease, it is mentioned accordingly in the following chapters. The details present in the literature concerning each of these minerals are described in appropriate references.

3.2.1. Calcium[69-71]

This is the most abundant element in the human body. The RDA for males and females is between 800 and 1200 mg/day. Milk and milk products are the highest sources of calcium in the diet of the Western world. However, with the changing dietary pattern, it is important for the clinician to be aware of other foods that supply considerable amounts of calcium. Vegetables such as collards, turnips, kale, broccoli, cabbage, and carrots have moderate amounts. Some fish products such as salmon, clams, oysters, and shrimp are other sources. Both molasses and eggs also contain moderate amounts of calcium.

Calcium is absorbed from the intestine by active transport against a concentration gradient. Passive absorption occurs throughout, but the main active process is in the duodenum. The level of calcium absorption is affected by parathyroid hormone, which stimulates the synthesis of 1,25-dehydroxyvitamin D, which in turn activates a Ca-binding protein for calcium absorption in the intestine as well as reabsorption of calcium from bone.[10,11] When the body is depleted, more parathyroid hormone is stimulated, and conversely, when there is an excess of calcium, parathyroid secretion is depressed, vitamin D being an integral part of the feedback mechanism. Approximately 600 mg is secreted into the gut during various absorptive processes. Depending on dietary intake, approximately 900 mg is lost in the stool per day and the remainder absorbed. The final calcium balance is regulated in conjunction with the kidney, where approximately 100 mg is lost per day. The body retains approximately 100 mg for its function in its final balance. The cal-

Table 3-2. Factors Affecting Calcium Absorption

Increase	Decrease
Vitamin D	Alkalinity in intestine
Parathormone	Phosphates in intestine
Hypocalcemia	Benzoic acid
Acidity in intestine	Excess fatty acids (LCT)
Growth hormone	Oxalates in intestine
1-Lysine	
Citrates	
Lactose	

cium:phosphorus ratio is important and is maintained in body homeostasis at a ratio of approximately 2:1. The body attempts to maintain this ratio for normal bone structure. Table 3-2 lists the factors that increase and decrease calcium absorption.

Calcium depletion is seen in severe malabsorptive disorders, pancreatitis, and malnutrition. It is described in more detail in those sections of this book. Associated with calcium depletion is osteoporosis, but the mechanism is poorly understood as it has been difficult to relate osteoporosis to calcium intake in man.[10]

Increased calcium uptake is noted with antacids. Hypercalcemia, seen in certain malignant diseases as well as sarcoidosis, causes symptoms of depressed neurogenic activity, contrasting to the hyperactivity and tetany of hypocalcemia.

Severe depletion of calcium is treated by intravenous replacement therapy. Long-term calcium deficiency is treated by increased oral intake.

3.2.2. Phosphorus[10,72-75]

The RDA for phosphorus is between 800 and 1200 mg/day for both males and females. It makes up approximately 1% of the body weight and is the key element for almost all body functions.

It is absorbed readily in the intestine and is affected by vitamin D, but the exact mechanisms are still unclear. Phosphorus is found in abundant amounts in dairy and meat products and in unprocessed cereals and grains, primarily as phytic acid. However, the availability of phytic acid for absorption is open to question as it may combine with calcium to form an insoluble product. Once absorbed, phosphorus has many significant roles in the body. It is essential in DNA and RNA, as part of phospholipids, in phosphorylation reactions, in all of the important energy-storage systems of ADP/ATP and in the oxidation–reduction system of NADP/NADPH. Furthermore, it is important as an organic phosphate in the body fluid buffering system.

Hyper- and hypophosphotemia are largely caused and related to either renal disease or disturbances in vitamin D metabolism.[10,11] Phosphorus deficiency can be seen in severe weight loss and in alcoholics. Mild deficiency can be treated with milk.

3.2.3. Magnesium[76-80]

The RDA for males is 350 to 400 mg/day and for females 300 mg/day. It is found in a variety of foods such as beans, peas, soybeans, nuts, and green leafy vegetables. It is also found in moderate amounts in dairy products as well as in all cereals. The average diet contains 10 mmol.

Approximately 55% of body magnesium is present in bone and about 27% in muscle. The body contains approximately 2000 meq. There appears to be a balance maintained between gastrointestinal absorption and renal secretion. Magnesium is passively absorbed throughout the small intestine, and its absorption varies greatly depending upon intraluminal factors such as foods and precipitating elements. High calcium intake appears to increase the requirement for magnesium. The kidneys actively excrete magnesium, and it is reabsorbed by the renal tubule. Magnesium in feces represents nonabsorbed magnesium. It is apparent that magnesium, calcium, parathyroid hormone, and bone metabolism are all related. Hypermagnesia is rarely seen. Magnesium deficiency is noted in malabsorption disorders, severe cirrhosis, and in association with calcium–parathyroid disturbances. It is treated by oral or parenteral replacement. Normal serum level is between 1.5 and 2.1 meq/liter.

3.2.4. Sodium[81-89]

There is no RDA for sodium. Estimates at the present time indicate that 10 to 60 meq of sodium or 600 to 3500 mg of sodium chloride would be an adequate intake to replace the amounts lost in feces, urine, and sweat. However, Western societies have grown accustomed to large amounts of sodium in the diet. This is clearly evident from the fact that foods such as cheese, bacon, corn flakes, ham, and corned beef have anywhere from 1500 to 5000 mg of sodium per 100 g of food. Fresh fruits, nuts, cereals, some fishes and meats have far less, ranging from 20 to 300 mg/100 ml.

Sodium is absorbed throughout the small intestine and in the colon. It is an integral ion in the transfer of osmolarity in the small bowel and, as such, is in constant flux throughout the small bowel. Large amounts of sodium passing into the right side of the colon are reabsorbed. The details of sodium exchange will be discussed in the sections on the small intestine and diarrhea.

Sodium deficiency certainly can occur in the presence of severe diarrhea, excessive sweating, and endocrinopathies. It is easily replaced, either orally or parenterally.

Sodium excess may occur with heavy overloading. However, the true toxicity of sodium is yet to be determined as it is under great suspicion as the cause of chronic vascular and renal disease. For all we know at the present time, the 3000-mg figure for basic replacement may represent the upper limit of normal tolerance.

3.2.5. Potassium[81,82,87,90]

There is no RDA for potassium. As noted in Table 3-1, potassium is more plentiful in the body than sodium. Both ions are positively charged and have similar qualities, potassium being more plentiful in intracellular fluids and sodium more plentiful in extracellular fluids.

Good sources of potassium in foods are soy flour, fruits (dried fruits are very high), potatoes, nuts, beets, fish, and some green vegetables.

The body contains about 5% of its ash in potassium and from 5 to 15 meq is lost per day in feces. There is a great range in urinary potassium loss depending on the intake. The average daily Western diet contains approximately 50 to 100 meq (3.7 to 7.4 g) of potassium chloride. Potassium fluxes back and forth in the small bowel, but much less so than sodium. Contrary to sodium, it is secreted in the colon rather than absorbed. Potassium appears to have some protective action against high sodium intake, but the relationship is not completely understood.

Potassium excess rarely occurs except in the presence of drug intake or renal failure. Deficiency also rarely occurs except in the presence of certain endocrine disorders, marked diarrhea or vomiting where there is excessive loss, or due to diuretic therapy. Occasional renal potassium loss also occurs. Probably the most common cause of potassium deficiency today is exogenous diuretic therapy. Oral replacement is usually adequate, and intravenous replacement is needed only where there is a rapid life-threatening depletion.

3.2.6. Chlorine[82,84-86,88,89]

Chlorine makes up 3% of body ash. There is no RDA for this anion. However, almost all of the sodium and potassium ingested in our diets is in the form of chloride. It is in a state of flux in the small bowel, and absorption closely parallels that of sodium. However, chloride can be absorbed against an electrical gradient; as a consequence, this can be helpful in certain diarrhea states. Chloride absorption probably occurs in exchange for bicarbonate.

Chloride is absorbed in the colon, but approximately 2 meq is lost in the stool. It is also lost in the kidney. Hypochloremia occurs in severe vomiting, in diarrhea, and in renal disease. It is easily corrected by intravenous therapy, depending on the cause and the electrolyte picture.

3.2.7. Sulfur[91]

There is no RDA for sulfur, although it makes up 4% of body ash and approximately 0.25% of body weight. It is present in methionine and cystine. The exact daily requirements are unknown.

In the body it is a part of thiamine and biotin and is present in connective tissue, skin, and hair. It is essential in certain amino acids and is an important part of body mucopolysaccharides.

Sulfates are excreted in the urine and are increased in high-protein and decreased in low-protein diets.

3.2.8. Iron[92-99]

The RDA for iron is 10 to 18 mg for males and females per day. Major sources of iron are beets, deep-green leafy vegetables, and whole-grain or enriched cereals and breads. Although controversial, iron has been added to breads to increase its availability in foods.

Iron is absorbed in the small intestine, primarily in the duodenum and jejunum. The ability of mucosal cells to absorb iron appears to be determined by their iron content. Iron can be transferred through the intestinal cells or deposited in the form of ferritin. Ionic iron-specific receptors in the brush border actively transfer iron to amino acids in the cytoplasm where it is in equilibrium with iron-poor ferritin. The absorption increases when there is a shortage and decreases when the body is overloaded. Absorption is aided by vitamin C. Serum ferritin is an accurate measure of body iron. Iron circulates in the plasma attached to a β-globulin, transferrin. Transferrin is usually 20 to 40% saturated, so that free iron is measured in relation to the total iron-binding capacity. When there is marked iron loss or depletion, bound iron falls below 15 to 20% and there is a relative rise in the total binding capacity. Most iron is found in the circulating red blood cell mass, but stored iron is in the form of ferritin in the marrow, intestine, and muscle.

Males and postmenopausal women lose approximately 800 to 1000 mg of iron per day into the gut that is ordinarily reabsorbed. The menstruating female may lose on an average up to 2 mg of iron per day. Iron deficiency may occur when there is blood loss, or decreased absorption, or a relative increased loss as compared to decreased availability of nutritional iron. The resulting decrease in iron causes a typical microcytic hypochromic iron deficiency anemia.

Iron overload may occur for many pathologic reasons, or because of massive increased intake in which the intestinal inhibitory system fails and permits an increase in absorption over the need. It is discussed in more detail in the chapter on liver disease.

3.2.9. Zinc[100-110],[116]

The RDA is 15 mg/day. Adolescent females require at least 11 mg/day. Most of our foods have only small amounts of zinc, except for animal protein.

Zinc is thought to be poorly absorbed from the small intestine. The ileum is the major site of absorption. Often, high calcium and phytate concentrations interfere with its absorption by forming insoluble compounds. High-fiber foods such as corn and black beans interfere with zinc availability. Zinc is an important part of many enzymes including alkaline phosphatase, alcohol dehydrogenase, carbonic anhydrase, lactic dehydrogenase, glutamic dehydrogenase, and carboxy dehydrogenase. Experimental zinc deficiency profoundly affects DNA and RNA metabolism. Zinc deficiency was thought to be extremely rare but has been documented, primarily in Iran, Egypt, and Central America, because of severe zinc-deficient diets. It is observed in malabsorption states, as in regional enteritis and post-bypass surgery for obesity. Zinc deficiency is also noted as a problem in alcoholism and cirrhosis. It has been established as an important factor in achrodermatitis enteropathica, where the disease is related to a hereditary defect in zinc absorption. Zinc deficiency can be corrected by oral or parenteral replacement. Experimental human deficiency is related to falls in serum levels, decrease in essential enzymes, weight loss, and an adverse effect on protein metabolism.

Zinc can be assessed by taste acuity and plasma, serum, RBC, hair, urine, and enzyme levels. From these available tests, it is obvious that no one test is preferred. Plasma zinc ranges from 75 to 120 μg/100 ml.

3.2.10. Iodine[111]

The RDA for iodine is 150 μg for males and females. Iodine is readily available in seafoods, vegetables, beef products, eggs, some dairy products, and bread. Variations are wide, depending on iodine content in the environment. Therefore, it is common for salt to be iodized with sodium or potassium iodide. Seafood contains 54 μg/100 ml and dairy products approximately 14 μg/100 ml.

Iodine is absorbed as inorganic iodides after being split from organic compounds in the small intestine. It is transported as free or protein-bound iodine in the circulation. The interested reader is referred to texts on thyroid function to understand the relationship of iodine and the thyroid gland. The cycle of pituitary thyroid-stimulating hormone–thyroid activity–thyroglobulin is the realm of iodine metabolism. Iodine is approximately 2500 times greater in concentration in thyroid tissue. There is approximately 25 to 50 mg in the thyroid gland.

Iodine deficiency is reflected in endemic goiter and may result in hypothyroidism. The iodine deficiency can easily be replaced by oral therapy, but the results of hypothyroidism or creatinism are not so easily corrected.

3.2.11. Copper[102-105,112,117]

There is no RDA for copper. Food sources are usually meats, shellfish, nuts, and cereals, and the copper content is dependent somewhat on the environment. Approximately 2 to 5 mg is in the average diet. Approximately 40% of the daily intake is absorbed from the stomach and small intestine. The exact mechanisms and control of absorption are not clearly understood. Once in the circulation, it is bound to a protein, ceruloplasmin, which is synthesized in the liver. The normal concentration is 34 ±4 mg/100 ml. Copper is primarily stored in the liver.

It has a wide variety of functions in the body including involvement in collagen, the central nervous system, bone development, hemoglobin formation, and several enzyme activities.

Copper deficiency is seen in severe malnutrition but is not described frequently in clinical medicine. Copper overload is more commonly seen in Wilson's disease and is discussed in the chapter on liver diseases. Excessive amounts of copper may be toxic and cause hemolytic anemia.

Zinc and copper have been implicated as possible factors in the etiology of coronary heart disease.[102] A suspected imbalance of the ratio of these two elements results in hypercholesterolemia and increased mortality. The reason for the imbalance is not understood, and the work has yet to be substantiated, but it certainly is interesting.

3.2.12. Trace Elements[102],[112-115]

All of these elements are considered essential in humans because they serve some pivotal role in homeostasis or body function.

Fluorine has a major role in preventing tooth decay. An overload can result in mottling of teeth when drinking water has more than 1.5 ppm. Tooth enamel can become dull and pitted, and certainly when fluorine is in excess of 2.5 ppm the incidence of mottling increases. When there is a large excess in the water (over 8 ppm), bone deposition occurs and may be associated with arthritis. Most foods have a very small amount of fluorine.

Manganese is present in the liver. It is important for normal growth and in bone, cartilage, brain, and lipid metabolism. In experimental animal studies, it has been shown to be a part of several enzymes. The minimum daily requirement is as yet not understood. It is present in many foods, totally absent in milk, and high in tea.

Selenium has been shown to be essential in experimental animals. Gluthione is a seleno enzyme. When mammals are deficient, the enzyme is depressed. Selenium is, as gluthione peroxidase, protective against peroxide activity. This is similar to vitamin E activity, and it is estimated that combined activities are needed at times to protect body homeostasis. Selenium is also present in some body proteins.

Cobalt appears in trace amounts in man and is essential for body function. Its role in cyanocobalamin (vitamin B_{12}) metabolism is well known and is discussed under that vitamin. Vegetarians may develop deficiencies as cobalt is present only in animal tissues and humans consume it only in the form of B_{12}.

Molybdenum, vanadium, nickel, silicon, and tin also appear to be essential trace elements, but their true roles in human metabolism need to be further elaborated.

REFERENCES

1. Nomenclature policy: Generic descriptors and trivial names for vitamins and related compounds. *J Nutr* 7, 1979.
2. Olsen JA: The metabolism of vitamin A. *Pharm Rev* 19:559, 1967.
3. Rodriguez MS, Irwin MI: A conspectus of research on vitamin A requirements of man. *J Nutr* 102:909, 1972.
4. Smith FR, Lindenbaum J: Human serum transport in malabsorption. *Am J Clin Nutr* 27:700, 1974.
5. Smith FR, Goodwin DS, Zaklama MS, *et al:* Serum vitamin A, retinal-binding protein and pre-albumin concentrations of protein-calorie malnutrition. 1. A functional defect in hepatic retinal release. *Am J Clin Nutr* 26:973, 1973.
6. Mahalanahis D, Simpson TW, Chakraborty ML, *et al:* Malabsorption of water miscible vitamin A in children with giardiasis and ascaris. *Am J Clin Nutr* 32:313, 1979.
7. Mansour MM, Mikhail MM, Farid Z, *et al:* Chronic salmonella septecemia and malabsorption of vitamin A. *Am J Clin Nutr* 32:319, 1979.
8. Russell RM, Morrison SA, Smith FR, *et al:* Vitamin A reversal of abnormal dark adaptation in cirrhosis. *Ann Intern Med* 88:622, 1978.

9. Muenter MD, Perry HO, Ludwig J: Chronic vitamin A intoxication in adults. Hepatic, neurologic and dermatologic complications. *Am J Med* 50:129, 1971.

10. Haussler MR, McCain TA: Basic and clinical concepts related to vitamin D metabolism and action. *N Engl J Med* 297:974, 1041, 1977.

11. DeLuca HF: Vitamin D metabolism and function. *Arch Intern Med* 138:836, 1978.

12. Skinner RK, Sherlock S, Long RG, *et al:* 25-Hydroxylation of vitamin D in primary biliary cirrhosis. *Lancet* 1:720, 1977.

13. Haddad JG: Vitamin D economy in gastrointestinal disease. *Ann Intern Med* 87:629, 1977.

14. Sitrin M, Meridith S, Rosenberg IH: Vitamin D deficiency and bone disease in gastrointestinal disease. *Arch Intern Med* 138:886, 1978.

15. Hazards of overuse of vitamin D. *Am J Clin Nutr* 28:512, 1975.

16. Hoywitt M: Status of human requirements for vitamin E. *Am J Clin Nutr* 27:1182, 1974.

17. Witting LA, Lee L: Dietary levels of vitamin E and polyunsaturated fatty acids and plasma vitamin E. *Am J Clin Nutr* 28:571, 1975.

18. Witting LA, Lee L: Recommended dietary allowance of vitamin E: Relation to dietary, erythrocyte and adipose tissue linoleate. *Am J Clin Nutr* 28:577, 1975.

19. Lehmann J, Marshall MW, Slover HT, *et al:* Influence of dietary fat level and dietary tocopherole on plasma tocopherole of human subjects. *J Nutr* 107:1006, 1977.

20. Binder HJ, Herting D, Herst D, *et al:* Tocopheral deficiency in man. *N Engl J Med* 273:1289, 1965.

21. Bieri JG: Vitamin E. *Nutr Rev* 33:161, 1975.

22. Herman RH, Stifel FB, Greene HL: Vitamin K and Vitamin K deficiency, in Dietschy JM (ed): *Disorders of the Gastrointestinal Tract, Disorders of the Liver, Nutritional Disorders.* New York: Grune & Stratton, 1976.

23. Suttie JW: The metabolic role of vitamin K. *Fed Proc* 39:2730, 1980.

23a. Gallop PM, Lian JB, Hauscha PV: Carboxylated calcium-binding proteins and vitamin K. *New Engl J Med* 302:1460, 1980.

24. Hodges RE, Baker EM, Hood J, et al: Experimental scurvy in man. *Am J Clin Nutr* 22:535, 1969.

25. Hodges RE, Hood J, Canham JE, *et al:* Clinical manifestations of ascorbic acid deficiency in man. *Am J Clin Nutr* 24:432, 1971.

26. Pauling LC: *Vitamin C and the Common Cold.* San Francisco: WH Freeman and Co, 1970.

27. Anderson TW, Suranyi G, Beaton GH: The effect on winter illness of large doses of vitamin C. *Can Med Assoc J* 111:31, 1974.

28. Miller DZ, Nance WE, Norton JA, *et al:* Therapeutic effect of vitamin C. A co-twin control study. *JAMA* 237:248, 1977.

29. Lynen F: The role of biotin-dependent carboxylations in biosynthetic reactions. *Biochem J* 102:381, 1967.

30. Baugh CM, Marline JH, Butterworth CE Jr: Human biotin deficiency: A case history of biotin deficiency induced by raw egg consumption in a cirrhotic patient. *Am J Clin Nutr* 21:713, 1968.

31. MacCormick TB: Biotin, in *Nutrition Reviews' Present Knowledge in Nutrition.* New York: Nutrition Foundation, Inc, 1976.

32. Herbert V: Vitamin B_{12}, in Goodman LS, Gilman A (eds): *Pharmacological Basis of Therapeutics,* ed 5. New York: Macmillan, 1975.

33. Toskes PP, Daren JJ: Vitamin B_{12} absorption and malabsorption. *Gastroenterology* 65:622, 1973.

34. Herbert V: Vitamin B_{12}, in *Nutrition Reviews' Present Knowledge in Nutrition,* ed 4. New York: Nutrition Foundation, Inc, 1976.

35. Carmel R: Nutritional vitamin B_{12} deficiency: Possible contributory role of subtle vitamin B_{12} malabsorption. *Ann Intern Med* 88:647, 1978.

36. Donaldson RM Jr: "Serum B_{12}" and the diagnosis of cobalamin deficiency. *N Engl J Med* 299:827, 1978.

37. Kolhouse JF, Kondo H, Allen NC, *et al:* Cobalamin analogues are present in human plasma and can mask cobalamin deficiency because current radioisotope dilution assays are not specific for true cobalamin. *N Engl J Med* 299:785, 1978.

38. Cooper BA, Whitehead VM: Evidence that some patients with pernicious anemia are not recognized by radiodilution assay for cobalamin in serum. *N Engl J Med* 299:816, 1978.
39. Herbert V: Biochemical and hematologic lesions in folic acid deficiency. *Am J Clin Nutr* 20:562, 1967.
40. Herbert V: The five possible causes of all nutrient deficiencies: Illustrated by deficiencies by vitamin B_{12} and folic acid. *Am J Clin Nutr* 26:77, 1973.
41. Stokstad ELR, Koch J: Folic acid metabolism. *Physiol Rev* 47:83, 1967.
42. Blair JA, Matty AJ: Acid microclimate. *Clin Gastroenterol* 3:183, 1974.
43. Rosenberg IH, Godwin HA: Digestion and absorption of dietary folate. *Gastroenterology* 60:445, 1971.
44. Daniel, Jr. WA, Gaines EG, Bennett DL: Dietary intakes and plasma concentrations of folate in healthy adolescents. *Am J Clin Nutr* 28:360, 1975.
45. Malin JD: Folic acid. *World Rev Nutr* and *Diet* 21:198, 1975.
46. Rodriguez MS: A conspectus of research on folocin requirements of man. *J Nutr* 108:1983, 1978.
47. Darby WJ, McNutt KW, Todhunter EN: Niacin, in *Nutrition Reviews' Present Knowledge in Nutrition*, ed 4. New York: Nutrition Foundation, Inc, 1976.
48. Editorial: Conversion of tryptophan to niacin in man. *Nutr Rev* 32:76, 1974.
49. The coronary drug project research group: Clofibrate and niacin in coronary heart disease. *JAMA* 321:360, 1975.
50. Kuo PT, Kostis JB, Moreyra AE, *et al:* Effective control of familial hypercholesterolemia with diet and bile acid sequestrant plus nicotinic acid. *Am J Clin Nutr* 32:941, 1979.
51. Wright LD: Pantothenic acid, in *Nutrition Reviews' Present Knowledge in Nutrition*. New York: Nutrition Foundation, Inc, 1976.
52. Glusman M: The syndrome of the "burning feet" as a manifestation of nutritional deficiency. *Am J Med* 3:211, 1947.
53. Hodges RE, Ohlson MA, Bean WB: Pantothenic acid deficiency in man. *J Clin Invest* 37:1642, 1958.
54. Sauberlich HE, Canham JE, Baker EM: Biochemical assessment of the nutritional status of vitamin B_6 in the human. *Am J Clin Nutr* 25:629, 1972.
55. Raskin NH, Fishman RA: Pyridoxine-deficient neuropathy, due to hydralazine. *N Engl J Med* 273:1182, 1965.
56. Harris JW, Horigen DL: Pyridoxine-responsive anemia, prototype variations on the theme. *Vitamin Horm NY* 22:721, 1964.
57. Stone WJ, Wainock LK, Wagner C: Vitamin B_6 deficiency in uremia. *Am J Clin Nutr* 28:950, 1975.
58. Leklem JE, Brown RR, Rose DP, *et al:* Metabolism of tryptophane and niacin in oral contraceptive users receiving controlled intakes of vitamin B_6. *Am J Clin Nutr* 28:146, 1975.
59. Rose DP, Leklem JE, Brown RR: Effect of all contraceptives and vitamin B_6 deficiency on carbohydrate metabolism. *Am J Clin Nutr* 28:872, 1975.
60. Rivlin RS: Riboflavin metabolism. *N Engl J Med* 283:463, 1970.
61. Heller S, Sokel RM, Korner WF: Riboflavin status in pregnancy. *Am J Clin Nutr* 27:1225, 1974.
62. Rivlin RS: *Riboflavin.* New York: Plenum Press, 1975.
63. Randi G, Ventura U: Thiamine intestinal transport. *Physiol Rev* 52:821, 1972.
64. Thomson AD, Baker H, Levy CM: Folic-induced malabsorption of thiamine. *Gastroenterology* 60:756, 1971.
65. Baker H, Frank O, Zettalman RK, *et al:* Inability of chronic alcoholics with liver disease to use food as a source of folates, thiamin and vitamin B_6. *Am J Clin Nutr* 28:1377, 1975.
66. Tandhaichiter V, Vimokesant SL, Dhanamitta S, *et al:* Clinical and biochemical studies of adult beriberi. *Am J Clin Nutr* 23:1017, 1970.
67. Williams RR: The world beriberi problem. *J Clin Nutr* 1:513, 1953.
68. Tandhaichiter V: Thiamine, in *Nutrition Reviews' Present Knowledge in Nutrition,* ed 4. New York: Nutrition Foundation, Inc, 1976.

69. Mitchell HS, Rynbergen HJ, Andersen L, *et al: Nutrition and Health in Disease,* ed 16. Philadelphia: JP Lippincott Co, 1968.
70. D'Souza A, Floch MH: Calcium metabolism and pancreatic disease. *Am J Clin Nutr* 26:352, 1973.
71. Linksweiler HM: Calcium, in *Nutrition Reviews' Present Knowledge in Nutrition.* New York: Nutritional Foundation, Inc, 1976.
72. Tewell JE, Clark HE, Hill JM: Phosphorus balances of adults fed rice, milk and wheat flour mixtures. *J Am Diet Assoc* 63:530, 1973.
73. Harrison HE, Harrison HC: Intestinal transport of phosphate, action of vitamin D, calcium and potassium. *Am J Physiol* 201:1007, 1961.
74. Lotz M, Zisman EN, Bartter FC: Evidence for a phosphorus-depletion syndrome in man. *N Engl J Med* 278:409, 1968.
75. Knochel JP: The pathophysiology and clinical characteristics of severe hypophosphotemia. *Arch Intern Med* 137:203, 1977.
76. Hersh T, Siddiqui DA: Magnesium in the pancreas. *Am J Clin Nutr* 26:362, 1973.
77. Wacker W, Parisi AF: Magnesium metabolism. *N Engl J Med* 278:712, 1968.
78. Shils ME: Magnesium deficiency in parathyroid hormone levels in man. *Am J Clin Nutr* 28:421, 1975.
79. Shils ME: Magnesium, in *Nutrition Reviews' Present Knowledge in Nutrition,* ed 4. New York: Nutrition Foundation, Inc, 1976.
80. Thoren L: Magnesium metabolism. *Prog Surg* 9:131, 1971.
81. Meneely GR, Battarbee HD: Sodium and potassium, in *Present Knowledge in Nutrition,* ed 4. New York: Nutrition Foundation, Inc, 1976.
82. Fortran JS, Ingelfinger FJ: Absorption of water, electrolytes and sugar from the human gut, in Code CF, Heidel W. (eds): *Handbook of Physiology, Section 6: The Alimentary Canal.* Washington DC: American Physiologic Society, 1968, vol 3.
83. Fortran JS, Rector FC Jr, Carter NW: Mechanisms of sodium absorption in the human small intestine. *J Clin Invest* 47:884, 1968.
84. Schultz SG, Currin PF: Coupled transport of sodium and organic solutes. *Physiol Rev* 50:637, 1970.
85. Turnberg LA: Electrolyte absorption from the colon. *Gut* 11:1049, 1970.
86. Dahl LK: Salt and hypertension. *Am J Clin Nutr* 25:231, 1972.
87. Gross OS, Weller JM, Hobler SW: Relationship of sodium, potassium and zinc to blood pressure. *Am J Clin Nutr* 24:605, 1971.
88. Schultz SG, Currin PF: Intestinal absorption of sodium chloride in water, in Code, CF, Heidel W (eds): *Handbook of Physiology, Section 6: The Alimentary Canal.* Washington, DC: American Physiologic Society, 1968, vol 3.
89. Turnberg LA: Intrarelationships of chloride, bicarbonate, sodium and glucose transport in the human ileum. *J Clin Invest* 49:557, 1970.
90. Turnberg LA: Potassium transport in the human small bowel. *Gut* 12:811, 1971.
91. Comar CL, Bronner F: *Mineral Metabolism.* New York: Academic Press, 1962.
92. Increased iron fortification of foods. *Med Lett* p. 1481, 1972.
93. Worwood M: The clinical biochemistry of iron. *Semin Hematol* 14:1, 1977.
94. Isen T, Brown ED: The iron-binding function of transferrin in iron metabolism. *Semin Hematol* 14:31, 1977.
95. Harrison PM: Ferritin: An iron-storage molecule. *Semin Hematol* 14:55, 1977.
96. Linder MC, Munro HN: The mechanism of iron absorption and its regulation. *Fed Proc Fed Am Soc Exp Biol* 36:2017, 1977.
97. Croshy WH: Current concepts in nutrition. Who needs iron? *N Engl J Med* 297:543, 1977.
98. Croshy WH: Iron deficiency anemia in a nutritionally complex situation. *Am J Clin Nutr* 32:715, 1979.
99. Editorial: Serum-ferritin. *Lancet* 1:533, 1979.
100. Editorial: Zinc in human medicine. *Lancet* 2:51, 1975.
101. Sandstead HH: Zinc in nutrition in the United States. *Am J Clin Nutr* 26:1251, 1973.

102. Burch RE, Hahn HK, Sullivan JF: Newer aspects of the roles of zinc, manganese and copper in human nutrition. *Clin Chem (NY)* 21:501, 1975.
103. Klevay LM: Coronary heart disease: Zinc/copper hypothesis. *Am J Clin Nutr* 28:764, 1975.
104. Greger JL, Zaikis SG, Abernathy RP, et al: Zinc, nitrogen, copper, iron and manganese balance in adolescent females fed two levels of zinc. *J Nutr* 108:1449, 1978.
105. Solomons NW: On the assessment of zinc and copper nutriture in man. *Am J Clin Nutr* 32:856, 1979.
106. Antonson DL, Barak AJ, Vanderhoff JA: Determination of the site of zinc absorption in rat small intestine. *J Nutr* 109:142, 1979.
107. Duncan JR, Hurley LS: Intestinal absorption of zinc: A role for a zinc-binding ligand in milk. *Am J Physiol* 235:E559, 1978.
108. Prasad AS, Rabbani P, Abbasii A, et al: Experimental zinc deficiency in humans. *Ann Intern Med* 89:483, 1978.
109. Atkinson RL, Dahms WT, Bray GA, et al: Plasma zinc in obesity and after intestinal bypass. *Ann Intern Med* 89:491, 1978.
110. Solomons N, Viteri F, Pineda O, et al: The effect of the Guatemalian diet on zinc bioavailability. *Am J Clin Nutr* 32:932, 1979.
111. Scrimshaw NS: Endemic goiter. *Nutr Rev* 15:161, 1957.
112. Reinhold JG: Trace elements: A selective survey. *Clin Chem (NY)* 21:476, 1975.
113. Sandstead HH, Burk RF, Booth GH Jr, et al: Current concepts on trace metals. *Med Clin North Am* 54:1509, 1970.
114. Burk RF: Selenium, in *Nutrition Reviews' Present Knowledge in Nutrition,* ed 4. New York: Nutritional Foundation, Inc, 1976.
115. Ulmen DD: Trace elements. *N Engl J Med* 297:318, 1977.
116. Freeland-Graves JH, Ebagit ML, Hendrikson PJ: Alterations in zinc absorption and salivary sediment zinc after a lacto-ovo-vegetarian diet. *Am J Clin Nutr* 33:1757, 1980.
117. Klevay LM, Reck SJ, Jacob RA: The human requirement for copper. I. Healthy men fed conventional, American diets. *Am J Clin Nutr* 33:45, 1980.

Fiber and the Intestinal Microflora

4.1. FIBER

During the past decade, the importance of fiber substances in the diet has been reawakened by epidemiologists. The theories and information brought forth by Burkitt and Trowell,[1] Cleave *et al.*,[2] Spiller and Amen,[3] and Walker[4] have postulated that fiber in the diet is essential in the prevention of a series of degenerative diseases that include constipation, carcinoma of the colon, diverticular disease of the colon, appendicitis, diabetes mellitus, hiatus hernia, cholelithiasis, varicose veins, and hemorrhoids. This is truly a diverse list that can only be tied together by the theories postulated by Birkitt and Trowell. There are four texts published within the past decade that contain almost all of the recent information regarding these theories and the factual information in the field.[1-3,5] In this section we can only outline available and pertinent information. There is enough information available now to indicate that fiber in the diet is an essential factor for normal intestinal function, which in turn may be essential for the prevention of a whole series of degenerative diseases. Early in this century, Metchnikoff[6] proposed that toxic products within the intestine caused much of our disease. The present work on fiber substances may well prove him correct within the coming decades.

4.1.1. Definition of Fiber

Scientists disagree on the definition of fiber. At the present time there is no definition that can fully cover all of the theories and information available on the subject. Several definitions are available, and needed, to understand the present field.

Crude fiber is the residue of plant food that is left after extraction by dilute acids and dilute alkalis.[7]

Dietary fiber is the residue of plant food after hydrolysis by human gut enzymes.[7] Dietary fiber is composed of cellulose, hemicellulose, and lignin. In human nutrition fiber can be thought of as foods of plant origin that are neither usually digested nor absorbed by humans in the upper gastrointestinal tract.[3]

However, some studies do show slight digestion of both cellulose and hemicellulose in the small and large intestines, probably from bacterial action.[8]

As chemical techniques develop, clearer definitions emerge. Southgate's[9] detailed techniques measure pectin, gums, hemicelluloses, and polysaccharides. The techniques developed by Van Soest and McQueen[10] employ neutral detergents that measure more fiber substances in the residue than determined in Southgate's methods and are referred to as *neutral detergent fiber* (NDF). Others have developed enzymatic techniques that measure a similar amount of fiber substances.[11] As yet, there is no clear-cut agreement among scientists as to which technique is to be adopted to assay foods. It is clear, though, that specific techniques will have to be employed to evaluate a specific fiber substance. One technique may prove better for cellulose, another for lignin, another for pectin, etc.

As chemical methods improve it is apparent that plant fibers can be classified as follows[12,13]:

Biological
  1. Structural fibers (cell walls, cellulose, hemicelluloses, lignins, and pectins).
  2. Gums and mucilages.
  3. Storage polysaccharides (usually digestible).

Analytical
  1. Insoluble—NDF (method of Van Soest and McQueen): cellulose, hemicelluloses, lignin.
  2. Soluble (method of Southgate): hemicelluloses, pectins, gums, mucilages, polysaccharides (storage and algal).

All fiber substances are a mixture of high-molecular-weight polymers. The basic structural unit of the carbohydrate is a variety of monosaccharides. Lignin is primarily a phenylpropane unit. *Cellulose* is composed of straight chains of approximately 3000 1,4-linked D-glucose molecules. The *hemicelluloses* can be grouped into those that do or do not contain uronic acids and usually contain two to four different sugars, a common xylose, and approximately 150 to 250 pentose and hexose sugars. There are over 250 different polymers.[12] *Pectins* are polymers of galacturonic acid with varying degrees of methyl or acetyl esterification. Ninety-five percent of all pectins are digested completely in the colon.[5,12] *Gums* and *mucilages* are used by plants to repair their structure. They are complex sticky polysaccharides containing glucuronic and galacturonic acids in a branched structure with sugars.[12] Table 4-1 lists the primary units of the common components of most fiber substances.

4.1.2. Fiber Content of Foods and Intake

Table 4-2 is a comparison of the fiber content of some common foods based on different methods.

Standard texts, such as Church and Church,[16] have listed the fiber content of foods based on previous crude fiber analysis. Paul and Southgate's book[17] is

Table 4-1. Common Components of Dietary Fiber[a]

Main group	Class	Primary units
Carbohydrate	Cellulose	Glucose
	Hemicellulose A	Xylose Galactose Glucose Mannose Arabinose
	Hemicellulose B	Half-uronic acid, also equals "acid" hemicellulose
	Pectin substances	Mainly galacturonic acid
Noncarbohydrate	Lignin	Phenylpropane

[a]Adapted from Cummings.[14]

the latest listing of foods based on the newer analytical fiber methods (Chapter 20, diet 26). By the time this text is in print, there undoubtedly will be newer lists. Although the exact analytical fiber content is needed, the important considerations are the physiologic effects that the various foods have on man. It is of academic and research interest as to what the actual fiber content of those foods is. If one uses the older definitions of crude fiber, then in the United States, particularly areas such as Connecticut, man ingests between 2 and 4 g of crude fiber per day,[18] while in Great Britain it is estimated that the intake is between 4 and 8 g of crude fiber per day.[19,20] Using the same standards, vegetarian diets would contain between 12 and 24 g of crude fiber.[21] A careful survey of British subjects revealed an average intake of 19.9 ± 5.3 g of dietary fiber per day.[22] The range was from 8 to 32 g/day with 41.3% of the intake from vegetables, 30.5% from cereals, and 28.2% from fruits and mixed sources. Chemical analysis revealed that intake of noncellulose polysaccharide, cellulose, lignin, hexoses, pentoses, and uronic acid were respectively 13.8, 4.7, 1.4, 3.3, and 3.0 g/day.[22] As dietary patterns change with awareness of the benefits of fiber, it is easy to speculate that the Western diet will increase fiber intake.

An intake of between 10 and 20 g of crude fiber should produce an optimal fecal output in order to correct the so-called fiber deficiency disease as postulated by Burkitt and Trowell.[1] However, standardization is necessary. Appropriate analysis and evaluation of foods for their physiological effects are needed before more stringent dietary recommendations can be made. We need to know the specific effects plant foods have in humans.

In Western societies, bran is the main additive for bulk. In other countries, maize is the stable bulk former.[1] Kelsay and associates[23] have shown that a diet consisting of fruits and vegetables (citris fruits, dates, corn, pineapple, carrots, cabbage, blackberries, blueberries, squash, broccoli, raisins, apples, and peas) without whole grain but consisting of approximately 20 g of neutral detergent fiber will significantly increase the number of defecations and de-

Table 4-2. Fiber Contents of Some Common Foods (g/100 g) as Analyzed by Different Methods[a] (see Chapter 20 for complete list)

Food	Crude fiber	Southgate[9] method			Van Soest and McQueen[10] method		
		Cellulose	Hemi cellulose	Lignin	ADF[b]	Lignin	Indigestible residue
Apple	1.0	0.47	0.66	0.18	0.99	0.18	
	0.66						1.15[c] 1.7–2.4[d]
		0.67	0.52	0.40			
Banana	0.5						3.4[d]
		1.00	6.10	0.20			
Beans,	1.5						12.1[c]
brown cooked	2.2						5.1[d]
Cabbage	0.8						
		1.04	0.92	0.24	1.19	0.14	
	1.46						1.64[c]
		0.81	0.20	0.08			2.2–3.4[d]
Carrot	1.0						1.64[c]
	1.01						2.9–3.1[d]
Celery	1.0						
		0.91	0.65	0.16	0.91	0.16	
							1.8[d]
Rolled oats	1.2						
oats, dry	1.5						6.9[c]
Peas	2.0						
	0.57						5.2[d]
		3.06	1.42	0.33			
Turnip	0.9						
		0.66	0.41	0.11	0.72	0.14	
	0.98						1.3[c] 2.2–2.8[d]
Wheat bran	9.1						
		9.3	21.7	4.3	12.1	2.8	
	9.1						48.8[c]

[a] Adapted from Kelsay.[15]
[b] Acid detergent fiber.
[c] Enzymatic determination. Values have been calculated for wet weight using figures for indigestible residue and percentage dry matter.
[d] Roughage "by difference," after subtraction of fat, available carbohydrate, and protein.

crease bowel transit time while increasing fecal weight. Therefore, one can assume that it is not only the cereals but a broad intake of bulk formers that can effectively increase fecal weight.[23] Chapter 20 (diets 23, 26, and 27 and Table 20-7) lists the foods commonly used that have a relatively high fiber content. In addition, one should consult the standard references (crude fiber,[16] neutral detergent fiber[17]) when more detail is desired.

4.1.3. Physical and Biochemical Properties of Fiber

The physical and biochemical properties of fiber substances should be discussed for each specific fiber substance as the term fiber includes such a large number of chemical compounds. Research into the properties of fiber substances has been minimal. Therefore, we can only broadly group their properties as follows:

1. Water holding
2. Binding
   a. Minerals and electrolytes
   b. Bile salts
   c. Bacteria and toxins
3. Metabolic–systemic effects
   a. Blood lipids, glucose metabolism, bile salt physiology
   b. Fecal nitrogen and fat loss

4.1.3.1. Water-Holding Property

The different polysaccharide structures of fibers found in foods vary in their ability to hold water. Most are capable of forming gels. Table 4-3 lists the weight of water that can be demonstrated *in vitro* to combine with acetone dry powder from various foods. Remember that each one of these foods contains a combination of fiber substances. At this point, there is little scientific data available on the water-holding physical properties of all of the various food polysaccharides. Nevertheless, it can be helpful clinically if we know the water-binding capacity of foods, and therefore Table 4-3 supplies realistic data for diet therapy. Foods such as turnips and potatoes will bind approximately

Table 4-3. Approximate Weight of Water Held by Fiber from 100 g of Fruit or Vegetable[a,b]

25–100 g	H_2O
Turnips, potatoes, rhubarb, banana, cauliflower, tomatoes, broad beans, cucumber, celery	
100–150 g	H_2O
Onion, pea, cabbage, lettuce, green bean, pear, orange, maize, aubergine	
150–200 g	H_2O
Oatmeal, brussels sprout, apple	
200–300 g	H_2O
Carrot	
300–400 g	H_2O
Mango	
400–500 g	H_2O
Bran	

[a] Adapted from McConnell.[24]
[b] Keep in mind results should vary greatly depending on where vegetable or fruit is grown.

one-tenth of the water that dry bran does.[37] This information is used in the treatment of constipation, where the goal is to increase the bulk and wet content of the stool. Cummings and associates[25] fed bran foods to six healthy volunteers for 3 weeks, increasing their dietary fiber intake from 17 to 45 g of neutral detergent fiber per day. Concomitantly, the fecal weight increased from 79 ± 6.6 g to 228 ± 29 g/day. Both the stool solids and the stool size increased. In a similar series of experiments, Fuchs and colleagues[26,27] demonstrated that fecal weight increased from an average of 103 g/day to 226 g/day in six subjects fed bran for 6 weeks. And although both fecal solids and water increased significantly, the percentage of water increased in direct relationship to the solids (Fig. 1).

Kelsay *et al.*[23] studied 12 men on a high-fruit and high-vegetable diet exclusive of bran in which neutral detergent fiber intake was increased from 3.6 to 20 g/day. They were able to demonstrate an increase in wet fecal weight from approximately 90 to 200 g/day with the percentage of water remaining the same but increasing in proportion to the increase in the solid content of the feces. These studies clearly demonstrate that it is the total fiber intake as well as the quality of the fiber that can increase the water content of the stool and hence increase the effect of water binding in the human gut. Cummings *et al.*[28] used 20 g of concentrated dietary fiber in 19 healthy volunteers and reported that fecal weight increased approximately 127% with bran, 69% with cabbage, 59% with carrot, 40% with apple, and 20% with guar gum. They correlated the stool with the pentose content of the fiber polysaccharide eaten. More studies

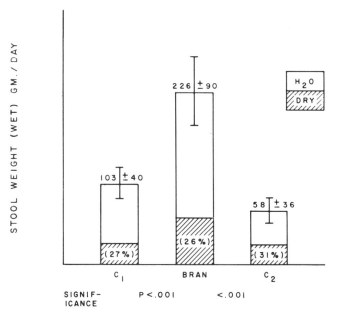

Figure 4-1. Bar graph demonstrates the increased size of stool produced on a high-bran (3 oz of All-Bran daily) diet. Note that the percentage of water remains insignificantly changed during control periods (C_1 and C_2), but the stool weight is greatly increased in the bran period.

will be needed to establish the exact chemical makeup of each fiber substance that determines water-holding and gel-forming capacities. Even though the physics of these food properties are not yet clear, it is clinically valuable that a combination of cereals, vegetables, fruits, and fiber substances is capable of increasing the water and total fecal weight content of human excrement (see Chapters 14 and 20 for high-fiber treatment of constipation).

4.1.3.2. Binding of Minerals and Chemicals

a. Minerals. Calcium, magnesium, zinc, phosphorus, and iron may all be significantly bound to fiber substances so that their absorption is decreased.

Calcium deficiency has been associated with diets rich in oatmeal and whole-meal wheat bread.[29-31] Furthermore, when cellulose is added to low- or high-fiber diets, calcium excretion may be increased.[32] The work of James and colleagues[33] reveals that both phytate and nonphytate components of fiber significantly bind calcium. The nonphytate calcium-binding fiber can be correlated with uronic acid content. Thus the amount of calcium expected to be bound and excreted in the stool can be calculated on the basis of the uronic acid millimol content of dietary fiber. In residents of Copenhagen who consumed 17.2 ± 5.1 g of dietary fiber per day that contained 12.3 ± 4.1 mmol uronic acid, they were able to predict that 3.8 ± 1.3 mmol (152 ± 15 mg of calcium) could be bound per day. From these data, one could expect societies that consumed 50 to 150 g of dietary fiber could easily bind almost 800 mg of calcium per day and consequently be in negative calcium balance and become calcium deficient.[33] But bacterial fermentation of fiber in the colon is primarily of the noncellulose polysaccharide, which results in fermentation of 80% of the uronic acid. This would free the calcium for potential absorption in the colon. In effect, Heaton and Pomare[34] did show that a high-bran diet can actually be associated with lower serum calcium values. This correlates with the observation that bran increases calcium excretion.[36] Magnesium excretion is also increased with high-fiber intake, but its metabolism has been less well studied. The demonstration that high fiber intake decreases both calcium and magnesium absorption is potentially of great importance in nutrition and may indicate that calcium intake would have to rise sharply if fiber intake is increased. Certainly, this aspect of increased fiber intake must be studied in more detail. (Chapter 20 lists foods associated with increased calcium excretion.)

There appear to be only minimal increases in sodium, chloride, and potassium excretion with high-fiber diets.[28] But iron and zinc deficiencies have been attributed to high intake of breads made from flours of high extraction.[35] One theory explains iron deficiency, while total intake of iron is adequate, on the binding of iron by phytate that leads to a decreased availability for intestinal absorption.[35] Further studies employing wheat bran reveal that phytate is definitely capable of binding as much as 72% of iron and 82% of zinc at a pH of 6.5 to 6.8.[35] Lignin and the cellulose fractions of wheat bran also have a high binding capacity with zinc. These authors postulate that bivalent metals bind to indigestible fiber components and become unavailable for gastrointestinal absorption.[35] Obviously, these observations have to be elaborated upon before they can be fully accepted. Many societies have a high-fiber diet and yet iron and zinc deficiencies are not obvious. Although Iranian children eating home-made bread can develop iron and zinc deficiencies, this fact cannot be extrapo-lated to all societies.[35,36] More study is needed in this area to determine which foods may have significant malabsorptive effect due to their binding capacities.

b. Bile Salts. As noted in Chapter 11, bile acid excretion appears to be influenced by increased fiber intake. Diets containing wheat and vegetables, maize, bagasse, cholesterol plus cellulose, psyllium, or cellulose may increase bile acid excretion to varying degrees.[15]

Eastwood and associates[37,38] have demonstrated that bile acids are ad-sorbed to dietary fiber. Dihydroxy bile acids are adsorbed more strongly than trihydroxy bile acids. Free bile acids are adsorbed more strongly than taurine or glycine conjugates. When food residues are added to bile salts *in vitro,* it can be shown that celery, corn, lettuce, potato, and string bean can potentially adsorb large amounts of bile salts, but hemicellulose alone adsorbs a small amount.[39]

Physiologically conjugated bile salts are present in the upper small intes-tine with fatty acids, mono-, di-, and triglycerides, lipase, cholesterol, and phospholipids. These components form micelles. Once bile salts are incorpo-rated into micelles, there is less adsorption to fiber substances. The extent to which adsorption is diminished appears to be affected by the fatty acid concen-tration, chain length, and unsaturation. Therefore, adsorption should occur less in the upper jejunum when bile acids are out of the micelle phase, such as occurs in the distal jejunum, more so in the ileum, and most in the colon.[37] Lignin has the greatest binding potential.[37] The entire phenomenon of food binding bile salts is most complex when one considers the great variability in food intake.[40] However, there may be dietary therapeutic implications in bile salt disturbances, such as cholelithiasis. This is discussed in detail in Chapter 11. It is possible that selected fiber diets may be able to cause increased excre-tion of selective bile salts and thereby have an effect on cholesterol and bile acid metabolism. Much more research is needed in this field before any definite clinical therapeutic benefit is established.

c. Binding of Bacteria and Toxins. Bacterial reactions occur in four gen-eral areas:

(1) *Gas formation* from fermentation of nonabsorbed carbohydrates with the production of carbon dioxide, hydrogen, and methane that may clinically produce more or less flatus and more or less noxious odors.
(2) *Acid production* of lactic acid and volatile fatty acids from nonab-sorbed carbohydrates that clinically may cause an increase in secretion of fluid in the bowel and may result in diarrhea when increased amounts are present.
(3) *Enzymatic reactions* by bacterial enzymes such as conversion of con-jugated to deconjugated bile acids and primary to secondary bile acids (in addition, the enzymatic assistance of food breakdown of proteins

into ammonia, the conversion of cholesterol to its breakdown products such as coprostanol, etc.).

(4) *Binding to bile acids* by bacteria that may result in either entrapment of bile acid or control of bacterial populations.[41,42]

Dietary intake has variable effects on the colonic flora in man.[43] Although the chemical nature of the diet should affect the flora, investigations into the quantitative and qualitative flora in man have not demonstrated clear results. When the theory of bacterial involvement in the production of carcinogens was brought forth, it was assumed that the flora varied in relation to dietary intake. The early evidence revealed that there was a difference in the broad spectrum of colonic bacteria of subjects living in underdeveloped countries compared with those living in highly developed countries consuming a diet consisting of more animal fat and less fiber. Studies in this field have yielded inconsistent results, as is discussed in carcinoma of the colon. Slight changes in the flora have been demonstrated with increasing intake of cellulose, bran, lactose, and high meat.[26,27,43-45] Fluctuations in the flora have been demonstrated in subjects placed on chemically defined, fiber-free diets,[43] and small but definite variations in flora and bacterial enzymes can be shown in vegetarians and Adventists.[46,47] Little work has been done on studying the effect of selective dietary factors, rather than food groups, in man. Most studies have been in rats, and the species variation is too great to make any clear analogies. Such factors as the bile-salt-binding capability of lignin versus the digestion of pectin by bacteria must have an influence on the intestinal flora. Future studies employing selective foods and food components should prove fascinating and may lead to therapeutic advances.

Eastwood and co-workers[37,38] found that hydrated fiber, after passing through the cecum, forms a cellular interior or mesh rather like a sponge. They suggest that conditions on the interior of the fiber create a different physiochemical environment so that the bacteria inside, or outside, of the fiber matrix may act entirely differently. Little is known about the effects of fiber on bacterial physiology, but from *in vitro* experiments it is suspected that bacteria may well align themselves via adsorption on the surface of the fiber. If they are adsorbed onto the fiber, then it can be postulated that the variations in solutes and pH would have a great effect on bacterial metabolism, and hence bacterial populations. What clinical significance this has is not yet apparent.

Although Metchnikoff's[6] theories of intestinal toxins causing disease did not gain wide acceptance, the understanding that fiber can adsorb chemicals reawakens the possibilities that it might be protective when eaten in appropriate amounts in man.

4.1.3.3. Metabolic–Systemic Effects

Experiments on the systemic effects of fiber foods in human metabolism have been few and far between. The major contributions can be grouped into the effects on (a) serum lipids including cholesterol, (b) glucose tolerance in diabetes mellitus, and (c) bile salts in cholelithiasis.

a. Serum Lipids. Although there have been many experiments on diet and atherosclerosis in animals, there are few references concerning experiments in man. This is surprising considering the great interest in atherosclerosis and its proven relationship in our society to dietary factors. It is well accepted that high saturated fat intake, often associated with high cholesterol and high simple sugar intake, and elevated serum cholesterol levels are related to increased coronary artery disease. Societies that have low intakes of these substances tend to have lower serum cholesterol blood levels and concomitantly increased fiber intake. Jenkins and colleagues[48] demonstrated that 36 g of pectin or guar gum given to healthy volunteers over a 2-week period resulted in a significant decrease in serum cholesterol levels. This work has been confirmed in experiments by Kay and Truswell,[49] who demonstrated that 15 g of pectin added to the diet daily over a 3-week period resulted in a significant decrease in plasma cholesterol, while there was a significant increase in the fecal fat, neutral steroid, and fecal bile acid excretion. Both of these works confirm the original presentation on the effect of pectin in the diet by Keys and associates.[50] Although the work with pectin has been confirmed, there has been little done with guar. It is of interest in the United States that guar is readily added to some ice cream products, such as soft ice creams.

The effect of bran intake on blood lipids has been controversial.[15] Several authors have found no significant effect with the use of bran,[48,51] but our own laboratory did find significant decreases in individual serum levels on a 3-week cereal-bran diet. Unfortunately, all of these experiments did not use similar amounts or types of bran.[12] The difference in the origin of the cereal (wheat, oat, etc.) probably is significant, for several investigations have shown significant effects on serum cholesterol using a variety of cereals.[12,52,53] Our own work included a larger amount than was employed in most other experiments. In any event, it does seem that substances such as pectin and guar may have a more dramatic response than bran. The reason for this is obviously in the chemical makeup of the food. From all of these experiments it appears that the systemic lipid effect is on serum cholesterol, as triglycerides appear to remain normal. This agrees with the observation that vegetarian subjects, when compared to controls, have a significantly lower mean cholesterol level.[54] Concomitantly, they have lower low-density lipoprotein, lower very-low-density lipoprotein, and mean triglyceride levels and are thinner, but analysis reveals the lipid differences are related to food intake rather than to weight.[54] Obviously, there are multiple factors involved rather than simply the increased fiber intake that occurs in vegeterians. Clinical application of these observations has resulted in an attempt to treat type 2A hyperlipoproteinemia patients with a high-fiber diet. Eight subjects were treated with powdered cellulose or soy hulls or both for 6 months and compared to subjects treated with cholestyramine.[55] Mean serum cholesterol levels decreased on all regimens although most significantly with the use of cholestyramine. In this clinical study, the authors felt that there was some benefit to the use of a high-fiber diet for their patients with hyperlipoproteinemia.[55]

Although most epidemiologists and nutritionists attribute high animal fat intake to dietary hypercholesterolemia, there is some evidence to indicate that a shift from animal protein to vegetable protein has a serum cholesterol lowering effect that is greater than can be explained by a reduction of the intake of saturated fats and cholesterol.[56,57] Obviously, serum cholesterol can be affected by all dietary factors. Much more experimental work is needed in the field to clearly demonstrate which dietary factors are most important and which are protective, if at all, against elevated serum lipids that result in high risk of atherosclerosis.

b. *Glucose Tolerance in Diabetes Mellitus.* The systemic effect of intake of fiber substances on glucose metabolism is discussed in the section on diabetes mellitus. Of note is that high carbohydrate fiber diets have been shown to improve glucose tolerance in diabetic subjects. Patients were able to do without sulfonylurea administration or decrease their requirements for insulin.[58,59] The effect on plasma glucose and serum insulin appears to be related to the consistency of the fiber food.[60] Haber and associates[60] showed that subjects fed whole apples as opposed to apple juice or applesauce had less of the rebound fall in plasma glucose and less of a rise in serum insulin. The reason for this is not clear. One could postulate that it is related to the important role of glucagon in diabetes, and its relationship with hormones such as somatostatin, which may be affected by dietary intake. More research is needed in this field to relate the diet to hormones related to diabetes mellitus and concomitantly to glucose tolerance.[61]

c. *Bile Salts in Cholelithiasis.* The effect on bile salt metabolism is discussed in Chapter 11, as is the effect of fiber substances on cholelithiasis.

Of note concerning wheat bran administration, the size of the bile salt pool is not significantly changed, but bran supplements are capable of decreasing the cholesterol saturation of bile while increasing chenic acid synthesis and the size of the chenic acid pool and decreasing the size of the deoxycholic pool. This effect has not been tested clinically, but it holds great promise.[62]

The metabolic effect of fiber can be related to its effect on absorption and digestion of other nutrient materials. This becomes clear from experimental works published by Cummings *et al.*[25,28] When healthy volunteers were given approximately 45 g of dietary fiber intake compared with their low fiber intake of 17 g, there was a significant increase in fecal fat, nitrogen, and calcium output. These chemical phenomena may be related to the binding effect, as discussed above, but probably have a significant long-term impact on total nitrogen and energy balance of the body. The increased excretion with high-fiber diets should then be related to systemic effects. What the meaningful effect in humans is, is not clear. Much more research is needed concerning the effect of specific fiber foods.

4.1.4. Fiber and Colon Cancer

As discussed previously, epidemiologic evidence reveals definite differences in the distribution of the incidence of carcinoma of the colon.[1] One of the clear differences between the populations that develop carcinoma of the colon and those that do not is the diet. Fiber and plant foods are eaten in large quantities in those societies that have a low incidence of carcinoma of the colon. Concomitant with this is the lesser consumption of animal fats and proteins. Some postulate the high intake of fiber protects against colon carcinoma, and the mechanisms for this can be outlined as follows.

1. By holding onto water, fiber acts to dilute potential carcinogens within the gut.
2. By decreasing transit time, fiber reduces the time potential carcinogens are exposed to the gut lining.
3. By its influence on bile salt metabolism, fiber alters the availability of certain sterols for the formation of carcinogens.
4. By its physical property of binding, fiber may bind potential carcinogens or carcinogens themselves.
5. And, finally, by its influence on intestinal microflora, fiber may alter the production of enzymes necessary to convert metabolites to carcinogens.

The evidence to indicate that fiber's bulking or water-binding effect has any influence on the cause of cancer of the colon is not strong. When Japanese men living in Hawaii were compared to Caucasian men living in Hawaii, it was found that there was a definite disparity in transit time, but both had the same risk of cancer of the colon,[63,64] whereas Japanese men in Japan do have a lower risk of carcinoma of the colon. This casts doubt on the effect of decreased transit time as a preventive factor. This impression was verified in a study in Scandanavia that compared high-risk-colon-cancer Danish men on high-meat intake to low-risk Finnish farmers on high-fiber milk diets where risk was unrelated to transit time.[65] This was further enforced by the work of Graham *et al.*[66] that compared the diets of 1182 rectal and colon cancer patients with 1411 matched normals and found no increased risk associated with meat intake, but a decreased risk in frequency of vegetable intake.[66] The frequent intake of particularly cabbage, brussels sprouts, and broccoli was associated with decreased risk.[66]

The theory that the bulking effect of fiber decreases the concentration of carcinogens in the gut is still open to question.[67-69] It was postulated that subjects in England with a high risk for colon cancer had a higher concentration of fecal steroids. This was also confirmed in the United States.[70,71] Patients with carcinoma of the colon and those with a high risk for carcinoma of the colon, such as ulcerative colitis and familial polyposis subjects, appear to have higher concentrations of fecal steroids. Depending on the study quoted, neutral (cholesterol metabolites) and/or acid (bile salts) steroids have been demonstrated to be increased in subjects with high colon cancer risk.[67-72] Our own laboratory studying carcinoma of the colon subjects in Connecticut who have poor fiber intake did not demonstrate an increase in either acid or neutral steroids.[73]

A very distinct possibility is that fiber has an effect on intestinal luminal steroid metabolism so that it will affect potential carcinogens or cancer metabolism of steroids. The effect of fiber on bile salts by its binding phenomenon is clear. Experiments have been performed to determine whether fiber addition to so-called Western diets has an effect on the concentration of colon steroids. A study among varied Hong Kong socioeconomic groups again clearly revealed that fecal steroids are more highly concentrated in the stools of subjects with a higher risk for colon cancer.[74] But the results vary, as indicated in a review by Hill.[75] It is clear that fiber has a definite effect on acid (bile salt) steroid metabolism. What relationship this has to cancer risk remains to be fully assessed.

The binding effect of fiber has been discussed above. It is well to remember that fiber can bind both bacteria and bile salts. As both of these are postulated as important in carcinoma of the colon, it is reasonable to suspect that this physical property will prove important in the final assessment of fiber's role in cancer of the colon. What is not yet known is how fiber can bind other potential carcinogens *in vitro* or *in vivo*.

4.2. INTESTINAL MICROFLORA

Finally, the role of the *intestinal microflora* in the production of carcinogens appears to be extremely important.[67,74-78] Quantitative counts of microflora in the stomach are usually less than 10^3 or 10^4 and consist of lactobacilli, streptococci, and fungi.[69] Bacterial counts and flora in normal humans remain low in the proximal small intestine due to cleansing acid production by the stomach and normal motility. When these are disturbed, bacterial overgrow occurs (see Chapter 13). Colon bacterial flora begins to appear in the ileum, where anaerobic counts begin to match aerobic counts, and finally in the colon anaerobes outnumber aerobes 100–1000:1 with total counts at 10^{11} to 10^{13} organisms per gram of feces. There is great variability of flora species harbored by humans. Approximately 400 species and strains of bacteria have been identified, the predominant ones from bacteroides, eubacteria, lactobacilli, klebsiella, *Escherichia*, streptococci, and peptostreptococci.[69,78]

Bacteria are the major contributors of enzymes within the gut lumen. These enzymes are assumed to be the major factor in the ability to convert food products or metabolites to carcinogens. Bacteria produce β-glucuronidase, nitroreductase, azoreductase, and probably most importantly 7α-dehydroxylase, the dehydroxylase capable of converting most bile salts. *Clostridium paraputrificum* does produce enzymes that are capable of introducing a double bond into the steroid nucleus as well as dehydroxylation. This metabolic transformation produces substances that can be carcinogenic for man.[75] On this note, Hill and associates[76] have demonstrated that 44 patients with cancer of the large bowel had both increased fecal bile acid concentrations as well as a higher incidence of Clostridia able to dehydrogenate. Although the British have confirmed this repeatedly in their own laboratories, our study on cancer of the colon subjects did not confirm this finding,[78] nor was it confirmed in the laboratories of Finegold *et al.*[79] studying high-risk colon cancer subjects.

The relationship of diet to alterations in the microflora has also been noted.

Initial investigators suspected that a high-fiber diet was related to a decreased number of anaerobic organisms and consequently a decreased anaerobic to aerobic ratio.[67] Our own studies reveal high-fiber (bran) intake tended to increase the anaerobic to aerobic ratio,[26] and our most recent study on subjects with cancer of the colon similarly showed an increased anaerobic ratio.[78] This area of study is further muddled by the finding that a high-meat diet (which would be contrary to a high-fiber diet) results in an increased number of anaerobes.[80] Although there is some controversy on alterations in the bacterial flora that are associated with carcinoma of the colon, there is no question that bacteria play an important role in metabolites within the colon lumen. If and when the carcinogen that produces carcinoma of the colon is identified, future research will tell us which bacteria are important.

Although some may still doubt that these theories are important, animal experiments are adding further important strength to them. When bran is added to the diet of male rats, it has a definite protective effect against the production of colon carcinoma induced by subcutaneous dimethylhydrazine injection.[81]

Although fiber is the most popular factor in theories on carcinoma of the colon, other dietary factors have received similar popularity in the past and have been reviewed by Alcantara and Speckmann.[82] The amount and type of fat in the diet have been related to carcinoma. In animal experiments increasing the fat intake from roughly 5 to 20% increases the risk of tumor formation. Controversy has developed concerning the role of saturated and unsaturated fats in cancer risk. Some claim that unsaturated fats actually increase risk, while others claim they decrease risk.[82] Similarly, there are conflicting reports on the relationship of serum cholesterol to colon carcinoma. Although it is clearly associated with increased risk for coronary artery disease, epidemiologic studies do not reveal clear associations with colon carcinoma.[82] Similarly, controversy occurs concerning the role of protein and amino acids in carcinoma. Although there is some epidemiologic evidence to indicate that increased animal protein does produce increased risk, there is animal experimentation to indicate that it does not have a definite association. Just as the evidence concerning fats and proteins is not clear, so the evidence concerning a variety of vitamins and minerals is similarly not clear. Thus, the composite picture reveals no definite role for dietary factors, but a contemporary theory that fiber and intestinal lumen metabolism have a relationship to colon cancer.[83]

REFERENCES

1. Burkitt DP, Trowell HC: *Refined Carbohydrate Foods and Disease.* New York: Academic Press, 1975.
2. Cleave TL, Campbell GD, Painter NS: *Diabetes, Coronary Thrombosis and the Saccharine Disease.* ed 2. Bristol: John Wright, 1969.
3. Spiller GA, Amen RJ: *Fiber in Human Nutrition.* New York: Plenum Press, 1976.
4. Walker ARP: Colon cancer and diet, with special reference to intakes of fat and fiber. *Am J Clin Nutr* 29:1417, 1976.

5. Spiller GA: *Topics in Dietary Fiber Research.* New York: Plenum Press, 1978.
6. Metchnikoff E: *Prolongation of Life.* New York: GP Putnam, The Knickerbocker Press, 1908.
7. Trowell H: Definition of dietary fiber and hypotheses that it is a protective factor in certain diseases. *Am J Clin Nutr* 29:417, 1976.
8. Holloway WD, Tasman-Jones C, Lee SP: Digestion of certain fractions of dietary fiber in humans. *Am J Clin Nutr* 31:927, 1978.
9. Southgate DAT: Determination of carbohydrates in foods. II. Unavailable carbohydrates. *J Sci Food Agric* 20:331, 1969.
10. Van Soest PJ, McQueen RW: The chemistry and estimation of fibre. *Proc Nutr Soc* 32:123, 1973.
11. Hellendorn EW: Enzymatic determination of the indigestible residue (dietary fibre) content of human food. *J Sci Food Agric* 26:1461, 1975.
12. Anderson JW, Chen WL: Plant fiber. Carbohydrate and lipid metabolism. *Am J Clin Nutr* 32:346, 1979.
13. Cummings JH: What is fiber?, in Spiller GA, Amen RJ (eds): *Fiber in Human Nutrition.* New York: Plenum Press, 1976.
14. Cummings JH: Dietary fibre. *Gut* 14:69, 1973.
15. Kelsay JL: A review of research on effects of fiber intake on man. *Am J Clin Nutr* 31:142, 1978.
16. Church CF, Church HN: *Food Values of Portions Commonly Used,* ed 12. New York: JP Lippincott Co, 1975.
17. Paul AA, Southgate DAT: *McCance and Widdowson's The Composition of Foods,* ed 4. New York: Elsevier/North-Holland Biomedical Press, 1978.
18. Dorfman SH, Ali M, Floch MH: Low fiber content of Connecticut diets. *Am J Clin Nutr* 29:87, 1976.
19. Trowell HC: Ischemic heart disease and dietary fiber. *Am J Clin Nutr* 25:926, 1972.
20. Robertson J: Change in the fibre content of the British diet. *Nature (London)* 238:290, 1972.
21. Hardinge MG, Chambers AC, Crooks H, *et al:* Nutritional studies of vegetarians. III. Dietary levels of fiber. *Am J Clin Nutr* 6:523, 1958.
22. Bingham S, Cummings JH, McNeil NI: Intakes and sources of dietary fiber in the British population. *Am J Clin Nutr* 32:1313, 1979.
23. Kelsay JL, Behall KM, Prather ES: Effect of fiber from fruits and vegetables on metabolic responses of human subjects. I. Bowel transit time, number of defecations, fecal weight, urinary excretions of energy and nitrogen and apparent digestibilities of energy, nitrogen, and fat. *Am J Clin Nutr* 31:1149, 1978.
24. McConnell AA, Eastwood MA, Mitchell WD: Physical characteristics of vegetable foodstuffs that could influence bowel function. *J Sci Food Agric* 25:1, 1974.
25. Cummings JH, Hill MJ, Jenkins DJA, *et al:* Changes in fecal composition and colonic function due to cereal fiber. *Am J Clin Nutr* 29:1468, 1976.
26. Fuchs HM, Dorfman S, Floch MH: The effect of dietary fiber supplementation in man. II. Alteration in fecal physiology and bacterial flora. *Am J Clin Nutr* 29:1443, 1976.
27. Floch MH, Fuchs H-M: Modification of stool content by increased bran intake. *Am J Clin Nutr* 31:185, 1978.
28. Cummings JH, Southgate DAT, Branch W, *et al:* Colonic response to dietary fibre from carrot, cabbage, apple, bran, and guar gum. *Lancet* 1:5, 1978.
29. McCance RA, Widdowson EM: Mineral metabolism on dephytinized bread. *J Physiol* 101:304, 1942.
30. McCance RA, Widdowson EM: Mineral metabolism of healthy adults on white and brown bread dietaries. *J Physiol* 101:44, 1942.
31. McCance RA, Walsham CM: The digestibility and absorption of the calories, proteins, purines, fat and calcium in wholemeat wheaten bread. *Br J Nutr* 2:26, 1948.
32. Ismail-Beigi F, Reinhold JG, Faraji JG, *et al:* Effects of cellulose added to diets of low and high fiber content upon the metabolism of calcium, magnesium, zinc and phosphorus by man. *J Nutr* 107:510, 1977.
33. James WPT, Branch WJ, Southgate DAT: Calcium binding by dietary fiber. *Lancet* 1:638, 1978.

34. Heaton KW, Pomare EW: Effect of bran on blood lipids and calcium. *Lancet* 49, 1974.
35. Ismail-Beigi F, Faraji B, Reinhold JG: Binding of zinc and iron to wheat bread, wheat bran, and their components. *Am J Clin Nutr* 30:1721, 1977.
36. Reinhold JG: Zinc and mineral deficiencies in man: The phytate hypothesis, in Charez A, Bourges H, Basta S (eds): *Procedures of the 9th International Congress of Nutrition, Mexico,* 1975, vol 1, p 115.
37. Eastwood M, Mowbray L: The binding of the components of mixed micelle to dietary fiber. *Am J Clin Nutr* 29:1461, 1976.
38. Eastwood MA, Hamilton D: Studies on the absorption of bile salts to nonabsorbed components of diet. *Biochim Biophys Acta* 152:165, 1968.
39. Eastwood MA, Girwood RH: Lignin: A bile-salt sequestering agent. *Lancet* 2:1170, 1968.
40. Birkner HJ, Kern F: *In vitro* adsorption of bile salts to food residues, salicylazosulfaphyridine, and hemicellulose. *Gastroenterology* 67:237, 1974.
41. Hellendoorn EW: Fermentation as the principle cause of the physiological activity of indigestible food residue, in Spiller GA (ed): *Topics in Dietary Fiber Research.* New York: Plenum Press, 1978.
42. Floch MH, Gershengoren W, Elliot S, *et al:* Bile acid inhibition of intestinal microflora—A function for simple bile acids? *Gastroenterology* 61:228, 1971.
43. Drasar BS, Hill MJ: *Human Intestinal Flora.* New York: Academic Press, 1974.
44. Maier BR, Flynn MA, Burton GC, *et al:* Effects of a high beef diet on bowel flora: A preliminary report. *Am J Clin Nutr* 27:1470, 1974.
45. Ellington T, Conn HO, Floch MH: Lactulose in treatment of chemic portal-systemic encephalopathy. *N Engl J Med* 281:408, 1969.
46. Finegold SM, Sutter UL, Sugihara PT, *et al:* Fecal microbial flora in Seventh Day Adventist populations and control subjects. *Am J Clin Nutr* 30:1781, 1977.
47. Reddy BS, Wynder EL: Metabolic epidemiology of colon cancer: Enzymatic activity of fecal flora. *Am J Clin Nutr* 29:1455, 1976.
48. Jenkins DJA, Leeds AR, Newton C, *et al:* Effect of pectin, guar gum, and wheat fibre on serum-cholesterol. *Lancet* 1:1116, 1975.
49. Kay RM, Truswell AS: Effect of citrus pectin on blood lipids and fecal steroid excretion in man. *Am J Clin Nutr* 30:171, 1977.
50. Keys A, Grande F, Anderson JT: Fiber and pectin in the diet and serum cholesterol concentration in man. *Proc Soc Exp Med Biol* 106:555, 1961.
51. Truswell AS, Kay RM: Absence of effect of bran on blood lipids. *Lancet* 1:922, 1975.
52. Tarpila S, Miettinen TA, Metsaranta L: Effects of bran on serum cholesterol, faecal mass, fat, bile acids and neutral sterols, and biliary lipids in patients with diverticular disease of the colon. *Gut* 19:137, 1978.
53. Munoz JM, Sandstead HH, Jacob RA, *et al:* Effects of some cereal brans and textured vegetable protein on plasma lipids. *Am J Clin Nutr* 32:580, 1979.
54. Sacks FM, Castelli WP, *et al:* Plasma lipids and lipoproteins in vegetarians and controls. *N Engl J Med* 292:1148, 1975.
55. Palumbo PJ, Briones ER, Nelson RA: High fiber diet in hyperlipemia. *JAMA* 240:223, 1978.
56. Editorial: Serum-cholesterol and the soya bean. *Lancet* 1:291, 1977.
57. Sirtori CR, Agradi E, Conti F, *et al:* Soybean-protein diet in the treatment of type-II hyperlipoproteinemia. *Lancet* 1:275, 1977.
58. Kiehm TG, Anderson JW, Ward K: Beneficial effects of a high carbohydrate, high fiber diet on hyperglycemic diabetic men. *Am J Clin Nutr* 29:895, 1976.
59. Jenkins DJA, Leeds AR, Gassull MA, *et al:* Decrease in postprandial insulin and glucose concentrations by guar and pectin. *Ann Intern Med* 86:20, 1977.
60. Haber GB, Heaton KW, Murphy D, *et al:* Depletion and disruption of dietary fibre. *Lancet* 1:679, 1977.
61. Raskin P, Unger RH: Hyperglucagonemia and its suppression. *E Engl J Med* 9:433, 1978.
62. Pomare EW, Heaton KW, Low-Beer TS, *et al:* The effect of wheat bran upon bile salt metabolism and upon the lipid composition of bile in gallstone patients. *Am J Dig Dis* 21:521, 1976.

63. Glober GA, Klein KL, Moore JO, *et al:* Bowel transit-times in two populations experiencing similar colon-cancer risks. *Lancet* 2:80, 1974.

64. Glober GA, Nomura A, Kamiyama S, *et al:* Bowel transit-time and stool weight in populations with different colon-cancer risks. *Lancet* 2:110, 1977.

65. International Agency for Research on Cancer, Intestinal Microecology Group: Dietary fibre, transit-time, fecal bacteria, steroids, and colon cancer in two Scandinavian populations. *Lancet* 2:207, 1977.

66. Graham S, Dayal H, Swanson M, *et al:* Diet in the epidemiology of cancer of the colon and rectum. *J Natl Cancer Inst* 61:709, 1978.

67. Aries V, Crowther JS, Drasar BS, *et al:* Bacteria and the aetiology of cancer of the large bowel. *Gut* 10:334, 1969.

68. Hill MJ: The effect of some factors on the fecal concentration of acid steroids, neutral steroids and urobilins. *J Pathol* 104:239, 1971.

69. Drasar BS, Hill MJ: Intestinal bacteria and cancer. *Am J Clin Nutr* 25:1399, 1972.

70. Reddy BS, Mastromarino A, Wynder EL: Further leads on epidemiology of large bowel cancer. *Cancer Res* 35:3403, 1975.

71. Reddy BS, Mastromarino A, Gustafson C, *et al:* Fecal bile acids and neutral sterols in patients with familial polyposis. *Cancer Res* 38:1694, 1976.

72. Reddy BS, Martin CW, Wynder EL: Fecal bile acids and cholesterol metabolites of patients with ulcerative colitis, a high-risk group for development of colon cancer. *Cancer Res* 37:1697, 1977.

73. Moscovitz M, White C, Barnett RN, *et al:* Diet, bile acids, and neutral sterol excretions in carcinoma of the colon, control subjects and in liver metastases. *Dig Dis Sci* 24:246, 1979.

74. Crowther JS, Drasar BS, Hill MJ, *et al:* Fecal steroids and bacteria and large bowel cancer in Hong Kong by socio-economic groups. *Br J Cancer* 34:191, 1976.

75. Hill MJ: Metabolic epidemiology of dietary factors in large bowel cancer. *Cancer Res* 35:3398, 1975.

76. Hill MJ, Drasar BS, Williams REO, *et al:* Fecal bile-acids and clostridia in patients with cancer of the large bowel. *Lancet* 1:535, 1975.

77. Goldin B, Gorbach SL: Colon cancer connection: Beef, bran, bile and bacteria. *Viewpoints Dig Dis* 10:1, 1978.

78. Vargo D, Moskovitz M, Floch MH: Fecal bacteria flora in cancer of the colon. *Gut,* 1980 (in press).

79. Finegold SM, Attebery HR, Sutter VL: Effect of diet on human fecal flora: Comparison of Japanese and American diets. *Am J Clin Nutr* 27:1456, 1974.

80. Reddy BS, Weisburger JH, Wynder EL: Effects of high risk and low risk diets for colon carcinogenesis on fecal microflora and steroids in man. *J Nutr* 105:878, 1975.

81. Fleiszer D, Murray D, MacFarlane J, *et al:* Protective effect of dietary fibre against chemically induced bowel tumours in rats. *Lancet* 2:552, 1978.

82. Alcantara EN, Speckmann EW: Diet, nutrition and cancer. *Am J Clin Nutr* 29:1035, 1976.

83. Huang CTL, Gopalakrishna GS, Nichols BL: Fiber, intestinal sterols, and colon cancer. *Am J Clin Nutr* 31:516, 1978.

Weight Loss and Nutritional Assessment

5.1. WEIGHT LOSS

Weight loss occurs because of inadequate caloric intake. [Optimal body weights are described in Chapter 2 (Table 2-6).] Weight loss is a state of malnutrition. However, malnutrition also includes obesity and disorders that result in overnutrition as well as undernutrition. Furthermore, malnutrition includes selective deficiencies in proteins, vitamins, and minerals such as described in Chapters 2 and 3. A simple lack of food supply remains the greatest cause of decreased caloric intake in the world. Consequently, starvation, kwashiorkor, and marasmus are the most common diseases resulting from inadequate food supply and are seen frequently in underdeveloped societies. Obviously, these are due to economic and sociologic factors rather than body disorders or digestive diseases.

Table 5-1 outlines diseases associated with weight loss. Anorexia nervosa is the best example of a psychiatric disorder that results in severe weight loss. Any clinician will verify that depression, or simple anxiety associated with so-called "worry," will result in decreased food intake and weight loss. Neurologic disorders that interfere with the neuromuscular activity necessary for eating will result in weight loss. Examples of these are destruction of nerve supplies involved in swallowing, such as occur in cerebral vascular accidents involving the cranial nerves, or posttraumatic accidents that damage nerve or muscle tissue of the tongue, mouth, or pharynx.

Addison's disease, chronic cardiac disease, neoplastic disease, and inappropriate antidiuretic hormone secretion are systemic diseases that result in weight loss. Mechanisms such as decreased plasma osmolality and disturbances in circulating electrolytes have been postulated as causes of anorexia by decreasing activities of hypothalamic feeding centers. However, this is theoretic and others have proposed circulating anorexigenic chemicals produced in chronic or neoplastic disease, but none have been identified. The exact cause of the loss in severe advanced endocrine and neoplastic disease remains unknown.[1]

In severe hyperthyroid disease, it is assumed that increased metabolic activity results in increased energy requirements. In diabetes mellitus, it is assumed that increased metabolic breakdown results in weight loss, yet mild

Table 5-1. Diseases Associated with Weight Loss

A. Psychiatric
   Anorexia nervosa
   Depression or anxiety

B. Neurologic
   Diseases interfering with neuromuscular activity of eating

C. Endocrine and secondary to systemic disease
   Neoplastic ("cancer cochexia"), Addison's disease, diabetes mellitus, hyperthyroid, chronic
   renal, and cardiac disease

D. Postsurgery
   Postoperative period

E. Gastrointestinal
   Following surgical resection and blind loops
   Due to primary and secondary small-bowel disease, gluten enteropathy, etc. (Chapter 6)
   Due to pancreatic insufficiency (Chapter 7)
   Due to liver disease (Chapter 5)

F. Starvation
   Kwashiorkor and marasmus

diabetes mellitus is associated with hyperphagia and may actually result in weight gain. The exact mechanisms of weight loss and weight gain are not completely understood in this disorder. These endocrine disorders appear to be associated with increased appetite, although it may well result in weight loss.[1]

The weight loss that occurs during the postoperative period is obviously due to decreased intake, but yet it is closely associated with increased energy demands of the stress of surgery.

The causes of weight loss associated with digestive diseases are outlined in Table 5-1, "Gastrointestinal." Those due to surgical resection of the gut, surgical complication such as blind loop, either primary or secondary small-bowel disease, pancreatic insufficiency, or liver disease will be discussed elsewhere in the text under those organs.

Evaluations of hospital patients have revealed a dramatic incidence of malnutrition in both surgical and medical patients.[2-6] Seventeen to forty-five percent of medical service patients had findings below the lower limits of normal for serum albumin, levels of vitamins A, C, and E, hemoglobin, hematocrit, and body weight when compared to hospital employees and patients hospitalized for vocational rehabilitation.[2] Thorough evaluations by Bistrian et al.[4] in the United States and by Hill et al.[5] in England reveal that there is a high incidence of malnutrition in patients on general medical services as well as pre- and postoperative patients. Butterworth and colleagues[3,6] have stressed this fact in a prospective evaluation of 134 consecutive admissions to a general medical service. They evaluated 8 nutrition-related parameters: serum folate, vitamin C, triceps skin fold, weight/height, arm muscle circumference, lymphocyte count, serum albumin, and hematocrit. It was found that on admission to the hospital, 48% of the patients had a high likelihood of malnutrition, which increased with the hospital stay and correlated with an increased mortality rate.[6] It was found that the longer the patient remained in the hospital, the greater the likelihood of malnutrition.[3,6] Although it is obvious that the sicker patient will require longer hospitalization and that the illness causing the sickness may well be causing the malnutrition, it is of utmost importance for the clinician to remember that nutritional support may limit weight loss and enhance recovery from the illness.

5.2. FINDINGS ASSOCIATED WITH WEIGHT LOSS

Anorexia is the loss of appetite or lack of desire for food. It is different from satiety, which is the loss of desire for food after ingestion of food. Appetite and satiety are discussed in detail in Chapter 2. Anorexia occurs commonly in chronic illnesses, and exact causes are unknown. It occurs commonly in depression and psychiatric disorders and is thought to be functional or psychogenic in these cases. Its cause in chronic systemic illness is thought to be physiologic, although the exact mechanism is not understood. Anorectic circulating substances as well as physiologic mechanisms such as hyperosmolality have been sought to explain anorexia.[1] One has only to look at the vast amount of research funds expended by drug companies to find an anorectic drug to understand the scientific feeling that there exists some chemical to produce anorexia. Certainly amphetamines do have a pharmacologic anorectic action, but their physiologic role is not understood.

Although exact mechanisms are uncertain, recent work has suggested that intermediary metabolites (peptides and oligonucleotides) produced by cancers may be responsible for the genesis of severe anorexia.[7]

Nausea and *vomiting* are most often due to organic causes. Those associated with diseases resulting in weight loss will be discussed in subsequent chapters. However, nausea and vomiting may be due to psychologic causes. When nausea is present for long periods of time and there is no associated physical finding, it is assumed that it can be due to psychologic causes. At times patients will vomit recurrently from so-called functional or psychiatric cause. Often this type of patient has multisystem complaints characteristic of hypochondriasis.[8] These patients may actually not lose weight, which is always puzzling in relation to the symptomatology, or they may deteriorate and lose significant amounts of weight. Psychiatric management is most often necessary when the problem becomes severe.[8]

Voluntary regurgitation is a clearer form of psychogenic vomiting. It occurs in retarded humans and may result in severe weight loss. Modified feeding techniques and forced feedings that result in satiation have stopped the vomiting and increased weight in selected cases.[9,10]

Sanders and colleagues[11] have described a patient with recurrent nausea and vomiting for 1 year that resulted in a 100-lb weight loss in whom they demonstrated increased gastric motor activity and suggested that it was related to increased production of prostaglandin E_2. These initial observations are not substantiated as yet, and the search for the systemic cause of nausea and vomiting continues.

The physical signs indicating weight loss and poor food intake are numerous. *Hair* may become dull, dry, thin, or sparse. The hair will feel fine or silky and may lose its natural curl. Color changes occur that may be referred to as a so-called flag sign. When severe deficiency occurs, there is overt hair loss. The *conjunctivae* may become pale or red and inflamed. The corners of the eyelids may become fissured and red, and the cornea may be affected depending upon selective vitamin deficiency so that it is dull, soft, or actually scarred. The lips may become swollen and red so that there is a cheilosis, or they may become fissured at the corners so that there is an angular stomatitis. The *tongue* may also present a varied appearance. It may become swollen and hyperemic without loss of papillae so that there is a glossitis, or it may become smooth with atrophic papillae. In severe malnutrition *teeth* may be missing or markedly retarded. In fluorosis there are actually gray or black spots. The gums surrounding the teeth may become recessed, or may be soft and bleed easily. The general appearance of the *face* will vary with the findings in the hair, eyes, lips, tongue, teeth, and gums. At times in the markedly malnourished patient there will be the so-called sunken facies with increased pigmentation around the eyes and over the cheeks. At other times with selective vitamin deficiencies, the face may appear swollen and be erythematous or red.

The *skin* will vary in weight loss. With minor loss, there will be no obvious defects or there may be selective abnormalities such as dryness of skin (xerosis), or a rough feeling such as noted in follicular hyperkeratosis. In marked niacin deficiency there will be actually swollen or pigmented skin. Eventually when severe weight loss occurs, the skin can easily be lifted off the muscular and bony structures with obvious deterioration of the fat.

The *hands* may demonstrate brittle or ridged *nails*. In selective type deficiencies, e.g., iron, the nail may become spoon shaped. As weight loss progresses, there is a wasted appearance and the small muscles of the hand will decrease so that spaces between the tendons of the hand are markedly depressed. As weight loss progresses, all of the muscles of the body may waste and lose tone so that the face appears skull-like and the extremities thin. Ends of bones may swell so that bumps may appear on the chest wall due to swelling of the ends of the ribs. Ends of long bones may swell so that the body becomes further distorted.

Systematic findings of severe weight loss might cause internal organs to become affected so that there is enlargement of the heart, hepatomegaly with or without splenomegaly, obvious tachycardia, and then the findings of mental confusion with marked neurologic defects such as loss of reflexes and loss of position and vibratory senses.

5.3. NUTRITIONAL ASSESSMENT

With acknowledgment that malnutrition may exist in as many as 20–60% of hospitalized patients, it becomes important to fully assess the nutritional status of almost all patients entering the hospital.[2-6,12] Although the average clinician has not adopted nutritional assessment in practice, the awareness that malnutrition is so high in the pre-hospitalized patient should create a situation where nutritional assessment becomes more readily used for the ambulatory patient.[13] Blackburn and associates[13] published the first clear outlines on a full assessment of the nutritional status of the hospitalized patient.

Table 5-2 lists the more common clinical laboratory parameters that can be used in assessing patients.

Ideal height and weight is listed in Table 2-6 as set by the Metropolitan Life Insurance standards. These heights and weights relate to body frame. The clinician must take into consideration the mesomorph, endomorph, and so-called variations in bony structure when he decides upon the ideal body weight

Table 5-2. Clinical and Laboratory Nutritional Parameters

Anthropometrics
  Height
  Weight
  Frame—small, medium, large
  Ideal body weight[a]
  Weight as percentage of ideal weight
  Triceps skin fold (TSF)
  Arm circumference (AC)
  Arm muscle circumference in cm = AC − (0.314 × TSF × 10)
  Triceps skin fold as percentage of standard[b]
  Arm circumference as percentage of standard[b]
  Arm muscle circumference as percentage of standard[b]

Laboratory
  Serum albumin
  Total iron-binding capacity (TIBC)
  Serum transferrin in mg/100 ml = (0.8 × TIBC) − 43
  Lymphocytes (L)
  White blood cell (WBC) count
  Total lymphocyte count = (L × WBC)/100
  From 24-hr urine collection:
    Urine urea nitrogen (UUN)
    Urinary creatinine (UC)
    Creatinine height index (CHI) as percentage of standard[c]
      e.g., male, ht = 177.8 cm; UC = 1436 mg;
      expected creatinine from table = 1596 mg;
      CHI = 1436/1596 × 100 = 90%.

Diet and nutrition
  Protein intake (PI) (g)
  Caloric intake (kcal)
  Nitrogen Balance in g = (PI/6.25) − (UUN + 4)

Skin test results in mm
  Normal = greater than 5 mm in diameter in 24 or 48 hr

[a]Table 5-3.
[b]Table 5-4.
[c]Table 5-5.

Table 5-3. Ideal Average Weight for Height[a]

Height (cm)	Weight (kg)	Height (cm)	Weight (kg)	Height (cm)	Weight (kg)
			Males		
145	51.9	159	59.9	173	68.7
146	52.4	160	60.5	174	69.4
147	52.9	161	61.1	175	70.1
148	53.5	162	61.7	176	70.8
149	54.0	163	62.3	177	71.6
150	54.5	164	62.9	178	72.4
151	55.0	165	63.5	179	73.3
152	55.6	166	64.0	180	74.2
153	56.1	167	64.6	181	75.0
154	56.6	168	65.2	182	75.8
155	57.2	169	65.9	183	76.5
156	57.9	170	66.6	184	77.3
157	58.6	171	67.3	185	78.1
158	59.3	172	68.0	186	78.9
			Females		
140	44.9	150	50.4	160	56.2
141	45.4	151	51.0	161	56.9
142	45.9	152	51.5	162	57.6
143	46.4	153	52.0	163	58.3
144	47.0	154	52.5	164	58.9
145	47.5	155	53.1	165	59.5
146	48.0	156	53.7	166	60.1
147	48.6	157	54.3	167	60.7
148	49.2	158	54.9	168	61.4
149	49.8	159	55.5	169	62.1

[a]This table corrects the 1959 Metropolitan Standards to nude weight without shoe heels. Adapted from Jelliffe DB: *The Assessment of the Nutritional Status of the Community.* WHO, Geneva, 1966.

(Table 5-3). The percentage of ideal body weight can then be calculated. In the malnourished patient, the height and weight are often enough of a clinical evaluation to begin therapy.

A full assessment of the patient may be more helpful when there is a marginal picture or when it is necessary to follow the course of nutritional therapy. When this is necessary, anthropometric measurements may be elucidative (Table 5-4). This becomes evident in evaluations of hospitalized patients who were not suspected of malnutrition from gross appearance.[12] It is also of help when managing nutritional therapy in the severely ill patient.[13,14] The triceps skin fold and arm circumference measurements are obtained and the arm muscle circumference is derived by formula (Table 5-2). These measurements can be matched against accepted standards to assess relative nutritional status. The standards have been set in accordance with optimal weights. Frisancho[15] reported measurements in 12,396 white subjects between the ages of 0 and 44 years living in the United States between 1968 and 1970. Table 5-5 lists the arm circumference and triceps skin fold for the various percentiles at the various ages, and Table 5-6 lists the derived arm muscle circumference for the 5th through 95th percentiles for those same ages. These tables can be of help in assessing the nutritional status of a puzzling patient. They have been of clear value and proved both sensitive and specific when used as indicators in identifying a malnourished population.[16] Although the arm circumference and arm muscle circumference have been little used by physicians, it is obvious from the widespread interest that they can be helpful diagnostically and certainly can prove helpful in measuring the therapeutic response. Often it may be difficult to obtain a weight, but it is certainly easy to measure the arm circumference.

It should be kept in mind that arm circumference is determined at the midpoint of the upper nondominant arm, between the acromial process of the scapula and the olecranon process of the ulna. The triceps skin fold is measured on the same arm with calipers.

Laboratory tests that may be of value in assessing a patient are listed in Table 5-2. Upon occasion the clinician may miss a deficiency but be alerted by a low serum albumin and a low lymphocyte count. Of importance in assessing the patient's course are determination of the nitrogen balance and estimation of the lean body mass. The creatinine height index is most helpful (Table 5-7). Creatinine excretion is very stable, and when calculated against the standard height, an index is obtained that is valuable in assessing the muscle mass. When the lean body mass as determined by the creatinine height index falls below 60%, it becomes urgent to vigorously treat the patient with nutritional replacement.[13]

Estimations of the nitrogen balance are simple to determine when protein intake is calculated. A 24-hr urine specimen is necessary. The urea nitrogen is obtained, and by simple formula (Table 5-2) positive or negative nitrogen bal-

Table 5-4. Standards of Adult Arm Circumference, Triceps Skin Fold, and Arm Muscle Circumference

Sex	Standard	90% Standard	80% Standard	70% Standard	60% Standard
Arm circumference (cm)					
Male	29.3	26.3	23.4	20.5	17.6
Female	28.5	25.7	22.8	20.0	17.1
Triceps skin fold (mm)					
Male	12.5	11.3	10.0	8.8	7.5
Female	16.5	14.9	13.2	11.6	9.9
Arm muscle circumference (cm)					
Male	25.3	22.8	20.2	17.7	15.2
Female	23.2	20.9	18.6	16.2	13.9

Table 5-5. Percentiles for Upper Arm Circumference and Triceps Skin Fold for Whites of the Ten-State Nutrition Survey of 1968–1970[a]

Age group	Age midpoint (yr)	Number	Arm circumference percentiles (mm)					Triceps skin fold percentiles (mm)				
			5th	15th	50th	85th	95th	5th	15th	50th	85th	95th
Males												
0.0–0.4	0.3	41	113	120	134	147	153	4	5	5	12	15
0.5–1.4	1	140	128	137	152	168	175	5	7	9	13	15
1.5–2.4	2	177	141	147	157	170	180	5	7	10	13	14
2.5–3.4	3	210	144	150	161	175	182	6	7	9	12	14
3.5–4.4	4	208	143	150	165	180	190	5	6	9	12	14
4.5–5.4	5	262	146	155	169	185	199	5	6	8	12	16
5.5–6.4	6	264	151	159	172	188	198	5	6	8	11	15
6.5–7.4	7	309	154	162	176	194	212	4	6	8	11	14
7.5–8.4	8	301	161	168	185	205	233	5	6	8	12	17
8.5–9.4	9	287	165	174	190	217	262	5	6	9	14	19
9.5–10.4	10	315	170	180	200	228	255	5	6	10	16	22
10.5–11.4	11	294	177	186	208	240	276	6	7	10	17	25
11.5–12.4	12	294	184	194	216	253	291	5	7	11	19	26
12.5–13.4	13	266	186	198	230	270	297	5	6	10	18	25
13.5–14.4	14	207	198	211	243	279	321	5	6	10	17	22
14.5–15.4	15	179	202	220	253	302	320	4	6	9	19	26
15.5–16.4	16	166	217	232	262	300	335	4	5	9	20	27
16.5–17.4	17	142	230	238	275	306	326	4	5	8	14	20
17.5–24.4	21	545	250	264	292	330	354	4	5	10	18	25
24.5–34.4	30	679	260	280	310	344	366	4	6	11	21	28
34.5–44.4	40	616	259	280	312	345	371	4	6	12	22	28
Females												
0.0–0.4	0.3	46	107	118	127	145	150	4	5	8	12	13
0.5–1.4	1	172	125	134	146	162	170	6	7	9	12	15
1.5–2.4	2	172	136	143	155	171	180	6	7	10	13	15
2.5–3.4	3	163	137	145	157	169	176	6	7	10	12	14
3.5–4.4	4	215	145	150	162	176	184	5	7	10	12	14
4.5–5.4	5	233	149	155	169	185	195	6	7	10	13	16
5.5–6.4	6	259	148	158	170	187	202	6	7	10	12	15
6.5–7.4	7	273	153	162	178	199	216	6	7	10	13	17
7.5–8.4	8	270	158	166	183	207	231	6	7	10	15	19
8.5–9.4	9	284	166	175	192	222	255	6	7	11	17	24
9.5–10.4	10	276	170	181	203	236	263	6	8	12	19	24
10.5–11.4	11	268	173	186	210	251	280	7	8	12	20	29
11.5–12.4	12	267	185	196	220	256	275	6	9	13	20	25
12.5–13.4	13	229	186	204	230	270	294	7	9	14	23	30
13.5–14.4	14	184	201	214	240	284	306	8	10	15	22	28
14.5–15.4	15	197	205	216	245	281	310	8	11	16	24	30
15.5–16.4	16	187	211	224	249	286	322	8	10	15	23	27
16.5–17.4	17	142	207	224	250	291	328	9	12	16	26	31
17.5–24.4	21	836	215	233	260	297	329	9	12	17	25	31
24.5–34.4	30	1153	230	243	275	324	361	9	12	19	29	36
35.5–44.4	40	933	232	250	286	340	374	10	14	22	32	39

[a]Taken from Frisancho.[15]

Table 5-6. Percentiles for Upper Arm Diameter and Upper Arm Circumference for Whites of the Ten-State Nutrition Survey of 1968–1970[a]

Age midpoint (yr)	Arm muscle circumference percentiles (mm)				
	5th	15th	50th	85th	95th
Males					
0.3	81	94	106	125	133
1	100	108	123	137	146
2	111	117	127	138	146
3	114	121	132	145	152
4	118	124	135	151	157
5	121	130	141	156	166
6	127	134	146	159	167
7	130	137	151	164	173
8	138	144	158	174	185
9	138	143	161	182	200
10	142	152	168	186	202
11	150	158	174	194	211
12	153	163	181	207	221
13	159	169	195	224	242
14	167	182	211	234	265
15	173	185	220	252	271
16	186	205	229	260	281
17	206	217	245	271	290
21	217	232	258	286	305
30	220	241	270	295	315
40	222	239	270	300	318
Females					
0.3	86	92	104	115	126
1	97	102	117	128	135
2	105	112	125	140	146
3	108	116	128	138	143
4	114	120	132	146	152
5	119	124	138	151	160
6	121	129	140	155	165
7	123	132	146	162	175
8	129	138	151	168	186
9	136	143	157	176	193
10	139	147	163	182	196
11	140	152	171	195	209
12	150	161	179	200	212
13	155	165	185	206	225
14	166	175	193	221	234
15	163	173	195	220	232
16	171	178	200	227	260
17	171	177	196	223	241
21	170	183	205	229	253
30	177	189	213	245	272
40	180	192	216	250	279

[a]Taken from Frisancho.[15]

Table 5-7. Ideal Urinary Creatinine Values

Height (cm)	Ideal creatinine (mg)	Height (cm)	Ideal creatinine (mg)
Men[a]		Women[b]	
157.5	1288	147.3	830
160.6	1325	149.9	851
162.6	1359	152.4	875
165.1	1386	154.9	900
167.6	1426	157.5	925
170.2	1467	160.0	949
172.7	1513	162.6	977
175.3	1555	165.1	1006
177.8	1596	167.6	1044
180.3	1642	170.2	1076
182.9	1691	172.7	1109
185.4	1739	175.3	1141
188.0	1785	177.8	1174
190.5	1831	180.3	1206
193.0	1891	182.9	1240

[a]Creatinine coefficient (men) = 23 mg/kg ideal body weight.
[b]Creatinine coefficient (women) = 18 mg/kg ideal body weight.

ance can be determined. It is best to assess this over a several day period so that errors in measurements can be evaluated and a trend established. When negative nitrogen balance exists, it is a clear indication to add nutritional support. The nitrogen balance is also an effective method of following therapy.

When malnutrition is obvious and the patient is in a marginal immunologic state, it may be helpful to perform skin tests to assess immunologic responsiveness. Skin test antigens that are commonly used are candida, mumps, streptokinase–streptodornase, DPT, and trichopytin.[17-20] The antigen is injected in the forearm and a positive response to one or more of the antigens as measured by an induration of at least 5 mm in diameter within 24 to 48 hr is considered adequate immunologic responsiveness.

Depending on the compulsiveness of the physician and the availability of certain tests, the nutritional assessment will vary. There are all too few studies available to tell us which test may be the best indicator of successful therapy or prognosis in disease. Therefore, it still remains for the clinician to determine which group of tests he will use; a few or the entire spectrum in either evaluating a patient or following patients for effective therapy. Most nutritionists do a study that will be sure to include at least a serial evaluation of weight, arm muscle circumference, lymphocyte and white blood counts, serum albumin, and intermittent evaluation of nitrogen balance. Whenever immunity is in question, skin tests can easily be performed. This seems like a minimal battery of tests when a nutrition problem exists. The physician should be encouraged to use them as one would in any organ or system evaluation.

REFERENCES

1. Hall RJD: Progress Report—Normal and abnormal food intake. *Gut* 16:744, 1975.
2. Bollet JB, Owens S: Evaluation of nutrition status in selected hospitalized patients. *Am J Clin Nutr* 26:931, 1973.
3. Butterworth CE: The skeleton in the hospital closet. *Nutr Today* 9:4, 1974.
4. Bistrian BR, Blackburn GL, Vitale J, *et al:* Prevalence of malnutrition in general medical patients. *JAMA* 235:1567, 1976.
5. Hill GL, Pickford I, Young GA, *et al:* Malnutrition in surgical patients. An unrecognized problem. *Lancet* 1:689, 1977.
6. Weinsier RL, Hunker EM, Krumdieck CL, *et al:* Hospital malnutrition. A prospective evaluation of general medical patients during the course of hospitalization. *Am J Clin Nutr* 32:418, 1979.
7. Theologlides A: Anorexia-producing intermediary metabolites. *Am J Clin Nutr* 29:552, 1976.
8. Swanson DW, Swenson WM, Huizenga KA, *et al:* Persistent nausea without organic cause. *Mayo Clin Proc* 51:257, 1976.
9. Ball TS, Hendricksen H, Clayton J: A special feeding technique for chronic regurgitation. *Am J Ment Defic* 78:486, 1974.
10. Jackson GM, Johnson CR, Ackron GS, *et al:* Food satiation as a procedure to decelerate vomiting. *Am J Ment Defic* 80:223, 1975.
11. Sanders K, Menguy R, Chey W, *et al:* One explanation for human antral tachygastrin. *Gastroenterology* 76:1234, 1979.
12. Bushman L, Russell RM, Curry G, *et al:* Malnutrition among patients in an acute care veterans facility: Anthropometric measurements. *Am J Clin Nutr* 32:948, 1979.
13. Blackburn GL, Bistrian BR, Maini BS, *et al:* Nutritional and metabolic assessment of the hospitalized patient. *J Parent Ent Nutr* 1:12, 1977.
14. Bethel RA, Jansen RD, Heymsfield SB, *et al:* Nasogastric hyperalimentation through a polyethylene catheter: An alternative to central venous hyperalimentation. *Am J Clin Nutr* 32:1112, 1979.
15. Frisancho AR: Triceps skin fold and upper arm muscle size norms for assessment of nutritional status. *Am J Clin Nutr* 27:1052, 1974.
16. Trowbridge FL, Staehling NS: Sensitivity and specificity of arm circumference indicators in identifying malnourished children. *Am J Clin Nutr* 32:937, 1979.
17. Law DK, Dudrick SJ, Abdon NI: Immunocompetence of patients with protein-calorie malnutrition. *Ann Intern Med* 79:545-550, 1973.
18. Bistrian BR, Blackburn GL, Scrimshaw NF, *et al:* Cellular immunity in semistarved states in hospitalized adults. *Am J Clin Nutr* 28:1148-1155, 1975.
19. Meakins JL, Pietsch JB, Bubenick O, *et al:* Delayed hypersensitivity: Indicator of acquired failure of host defenses in sepsis and trauma. *Ann Surg* 186:241-249, 1977.
20. Copeland EM, MacFadyen BU Jr, Dudrick SJ: Effect of intravenous hyperalimentation on established delayed hypersensitivity in the cancer patient. *Ann Surg* 184:61-64, 1976.

II

Gastrointestinal Pathophysiology, Diseases, and Nutritional Recommendations

Starvation: Anorexia Nervosa, Marasmus, and Kwashiorkor

Mild malnutrition results in weight loss. The characteristics of weight loss and its nutritional assessment are described in Chapter 5. The loss of weight in the hospitalized patient or in association with mild chronic disease is often not severe, but it is significant. The weight loss that is seen with starvation in such clinical entities as anorexia nervosa and childhood marasmus and kwashiorkor is described in more detail in this chapter.

6.1. ANOREXIA NERVOSA

Anorexia nervosa is a disorder characterized by restriction of food intake, amenorrhea, and weight loss. It is relatively rare and occurs most commonly between the ages of 13 and 20.[1] The cause is unknown; however, most clinicians and scientists agree that is probably a psychiatric disorder. Epidemiologic and genetic investigations tend to indicate no clear pattern other than an environmental family association.

Clinical presentation varies, but there is always progressive weight loss associated with amenorrhea. The incidence ranges from 10:1 to 20:1 predominantly in females.[1] The average age of onset is 17, but most often it begins just after puberty. The patient is often chubby or obese and progresses from a pattern of mild dietary restriction to severe dietary restriction. There are two rather typical presentations. The patient may be a young model child from a so-called "good" family, but slightly obese. She is often active and begins voluntary diet restriction with an excellent knowledge of caloric values of foods. The family frequently is unaware of the onset of the illness, which progresses insidiously until it becomes obvious that the child has lost more weight than is desired. At this point, the family becomes upset. All urging to eat results in a rather cold response from the child. Starvation continues, with an obvious defiance by the child, until she is almost literally dragged to the physician for help. The second characteristic presentation is the fad eater and dieter. In this situation, the child tends to be obese but will go on a regulatory diet, alternated by periods of binge eating. During one of the diet periods, weight reduction is accomplished but continues into a clear disease pattern of severe weight loss.

As weight loss progresses, menses cease. In addition, the patient might become intensely active although appearing thin and drawn. The child's behavior pattern and relationship to the family may become bizarre. As the child becomes emaciated and thin, the family becomes more and more disturbed. The relationship between the child and the mother at onset is often not a good one[2] and becomes aggravated. Abnormal behavior of the patient is often noted with peculiar habits of hiding food, eating and perhaps vomiting food after eating, and all sorts of "shenanigans."

The early onset of the disease is seen in a chubby child who has been teased by the family, in a child who has a poor relationship with one of the parents, or in a child who is a food expert or calorie counter. However, these sociologic settings are so common and the disease is so rare that it is difficult to accept any of these clinical settings as predictive of the onset of the disease.

The physical findings may be varied once the patient has had severe weight loss. There may be evidence of vitamin deficiency, such as angular stomatitis or thinning of hair. Peripheral edema may be present. Blood pressure may be extremely low and there might be mild hypothermia. The patient should be evaluated to make sure that there is no underlying intestinal disease, such as regional enteritis or gluten enteropathy, and that no endocrinopathy is present. Routine tests, including barium studies of the gut, will easily rule out these disorders.

A detailed clinical and endocrine evaluation[3] of 101 anorexia patients (95% females) at the Mayo Clinic has been reported. In 94% of the patients, the anorexia nervosa began before age 30. Evidence of gonadal dysfunction was the predominant manifestation. Amenorrhea occurred before or with the onset of weight loss in 65% of the women. The average weight loss was 28% before the illness began. In 11% the disease began before menarche. The mean age of menarche in patients with secondary amenorrhea was 13 yr. Urinary excretion of pituitary gonadotropin was undetectable in 44 of 65 patients and was below 19 rat units/24 hr in the remaining patients. Serum luteinizing hormone level was below 8 μg/dl in 15 of 27 patients studied and serum follicle-stimulating hormone was below 10 μg/dl in 7 of 27 patients studied. Mean serum and/or urinary estrogens were low in more than 50% of the patients, and elevation of serum corticosteroids and/or loss or reversal of diurnal variation were noted in 50%. Fasting serum growth hormone levels were elevated in 45%. Mean total and free serum thyroxine, thyroid-stimulating hormone, and triiodothryonine levels were low.[3]

Similar endocrine evaluations at Montefiore Hospital revealed subnormal plasma triiodothryonine concentrations and decreased 11-hydroxysteroid dehydrogenase and 5α-reductase enzyme activity.[4] This results in the hypothyroidlike abnormalities of cortisol and androgen metabolism. The Mayo and Montefiore studies suggest nutritional hypothalamic–pituitary hypofunction as an adaptive mechanism against chronic starvation.

When self-induced vomiting is part of the clinical syndrome, alkalosis may be severe. Often, with severe weight loss, hypokalemia and hypoglycemia may be noted.

The prognosis depends on the success of the treatment. The extent and length of the weight loss may vary from a situation where it is mild and almost goes undiagnosed to that of death. Mortality statistics vary from 5 to 20%.[1,2]

Once the diagnosis is made, it is best to obtain psychiatric consultation to help determine the best course of treatment. In mild disease, treatment may be accomplished as an outpatient. Once severe weight loss ensues with evidence of metabolic disturbance, an authoritative approach is best adopted and the patient admitted to the hospital.[5,6] Once admitted, the patient cannot be forced to eat, but often the threat of tube feedings results in a return to eating. If eating begins, then diets are gradually increased to demonstrate that there is adequate weight gain. The patients begin to gain weight at the rate of approximately 5 lb/week when they consume in the area of 4000 to 4500 cal/day.

If food intake does not begin in the hospital, then hyperalimentation can be started. Depending on the patient, tube feeding or TPN is employed. Because the gut is functioning, it is best to consider tube feeding first (see Chapter 18). It is a rare patient who will need long-term TPN. Remembering that the disease can be fatal, TPN can be a lifesaving measure (see Chapter 19). If the patient is absolutely resistant to tube feeding, TPN should be used and can be lifesaving by reversing the course of the disease.

Once the course is reversed and menses return, the prognosis is relatively good and relapses are relatively rare.[6] However, it often requires some family therapy. Surprisingly, psychoanalysis has not been extremely helpful.[6] The entire process, from beginning to either recovery or death, is most destructive to the family situation and often requires psychotherapy for members of the family. An approach of both behavior modification of the patient and family therapy has been tried and recommended as successful.

When the course has not been reversed by hospitalization and the patient has not begun to take in food, death occurs. It also occurs when the course of the disease has not been interrupted by hospitalization.

If recovery occurs, the patient frequently goes on to have a very fruitful and productive life, although there are few long-term studies that follow up on these patients. A study of 100 females with anorexia nervosa, followed for 4 to 8 yr after the onset, revealed that 48 had an excellent outcome, 30 an intermediate outcome as evidenced by continued nutritional, mental, or socioeconomic problems, 20 a very poor outcome, and 2 died.[7]

With the advent of enteral and parenteral nutrition, the short-term outcome should be improved, but advances in the psychiatric component of the disease must be made in order to improve the long-term outcome.

6.2. MARASMUS

Marasmus is severe malnutrition resulting from decreased total caloric intake. This decrease also includes protein malnutrition; as a result, marasmic children have protein malnutrition. Marasmus is distinguished from kwashiorkor in that the latter is primarily protein malnutrition.

Protein–calorie malnutrition can be classified into two types: The primarily caloric-deficient subjects, who are usually nonedematous, and the primarily protein-deficient subjects, who are edematous. Clinically, there is a great overlap between these two groups. The grade of protein deficiency varies so that there may be mild edema in relationship to the grade of total caloric starvation.

The classic marasmus is a worldwide disease. It usually occurs in infants, who appear to adapt to stress and infection but respond poorly to treatment, and the prognosis for a long life is only fair. This apparently occurs because of the long-term effects of poor mental and physical development.

The typical clinical picture is one of thin hair, hepatomegaly, and severe wasting of muscular tissue. Anemia is common and often severe. Edema and dermatologic disease are rare.

Laboratory studies reveal few abnormalities in serum protein, but decreased insulin and glucose levels, and a low or normal growth hormone level.[8]

The treatment of the marasmic child includes adequate total caloric replacement, including protein, vitamins, and minerals. As mentioned above, once growth retardation has begun, the institution of adequate nutritional replacement does not guarantee a complete recovery. If the deleterious adaptive changes to malnutrition have not taken place, total caloric replacement may result in complete recovery.[9]

Marasmus is a rare disease in the United States, but it is seen in children deprived of food intake and, unfortunately, it is still present in underdeveloped countries of the world.

6.3. KWASHIORKOR

Kwashiorkor is protein–calorie malnutrition with edema as its primary symptom. It can occur when adequate calories are derived from carbohydrates but protein intake is extremely low. It is most commonly seen in infants between the ages of 1 and 3 yr in Asia and Africa.

In contrast to marasmus, children suffering from kwashiorkor respond poorly to stress, but usually very well to nutritional treatment. There may be retarded growth, but once treatment begins, there is little long-term mental or metabolic damage. The cardinal sign of the disease is edema, which may be deforming. Although the infant may be malnourished, the edema gives a characteristic appearance, is associated with frequent skin lesions, thinning of the hair, hepatomegaly, and a variety of symptoms relating to vitamin deficiencies (see Chapter 3).

Laboratory results reveal impairment of a variety of functions resulting in depleted body potassium, malabsorption, liver function abnormalities, bone loss, and renal malfunction.[8-13] The serum albumin level becomes strikingly low, as are most of the enzymes. Surprisingly, the ratio of nonessential to essential amino acids, as well as growth hormone in the serum, is high. Serum transferrin may be helpful in differentiating protein–calorie deficiency,[12] and is a good indicator of the prognosis. The dietary management of this disease is one of adequate protein and caloric replacement. Administration of any one of many protein foods, flour, skim milk, etc., can adequately correct the deficiency. The replacement should be slow and cautious so that the body is not overwhelmed. Often, there is a dangerously low level of serum potassium. In these children it is advisable to replace minerals first, and slowly, so that vital circulatory and cardiac functions are maintained. It has been shown that vigorous iron therapy may cause death when there is low serum transferrin.[12] It has also been shown that adequate doses of magnesium can initially cause a remarkable clinical improvement. It is therefore wise to replace the vital minerals, such as potassium, calcium, and magnesium, before one approaches vigorous nutritional replacement.[12,14]

REFERENCES

1. Theander S: Anorexia nervosa: A psychiatric investigation of 94 female patients. *Acta Psychiatr Scand Suppl* 214, 1970.
2. Dally P: *Anorexia Nervosa.* New York: Grune & Stratton, 1969.
3. Hurd HP, Paumbo PF, Gharib H: Hypothalamic-endocrine dysfunction in anorexia nervosa. *Mayo Clin Proc* 52:711, 1977.
4. Boyar RJ, Hellman LD, Roffwarg H, *et al:* Cortisol secretion and metabolism in anorexia nervosa. *N Engl J Med* 296:190, 1977.
5. Silverman JA: Anorexia nervosa: Clinical observations in a successful treatment plan. *J Pediatr* 84:68, 1974.
6. Stunkard A: Anorexia nervosa, in Dietschy JM (ed): *Disorders of the Gastrointestinal Tract.* New York: Grune & Stratton, 1976.
7. Hsu LKG, Crisp AH, Hardina B: Outcome of anorexia nervosa. *Lancet* 1:13, 1979.
8. McLaren DS: *Nutrition and Its Disorders.* Baltimore: Williams & Wilkins Co, 1972.
9. Rao KS: Evolution of kwashiorkor and marasmus. *Lancet* 1:709, 1974.
10. Brock JF, Hansen JDL: The role of amino acids in kwashiorkor. *Am J Clin Nutr* 4:286, 1956.
11. Jelliffe DB: Protein-calorie malnutrition in tropical preschool children: A review of recent knowledge. *J Pediatr* 54:227, 1957.
12. McFarlane H, Reddy S, Adcock KJ, *et al:* Immunity, transferrin and survival in kwashiorkor. *Br Med J* 4:268, 1970.
13. Adibi SA, Allen E, Raworth R: Impaired jejunal absorption of essential amino acids induced by either dietary calorie or protein deprivation in man. *Gastroenterology* 59:404, 1970.
14. Caddell JL: Studies of protein-calorie malnutrition. II. A double blind clinical trial to assess magnesium therapy. *N Engl J Med* 276:535, 1967.

Esophagitis and Esophageal Disease

7.1. ACUTE ESOPHAGITIS

Acute esophagitis is a rare disorder due either to irritating agents or viral inflammation. The course may vary from a few days' duration in viral inflammation to one of life-long treatment after a caustic burn.

The dietary treatment is limited at first to antacids. If these are tolerated, liquid, nonirritating foods are then prescribed. Soft, bland foods are given as soon as the patient can tolerate them. (See Chapter 20, diet 2.)

During the healing phase, which may be anywhere from a week to months, the diet is gradually increased toward regular intake. The liquid foods employed during the acute phase are usually bland and include skim milk, whole milk, noncitric juices, and only slightly warmed soups. All irritating hot and cold liquids that contain any alcohol or orange, grapefruit, pineapple, or tomato juice should be avoided. As the esophagus heals from the acute insult, feeding as in chronic esophagitis should be employed.

Usually, healing of mild viral or idiopathic esophagitis is rapid and the patient will rapidly learn what can and cannot be tolerated. After the liquid phase, soft foods may be consumed, once again avoiding all irritating agents as mentioned above, and limited to mashed or pureed foods that require less mastication. Large boluses of food and those foods requiring chewing, such as meats and dry breads, should be avoided at the onset of eating solid foods.

Tube feeding may be required in severe caustic burns of the esophagus, and intravenous feeding if the burn is severe and the esophagus occluded. In these cases it is probably wise to start TPN early as the course of the disease will be prolonged. If there is any doubt, peripheral vein amino acid therapy can be started until the course of the disease is clearer.

7.2. CHRONIC ESOPHAGITIS AND REFLUX

7.2.1. Pathophysiology

Although common knowledge associates hiatus hernia with esophagitis, the pathophysiology of chronic esophagitis is clearly related to reflux from the stomach. Reflux correlates well with the competence of the lower esophageal sphincter.[1] Normally, the basal sphincteric pressure is 10 to 35 mm Hg higher than the intragastric pressure. As described in Chapter 2, after swallowing, the lower esophageal sphincter pressure falls to permit the bolus of food to pass into the stomach. Following passage of the food into the stomach and cessation of eating, the lower esophageal sphincter pressure should remain high enough to prevent gastric contents from passing back up into the esophagus. When the sphincteric pressure is low, then reflux can occur. When reflux is chronic, chronic inflammation and disease of the esophagus can occur.[1-5] It is not clear whether chronic ingestion of irritating substances such as alcohol can, in themselves, cause chronic esophagitis. However, it is clear that such agents are associated with an increased incidence of carcinoma of the esophagus.

Neurogenic, hormonal, and drug factors that affect the lower esophageal sphincter (LES) are listed in Table 7-1. The relative significance of all these factors is not clear. There is no question that there is a marked interplay of neurogenic and hormonal influences in control of the sphincter. However, which are physiologically paramount at any given moment is not certain. In all probability, further research will reveal that all of these factors are important

Table 7-1. Substances That Alter Lower Esophageal Sphincter Pressure[a]

Action	Autonomic drugs	Hormones	Others
Increase in LES pressure	Cholinergic (muscarinic) agents (bethanechol,[b] methacholine[b]) Anticholinesterase (edrophonium[b]) α-Adrenergic agents (norepinephrine, phenylephrine) β-Adrenergic blocking agents	Gastrin,[b] pentagastrin[b] Caerulein[c] Bombesin Motilin	Metoclopramide[b] Betazole[b] Histamine (H$_1$ receptor) Serotonin Prostaglandin F$_2\alpha$[b]
Decrease in LES pressure	Ganglionic stimulants (nicotine[b]) Atropine[b] (large doses) β-Adrenergic agents[b] (epinephrine, isoproterenol) β_2-Adrenergic agents (salbutamol) α-Adrenergic blocking agents Dopamine	Secretin[b] Cholecystokinin[b] Glucagon[b] Progesterone and estrogen[b] Vasoactive intestinal peptide[c]	Smooth muscle relaxants (dipyridamole; nitrites: nitroglycerin[b] Theophylline Ethanol[b] Histamine (H$_2$ receptor) Prostaglandins E$_1$, E$_2$[b], A$_2$

[a]Modified from Goyal.[2]
[b]Effects demonstrated in human subjects
[c]Unpublished data.

Table 7-2. Foods That Alter Lower
Esophageal Sphincter Pressure

Increase LES
Proteins: skim milk
Decrease LES
Fats: whole milk
Chocolate
Orange juice
Tomato juice

and that when an excess of any one factor occurs, dysfunction of the LES may then occur, resulting in reflux or spasm.

Knowledge of the hormonal factors helps explain some of the effects of various foods. Table 7-2 lists the effects of foods on the LES.[6-10] By testing the ingestion of various foods on human subjects while simultaneously measuring the pressure in the lower esophageal sphincter, workers have established that liquids containing fats and chocolate reduce the pressure in the lower esophageal sphincter significantly. Decreased pressures occur from whole milk versus skim milk and from chocolate. This work has led to the recommendation that persons suffering from reflux limit the intake of fats. Protein foods most significantly stimulate the production of acid and therefore gastrin.[9-11] Although the physiologic role of gastrin is still in question, most agree that gastrin does have some effect on the lower esophageal sphincter; consequently, increased protein intake would, if anything, have a beneficial effect on maintaining LES pressure.[11,12] Antacids also have been observed to increase the LES, with the mechanism thought to be hormonal.[2]

The mechanism by which protein increases lower esophageal pressure is not clearly understood at the present time, but it is postulated that it relates to an increased production of gastrin.

The mechanism by which fats reduce the lower esophageal sphincter pressure is also uncertain at this time. It is postulated that fats increase the production of cholecystokinin, and perhaps that hormone has a negative effect on the LES. Why chocolate reduces the sphincteric pressure is also unclear but could be explained on the basis of its fatty acid content.

Tomato juice causes a direct symptomatic effect in the esophagus, even when it does not cause a drop in the LES.[13]

The experiments of Babka and Castell[6] are a clear demonstration of how protein may be beneficial and fat harmful. Both skim milk and whole milk may initially cause a slight drop in the LES basal pressure, but after 0.5 hr the skim milk, whose content is primarily protein, will cause a rise in LES whereas the whole milk, which contains significant amounts of fat (4%), will cause a drop in the LES. This raises the question of the effect of mixed meals on the LES. When there is a low basal LES, meals should contain small amounts of fat and very significant amounts of protein if one is to adopt the physiologic approach to the treatment of this disease. The addition of protein to fat helps protect against the lowering of the LES by the fat.[7] Therefore, some might assume that a mixture of protein with fat will maintain normal basal sphincter pressures. However, in dealing with a patient who starts with a low LES, the clinician should assume that increased amounts of protein and decreased amounts of fat are of benefit.

7.2.2. Clinical Picture

The patient with esophagitis most often presents with retrosternal pain. It may be characterized as burning in nature and consequently vividly described as heartburn. Others may merely feel a dull ache, while others may experience a severe oppressive retrosternal pain that is readily confused with cardiac pain. The pain may be aggravated by intake of citric juices or alcohol. At times the patient will describe a classic water brash in which there is reflux of acid contents felt in the throat. Rarely will esophagitis present with bleeding. Once severe dysphagia occurs, the complications of stricture are often present, and treatment of this will be discussed below.

The clinical impressions are usually confirmed by barium contrast studies, esophagoscopy, and physiologic evaluation by esophageal manometric studies and either pH monitoring for reflux or acid perfusion evaluation to demonstrate the pain.

Once the diagnosis is confirmed, the clinician is challenged in the attempt to make his patient asymptomatic. The patient should also be serially observed for the development of stricture or carcinoma, both of which occur with increased incidence in chronic esophagitis.

7.2.3. Treatment

7.2.3.1. Dietary

Table 7-2 lists the foods that affect the lower esophageal sphincter pressure. On the basis of that information the following dietary recommendations are in order.

1. During the acute phase the diet should be limited to liquid and soft foods, as described in Section 7.1.1. This stage may be transient. (See Chapter 20 for detailed diets.)

2. During the chronic phase, if there is no stricture, solid foods may be ingested providing there is careful mastication. The following are recommended (also see Chapter 20 for detailed diet plans):

    a. Low fat intake.
    b. High protein intake.
    c. Elimination from the diet of chocolate, orange juice, grapefruit juice, other citric juices, tomato juice.
    d. Elimination of all alcoholic beverages and alcohol-containing foods.
    e. As yet, there is no experimental evidence that sharp spices such as those in garlic and pepper are harmful to persons with esophagitis.

However, caution in using these substances is in order. All one has to do is keep one of these spices on the buccal mucosa for a few seconds to see how irritating it can be. Therefore, I caution against their use by chronic esophagitis patients.

f. Most observers caution against the use of coffee and tea. However, the work of Cohen and Booth[14] raises a question as to whether coffee and caffeine are actually helpful or harmful.[20]

7.2.3.2. Supportive[15]

Antacids have been shown to be helpful. It is recommended that they be given hourly during the acute phase in significant amounts so that the patient receives 1 oz/hr while awake.

The patient should be urged not to bend or slouch so that intraabdominal pressure is not increased. Concomitantly, if the patient is obese, optimal weight should be an objective of the therapeutic regimen. Furthermore, during sleeping hours, when the lower esophageal sphincter pressure is lowest, the patient should be in a position where the chest is above the level of the stomach, i.e., not slouching. The best way to achieve this is to eliminate the use of several pillows and elevate the head of the bed on either wooden blocks or bricks.

The patient should be cautioned against the use of nicotine, which decreases the LES pressure. This means smoking should be eliminated.

If these modalities do not succeed, the use of drugs to strengthen the LES, such as bethanechol and metoclopramide, are recommended.[10] Twenty-five milligrams of bethanechol can be tried three to four times a day. However, there have been only scattered reports of these drug regimens being successful.

Atropine and anticholinergics actually decrease the LES. Consequently, they are contraindicated when esophagitis is present.[1-5]

The treatment of mild esophagitis is most often successfully accomplished by the patient without physician referral. However, once esophagitis becomes chronic and the patient seeks physician assistance, the treatment can often be difficult. It may require repeated visits with repeated reassurance. The exact causes for the beginning and prolongation of symptoms have always been obscure. When an emotional, stressful situation can be ascertained, treatment of that situation can often be helpful. Most often, when severe chronic esophagitis is symptomatic, the treatment must be vigorous and it taxes the patience of both patient and physician. When symptoms are not under control within a reasonable amount of time, or when episodes of symptoms recur so that they cannot be managed by the patient, surgical repair of the LES must be considered. Surgery has had limited success, but in selected cases it may be very beneficial. The exact type of repair depends on the surgical experience at the institution in which the physician practices. Surgery should be a last resort in the treatment of this disease as postoperative recurrences are common.

In summary, treatment of reflux esophagitis entails cessation of all offending factors and use of a low-fat, high-protein diet with elimination of chocolate, citrus juices, tomato juice, nicotine, and anticholinergic agents. If necessary, drugs such as bethanechol and metoclopramide may be added and surgery employed as a last resort.

7.3. ESOPHAGEAL STRICTURE AND LOWER ESOPHAGEAL RINGS

When a stricture of the esophagus develops, the dietary management is similar to that used in chronic esophagitis, but dilation of the esophagus is necessary.

Lower esophageal mucosal and muscular rings are diagnosed once the symptom of dysphagia occurs, or as an incidental finding of barium studies. This similarly requires dilation; once accomplished, the patient can often eat a regular diet. Dilation is essential in relieving the symptoms and the clinician should not have the patient believe that dietary management will cure the symptom when dilation is actually necessary.

Esophageal spasm requires drug therapy and is also not readily managed by diet therapy. However, the patient who does suffer from esophageal spasm may often induce an episode by eating hurriedly or by indulging in excessive amounts of alcohol or other irritating factors. Consequently, they should be cautioned against dietary indiscretion and hurried eating.

7.4. CARCINOMA OF THE ESOPHAGUS[17]

Carcinoma of the esophagus is usually of the epidermoid type and rarely adenocarcinoma (less than 5%). The latter occurs primarily at the cardioesophageal junction. The incidence of the disease varies throughout the world. It is most common in Asia and least common among certain ethnic groups.[18] In the United States the incidence is relatively low. The disease is associated with certain predisposing factors such as chronic alcoholism, the Plummer–Vinson syndrome, previous lye strictures, and possibly with smoking. The disease may be localized to the esophagus, spread through the wall, or have distant metastases.

The clinical presentation is dysphagia associated with weight loss. Rarely, a delayed visit to the physician may cause the patient to present with a fistula through to the respiratory tree, and consequently coughing or an aspiration pneumonia. Early in the disease, examination reveals little; late into the disease, the obvious malnutrition is present. The diagnosis is made by barium contrast radiographic study followed by esophagoscopy with multiple biopsies and cytology. The combination of these tests yields an accuracy of over 98%.

Once the diagnosis is made, the appropriate therapy for adenocarcinoma involving the distal third is surgery if there are no obvious metastases. If the lesion is in the upper or middle third, then the decision to resect or to treat with radiation depends on the patient's age, any signs of spread, and the expertise that exists in the hospital staff. The prognosis for life is not good because less than 10% survive 5 yr and most are dead within 1 yr. If there is any sign of fistulization, or fistulization occurs with therapy, then therapy must be a tube bypass and possibly gastrojejunostomy for feeding. However, it should be stressed that all patients should be treated. An obstructing lesion of the esophagus leads to inability to swallow, coughing, and a horrible choking existence. Treatment does permit a more comfortable life. The nutrition of these patients has been neglected in the past. Modern modalities of hyperalimentation are yet to be thoroughly tested in their ability to make life more bearable.

The nutritional status of the patient should be assessed at each stage of the disease, that is, prior to treatment, during treatment, and terminally.

7.4.1. Diet Prior to Treatment

While the decision for medical or surgical treatment is being made, if the patient has lost weight he should immediately be placed on some form of supplementation. As described in Chapters 18 and 19, intravenous amino acid therapy or hyperalimentation will reverse the negative nitrogen balance and permit more rapid postoperative healing. If the patient can tolerate oral feedings, then soft foods or elemental diets should be instituted with appropriate documentation of intake and of weight. Vitamins and minerals should be added to the diet. Remember, the cancer patient requires more calories because of the disease. The weight is essential to determine whether there is adequate energy intake. A patient who ordinarily requires 2000 cal/day may well require double that amount in order to maintain weight during the treatment phase. If the case is seen late and is totally obstructed, then certainly hyperalimentation may be considered so that adequate energy and body requirements can be maintained until some form of therapy can be instituted. However, once therapy has been concluded, controversy exists as to whether TPN should be used to prolong life.

7.4.2. Diet during Treatment

If surgery is to be performed, intravenous supplementation is in order. However, one of the most accepted forms of therapy is radiation. Four thousand to six thousand radians will melt away most lesions. If a fistula, which is rare, does not develop due to radiation, the patient's esophagus can be cleansed of the tumor. Unfortunately, radiation has not proven to prolong life, although it makes it much more manageable. During the radiation or immediately after, the patient may experience difficulty in eating because of the resulting edema. During this phase the patient can be placed on oral elemental supplementation if liquids can be tolerated (see Chapter 18) or may do well with merely eating blended or soft foods. This period of treatment may last from a week to a month. Once it is over, the patient can usually tolerate feedings, and nutrition is not much of a problem. If a fistula develops, radiation must be stopped and, consequently, a celestine tube[19] or some form of gastric tube may be placed in the esophagus to bypass the fistula.

7.4.3. Diet after Treatment

Following radiation a stricture may develop. It is essential that contrast X-ray and endoscopy be used to thoroughly evaluate the patient in order to maintain adequate nutrition. If it is a recurrence of tumor, then dilatation may be of temporary assistance; but if it is a stricture, dilatation by either mercury-weighted bougies or metal dilators will give marked relief. Strictures are not an uncommon complication. If it is recurrence of disease, dilatation may be a temporary help, but invariably a tube will be necessary. Patients who reach a terminal stage following radiation surprisingly do not expire from obstruction. The cause of their demise is often uncertain.

Celestine tubes are employed occasionally to give the patient relief for long periods of time.[19] Success depends on the experience of the physician. Once there is total obstruction and neither surgery nor radiation is any longer possible, or once a fistula has developed, a tube for feeding can be placed in the esophagus by one of the many techniques available. This relief, naturally, is only temporary and may vary from weeks to as long as a year. However, it does permit the patient to swallow.

With the advent of intravenous hyperalimentation (TPN), the use of gastrotomous feeding has decreased. Nevertheless, at times a patient may be a good candidate for gastrotomy and the feeding of foods through a gastrotomous tube is feasible when cooperative patients are involved.

The use of long-term, home TPN in these patients has not yet been fully evaluated. However, this technique may be able to give nutrition and add comfort and a better life situation to the debilitated, malnourished patient who is not deteriorating rapidly. The techniques for oral supplementation and intravenous hyperalimentation are described in Chapters 18 and 19.

While the patient is undergoing evaluation or during therapy or the post-therapy period, some observers have stressed that nasogastric tube feedings may be of help. In these instances a fine tube is passed into the stomach through the nose. Number five French polyethylene tubes are of small size and readily available from various companies. Once the tubes are in the stomach, feeding can be accomplished with various formulas. Relatively large amounts can be given by drip or machine-regulated monitoring. These patients must be watched because of the danger of aspiration. The patients should be tube fed while in at least a 30° upright position. They should not lie down until the stomach has been given a chance to empty. In our own experience, we have had little success at prolonging a comfortable life with nasogastric tube feeding. Using this as a long-term method of therapy is still questionable. Short-term use in debilitated elderly people often results in aspiration or prolongs the misery of their existence.

In summary, carcinoma of the esophagus is a debilitating disease with poor prognosis that involves severe weight loss and often obstruction of the esophagus. Depending upon the phase of the disease and the treatment phase, the nutritional supplementation will vary. The physician must be ready to prescribe oral or intravenous supplementation. Furthermore, the diet will have to be pureed or soft during the various phases of the disease. Nasogastric tube feeding is difficult to employ and often unrewarding. However, the clinician will often have a more comfortable patient and a happier family if the various aspects of nutritional support and maintenance of energy sources are closely observed so that weight is maintained. Therefore, some form of oral or intravenous nutritional support should be prescribed during the various phases of the course of the disease. For more details on the use of supplementation, see Chapters 18, 19, and 20 (diets 2 and 3).

REFERENCES

1. Cohen S: Motor disorders of the esophagus. *N Engl J Med* 301:184, 1979.
2. Goyal R: The lower esophageal sphincter. *Viewpoints Dig Dis* 8:3, 1976.
3. Ingelfinger FJ: Esophageal motility. *Physiol Rev* 38:533-584, 1958.
4. Castell DO: Lower esophageal sphincter: Physiologic and clinical aspects. *Ann Intern Med* 83:390-401, 1975.
5. Goyal RK, Rattan S: Mechanism of the lower esophageal sphincter relaxation. *J Clin Invest* 52:337-341, 1973.
6. Babka JC, Castell DO: On the genesis of heartburn. The effects of specific foods on the lower esophageal sphincter. *Am J Dig Dis* 18:391-397, 1973.
7. Nebel OT, Castell DO: Inhibition of the lower esophageal sphincter by fat—A mechanism for fatty food intolerance. *Gut* 14:270-274, 1973.
8. Nebel OT, Castell DO: Kinetics of fat inhibition of the lower esophageal sphincter. *J Appl Physiol* 35:6-8, 1973.
9. Nebel OT, Castell DO: Lower esophageal sphincter pressure changes after food digestion. *Gastroenterology* 63:778-783, 1972.
10. Wright LE, Castell DO: The adverse effect of chocolate on lower esophageal sphincter pressure. *Dig Dis Sci* 20:703-707, 1975.
11. Csendes A, Oster M, Brandburg J, *et al:* Gastroesophageal sphincter pressure and serum gastrin: Reaction to food stimulation in normal subjects and in patients with gastric or duodenal ulcer. *Scand J Gastroenterol* 13:879, 1978.
12. Cohen S: Recent advances in management of gastroensophageal reflux. *Postgrad Med* 57:97, 1975.
13. Price SF, Smithson KW, Castell DO: Food sensitivity in reflux esophagitis. *Gastroenterol* 75:240, 1978.
14. Cohen S, Booth GH Jr: Gastric acid secretion and lower-esophageal sphincter pressure in response to coffee and caffeine. *N Engl J Med* 293:897, 1975.
15. Editorial. Medical Management of Heartburn. *Br Med J* II:64, 1974.
16. Farrell RL, Roling GT, Castell DO: Cholinergic therapy of chronic heartburn. A controlled trial. *Ann Intern Med* 80:573, 1974.
17. Spiro H: *Clinical Gastroenterology.* New York: Macmillan Publishing Co, 1977, pp 131-143, 248-291.
18. Hormozdiari H, Day NE, Aramesh B, *et al:* Dietary factors and esophageal cancer in the Caspian Littoral of Iran. *Cancer Res* 35:3493-3498, 1975.
19. Sanfellipo PM, Bernatz PE: Celestine-tube palliative for malignant esophageal obstruction. *Surg Clin North Am* 53:921, 1973.
20. Cohen S: Pathogenesis of coffee-induced gastrointestinal symptoms. *N Engl J Med* 303:122, 1980.

8

Gastric and Duodenal Diseases

These diseases are probably the cause of the major amount of epigastric distress experienced by humans. The reason the word "probably" must be stressed is that academicians still remain confused concerning the significance of abnormalities observed on both endoscopic and histologic evaluation of gastric mucosa in patients suffering from the symptoms. The clinician attempting to treat these diseases should understand some of the pathophysiology.

8.1. GASTRITIS

8.1.1. Pathophysiology

Injury to the mucosa can occur due to a direct irritating factor such as caustic and corrosive agents, drugs (e.g., aspirin), and food irritants. It is not possible in this text to discuss all of the causes of acute injury to the gastric mucosa and the interested reader is referred to inclusive discussions in the textbook by Bockus.[1] Whether foods cause injury to the mucosa is open to question. Surprisingly, there are very few studies on the subject. Tea has been associated with gastritis, but the injury to the mucosa in the chronic reaction has been attributed to the high temperatures at which the tea is ingested.[2] Animal studies show that a minimal exposure of 10 sec to temperatures above 122°F injures the mucosa.[3]

There is some controversy as to whether spices increase gastric secretion. The work of Solanke[4] showed red pepper induced significant acid secretion. The classic work of others such as Lennard-Jones and Babouris[5] revealed no significant difference in acid secretion between duodenal subjects who eat foods of their own choice and those who eat so-called bland foods. However, the experiments[5] did not test the effect of individual foods.

Anyone who has held a hot spice in his mouth can testify that it causes erythema and local irritation. Although academia has not accepted this as the cause of injury to the stomach, there have been gastroscopic demonstrations that hot pepper, chili, and mustard will cause erythema, edema, and damage of the gastric mucosa.[6,7] Surprisingly little work has been done in this field, and much more research will be needed to determine which foods, if any, cause possible chronic damage to the mucosa.

The actual mechanism of injury to the mucosa, and its breakdown to cause either acute or chronic ulceration (and possibly chronic inflammation), has been demonstrated. Bile salts and certain pharmacologic agents, such as aspirin and antiinflammatory drugs, will break the so-called "gastric mucosal barrier" (glycocalyx) and permit hydrogen ions that are present within the mucosa to enter epithelial cells and glands in a reflux manner so that the mucosa becomes injured. The classic demonstration of this[8,9] reveals that when hydrochloric acid is instilled into the stomach, it is ordinarily removed with no loss. However, when "mucosal-barrier-breaking substances" are exposed to the mucosa, there is a net loss of hydrogen ion, indicating that it is reabsorbed into the stomach after exposure to the irritating factor (Fig. 8-1).

No discussion of injury to the mucosa is complete without mention of alcohol. Because as much as 7% of the calories consumed in the United States are derived from alcohol, and because alcohol does break the gastric mucosal barrier, one can appreciate the extent of injury to the gastric mucosa in our society.

Inflammation follows injury to the mucosa and results in acute or chronic gastritis. *Acute gastritis* can be caused by the ingestion of pharmaceutical agents or chemicals, corrosive agents, thermal or food irritants, radiation, associated with infectious disease, allergic disorders, or so-called central nervous system diseases and stress. Acute gastritis can result in acute ulceration of the stomach with erosions that can even cause massive hemorrhaging of the stomach.

Chronic gastritis can be the result of repeated insults by the agents causing acute gastritis, or it can be of idiopathic origin. When inflammation involves the superficial mucosa, the disease is called superficial gastritis. As the inflammation continues, it is associated with atrophy of specific cellular elements of the mucosa, parietal and chief cells. Partial atrophy may progress over years so that it becomes almost complete and results in decreased acid, pepsin, and intrinsic factor production. Gastric atrophy can be so-called type A and hereditary, as seen in classic Anderson's pernicious anemia, or it can result from prolonged inflammation.

The clinical symptomatology in acute and chronic gastritis will dictate the therapy. The patient with mild acute gastritis or chronic gastritis may be rela-

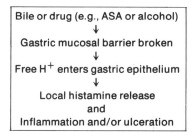

Fig. 8-1. Mechanism causing gastritis or ulcer.

tively asymptomatic, or may complain of epigastric burning pain. Food often may relieve the pain or may actually aggravate the symptoms. Certainly, the patient with significant gastritis cannot tolerate large amounts of food. The inflamed stomach rebels against distension. Physical examination reveals little information except when there is massive erosive gastritis that results in hemorrhage. In that instance, hospitalization is required, and all treatment is directed to control the upper gastrointestinal bleeding.

A diagnosis of acute gastritis is most often made clinically and substantiated by gastroscopy. Establishing the diagnosis of chronic gastritis is difficult and clinically controversial. Because gastritis is a diffuse inflammation of the stomach, one must have a histologic specimen to substantiate the diagnosis. Whenever gastric atrophy is suspected, the diagnosis should be substantiated, as treatment for vitamin B_{12} deficiency is essential to prevent neurologic complications. All too often the clinician suspects gastritis, or the academician slights clinical significance of the diagnosis, and the patient continues on in intermittent misery and is potentially vitamin B_{12} deficient if gastric atrophy evolves from chronic gastritis. A simple suction biopsy of the gastric mucosa can reliably confirm the diagnosis.[10,11]

Treatment of acute gastritis depends on the cause. Chemical and corrosive gastritis must be treated specifically in accordance with the agent causing the disease. The other acute gastridides are treated by removal of the agent causing the break in the mucosal barrier and then neutralization of the acid production by antacids and H_2 inhibitors (cimetidine).

8.1.2. Dietary Management

Dietary management should include elimination of any irritating foods, such as those containing alcohol or spices. In addition, the patient should receive an hourly feeding. There are several regimens that are acceptable. A typical feeding regimen alternates antacids and milk (see Chapter 20, diet 4). Depending on the severity of the gastritis, the patient may rapidly progress to more-solid foods. However, the foods should not include substances difficult to digest, e.g., fatty foods that are retained for long periods of time in the stomach, hot beverages that may produce thermal injury, or spicy or alcohol-containing drinks. The acute stage of the gastritis may heal within 5 to 7 days and the diet progress to traditional bland diets (see Chapter 20, diets 1–3).

Treatment of chronic gastritis is less well understood. Actually, when an etiologic agent can be demonstrated, it should be removed. However, most often there is no definite etiology and in these cases corticosteroid therapy has been successful in reversing a chronic gastritis with elements of atrophy. However, long-term corticosteroid therapy is not recommended.[12,13]

Patients should be observed for the development of atrophy once gastritis has been documented. It is important to remember that B_{12} stores may take 2 to 3 yr to become depleted; consequently, the clinician should warn his patient to remain in follow-up care.

Dietary management of chronic gastritis is not clear. Patients who complain of epigastric distress and distension following large meals should be limited to smaller feedings and, as in acute gastritis, avoid the intake of greasy foods, hot beverages, harsh spices such as pepper, mustard, chili, and alcohol (see Chapter 20, diets 1–3). They should also be cautioned against the use of pharmacologic agents that can cause gastritis. As secretory studies have shown that red pepper can increase acid secretion,[4] there is some rationale to avoid these foods. However, in the often quoted study of Lennard-Jones and Babouris,[5] bland-diet feeding appears to have little effect in relation to ad lib food feeding. The scientific approach to dietary management of chronic gastritis remains controversial. Nevertheless, the patient should be counseled and instructed to eliminate any potential sources of inflammation to the stomach as well as any foods the individual determines to be irritants.

Eosinophilic gastroenteritis[14,15] is often associated with allergies. However, allergic testing for these patients has not resulted in methods of successful therapy. The patient with eosinophilic granulomas involving the pylorus may require surgery if the granuloma forms a polyp that is large enough to obstruct the pylorus. The patient with intermittent symptoms due to eosinophilic infiltration of the stomach and small intestine may benefit from intermittent corticosteroid therapy. Most often, these patients have acute exacerbations of eosinophilic gastroenteritis and then spontaneously, or after corticosteroid therapy, will become asymptomatic. Diet therapy during the acute phase should be as in acute gastritis and any obvious allergens removed from the diet or environment. As symptoms abate, patients can be returned to a regular diet with care to avoid any food they feel is an offending agent.

8.2. GASTRIC ULCER

Although the dietary treatment of peptic ulcer does not differentiate between gastric and duodenal ulcer, there is no question that each of these diseases has its own characteristics. Consequently, peptic ulceration in each area will be discussed separately,' and the dietary management for both will be discussed in Section 8.4.

8.2.1. Incidence

The incidence of gastric ulcer is less than duodenal ulcer when the diagnosis is based on clinical findings, but equal at autopsy studies. The incidence is slightly higher in men and in those past middle age. It appears to be higher in those with type A blood. Other predisposing factors from epidemiologic evaluations appear to be smoking,[16,17] alcohol in large amounts, intake of significant amounts of aspirin,[18,19] and dietary factors.

8.2.2. Food Associations

A rough association of decreased incidence of gastric ulcer in subjects eating high-fiber diets suggests that fiber foods may be protective.[20,21] Although high fiber may be protective, peptic ulceration does occur in certain high-fiber-diet societies. This may be associated with an increased intake of highly spiced foods such as is observed in South Nigeria,[5] where red pepper is used to excess, and in areas of the world where the Japanese pickle is associated with severe gastritis.[22]

8.2.3. Pathophysiology

Advances in the understanding of gastric physiology have led to hypotheses that are now often accepted, explaining the pathophysiology and cause of gastric ulcer.[23] One suggests there is damage to the gastric mucosa that disrupts the mucosal barrier so that back-diffusion of hydrogen ions can occur. The diffusing hydrogen ions cause histamine release and damage to the mucosa (Fig. 8-1). This hypothesis has gained strength with the finding that bile reflux can occur.[24] The breaking of the mucosal barrier can be caused not only by bile salts, but by many other agents including aspirin, butazolidin, indocin, and other antiinflammatory agents.[25] Following a break in the barrier and damage to the mucosa, either an ulcer or the inflammation and small erosions of chronic gastritis can occur. Chronicity of the inflammatory response may even advance to atrophy. The decreased acid production often noted in gastric ulcer patients is explained either by the mild atrophy or by a loss of hydrogen ion for collection due to its back-diffusion if the mucosal barrier is broken.[26] Although this theory of general mucosal damage may explain the cause of benign gastric ulcer and is popular today, it does not explain the reason for deep localized ulceration without the presence of an associated gastritis.

Another hypothesis to explain gastric ulcer that is less well accepted is that of chronic motility disturbance associated with a low-grade pyloric obstruction.[27] The obstruction results in distension of the antrum, which in turn causes increased gastrin excretion, which causes increased acid production and gastric ulcer. Pyloric obstruction is difficult to demonstrate in all patients with gastric ulcer. Hypersecretion is rarely demonstrated in these subjects, whereas hyposecretion is common. Consequently, this theory is less widely accepted.

The diagnosis of gastric ulcer is established by barium contrast radiographic study or fiberoptic endoscopy. In this day of the "golden" fiberscope, almost all gastric ulcer patients are evaluated by endoscopy to ensure that the ulcer is benign. However, a barium contrast study is statistically accurate in young females so that it is reassuring for initial observation and therapeutic management. However, the radiographic criteria for benignity must be clearly present. If there is the slightest doubt, then endoscopy should be performed with cytology and biopsy to clearly rule out any potential malignancy. Certainly in countries where the incidence of gastric carcinoma is so high, there is no excuse not to use endoscopy and establish firm histologic criteria for a benign ulcer. Acid secretory studies are of little value except when there is a suspicion of an endocrine tumor. In these cases, both serum gastrin and secretory studies should be obtained to thoroughly evaluate the patient.

8.2.4. Treatment

1. Dietary management is discussed in Section 8.4.

2. There has been some initial evidence that patients placed at rest heal more quickly.[28] In addition, there is benefit from decreased smoking.[23,29]

3. At this time, cimetidine (H_2 inhibitor) appears to be the drug of choice in the acute attack as it has the fewest side effects. Long-term use is still under study.[30,32]

4. Carbenoxolone sodium is apparently an effective treatment for gastric ulcer. Its action is not completely understood, but it does have side effects due to an aldosteronelike action that can result in fluid retention and renal damage. Although it is an effective drug, it must be used cautiously.[29,31] It cannot be used in the United States, but it is used effectively in other countries.[30] It is synthetically derived from glycyrrhitinic acid, which is extracted from licorice. The side effects of sodium and fluid retention may be counteracted by Aldactone. Deglycyrrhizinated licorice does not appear to be an effective agent.[30]

5. The patient with gastric ulcer should be taken off all drugs that have an effect on the mucosa, such as aspirin, phenylbutozone, indomethacin, and alcohol. Corticosteroids and ACTH have been shown to decrease epithelial cell turnover and possibly increase acid production. However, ulcerative disease heals while patients continue to take these medications, and if they are absolutely necessary, the ulcer is not a contraindication.[33]

6. Antacids are almost always recommended. However, their effectiveness for gastric ulcer treatment is still controversial.[34,35] Although they appear to be of great help to the clinician treating the ambulatory patient, there is a serious question as to whether they are any help in the hospitalized patient under close observation. One study has shown that they appear to have no effect over placebo.[35]

Hemorrhage, pyloric obstruction, or malignant ulcer must each be treated as complicated gastric ulcer. Malignant ulcer is a separate disease and will be discussed elsewhere; hemorrhage and pyloric obstruction most often require surgical consultation. Bleeding ulcer may often spontaneously stop and not require surgery, whereas a patient who has developed obstruction due to a pyloric ulcer rarely escapes some surgical procedure.

8.3. GASTRIC CANCER

Cancer of the stomach remains a frustrating challenge to the clinician and a mortal diagnosis for the patient. The field of nutrition holds great promise for a potential preventative triumph while being an important aspect of therapy. Approximately 8% of the cancers of the stomach are superficial spreading, and the rest are more focal lesions involved with ulceration or mass formulation. For a more detailed analysis of all of the factors in gastric carcinoma, the interested reader is referred to the latest in gastroenterologic texts.[33]

8.3.1. Incidence

The incidence of gastric cancer has been falling in Europe and the United States. Where it was previously 20 per 100,000 it is now 8 per 100,000. The reason for this is not apparent. In countries such as Japan, where the disease appears to have some different characteristics and responds much more to surgical therapy, the incidence has been stable and it accounts for 60% of all cancers in men and approximately 50% of all cancers in women. In Japan, the death rate for cancer of the stomach is five times that of the United States.

8.3.2. Etiology

The cause of cancer of the stomach is unknown, but the changes in incidence have been strong indications that environmental factors are involved. There have been some genetic trends, i.e., persons with type A blood may be more susceptible to the disease. However, most recent epidemiological surveys strongly implicate environmental factors. The modality most readily implicated in the environment is diet. Even in Japan, where it was thought that the incidence of stomach carcinoma was stable, a downward trend has been suggested and attributed to the change in the dietary patterns of the Japanese.[36] Although no specific factor has appeared, there is no question that the Japanese have tended to Westernize their diet and have significantly increased milk, milk products, and egg intake. Oiso,[37] who analyzed the dietary habits and nutritional trends of the Japanese, pointed out that rice and salty foods have been the staple diet. White rice (that with the husk removed and the bran layer taken off) makes up 40% of the Japanese diet. It was pointed out that large amounts of rice frequently distend the stomach and has an influence on digestion. In parts of Japan the morning meal consists of hot rice gruel with soybean sauce that has 18% salt as a condiment. This hot gruel, referred to as chagayu, has a definite correlation with stomach cancer. Surprisingly, there also has been an increase in the Japanese consumption of animal fats and a decrease in their traditional foods. It may all prove to be ironic in that the change in the Japanese diet may decrease cancer of the stomach but increase the incidence of cancer elsewhere in the body so that mortality is not affected. A brief report from Mexico City shows that the incidence of stomach cancer is related to ethnic and economic factors.[38] Stomach cancer was high in the poor, who have a low-protein high-fiber diet, and low in the affluent, who have a high-protein low-fiber diet. But the incidence of colon cancer rose sharply in the latter group.[38] These reports indicate nutritional factors may yet be isolated that could selectively be removed from the diet for cancer prevention.

The fact that alkyl nitrosamides can produce gastric cancer in different species of animals offers another hypothesis to explain the relationship of dietary factors to gastric cancer. Nitrates are readily available in drinking water, foods, and the soil, particularly in areas of the world such as Japan, Latin America west of the Andes, and some parts of the Caribbean and eastern

Europe where carcinoma of the stomach is very high.[39] The theory proposes that nitrates are converted to nitrites, and that nitrites, in the presence of a corresponding amine or amide, form a nitrosamine or a nitrosamide in the stomach that is carcinogenic.[39] Weisburger and Raineri[39] postulate that subjects susceptible to cancer are readily exposed to a high volume of nitrates that are converted to nitrites, then ingested, and converted by bacterial action to nitrosamines or nitrosamides. Their work also points out that ascorbic acid tends to inhibit the formation of nitrites and that the conversion of nitrates to nitrites does not occur at 2–4°C. Therefore, the hypothesis includes the evolution of the refrigerator as an important preventive modality in some societies. The refrigeration of food thereby prevents the formation and ingestion of large amounts of nitrites, whereas certain societies that do not refrigerate food would have ready conversion at room temperature and thereby be more susceptible to cancer. This is a fascinating and interesting theory that may prove true. It certainly seems wise at this point in our understanding of gastric cancer to eliminate consumption of any significant amounts of nitrite. The significance of the small amount that is in foods as a preservative is still open to question. Most believe that the insignificant amount does not cause malignancy.[39]

The *clinical picture* is all too often typical. The patient will present with abdominal pain, weight loss, anorexia, and less commonly anemia. On evaluation, there may already be evidence of malnutrition, or the screening radiographic study may reveal the classic lesions of malignant gastric ulceration, a mass, or linitus plastica. It is rare for an early cancer to be detected, but this does occur. In Japan, where the incidence of cancer of the stomach is high, the diagnostic acumen of the physicians is so sharp that early lesions are picked up with facility.

It is the endoscopist today who demonstrates the lesion and, via cytology or biopsy, confirms the barium contrast radiographic findings. Once the diagnosis is made, the all too often horrendous prognosis follows. Five-year survival rates vary tremendously between 4 and 21%. Smaller lesions apparently do better than larger lesions, and the infiltrating lesions do the worst. Possibility of survival is best with tumors that are resectable without evidence of metastasis. Lymphomas of the stomach do respond to radiation and chemotherapy and are not usually included in these statistics. If the lesion is resectable (small or large) and there is no metastasis or spread, then the patient can be treated nutritionally as described in Chapter 9. However, in the wide majority of patients, if no surgery is feasible, or if surgery is performed and the lesion spreads (which most often occurs within months to a year after surgical exploration), then the experience of a painful starvation follows. These patients can be made more comfortable by adequate nutrition. All too often they starve to death. In the present milieu of intravenous alimentation, it is possible to give them calories if one pursues the feeding with diligence. Patients who are in extreme pain with widespread metastases cannot be adequately helped. Patients who do not suffer severe pain and in whom the metastases do not seem to be overwhelming can benefit from assisted nutrition. Remember that all too often the only hope the patient has for comfort is the physician's approach to his treatment. When surgery, chemotherapy, and all other treatments fail, if the patient is to be treated further it is essential to keep the patient comfortable and help maintain adequate nutrition. The physician should not attempt to support a meaningless survival, but he is obligated to keep his patient comfortable, and adequate nutrition assists that goal.

Cachexia is seen in the cancer-riddled patient. Regardless of the physiology, these patients can benefit from nutritional supplementation, either oral or intravenous. Some suffer from a loss of taste. Recent demonstrations that zinc deficiency can correct the loss of taste and, consequently, mild anorexia and thereby increase food intake have been encouraging. Therefore, zinc supplements should be tried.[40] More research is needed in this particular aspect of the cancer-riddled patient. Why aversions to certain foods such as meat occur and how to correct them might well help the patient survive more comfortably.

8.3.3. Dietary Treatment

The following may prove helpful in the most difficult situations.

1. The patient who has no functional stomach and widespread metastases and is not responding to any therapy should be made comfortable. Adequate sedation and pain relief should be supplied. Some of these patients may be helped by fluid and electrolyte replacement. Few may be helped by TPN. The use of the latter depends on the philosophy of the family and the physician and the chance for a comfortable survival. When there is such, TPN should be instituted and can be maintained at home according to one of the courses outlined in Chapter 19. TPN can be expected to slow weight loss, but it is a rare patient who will gain weight in this situation.

2. The patient who still has a functional stomach and can adequately absorb and digest will benefit from a bland diet. All too often irritating foods such as spices and alcohol will further irritate a chronic lesion and increase the symptomatology. Therefore, a bland diet is recommended with mild spices. If the stomach is small in size, then frequent feeding can be of help, as opposed to large feedings. The patient should be watched closely for weight loss. Nutritional assessment by weight and N_2 excretion can be done. If weight loss ensues, then nutritional supplements in the form of liquid feedings may be tried (see Chapter 18). If these succeed in reversing weight loss, they can be maintained until weight loss again ensues and then TPN may be begun depending on the factors as described in (1).

3. Protein-losing gastropathy is often associated with gastric carcinoma or lymphoma. These patients have low serum protein levels. This is further aggravated by poor intake, which results in inadequate protein synthesis. These particular patients may be helped by albumin transfusion. Certainly, if oral protein intake is inadequate, TPN or intravenous amino acids may help synthesis. Nevertheless, if protein loss to the gut is significant, increased supplementation will be necessary to keep up with the loss.

4. Electrolyte and fluid balances must be maintained when there has been persistent vomiting or vomiting secondary due to metastasis. In a stomach that

is capable of producing hydrochloric acid, the associated vomiting will cause a hypochloremic alkalosis and replacement of adequate amounts of chloride and potassium is often necessary. However, the vomiting associated with an atrophic mucosa may merely require mixed electrolyte solutions. This type of patient is particularly best monitored by serum and urine electrolyte evaluation and replacement depending upon the findings. It is impossible to evaluate such a patient without adequate serum electrolyte determinations; however, the physician should not become compulsive. All too often, preterminal patients with this disease receive a daily bloodletting so that their condition is aggravated. Baseline electrolyte evaluation followed by fluid replacement and then frugal laboratory studies are often adequate to maintain the patient.

Once again, it is important to stress that these patients can benefit from the physician's therapeutic acumen. Antiemetics, encouragement, and nutritional supplements can create a more comfortable situation. Tumors require a great deal of energy, and additional caloric supply to the armamentarium may well help the patient to be more comfortable. Naturally, the use of nutritional supplement must be weighed against the chance of prolonged suffering. When that is the case, the physician and family and patient must accept the reality of the situation. However, when it is the desire of the patient or there are signs of potential improvement, vigorous nutritional supplementation should be considered.[41]

8.4. DUODENAL ULCER

Both gastric and duodenal ulcers are often included in discussions of peptic ulcer. However, ulceration in the duodenal bulb and postbulbar area has distinct characteristics. For the first half of this century, clinicians and the public have always associated duodenal ulcer with stress and have treated it with diet therapy. The decrease in the incidence of peptic ulcer and the addition of an effective acid inhibitor, cimetidine, may completely revolutionize our thinking about duodenal ulcer disease. Although cimetidine use has increased greatly, diet therapy and antacids have not become obsolete. There is the entire question as to the correct diet for the peptic ulcer patient between attacks. The old dictum "once an ulcer, always an ulcer" seems to be true for at least 50% of the patients. Therefore, their diet therapy, even should there be medication to cure the acute ulceration, may remain of importance.

8.4.1. Incidence

At the present time there is a decline in the incidence of peptic ulceration of the duodenal bulb. Until the 1950's, there was a gradual increase, and since that time there has been a rather rapid and persistent decline.[42,43] The cause for this is not certain, although I feel that it is partially due to a much more accurate diagnostic approach by practicing physicians, who now do not accept the radiologic report of duodenal irritability as evidence of duodenal ulcer.

Furthermore, the advent of the duodenoscope has greatly increased our ability to diagnose the disease. Nevertheless, epidemiologists do claim that the decline in incidence is real regardless of increased diagnostic acumen.[33,42] There also appears to be marked difference among certain racial and geographic groups and a marked decline in the complications of duodenal ulcer, such as perforation.[4,43] These facts all indicate a changing trend in the disease.

8.4.2. Etiology

Along with a poor understanding of the changing incidence of the disease, there is still no understanding of the cause of peptic ulceration of the duodenal bulb. There are family trends, which raise the importance of inheritance. Patients with duodenal ulcer are more than likely to have type O blood as well as be nonsecretors of blood antigens.

For ages, duodenal ulceration has been associated with increased stress. Yet how can we explain the declining incidence in these times when stress and the so-called hectic pace of society have increased? Although stress ulceration can be reduced in animal studies and emotional tension can increase physiologic factors that aid in the development of ulceration, it is clear that this is not the only element involved in the development of a peptic ulcer. The so-called mucosal barrier has been demonstrated to be important in the prevention of gastric ulceration. However, this protective effect of mucus is not a clear factor in the duodenum. Along this line, Hollander and Harlan[34] suggested that patients suffering from duodenal ulceration had inefficient buffering of acid by mucus, but this hypothesis has never been substantiated.

There is one clear physiologic requirement for duodenal ulcer—the presence of gastric acid secretion. As a group, patients with duodenal ulcer appear to have both higher basal and maximal acid secretions, whereas patients with gastric ulcer tend to have low acid secretion as compared to normal subjects. Therefore, it is assumed that the presence of increased amounts of acid in the duodenum is the major part of the pathophysiology of ulcer development. Although gastrin is the major stimulant for the production of acid, surprisingly it is not increased in patients with simple duodenal ulcer. It is increased only in subjects with recurrent ulceration due to gastrin-producing tumors as occur in Zollinger–Ellison syndrome.[44,45] If gastrin is not the cause of increased acid secretion in these subjects, then it would seem vagal–neurologic mechanisms are at cause, although this has not been clearly demonstrated. The entire role of gastrointestinal hormones in this disease is not completely understood. For example, secretin has an inhibitory effect on the gastric acid response to food, but this effect is not defective in duodenal ulcer patients.[46] Subjects with duodenal ulcer have abnormally rapid gastric emptying, especially for meals of high-energy density,[46] an interesting fact that has some therapeutic and dietary importance and may have etiologic importance. Studies that demonstrate increased and rapid emptying of the stomach correlate with a history of hunger in ulcer subjects. Along with the increased emptying of the stomach, there is an abnormally prolonged gastric secretory response to a meal that continues into a latent postprandial period. This acid response is usually quickly inhibited in healthy persons but not in subjects with duodenal ulcer.[48],[49] The failure of the mechanism of normal inhibition is not understood. In duodenal ulcer patients there is also a concomitant increase in peak postprandial secretory rates with increased emptying of the stomach so that more acid is delivered more rapidly to the duodenum. Therefore, there is a dual defect in duodenal ulcer subjects. They cannot retain food as long, and they appear to secrete acid for a longer period postprandially.

The diagnosis of duodenal ulcer is often made on clinical grounds. The adult patient presents with the classic symptoms of epigastric pain, often radiating to the right or to the back, awakening the subject several hours after falling asleep, or being relieved only by eating, and often aggravated by irritating foods. In children, the clinical presentation may be that of vomiting or gastrointestinal hemorrhage. Physical findings are often not characteristic. The diagnosis today is established by barium contrast studies but when X-rays are in doubt, it is established by fiberoptic duodenoscopy. Every patient with chronic ulceration should have a serum gastrin level to rule out hormone-producing tumors.

8.4.3. Treatment

The traditional therapy for peptic ulcer remains the use of *antacids.* Whenever the cause of a disease is not apparent, the therapy usually is controversial. Scientists attempting to evaluate the most effective means of treating peptic ulcer remain confused. The effects of cimetidine are initially not dramatically more significant than vigorous antacid therapy.[32] Antacid therapy has been a proven treatment of acute peptic ulcer.[34] Antacids are effective in neutralizing acids and consequently, when used judiciously, can have a therapeutic effect.[50] However, recent comparison to placebo indicates that antacids do not appear to relieve pain on the basis of their neutralizing effect.[51]

The H_2 inhibitor cimetidine is gaining wide acceptance as an inhibitor of acid production and consequently is of great use in treating duodenal ulcer.[30],[32] However, the experience in Europe has shown that it is not the panacea hoped for. Although of great help, the results of the wide use the drug will receive in the United States remain to be evaluated. There is no question that anyone with duodenal ulcer today should be treated with cimetidine. The question is, how effective will the drug be in long-term use?

A careful evaluation of random clinical trials[52] reveals that antacids, anticholinergics,[53-57] and giving up smoking all have significant effects on the treatment. Certainly, cimetidine will be added to these three. Both carbenoxolone[58],[59] and deglycyrrhizinated licorice[60],[61] have been proven to heal ulceration. These medications are used effectively in Great Britain, but they have not been approved for use in the United States because of their potential for sodium retention and its resultant complications.

8.4.4. Diet Therapy

Controversy still exists as to the benefits of diet therapy in acute and chronic peptic ulcer disease. Such noted clinicians as Spiro[33] and Bockus[62] support the clinical value of a bland diet therapy, while Peterson and Fordtran[63] and Palmer[5] point out the uncertainty of the benefit of diet therapy. Nevertheless, approximately one-half of all clinicians employ diet therapy in the treatment of ulcer disease.[65]

Table 8-1 is the position paper of the American Dietetic Association on the use of bland diets and the treatment of chronic duodenal ulcer disease and summarizes the present diet approach among clinicians.

Table 8-1. *American Dietetic Association Position Paper on Bland Diet in the Treatment of Chronic Duodenal Ulcer Disease*

I. [The ADA] recognizes that the rationale (chemically and mechanically nonirritating) for the bland diet is not sufficiently supported by scientific evidence.

   A. Spices, condiments, and highly seasoned foods are usually omitted on the basis that they irritate the gastric mucosa. However, experiments have indicated that no significant irritation occurs, even when most condiments are applied directly on the gastric mucosa. Exceptions are those items which do cause gastric irritation, including black pepper, chili powder, caffeine, coffee, tea, cocoa, alcohol, and drugs.

   B. Milk has been the basis of diets for duodenal ulcer for many years. One of the primary aims in dietary management of duodenal ulcer disease is to reduce acid secretion and neutralize the acid present. While milk does relieve duodenal ulcer pain, the acid-neutralizing effect is slight. Its buffering action could be overweighed by its ability to stimulate acid production. Most foods stimulate acid secretion to some extent; protein provides the greatest buffering action and is also the most powerful stimulus to acid secretion. The use of milk therapy has been greatly reduced over the past decade, owing to a better knowledge of its side effects and allergic reactions. The controversy regarding the use of milk still continues. There are those who still advocate the regular use of milk, primarily during the active stage of acute duodenal ulcer; however, strict insistence on its use during remission is unwarranted.

   C. Roughage, or coarse food, has been excluded from the diet on the basis that it aggravates the inflamed mucosal area. There is no evidence that such foods as fruit skins, lettuce, nuts, and celery, when they are well masticated and mixed with saliva, will scrape or irritate the duodenal ulcer. Grinding or pureeing of foods is necessary only when the teeth are in poor condition or missing.

   D. The effect of a bland diet on the healing of duodenal ulcer has been studied extensively. Investigations have compared various bland diets with regular or free-choice diets. The results indicate that a bland diet made no significant difference in healing the ulcer. One such study demonstrated that the acidity of the gastric contents was frequently lower when a free-choice diet was taken. Many foods have been incriminated as the cause of gastric discomfort and are subsequently eliminated from a patient's diet. Studies done on patients with and without documented gastrointestinal disease indicate that those with gastrointestinal disease cannot be distinguished by food intolerance. Symptoms of intolerance were more related to individual response than to intake of specific food or the presence of disease.

(continued)

Table 8-1 (Continued)

II. Believes that scientific investigation supports the validity of frequent, small feedings in the management of patients with duodenal ulcer disease. These have been found to offer the most comfort to the patient; additionally, acidity of the gastric contents is lower with small-volume, frequent feedings. It must also be recognized that rest, preferably in bed, rapidly reduces duodenal ulcer symptoms. This is a specially important factor in the healing of the ulcer.

III. Believes the following points should be of major consideration in developing a dietary plan for duodenal ulcer patients.

    A. Individualization of the dietary plan, since patients differ as to specific food intolerances, living patterns, life-styles, work hours, and education.

    B. Utilization of small-volume, frequent feedings.

    C. Provision of educational materials relative to dietary support.

IV. Advocates the continued pursuit of current research and recommends that valid information be utilized in updating dietary regimens.

V. Suggests that dietetic practitioners be cognizant of the possible harmful effects of a milk-rich bland diet in patients who have a tendency toward hypercalcemia and/or atherosclerosis.

    In a review of hospitals in the United States, bland diets were commonly used for peptic ulcer disease in 77% of the hospitals, milk routinely given in 55%, a very restricted bland diet in 72%, and patients discharged on a less restrictive bland diet in 50%. Of significant note was the tremendous variation of the "bland" diet. Some institutions actually included citric juices and caffeine-containing beverages, while others eliminated such bland foods as vegetables. This great variation in what is called a bland diet reflects the confusion that exists in diet therapy. Probably the greatest reason for maintaining diet therapy in the treatment of ulcer disease is the fact that eating helps relieve the acute symptoms. A clinical trial comparing antacids to placebos revealed that there was no difference between the two in *relieving acute pain* associated with the ulcer.[51] This well points to the fact that any bland substance in the stomach will help relieve ulcer pain, a fact known by clinicians for ages.

    The placebo employed in the above-named study contained cellulose. It would seem that if merely filling the stomach helps relieve symptoms, then perhaps fiber-containing foods may have some beneficial effect. This has also been suggested by epidemiologists, who noted that the incidence of duodenal ulcer tends to be lower in underdeveloped countries where subjects eat high-fiber or unrefined-carbohydrate diets. Although rice, bran, and unrefined-carbohydrate foods tend to buffer acid *in vivo,* they also increase acid production. What the role is in preventing chronic disease, or acute symptoms, is not completely understood.[66]

    The question as to the effect of diet on the *healing* of duodenal ulcer also remains an enigma. The story told repeatedly in gastroenterologic circles is that of Dr. Eddy Palmer, who treated his in-patient military duodenal ulcer subjects with chili sauce and their ulcers healed. What we do not know is whether the subjects would have healed if they were not removed from their environment. Other studies such as performed by Buchman and colleagues[67] have shown that duodenal ulcer heals as well with unrestricted as with bland diets. However, when one looks carefully at such studies, patients received antacids and were hospitalized, isolated from their environments. Almost all recent studies reveal no conclusive evidence concerning diet therapy.[67-72] Certainly, a bland diet does no harm, and, if anything, it is helpful in maintaining the patient under clinical observation and control. Most clinicians insist that their patients do better and are less symptomatic when placed on a bland diet. There is evidence to indicate that spices such as red pepper increase acid secretion[4] and that foods differ in their ability to stimulate acid production in duodenal subjects.[5] If this is so, then it is only reasonable to accept that certain foods will be less acid stimulating and consequently help limit the amount of acid delivered to the duodenum.

One of the reasons most clinicians insist on using a bland diet is that it gives good results. Very often, patients complain of an intolerance to spicy foods. Surprisingly, this complaint does not stand up statistically in peptic ulcer studies[72] because as many patients without ulceration complain of intolerance as do those with ulceration. Gastritis and duodenitis should be related to these intolerances, but there have not been studies that correlate the clinical picture with histologic findings. Bland diets eliminate foods that are supposedly irritating because of their mechanical effect or chemical or physical composition. Caffeinated beverages are usually eliminated because of their ability to stimulate acid production. However, it has been demonstrated that decaffeinated coffee can be as strong a gastric stimulant as regular coffee.[73] Facts such as this have caused confusion among nutritionists and, as a consequence, is reflected in the tremendous variation of so-called bland diets in hospital diet manuals. Perhaps the use of cimetidine will obviate the need for a bland diet, but that seems unlikely as epigastric distress associated with the history of peptic ulcer will move physicians and patients toward restriction of the diet when they are less acute and off the medication. Consequently, it behooves the clinician interested in nutrition to have the bland diet regimen available in his armamentarium of treatment for peptic ulcer.

The basic treatment eliminates gastric stimulants, foods that because of their physical or biochemical composition are locally irritating and any direct gastric secretory stimulants. The bland diet regimen should also be graded. One can then follow the very rigorous hourly feeding protocols outlined by Bockus or plan to use the more liberal approach professed by Fordtran.

The use of cimetidine[30,32] will modify the approach. If it is used, the clinician may elect to go directly to step 2 or 3, depending on the individual patient. For those interested in using the traditional diet therapy, the following outline may be of help (see Chapter 20, diets 1–3).

1. During the acute episode the aim is not to stimulate gastrin production, and therefore not to distend the stomach, and to keep acid neutralized as well

as sufficient calories available. Consequently, it is recommended that the patient receive hourly feedings of milk. Depending on the patient, this could range from skim milk to whole milk to half-milk/half-cream. These should be alternated with 30 ml of a palatable antacid. Magnesium antacids frequently cause diarrhea; therefore, during the acute phase it is recommended that the patient take an antacid containing aluminum hydroxide or calcium carbonate. For the cardiac patient, one should employ low-sodium antacids (such as Riopan). Depending on the patient's cooperation, he will be taking either an antacid or milk every hour while awake. This regimen can be quite rigorous, and it is much easier to employ in the hospital. For those who work, it can be difficult. Nevertheless, some compulsive patients do maintain the regimen while active. After relief of pain is obtained, step 2 is established.

2. The patient has six small but satisfying feedings per day, continuing to adhere to a bland diet. Antacids can be given. If the patient is feeling well after several weeks, step 3 follows.

3. Three regular meals of bland feeding with between-meal snacks are recommended for approximately 1 month until the patient is totally asymptomatic. A regular diet can then be instituted. As indicated above, subjects who have had repeated episodes of peptic ulceration will never return to a so-called regular diet and will always have to maintain bland feedings. These subjects should always avoid spicy, irritating meals.

Depending on the clinician and the patient's symptoms during steps 2 and 3, antacid may be added approximately 45 to 60 min following each feeding. It is important to remember that the major difference between steps 2 and 3 is the amount of feeding. In a bland diet there is an attempt to eliminate all spices that may act as gastric irritants or stimulants. It is also important to reiterate that scientific evidence is meager concerning the benefits of a restricted bland diet in peptic ulcer therapy. Nevertheless, clinicians interested in nutrition have successfully employed this modality of therapy for decades. It is also important to stress that they have succeeded in treating and healing peptic ulceration with a bland-diet therapy after failure has occurred with an unrestricted-diet therapy.

The virtues of bed rest, elimination of smoking, alcohol, aspirin, and antiinflammatory agents, and reassurance along with sedation are all important in the treatment. Naturally, their use depends on the individual patient and on the judgment of the physician.

If diet therapy fails, and there is failure with new agents such as cimetidine, and there is resultant intractable pain, bleeding, or obstruction from the ulcer, then surgery is necessary. Postoperative dietary management is discussed in Chapter 9.

8.4.5. Milk–Alkali Syndrome[33],[63]

Patients who drink significant amounts of milk in conjunction with absorbable antacids such as sodium bicarbonate and calcium carbonate may develop the symptoms and complications of the milk–alkali syndrome. The acute form may occur a few days after taking large amounts of absorbable antacids and consists of an acute alkalosis with temporary renal insufficiency, which is reversible when antacids are discontinued. There is a subacute form that is associated with hypercalcemia and elevated serum bicarbonate and low serum chloride levels. This form is also transient.

The more classic milk–alkali syndrome (Burnett's syndrome) may be irreversible. Following long-term ingestion of milk and antacids, the patient develops symptoms of hypercalcemia. There may be anorexia, ulcer pain may become intractable, and there may be evidence of calcinosis with band keratopathy and conjunctivitis. The laboratory will reveal high serum calcium, normal alkaline phosphatase, normal or slightly elevated serum phosphorus, and radiologic evidence of increased calcification. Often it is very difficult to differentiate this patient from the one who has hyperparathyroidism with associated duodenal ulcer.

Once the diagnosis is suspected, the patient should be taken off all absorbable antacids and placed on a low-calcium diet (see Chapter 20, diet 6). If the azotemia persists and elevated serum calcium persists after 1 month, the patient must be thoroughly evaluated for hyperparathyroidism. It is wise to remember that elevated calcium increases gastric stimulation and consequently aggravates any peptic ulceration. How these patients will be treated now that H_2 antagonists are available in the form of cimetidine is uncertain at this printing. Potentially, it should be helpful in treating hyperacidity that results from hypercalcemia. Once intractable renal disease occurs, that must be treated in itself.

The physician employing absorbable antacids should be aware of this milk–alkali syndrome and at the first sign of any symptoms a careful evaluation should be made.

REFERENCES

1. Bockus H: *Gastroenterology.* Philadelphia: WB Saunders Co, 1974, Vol I, pp 487-573.
2. Edwards FS, Edwards JH: Tea drinking and gastritis. *Lancet* 2:543, 1956.
3. Davis RE, Ivy AC: Thermal irritation in gastric disease. *Cancer* 2:138, 1949.
4. Solanke T: The effect of red pepper on gastric acid secretion. *J Surg Res* 15:385-390, 1973.
5. Lennard-Jones JE, Babouris N: Effect of different foods on acidity of the gastric contents in patients with duodenal ulcer. *Gut* 6:113, 1965.
6. Sanchez PE: Concept of the mucous barrier and its significance, changes in the gastric mucosa produced by the local actions of spices and other irritative agents. *Gastroenterology* 18:269, 1951.
7. Schneider MA, DeLuca V, Gray SJ: The effect of spice ingestion upon the stomach. *Am J Gastroenterol* 26:722, 1956.
8. Davenport HW: *Physiology of the Digestive Tract,* ed 4. Chicago: Year Book Medical Publishers Inc, 1977.
9. Ritchie WP: Acute gastric mucosal damage induced by bile salts, acid and ischemia. *Gastroenterology* 68:699, 1975.
10. Joske RA, Vinckh ES, Wood IJ: Gastric biopsy, a study of 1000 consecutive successful gastric biopsies. *Q J Med* 24:269, 1955.
11. Williams AW, Edwards FC, Lewis THC, *et al:* Investigation of non-ulcer dyspepsia by gastric biopsy. *Br Med J* 1:372, 1957.

12. Jeffries GH, Todd JE, Sleisenger MH: The effect of prednisolone on gastric mucosal histology, gastric secretion and vitamin B_{12} absorption in patients with pernicious anemia. *J Clin Invest* 45:803, 1966.

13. Siurala M, Varis K, Wiljasalo M: Studies on patients with atrophic gastritis: A 10 to 15 year follow-up. *Scand J Gastroenterol* 1:40, 1966.

14. Pitchumoni CS, Dearani A, Floch MH: Eosinophilic granuloma of the gastrointestinal tract. *JAMA* 211:1180, 1970.

15. Klein N, Hargrove RL, Sleisenger M, *et al:* Eosinophilic gastroenteritis. *Medicine* 49:299, 1970.

16. Langman, MJS: Changing patterns in the epidemiology of peptic ulcer. *Clin Gastroenterol* 2:217, 1973.

17. Doll R, Jones FA, Pygott F: Effect of smoking on the production and maintenance of gastric and duodenal ulcers. *Lancet* 1:657, 1958.

18. Cameron AJ: Aspirin and gastric ulcer. *Mayo Clin Proc* 50:565, 1975.

19. Gillies MA, Skyring AP: Gastric and duodenal ulcer. The association between aspirin ingestion, smoking and family history of ulcer. *Med J Aust* 2:280, 1969.

20. Cleave TL: *Peptic Ulcer.* Bristol: John Wright, 1962.

21. Dekkers HJN: Carbohydrate diet for duodenal ulcer, some remarks about the mechanism of pain in duodenal ulcer. *Gastroenterology* 86:496, 1965.

22. MacDonald WC, Anderson FH, Hashimoto S: Histological effect of certain pickles on the human gastric mucosa; A preliminary report. *Can Med Assoc J* 96:1521, 1967.

23. Rhodes J, Calcroft B: Aetiology of gastric ulcer with special reference to the roles of reflux and mucosal damage. *Clin Gastroenterol* 2:227, 1973.

24. DuPlessis DJ: Pathogenesis of gastric ulceration. *Lancet* 1:974, 1965.

25. Davenport HW: The gastric mucosal barrier. *Digestion* 5:162, 1972.

26. Davenport HW: Is the apparent hyposecretion of acid by patients with gastric ulcer a consequence of a broken barrier to diffusion of hydrogen ions into the gastric mucosa? *Gut* 6:513, 1965.

27. Dragstedt LR, Woodward ER: Gastric stasis: A cause of gastric ulcer. *Scand J Gastroenterol* 5:243, 1970.

28. Doll R, Pygott F: Factors influencing the rate of healing of gastric ulcers: Admission to the hospital, phenobarbitone and ascorbic acid. *Lancet* 1:171, 1952.

29. Doll R, Hill ID, Hutton C, *et al:* Clinical trial of a triterpenoid liquorice compound in gastric and duodenal ulcer. *Lancet* 2:793, 1962.

30. Editorial: Drugs for duodenal ulcer. *Lancet* 2:1012, 1977.

31. Baron JH: Effect of carbenoxolone sodium on human gastric acid secretion. *Gut* 18:721, 1977.

32. Winship D: Cimetidine in the treatment of duodenal ulcer. Review and commentary. *Gastroenterology* 74:396, 1978.

33. Spiro H: *Clinical Gastroenterology.* New York: Macmillan Publishing Co, 1977.

34. Hollander D, Harlan J: Antacids versus placebo in peptic ulcer therapy. *JAMA* 226:1181, 1973.

35. Butler ML, Gersh H: Antacid versus placebo in hospitalized gastric ulcer patients: A controlled therapeutic study. *Am J Dig Dis* 20:803, 1975.

36. Hirayama T: Epidemiology of cancer of the stomach with special reference to its recent decrease in Japan. *Cancer Res* 35:3460, 1975.

37. Oiso T: Incidence of stomach cancer and its relation to dietary habits and nutrition in Japan between 1900 and 1975. *Cancer Res* 35:3254, 1975.

38. Villabolos JJ: Editorial. Cancer of the stomach in Mexico. *Curr Concepts in Gastroent* 4:2, 1979.

39. Weisburger JH, Raineri R: Dietary factors and the etiology of gastric cancer. *Cancer Res* 35:3469, 1975.

40. DeWys W, Walters K: Abnormalities of taste sensation in cancer patients. *Cancer* 36:1888, 1975.

41. Shils M: *Nutrition in Neoplasia in Modern Nutrition in Health and Disease.* Goodhardt WM, Shils M (eds). Philadelphia: Lea & Febiger, 1973, chap 36.

42. Mendelhoff Al: What has been happening to duodenal ulcer? *Gastroenterology* 67:1020, 1974.

43. Smith MP: Decline in duodenal ulcer surgery. *JAMA* 237:987, 1977.

44. Trudeau WL, McGuigan JE: Serum gastrin levels in patients with peptic ulcer disease. *Gastroenterology* 59:6, 1970.

45. Zollinger RM, Moore FT: Zollinger–Ellison syndrome comes of age. *JAMA* 204:361, 1968.

46. Konturek SJ, Biernat J, Grzelec T: Inhibition by secretin of the gastric acid response to meals and to pentagastrin in duodenal ulcer patients. *Gut* 14:842, 1973.

47. Stubbs DF, Hunt JN: A relation between the energy of food and gastric emptying in men with duodenal ulcer. *Gut* 16:693, 1975.

48. Malagelada JR, Longstreth GF, Deering TB, *et al:* Gastric secretion and emptying after ordinary meals in duodenal ulcer. *Gastroenterology* 73:989, 1977.

49. Fordtran JS, Walsh JH: Gastric acid secretion rate and buffer content of the stomach after eating. *J Clin Invest* 52:645, 1973.

50. Fordtran JS, Morawski SG, Richardson CT: *In vivo* and *in vitro* evaluation of liquid antacids. *N Engl J Med* 288:923, 1973.

51. Sturdevant RAL, Isenberg JI, Secrist D, *et al:* Antacid and placebo produced similar pain relief in duodenal ulcer patients. *Gastroenterology* 72:1, 1977.

52. Christensen E, Juhl E, Tygstrup N: Treatment of duodenal ulcer: Randomized clinical trials of a decade (1964 to 1974). *Gastroenterology* 73:1170, 1977.

53. Sun DCH: Long-term anticholinergic therapy for prevention of recurrence in duodenal ulcer. *Am J Dig Dis* 9:706, 1964.

54. Goyal RK, Bhardwaj OP, Chuttani HK: Parasympatholytic agents in duodenal ulcer: Double-blind controlled trial with oxyphencyclimine hydrochloride. *J Indian Med Assoc* 50:365, 1968.

55. Sun DCH, Ryan ML: A controlled study on the use of propantheline and amylopectin sulfate for recurrences in duodenal ulcer. *Gastroenterology* 58:756, 1970.

56. Kaye MD, Rhodes J, Beck P, *et al:* A controlled trial of glycopyronium and l-hyoscyamine. *Acta Med Scand* 516:(suppl): 1, 1970.

57. Villabolos JJ, Kershenobich D, Mata MR: Antiacido adicionado a leche semidescremada en el tratamiento de la ulcera duodenal. *Rev Invest Clin* 22:183, 1970.

58. Montgomery RD, Lawrence IH, Manton DJ, *et al:* A controlled trial of carbenoxolone sodium capsules in the treatment of duodenal ulcer. *Gut* 9:704, 1968.

59. Cliff JM, Milton-Thompson GJ: A double-blind trial of carbenoxolone sodium capsules in the treatment of duodenal ulcer. *Gut* 11:167, 1970.

60. Feldman H, Gilat T: A trial of deglycyrrhizinated liquorice in the treatment of duodenal ulcer. *Gut* 12:449, 1971.

61. Misiewicz JJ, Russell RI, Baron JH, *et al:* Treatment of duodenal ulcer with glycyrrhizinic-acid-reduced-liquorice. A multicenter trial. *Br Med J* 3:501, 1971.

62. Bockus HL: Management of uncomplicated peptic ulcer, in Bockus HL (ed): *Gastroenterology.* Philadelphia: WB Saunders Co, 1974, pt 1, pp 674-709.

63. Peterson WL, Fordtran JA: Reduction of gastric acidity, in Sleisenger MH, Fordtran JS (eds): *Gastrointestinal Disease.* Philadelphia: WB Saunders Co, 1978, pp 891-913.

64. Palmer ED: *Functional Gastrointestinal Disease.* Baltimore: The Williams and Wilkins Co, 1967.

65. Welsh JD: Diet therapy of peptic ulcer disease. *Gastroenterology* 72:740, 1977.

66. Tovey FK: Aetiology of duodenal ulcer: An investigation into the buffering action and effect on pepsin of bran and unrefined carbohydrate foods. *Postgrad Med J* 50:683, 1974.

67. Buchman E, Kaung GT, Dolan K, et al: Unrestricted diet in the treatment of duodenal ulcer. *Gastroenterology* 56:1016, 1969.

68. Nicol BM: Peptic ulceration. Results of modern treatment. *Lancet* 2:466, 1942.

69. Lawrence JS: Dietetic and other methods in the treatment of peptic ulcer. *Lancet* 1:482, 1952.

70. Doll R, Friedlander P, Pygott F: Dietetic treatment of peptic ulcer. *Lancet* 1:5, 1956.

71. Truelove SC: Stilboestrol, phenobarbitone, and diet in chronic duodenal ulcer. *Br Med J* 2:559, 1960.

72. Koch JP, Donaldson RM Jr: A survey of food intolerances in hospitalized patients. *N Engl J Med* 271:657, 1964.

73. Cohen S, Booth GH Jr: Gastric acid secretion and lower-esophageal sphincter pressure in response to coffee and caffeine. *Engl J Med* 293:897, 1975.

Nutritional Problems Following Gastric and Small-Bowel Surgery

Weight gain is often difficult following gastric surgery. Many patients never regain their preoperative weight, and others have significant weight loss. Still others develop selected deficiencies that may result in anemia or osteoporosis. Many types of gastric surgery have been employed for peptic ulcer, neoplastic, and traumatic disorders.

Total gastrectomy entails total removal of the stomach. A loop of small intestine is anastomosed to the distal esophagus. This procedure is rare and is most often used for carcinoma of the stomach or severe traumatic damage to the stomach.

In *subtotal gastrectomy* 50 to 75% of the stomach is removed and the remaining stomach sutured directly to the duodenum. This is referred to as a Billroth I procedure. When the remaining stomach is sutured directly to the jejunum with closure of the duodenal stump, the procedure is referred to as Billroth II. Many variations of Billroth I and II procedures are employed and include anastomosing the jejunum to the anterior or posterior wall of the stomach in either an anterior or retrocolic position.

Esophagogastrectomy is employed to bypass or remove benign or malignant lesions of the proximal half of the stomach. This leaves the distal part of the stomach intact.

Vagotomy in conjunction with gastric resection, or drainage of the stomach, has become more widespread in the treatment of peptic ulcer disease. It may be *truncal,* which divides the two main vagal trunks; *proximal,* which interrupts only the branch to the acid-secreting part; or *selective,* which involves tedious section of all gastric fibers and maintenance of the celiac and hepatic innervation. A truncal procedure always needs an associated "drainage" (pyloroplasty).

The type of surgery employed for peptic ulcer disease varies with surgical experience and the present vogue. Consequently, the patient who presents with a nutritional problem after gastric surgery may have had any one of a variety of procedures. It is important to establish which surgery has been done as the symptoms and type of treatment may vary. For example, weight loss associated with dumping syndrome is treated differently than mild anemia and little weight loss. At present, more vagotomies are done than surgical resections of the stomach. Nevertheless, patients with surgical resection are plentiful in our population. Naturally, the patient who has had surgery for carcinoma will have resection as needed to remove the tumor plus all of the potential problems of weight loss from the primary disease. It is not within the scope of this text to describe all of the gastric surgical procedures, and the interested reader is referred to the latest surgical textbooks or monographs on the subject.[1]

After *subtotal gastrectomy*, the following problems may arise that can result in nutritional deficiencies which can be treated by diet management:

1. The dumping syndrome
2. Weight loss (decreased caloric intake or absorption)
3. Selective deficiencies due to absorption defects
   a. Steatorrhea
   b. Protein malnutrition
   c. Osteoporosis and osteomalacia
   d. Anemia—vitamin B_{12} deficiency
      Anemia—iron deficiency
   e. Hypoglycemia
   f. Vitamin A deficiency
4. Diarrhea
5. Appearance of latent diseases
6. Postoperative anatomic complications[1]
   a. Anastomotic ulcer
   b. Afferent loop syndrome
   c. Intussusception
7. Vagotomy

9.1. THE DUMPING SYNDROME

The dumping syndrome is the complex of intestinal and vasomotor symptoms that occur in a postgastrectomy patient after rapid passage of a hyperosmolar meal from the gastric pouch to the intestine.

The term "dumping" was first applied by Mix[2] in 1922 and merely refers to postgastrectomy rapid transit of food. Krieger *et al.*[3] and Abbott *et al.*[4] champion this definition and insist that dumping per se occurs frequently when the stoma size is greater than 2 cm. However, the term "dumping syndrome" is applied only when symptoms occur. Although dumping occurs frequently, the symptoms of the syndrome do not necessarily follow.

Machella[5] demonstrated that the syndrome could be caused by intrajejunal administration of hyperosmolar and hypertonic solutions. His work was confirmed by Roberts *et al.*[6] and Fischer *et al.*,[7] who also showed that a concomitant drop in plasma volume occurred with symptoms and that this could also be produced in normal people. Weidner *et al.*[8] and Scott *et al.*[9] further elaborated the etiology by showing that (1) the type of gastric resection is not related to the syndrome, (2) large quantities of hypertonic solution will produce the syndrome in all postgastrectomy patients, and (3) normal people develop the syndrome when these hypertonic solutions are fed directly into the jejunum, but not when given orally.[8,9] This syndrome occurs in anywhere from 1 to 75% (overall 25%) of sub-total gastrectomy patients regardless of age or sex.

The typical sequence of events is as follows. The susceptible patient eats a meal composed of hyperosmolar food (usually highly concentrated simple carbohydrates). The food rapidly enters the jejunum where it causes a sudden shift in fluid so that there is a measurable fall in plasma volume. Hyperperistalsis, bloating, and diarrhea may occur associated with varying degrees of vasomotor symptoms including weakness, dizziness, sweating, pallor, and tachycardia. At this stage the patient is usually hyperglycemic, but rebound hypoglycemia and hypokalemia have been recorded. Although most of the symptoms may be explained as vasomotor effects, disturbances of intestinal adaptation to food rapidly entering the jejunum or duodenum may be at fault. Ordinarily, food enters the duodenum slowly and permits appropriate motility and hormone activity to occur. Dumping into the proximal intestine may cause distension and inappropriate or inadequate release of intestinal hormones. It has been shown that patients with dumping syndrome do have significantly higher levels of enteroglycagon.[10]

The hypoglycemia is often referred to as the late aspect of the dumping syndrome. It is thought to occur because of the rapid rise in blood sugar followed by a rapid fall that may cause dizziness, fainting, nausea, and sweating.

Although most postgastrectomy patients may dump food into the intestine, symptoms may not develop. However, when challenged with hypertonic simple carbohydrate solutions, these same patients may develop symptoms of the dumping syndrome.[8,9]

Careful *dietary management* may prevent the dumping syndrome.[11-13] The recommended diet is usually limited simple-carbohydrate and fluid intake during meals (see Chapter 20, diet 8). This type of regimen is 80% successful. It was previously believed that patients who could not be controlled required surgical resection; however, high dietary fiber (poorly digested polysaccharides) may be of great help in preventing both the early and the late type of dumping symptoms.[13] Jenkins and colleagues have demonstrated that 10 g of pectin per day prevents recurrent postprandial hypoglycemic attacks in the most severe dumping syndrome patients. Whether other nonabsorbable polysaccharides will be helpful is not known. However, Jenkins and associates have demonstrated that 5-g sachets of pectin twice a day are an effective regimen in preventing the symptoms of the syndrome. Therefore, the present recommendation would be limited fluid and simple-carbohydrate intake during meals. If this does not prevent the symptoms, then pectin is added to the regimen. Foods high in pectin are listed in Chapter 20, diet 9. A protein dietary supplement consisting of casein and soy[14] has also been reported to be effective in combating severe exacerbations of the dumping syndrome. The results claimed for both dietary fiber and protein dietary supplement have not yet been reproduced by other investigators, but the rationale is good and they should prove effective for some patients. Should all of these regimens fail, then surgical correction or conversion is indicated.[4,15]

9.2. WEIGHT LOSS AND STEATORRHEA

The incidence of weight loss after gastric resection is reported to be anywhere from 10 to 50%.[16-21] The weight loss may be as much as 10 to 20% of body weight. It may occur rapidly after gastric surgery and then stabilize, or it may be a chronic problem. The degree of nutritional depletion is in some way related to the degree of surgical resection. Weight loss after total gastrectomy is always a problem, but nutritional problems after a partial gastrectomy vary. Some observers believe better results are obtained for ulcer surgery when the stomach is left intact.[22] Others believe there is no specific relationship to the type of operation, but rather to the extent of the resection.[20,21]

Weight loss is also related to postgastrectomy steatorrhea.[20] The loss of weight or the inability to maintain weight is related both to decreased caloric intake, which may be the result of decreased stomach size, and to malabsorption. The steatorrhea is clear evidence of malabsorption.

The reasons for steatorrhea are multiple. When the gastric stump is connected to the jejunum, it bypasses the duodenum. As described in Chapter 1, the food, or chyme, entering the duodenum is a pivotal point in digestion and absorption. It stimulates the production of gastrointestinal hormones and, consequently, maximal electrolyte and enzyme secretion. There are degrees of decreased absorption and resultant steatorrhea and nitrogen loss.[23-25] The poor mixing of duodenal contents with chyme that enters the jejunum results in defective absorptive mechanisms, including deficient micelle formation due to decreased bile and lipase, and actual decreases in the amounts of enzymes.[25]

In addition to poor mixing, there is occasional bacterial overgrowth in the jejunum due to decreased acid production by the stomach and, consequently, decreased cleansing of the chyme. The bacterial overgrowth in the upper small bowel can further add to the malabsorption.[26,27] The increased colonization of bacteria can be demonstrated both by direct counts from aspirates and by the breath-hydrogen test.[27] Once bacterial overgrowth occurs, the patient is subject to many of the problems that are seen in the so-called blind loop syndrome.

Treatment of this condition challenges the physician's clinical acumen. One maneuver which may be helpful is to start the hormonal processes and gut secretion 30 min before the major meal by having a small snack. If this does not work, it is wise to supplement the patient's food with enzymes. When bacterial overgrowth is demonstrated, treatment with antibiotics may be helpful. Most patients will stabilize and maintain a steady body weight on regular diets. However, those who do not can be aided initially by frequent small feedings, or, as a last resort, by the use of simple-protein foods such as may be found in elemental diets (see Chapter 18).

9.3. SELECTED DEFICIENCIES

Iron deficiency anemia is common following gastric resections. Hobbs[28] found that 50% of 242 subtotal gastrectomy patients developed it. When strict criteria of anemia are employed, statistics reveal that almost all females and half of the males will develop iron deficiency following subtotal gastrectomy.

The cause is usually a combination of initial blood loss (rarely continued blood loss), poor nutrition, and poor absorption of iron. In the female, these factors are superimposed on small iron stores that are unusual in a healthy male. The small bowel can absorb iron normally in postgastrectomy patients,[29,30] but poor absorption does occur due to bypass of the duodenum and rapid passage through the small bowel. It is not due to a mucosal malfunction. The iron deficiency anemia is usually mild and responds easily to oral iron replacement. Easily dissolved forms of iron should be employed because there is rapid transit and duodenal bypass. The presence of food in some subjects may decrease iron absorption and therefore iron supplements should be given between meals.[31] Vitamin C may be helpful in these patients. Consequently, some physicians add orange juice not only for its C content but also for its effectiveness in promoting iron absorption.

Megaloblastic anemia due to B_{12} or folic acid deficiency will develop after total gastrectomy[12] and may also occur after subtotal gastrectomy.

There are two mechanisms in the development of B_{12} deficiency: (1) lack of intrinsic factor (IF) and (2) competitive bacterial absorption in a blind loop of the intestine or with associated bacterial overgrowth. Intrinsic factor is produced in the neck cells of the stomach. After a total gastrectomy, all IF production ceases. However, after a subtotal gastrectomy, IF may or may not be absent depending on the integrity of the gastric stump. Radioactive B_{12} studies correlate well with the structural integrity of the stomach. With atrophy of the stump, mucosal B_{12} absorption is impaired and eventual development of megaloblastic anemia should occur.[33,34] Depletion of B_{12} stores may take as long as three years. The Schilling test is used to evaluate B_{12} absorption. If IF is absent no B_{12} is absorbed. When the test is repeated with added IF, absorption and secretion of B_{12} are normal, proving that the defect is not B_{12} absorption but absence or depletion of an intrinsic factor.

B_{12} depletion and megaloblastic anemia may occur because of bacterial overgrowth in an afferent loop, blind loop, or contaminated jejunum.[34,35] That the stagnant loop causes the anemia has been demonstrated by both surgical repair and antibiotic therapy.

Treatment of B_{12} deficiency megaloblastic anemia caused by lack of IF is easily corrected by a monthly dose of B_{12}. It should be anticipated in all gastrectomies and is certain following total gastrectomy. Whenever achlorhydria is present in any subtotal gastrectomy, the physician should be on guard for potential B_{12} deficiency. The problem of a blind loop or afferent loop megaloblastic anemia is more difficult. Although parenteral B_{12} and antibiotics will correct the anemia, overgrowth syndromes are troublesome and surgical correction may be necessary if a blind loop is demonstrable.

It is vital to discover any B_{12} deficiency as not only anemia but also irreversible neurologic complications may develop. Weir and Gatenby[36] demonstrated subacute combined degeneration of the cord after partial gastrectomy.

In one study, 8 of 13 patients with postsurgical megaloblastic anemia were definitely folic acid deficient and 3 had total folic acid depletion.[37] Conversely, in some patients with bacterial overgrowth, folic acid levels may be extremely high due to production of folic acid by some species of bacteria.[38] Treatment of deficiency requires folic acid replacement. Apparently, the small bowel is capable of absorbing folic acid when there is no latent celiac disease. Oral replacement doses should correct the deficiency.

Blood lipids appear to be lower in subjects following gastrectomy, and this leads to interesting epidemiologic theories. Japanese-American men between the ages of 45 and 69 with previous partial gastrectomies were compared to a control nongastrectomy population. They had significantly lower cholesterol and triglyceride serum levels in association with weight loss and decreased blood pressure. The depressed lipid and blood pressure values could not be entirely explained by the reduced weight. None of these differences appeared related to diet or living habits. Those operated on for gastric ulcer had, on the average, lower systolic pressures than duodenal patients and those with gastro-jejunal anastomosis had lower cholesterol levels than patients with a gastro-duodenostomy.[37] These findings need further elaboration as to their clinical significance, but they certainly correlate with observations on abnormal absorption.

The incidence of *osteomalacia* following gastrectomy is approximately 15%.[40] Fine cortical analysis studies indicate there may be a higher incidence of very subtle bone changes, which can be attributed to *vitamin D deficiency.* Bordier *et al.*[41] report that in several patients with no obvious vitamin D deficiency, early bone changes were nevertheless corrected by administration of vitamin D. It is difficult to determine whether the *osteoporosis* and osteomalacia that develop after gastrectomy are due to vitamin D deficiency, poor calcium absorption, or would have been present in any event. However, it is clear that these subjects do absorb calcium less readily[42] and also tend to have vitamin D deficiency. This may merely be nutritional depletion and needs further clarification.

Nevertheless, the patient with elevated serum alkaline phosphatase following gastrectomy is suspected of having a defect in bone metabolism and should be treated with appropriate vitamin D and calcium replacement therapy. Serum alkaline phosphatase and 5-nucleotidase are the best screening guides for a possible bone problem. When they are normal, it is highly unlikely that osteomalacia is developing in the patient. When it is abnormal, it is a distinct possibility, and therapy should be instituted unless the physician is ready to perform a bone biopsy to ensure that there is no impending lesion. As discussed elsewhere, the average diet of patients requiring gastrectomy is often deficient in calcium because of their age, and these patients in particular require more calcium and vitamin D in their dietary management. (See Chapter 20 for high-calcium foods.)

Hypoglycemia may be a problem following gastrectomy. It may be part of the so-called latent dumping syndrome. Shultz and colleagues[43] carefully studied glucose metabolism in nine postgastrectomy patients and demonstrated early hyperglycemia and late hypoglycemia after glucose ingestion. Insulin secretion was exaggerated. All of their data suggested that postgastrectomy hypoglycemia is caused by an inducible gastrointestinal factor that potentiates glucose-mediated insulin release. Since the enteroglucagon-producing cell has been located in the intestinal mucosa, Bloom and associates[10] have been able to demonstrate that levels of enteroglucagon are significantly higher in subjects who developed hypoglycemia following gastrectomy.

Treatment of the hypoglycemia is accomplished by dietary management, as mentioned above in the dumping syndrome, and should not require further surgery (see Chapter 20, diets 8 and 9).

Vitamin A may become depleted in total gastrectomy patients and is absorbed poorly after subtotal gastrectomy.[44,45] The deficiency is related to an increase in cholinesterase, poor pancreatic secretion, and steatorrhea. One should be on guard for vitamin A deficiency and oral supplements should be supplied.

9.4. DIARRHEA

Diarrhea following gastric resection is a nonspecific symptom and can be due to several mechanisms, some of which are clearly associated with subtotal gastrectomy. The dumping disorder and steatorrhea are causes of diarrhea. Furthermore, these patients are more prone to develop Salmonella infections. Complications such as gastrocolic fistula, anastomotic ulcer, blind loop syndrome with malabsorption, and latent diseases such as gluten enteropathy and lactose intolerance are all more common in the patient who has undergone gastrectomy than in the average population. Naturally, treatment of the diarrhea is dependent on the cause. If one of the above is found, then it must be treated accordingly.

Some subjects will develop diarrhea following subtotal gastrectomy without any apparent cause. These must be handled symptomatically. Nutritional variations may help at times. Studies such as those performed by Jenkins *et al.*,[13] where pectin is added to the diet, may prove to be of some benefit for these patients.

9.5. POSSIBLE LATENT INTESTINAL DISEASE

Shiner and Doniach[46] have demonstrated intestinal mucosal atrophy and inflammation in postgastrectomy patients. It was suggested that some of these patients were suffering from latent gluten enteropathy. Whether these lesions preceded or followed surgery is not clear, but the relationship of an intestinal lesion to steatorrhea and malnutrition is important. Therefore, when steatorrhea is discovered following subtotal gastrectomy, it is important that a small-bowel biopsy be performed so that *gluten enteropathy* can be ruled out. It would be most unfortunate to treat a postgastrectomy weight loss symptomatically when there is a specific cause such as gluten enteropathy. The usual workup for malabsorption in these patients will delineate the various specific causes.

Milk intolerance due to *lactase deficiency* is a common finding. It may not be apparent prior to sub-total gastrectomy. Following surgery, there is a small gastric reservoir. The lactose that is present in dairy products enters the small intestine rapidly and lactase deficiency may become more evident and cause more symptoms.[47,48] This is treated by decreased lactose intake *or* by the addition of lactase to foods prior to oral intake. (See Chapters 13 and 20 for lactose intolerance and low-lactose and lactose-free diets.)

9.6. OTHER POSTOPERATIVE COMPLICATIONS

Complications of subtotal gastrectomy such as the afferent loop syndrome, anastomotic ulcer, and jejunogastric intussusception do not readily concern the nutritionist, but they should be carefully analyzed by the clinician before the troubled postgastrectomy patient is subjected to lengthy treatments.[1] The incidence of cancer of the stump is increased. Whether it is due to reflux of bile irritation or increased nitrites is not yet clear.[65,66]

9.7. VAGOTOMY

Today, surgery for peptic ulceration more commonly employs vagotomy, selective or otherwise, in association with a simple drainage or pyloroplasty procedure due to the decreased incidence of complications. But some persist. Diarrhea following vagotomy is often a problem, and the mechanism remains unknown. Although Temple *et al.*[49] have demonstrated hypertonicity and excessive osmotic loads in the small bowel are a possible cause of diarrhea, this has not been substantiated by others. Gastric emptying is actually slowed after vagotomy. Decreased emptying time from the stomach permits more efficient digestion and results in more efficient absorption so that less steatorrhea or malabsorption is seen in the vagotomy/pyloroplasty patient than in the subtotal gastrectomy patient.[50] Subtotal gastrectomy with gastrojejunostomy results in more rapid emptying and abnormally low concentrations of trypsin and bile salts in the intestinal lumen. This should not occur with vagotomy/pyloroplasty. Therefore, the latter procedure is preferred by some.[51]

Selective vagotomy, so-called proximal gastric vagotomy, or parietal cell vagotomy, has become popular as a procedure for decreasing acid production and therefore as a treatment of peptic ulcer. Although vagotomy and vagotomy/pyloroplasty have less complications than subtotal gastrectomy, they do show evidence of causing disturbances in digestion. Proximal vagotomy has been shown to cause decreased sensitivity of the pancreas to secretin; consequently, low doses of secretin were able to produce increased bicarbonate secretion by the pancreas. Such increased sensitivities have not been demonstrated to produce any clinical syndrome as yet. However, alteration in vagus function will require much more study to determine whether there are any nutritional consequences.[52] Although myriad papers have been written on the subject of vagotomy, there is little that is of use to the nutritionist in the present state of the art.

9.8. TOTAL GASTRECTOMY

Total gastrectomy is performed because of severe injury to the stomach or the presence of malignancy. Incapacitating nutritional debilitation may result from the procedure. A small bit of retained stomach may prevent nutritional crisis. Bradley and associates[53] studied ten patients who had undergone total gastrectomy without evidence of malignancy. They documented that the subjects had mild steatorrhea and increased nitrogen loss; in a controlled hospital situation, however, they consumed large amounts of calories and maintained body weight. But in their home environments, they did not consume enough calories to maintain weight. Careful nutritional support was needed to maintain the patients adequately. In a discussion of this presentation[53] Dr. Zollinger reported that 85% of his patients having total gastrectomy for Zollinger–Ellison syndrome maintained their optimal weights. Moore and colleagues[53] reported that several pediatric patients requiring total gastrectomy grew and developed normally. Although difficult, careful guidance can help the total gastrectomy patient. The program should include caloric intake assessment and use of food supplements as needed.

9.9. SMALL-BOWEL RESECTION

Patients with small-bowel resection are relatively rare. Until 1950, only 250 cases were reported and tabulated.[53-58] The greatest percentage of these are due to infarction of the small bowel. Reports of massive resections with surprisingly long survivals are beginning to accumulate.[54-59] For example, a patient studied by Linder and colleagues[56] survived 3.5 yr with only 7 in. of small intestine. And a patient of Levin and colleagues[58] survived 6 yr with little more than a duodenum.

Because the small bowel is the site of both digestion and absorption, the problems that evolve after resection are primarily nutritional deficiencies. The degree and type of deficiency that develops vary directly with the extent of the surgery and whether the jejunum or ileum is resected. It logically follows that patients with small resections will have less nutritional problems than those with more extensive resection.[57] It is possible to remove 1 to 2 ft of bowel with little or no consequence.

Booth and colleagues[60,61] showed that specific problems arise when certain sites are resected. For example, fat is absorbed more efficiently in the upper bowel; as a result steatorrhea is greatest when the duodenum and jejunum are lost. Similarly, B_{12} is best absorbed in the ileum. If the ileum is lost, megaloblastic anemia develops readily.[61]

When a significant length of bowel is resected (usually more than 3 ft), some degree of malabsorption occurs. In order to establish this, the patient must be observed for weight and deficiencies. However, when malabsorption is only minimal, then little need be done other than to ensure no vitamin deficiencies occur.

In massive resection, the problems are multiple and call for frequent supportive nutrition. Electrolyte depletion occurs and potassium nephropathy is encountered.[58] Blood chemistries should be checked for latent depletions; when present, supplements of potassium, calcium, and magnesium are prescribed. Management of a massive resection is a challenge to the physician, and it has been shown that these patients can survive for many years if carefully managed. Treatment of a primary illness must be concomitantly provided. For example, in small-bowel infarction, if cardiac disease exists, treatment of an arrhythmia and possible anticoagulant therapy may be needed.

The small bowel can adapt to a certain extent. Clayton and Cotton[62] reported a case of a ten-month-old infant where the entire jejunum and half of the ileum were resected. However, the child had subsequent normal growth and development. It was their impression that the prognosis of massive intestinal resections is better in the child than in the adult because natural growth permits better adaptation.[62] Actual increased cell populations of the small intestine can occur as well as adoption of function by the remaining bowel.[63,64]

The patient with massive resection of the small bowel should be under regular nutrition management. During the early postoperative stages, there need be weekly and then monthly analyses of weight, dietary intake, and all nutritional factors as noted in Chapters 2 and 3. The patient's diet should be adjusted with supplements to maintain adequate nutrition. Medium-chain triglycerides may help initially, but as the bowel adapts, patients can often tolerate regular foods. Vitamin and mineral supplements are often needed and their replacement should be guided by serum biochemical evaluations. Serum K, Mg, and Ca should be evaluated until maintenance therapy is established. Similarly, body proteins such as albumin and transferrin should be stable on the accepted dietary intake for the patient.

Initial management may require TPN and also home management on TPN (see Chapter 19). Long-term TPN should only be needed in massive, almost total resections. If the jejunum is present, enteral supplements (Chapter 18) will be helpful, but rapid progression to foods is advised. Each patient must be assessed carefully and managed individually as the degree of resection and the primary disease will modify each situation. (Therapies such as the binding of bile salts, for bacterial overgrowth, and the blind-loop syndrome will be discussed in Chapter 13.)

REFERENCES

1. Blum A, Siewert JR: Postsurgical syndromes. *Clin Gastroenterol* 8:235, 1979.
2. Mix CL: "Dumping stomach" following gastrojejunostomy. *Surg Clin North Am* 2:617, 1922.
3. Krieger H, Abbott WE, Bradshaw JS, *et al:* Correlative study of postgastrectomized patients. *Arch Surg* 79:333, 1959.
4. Abbott WE, Krieger H, Levey S, *et al:* The etiology and management of the dumping syndrome following gastroenterostomy and subtotal gastrectomy. *Gastroenterology* 39:12, 1960.
5. Machella TE: Mechanism of post-gastrectomy "dumping" syndrome. *Ann Surg* 130:145, 1949.
6. Roberts KE, Randall HT, Farr HW, *et al:* Cardiovascular and blood volume alterations resulting from intrajejunal administration of hypertonic solutions to gastrectomized patients; relationship of these changes to dumping syndrome. *Ann Surg* 140:631, 1954.
7. Fisher JA, Taylor W, Cannon JA: The dumping syndrome: Correlations between its experimental production and clinical incidence. *Surg Gynecol Obstet* 100:599, 1955.
8. Weidner MG, Scott HW, Bond AG, *et al:* The dumping syndrome. I. Studies in patients after gastric surgery. *Gastroenterology* 37:188, 1959.
9. Scott HW, Weidner MG, Shull HJ, *et al:* The dumping syndrome. II. Further investigations of etiology in patients and experimental animals. *Gastroenterology* 37:194, 1959.
10. Bloom SR, Royston CMS, Thomson JPS: Enteroglucagon release in the dumping syndrome. *Lancet* 789, 1972.
11. Hayes MA: Dietary control of post-gastrectomy "dumping syndrome." *Surgery* 37:785, 1955.
12. Robinson FW, Pittman AC: Dietary management of postgastrectomy dumping syndrome. *Surg Gynecol Obstet* 104:529, 1957.
13. Jenkins DJA, Gassull MA, Leeds AR, *et al:* Effect of dietary fiber on complications of gastric surgery: Prevention of postprandial hypoglycemia by pectin. *Gastroenterology* 72:215, 1977.
14. Alexander HC: A protein dietary supplement for the severe dumping syndrome. *Surg Gynecol Obstet* 141:863, 1975.
15. Wallenstein S: The dumping syndrome. III. Conversion of Billroth II to Billroth I gastrectomy for severe dumping syndrome. *Acta Chir Scand* 118:278, 1960.
16. Kiefer ED: Life with a subtotal gastrectomy. A follow-up study ten or more years after operation. *Gastroenterology* 37:434, 1959.
17. Landau E, Sullivan ME, Dwight RW, *et al:* Partial gastrectomy for duodenal ulcer. *N Engl J Med* 264:428, 1961.
18. Scott HW, Herrington JL, Shull ELW, *et al:* Results of vagotomy and antral resection in surgical treatment of duodenal ulcer. *Gastroenterology* 39:590, 1960.
19. Johnson AH, McCorkle HJ, Harper HA: The problem of nutrition following total gastrectomy. *Gastroenterology* 28:360, 1955.
20. Postlethwait RW, Shingleton NW, Dillon ML, *et al:* Nutrition after gastric resection for peptic ulcer. *Gastroenterology* 40:491, 1961.
21. Saxon E, Zieve L: Weight loss after gastrectomy: Comparative importance of residual pouch capacity, presence of innervated pylorus, fat excretion, and post-operative symptoms. *Surgery* 48:666, 1960.
22. Austen WG, Edwards HC: A clinical appraisal of the treatment of chronic duodenal ulcer by vagotomy and gastric drainage operation. *Gut* 2:158, 1961.
23. Lundh G: Intestinal digestion and absorption after gastrectomy. *Acta Chir Scand Suppl* 231, 1958.
24. Lawrence W, Venamee P, Peterson AS, *et al:* Alterations in fat and nitrogen metabolism after total and subtotal gastrectomy. *Surg Gynecol Obstet* 110:601, 1960.
25. Lundh G: The mechanism of postgastrectomy steatorrhea. *Gastroenterology* 42:637, 1962.
26. *Annotations.* Protein malnutrition after partial gastrectomy. *Lancet* 630, 1968.
27. Metz G, Gassull MA, Drasar BS, *et al:* Breath-hydrogen test for small-intestinal bacterial colonisation. *Lancet* 668, 1976.
28. Hobbs JR: Iron deficiency after partial gastrectomy. *Gut* 2:141, 1961.
29. Choudhury MR, Williams J: Iron absorption and gastric operations. *Clin Sci* 18:527, 1959.
30. Zingg W, Green PT, Thomas EJ: Studies of Fe absorption following partial gastrectomy. *Gastroenterology* 36:806, 1959.

31. Bothwell TH, Pirzio-Biroli G, Flinch CA: Iron absorption. I. Factors influencing absorption. *J Lab Clin Med* 51:24, 1958.
32. Gambill EE, Campbell PC, Balfour DC, *et al:* Pernicious anemia following total gastrectomy for giant hypertrophy of the gastric rugae (Mentrier's disease). *Gastroenterology* 39:347, 1960.
33. Badenoch J, Evans JR, Richards WCD, *et al:* Megaloblastic anemia following partial gastrectomy and gastroenterostomy. *Br J Hematol* 1:389, 1955.
34. Hoffman WA, Spiro HM: Afferent loop syndromes. *Gastroenterology* 40:201, 1961.
35. Neilly RW, Kirsner JB: The blind loop syndrome. *Gastroenterology* 37:491, 1959.
36. Weir D, Gatenby P: Subacute combined degeneration of the cord after subtotal gastrectomy. *Br J Med* 2:1175, 1963.
37. Gouch KR, Thirkettle JL, Read RE: Folic acid deficiency after gastric resection. *Am J Med* 34:1, 1965.
38. Hoffbrand AV, Tabaqchali S, Mollin DL: High serum plate levels in intestinal blind-loop syndrome. *Lancet* 1:1339, 1966.
39. Glober GA, Rhoads GG, Liu F, *et al:* Long-term results of gastrectomy with respect to blood lipids, blood pressure, weight and living habits. *Ann Surg* 179:896, 1974.
40. Garrick R, Ireland A, Posen S: Bone abnormalities after gastric surgery. *Ann Intern Med* 75:221, 1971.
41. Bordier P, Matrajt H, Hioco D, *et al:* Subclinical vitamin-D deficiency following gastric surgery. *Lancet* 437, 1968.
42. Deller DJ: Radiocalcium absorption after partial gastrectomy. *Am J Dig Dis* 11:10, 1966.
43. Shultz KT, Neelon FA, Nilsen LB, *et al:* Mechanism of postgastrectomy hypoglycemia. *Arch Intern Med* 128:240, 1971.
44. Althausen T, Vyeyama LK, Loren MR: Effects of alcohol on absorption of vitamin A in normal and in gastrectomized subjects. *Gastroenterology* 38:942, 1960.
45. Adams JF, Johnstone TM, Hunter RD: Vitamin A deficiency following total gastrectomy. *Lancet* 1:415, 1960.
46. Shiner M, Doniach I: Histopathologic studies in steatorrhea, in *Proceedings of the World Congress on Gastroenterology.* Baltimore: Williams & Wilkins Co, 1958.
47. Gryboski J, Thayer W, Gryboski W, *et al:* A defect in disaccharide metabolism after gastrojejunostomy. *N Engl J Med* 268, 1963.
48. Pirk F, Skala I, Vulterinova M: Milk intolerance after gastrectomy. *Digestion* 9:130, 1973.
49. Temple JG, Birch A, Shields R: Postprandial osmotic and fluid changes in the upper jejunum after truncal vagotomy and draining in man. *Gut* 16:957, 1975.
50. MacGregor IK, Martin P, Meyer JH: Gastric emptying of solid food in normal man and after subtotal gastrectomy and truncal vagotomy with pyloroplasty. *Gastroenterology* 72:206, 1977.
51. MacGregor I, Parent J, Meyer JH: Gastric emptying of liquid meals and pancreatic and biliary secretion after subtotal gastrectomy or truncal vagotomy and pyloroplasty in man. *Gastroenterology* 72:195, 1977.
52. Berstad A, Roland M, Petersen H, *et al:* Altered pancreatic function after proximal gastric vagotomy in man. *Gastroenterology* 71:958, 1976.
53. Bradley EL, Isaacs J, Hersh T, *et al:* Nutritional consequences of total gastrectomy. *Ann Surg* 182:415, 1975.
54. Berman LG, Vluvitch H, Haft HH, *et al:* Metabolic studies of an unusual case of survival following resection of all but 18 inches of small intestine. *Ann Surg* 132:64, 1950.
55. Althausen TJ, Doig RK, Weiden S, *et al:* Digestion and absorption after massive resection of the small intestine. *Gastroenterology* 16:126, 1950.
56. Linder AM, Jackson WPU, Linder GG: Small gut insufficiency following intestinal surgery. III. Further clinical and autopsy studies of a man surviving $3\frac{1}{2}$ years with 7 inches of small intestine. *S Afr J Clin Sci* 4:1, 1953.
57. Kalser MH, Roth JLA, Tumen H, *et al:* Relation of small bowel resection to nutrition in man. *Gastroenterology* 38:605, 1960.
58. Levin H, Zamcheck N, Gottlieb LS: Six-year survival following massive intestinal resection with eventual potassium depletion nephropathy. *Gastroenterology* 40:818, 1961.
59. Kinney JJ, Goldwyn RM, Barr TS: Loss of the entire jejunum and ileum, and the ascending colon. *JAMA* 179:529, 1962.

60. Booth CC, Alldis D, Read AE: Studies on the site of fat absorption. 2. Fat balances after resection of varying amounts of the small intestine in man. *Gut* 2:168, 1961.

61. Booth CC, Mollin DL: The site of absorption of vitamin B$_{12}$ in man. *Lancet* 1:18, 1961.

62. Clayton BE, Cotton DA: A study of malabsorption after resection of the entire jejunum and the proximal half of the ileum. *Gut* 2:18, 1961.

63. Trier JS: The short bowel syndrome, in Sleisweger MH, Fordtran J (eds): *Gastrointestinal Disease*, ed 2. Philadelphia: WB Saunders Co, 1978, pp 1137-1143.

64. Williamson RCN: Intestinal adaptation. *N Engl J Med* 298:1393, 1444, 1978.

65. Duane WC, Wiegand DM, Gilberstadt ML: Intragastric duodenal lipids in the absence of a pyloric sphinctor: Quantitation, physical state, and injurious potential in the fasting and postprandial states. *Gastroenterology* 78:1480, 1980.

66. Schlag P, Bockler R, Ulrich H: Are nitrite and *N*-nitroso compounds in gastric juice risk factors for carcinoma in the operated stomach? *Lancet* 1:727, 1980.

Liver Diseases

The liver is a pivotal organ in human nutrition. It modifies much of the nutrient material it receives through the portal circulation, it adds significantly to the intraluminal digestive process through bile secretion, and it has an important hormonal or humoral effect via feedback mechanisms such as exist in cholesterol and bile acid metabolism. There are undoubtedly many other nutritional systems it affects that are yet not fully understood. When the liver becomes diseased, these functions may be affected. A full text would be required to describe completely the biochemical role of the liver. In this chapter, an attempt is made to describe some major biochemical phenomena that are affected by disease and treatable by diet modifications.

10.1. NORMAL CARBOHYDRATE METABOLISM[1]

The liver is probably the main organ site in the control of human carbohydrate metabolism. Monosaccharides are absorbed from the small intestine and passed directly to the liver through the portal vein. Glucose is the pivotal monosaccharide as fructose and galactose are converted in the liver to glucose. Both can be converted to glycogen or used as glucose. Glucose is stored primarily as glycogen; when the stores are filled or there is a demand for other products, glucose can be converted to fatty acids or amino acids. Conversely, when there is a demand for glucose or glycogen storage, fatty acids or amino acids can be utilized to supply adequate carbohydrates. The following is a simple outline of these biochemical steps (PEP = phosphoenolpyruvate):

liver cell
membrane

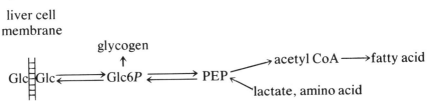

The major enzymes important in this system are glucokinase, which converts glucose (Glc) to glucose-6-phosphate (Glc6P), and glucose-6-phosphatase, which converts glucose-6-phosphate to glucose. Glucokinase catalyzes the same reaction as hexokinase and is present in the liver, but it is not inhibited by glucose-6-phosphate. The interrelations of these enzymes provide a sensitive regulatory mechanism for control of glucose release and storage in the liver.[1] Also involved in the system, as described in Chapter 1, are portal blood levels of glycogen, rates of glycogen synthesis, and probably hormonal levels of insulin and glucagon. The mechanism is extremely complex and as yet not completely understood. However, glucagon and insulin essentially exert opposite effects on glucose uptake and release by the liver. Insulin increases the maximal rates of glucokinase and glycogen synthesis and reduces that of gluconeogenesis. Glucagon increases glycogen breakdown and gluconeogenesis. As a final effect, glucose release is aided by low insulin or high glucagon levels.[2]

Through its glycogen storage mechanism and release of glucose into the blood, the liver has the primary role in maintaining adequate blood glucose levels. When a patient is starved or malnourished, glycogen stores are depleted and the liver then enters into body homeostasis by gluconeogenesis. It converts fatty acids and amino acids into glucose. Gluconeogenesis is also another mechanism for the removal of lactate from the blood.[3] There is evidence of a definite relationship between carbohydrate metabolism and fatty acid and protein metabolism. This is particularly evident during starvation, when carbohydrate stores are depleted and fat stores must be used to supply glucose for brain utilization. It is also evident during starvation that ketone bodies are used and produced in the liver by partial oxidation of fatty acids.[4]

10.2. ABNORMAL CARBOHYDRATE METABOLISM

When serious hepatocyte disease occurs, any of the above mechanisms may become disturbed. Impaired glucose tolerance occurs in patients with active and chronic viral hepatitis, fatty liver, and all forms of cirrhosis.[5-8] It has been proposed that this occurs due to a disturbance in the glucose uptake regulation in the liver and is associated with abnormal insulin and glucagon levels.

Patients with severe liver disease may also suffer from hypoglycemia due to their inability to adequately release glucose or to rapidly avail themselves of gluconeogenesis. This may be particularly evident in a liver patient who has syncope after not eating well and starves for a 24- to 36-hr period.[1] It is also a definite problem in liver disease patients undergoing surgery.

Patients with severe disease are also prone to lactic acid acidosis caused by their inability to remove lactic acid from the bloodstream through normal carbohydrate mechanisms.[3] Studies of patients with portocaval anastomosis reveal that there is definite resistance to endogenous insulin, which probably plays a major role in glucose intolerance. It was believed an increased islet cell response caused elevated blood insulin levels.[9,10] However, the finding that there is decreased insulin degradation in the cirrhotic liver helps explain elevated insulin levels.[11]

10.3. DIET THERAPY FOR CARBOHYDRATE DISTURBANCES

Glucose intolerance is rarely a severe problem in acute or chronic liver disease. In the acute state, the physician need not worry about it as a major problem. In chronic liver disease, the intolerance is most often controlled by a diabetic diet (see Chapter 20). Rarely is insulin needed in glucose intolerance due simply to liver failure. When insulin is needed, this usually is due to a concomitant diabetes mellitus or hemochromatosis associated with diabetes.

There is a particular need for glucose when a patient is starving. Furthermore, the need is urgent preceding and during surgery in the patient with liver disease so that the important limited glycogen stores of the liver are not depleted.

10.4. NORMAL FAT METABOLISM[12]

Within the hepatocyte, fat metabolism is a constant, dynamic process so that at any given time there are fatty acids, cholesterol, phospholipids, and triglycerides in flux. The fatty acids are increased in the liver due to (1) dietary fat, (2) adipose tissue fat, and (3) synthesis of fatty acids.

Fatty acids are removed from the liver by (1) esterification, (2) very-low-density lipoprotein formation, and (3) oxidation of fatty acids. Fats are absorbed through the intestine as described in Chapter 1. Long-chain fatty acids form triglycerides that are passed into the lymph and bloodstream as chylomicrons, arriving at the liver through the systemic circulation. Short-chain and medium-chain fatty acids form triglycerides that pass directly to the liver through portal blood. Fat that is mobilized from body tissues arrives at the liver through the circulation. Glucose and amino acids from the systemic circulation are also used by hepatocytes to form fats.

Triglycerides are reformed in the microsomes of the hepatocyte, pass into the smooth endoplasmic reticulum, where they are esterified, and are then coated with a β-apoprotein in the rough endoplasmic reticulum along with cholesterol and phospholipids to form very-low-density lipoproteins.

Fatty acids are oxidized in the mitochondria of the hepatocyte-liberating ketone bodies, which are not metabolized by the hepatocyte.

10.5. ABNORMAL FAT METABOLISM[12]

Plasma-free fatty acids are elevated in almost all liver diseases. Serum triglyceride levels are elevated in obstructive jaundice but are not elevated when hepatocytes are severely damaged, as in fulminant hepatitis.

The major disturbance of fat metabolism relating to the liver is fatty liver. Fat deposition occurs within hepatocytes. Fatty acids collect within the cytoplasm to distort the cell and push the nucleus to one side in alcohol-induced fatty liver, severe malnutrition such as kwashiorkor, severe obesity, following intestinal resection for obesity, and diabetes mellitus. A fine intracytoplasmic vacuolar distribution of fat without displacement of the nucleus occurs in fatty liver of pregnancy, drug toxicity (e.g., tetracycline), Reye syndrome, and in cholesterol storage disease.

Fatty liver also occurs in significant amounts in congenital disorders such as abetalipoproteinemia, Tangier disease, Wolman disease, and Niemann–Pick, Gaucher, and Hurler syndromes. Fatty liver may be seen in almost all hereditary metabolic disorders. The increase of fat in the liver occurs either due to an increase in any one of the processes that delivers fatty acids to the liver or to a decrease in any one of the processes that removes fatty acids from the liver. In each specific disease there appears to be a predominance of one of the defects. For example, in increased ethanol there is a major impairment of the oxidation of fatty acids even though there is an increased influx of fatty acids to the hepatocytes. In obesity the mechanism of increased fat in the liver still remains obscure although it seems to be related to increased storage. In diabetes there appears to be increased production of fat from carbohydrates as well as increased mobilized fat stored in the liver. The following list characterizes the mechanisms in specific diseases:

Effectively decreased very-low-density lipoprotein formation: kwashiorkor, tetracycline toxicity.
Increased mobilization of fat and carbohydrates: obesity, diabetes.
Decreased oxidation of fatty acids with some increased production: alcohol.

The increased fatty liver that occurs in severe malnutrition such as kwashiorkor is most fascinating. One would expect that the liver would become depleted. However, the mechanism that appears to cause this incongruous effect of malnutrition is a protein deficiency of the hepatocyte that does not permit very-low-density lipoprotein formation and subsequent release from the cell.

Treatment of fatty liver requires treatment of the disease responsible for its cause; for example, the feeding of a high-protein diet in kwashiorkor, or the removal of a toxin such as tetracycline. However, when the exact cause is not apparent or there is no specific treatment, symptomatic therapy is in order. In diabetes and obesity a low-sugar, high-protein diet and maintenance of optimal weight were considered essential to remove fat from the liver. However, theories of high polysaccharides in diabetic diet therapy need to be tested.[13] In severe morbid obesity the fatty liver is corrected if there is weight reduction. However, if jejunoileal bypass is the procedure to correct weight, fatty liver invariably develops, so that in reality there is no help for the fatty liver of obesity. In all conditions of fatty liver, a high-protein, low-sugar diet (see Chapter 20), with caloric intake established to maintain optimal body weight, is the diet of choice. At present, there is no definite evidence to indicate that a very-low-fat diet is of help in any of these conditions except in the acute fatty liver of alcoholism. The theories of high carbohydrates (polysaccharides) need to be tested.[13] (See Chapters 12 and 20 and diets 23 and 24.)

10.6. CHOLESTEROL AND BILE ACID METABOLISM

The details of steroid metabolism are discussed in Chapters 1 and 11. Cholesterol is absorbed from the small intestine into enterocytes where it may be secreted, synthesized, or passed on to the systemic metabolism. The major amount of endogenous cholesterol and its esters are formed in the liver. Similarly, primary and secondary bile acids are absorbed from the small intestine (small amounts from the colon) into the portal circulation and passed to the liver where they are conjugated and along with newly synthesized primary bile acids, cholic and chenic, are passed into the small intestine to continue the enterohepatic circulation. Synthesis of cholesterol in the liver is determined by levels of bile acid in the small bowel as well as by the amount of dietary cholesterol.

10.7. NORMAL PROTEIN METABOLISM

The normal requirements described in Chapter 2 hold true for normal liver function. However, in a patient with severe liver disease, protein metabolism becomes more complicated and is less well understood. The importance of adequate protein intake is emphasized by its protective action against hepatotoxins such as chloroform and its importance in the treatment of the cirrhotic patient.[14,15] However, there is not complete agreement that food protein is a key protective factor in alcoholic liver disease.[16] Amino acids and polypeptides are absorbed from the small intestine and pass into the systemic circulation where, as part of the amino acid pool, they are presented to the hepatocyte for its use.

The synthesis of serum proteins is one of the key homeostatic functions of the liver, producing albumin, fibrinogen, prothrombin, haptoglobin, and a whole series of protein complexes important in body metabolism, such as transferrin and ceruloplasmin. Its production of clotting factors is essential; in severe liver disease their deficiency becomes of paramount clinical importance.

10.7.1. Albumin

Albumin is probably the best studied of the proteins produced by the liver. A normal serum level of 3.5 to 5.0 g/100 ml is the result of a balance between synthesis, distribution in different body compartments, and degradation.[17] It is a major regulator of osmotic pressure and serves as the carrier protein for almost all important circulating chemicals (hormones, drugs, iron, and metabolites), and assists in the transport of substances for degradation.[18,19]

Albumin is diffuse throughout the body. Forty percent is distributed in the plasma and 60% in extraplasma tissues, with skin and muscle the largest storehouse and small amounts in the gut and liver. Albumin is secreted in bile, but it can be detected in all body secretions, e.g., feces, sweat, bronchial secretion, tears, exudates, and transudates.[7,20]

Albumin is distributed rapidly through the intravascular compartments and becomes equilibrated after two or three circulations. Movement through the extravascular pools is much slower, and equilibriation between the intravascular and extravascular pools may be as long as 7 to 10 days. Furthermore, there appears to be differential equilibration of albumin depending on its location. Major amounts of extravascular albumin are in the interstitial spaces and small lymphatic radicals, whereas a small amount exists in the larger lymphatic ducts.[17,21]

Albumin is synthesized within the hepatocyte. Amino acids are delivered to the small ribosomal subunits. A polysome, consisting of mRNA and rRNA, forms the albumin molecule, which is transported through larger ribosomal units into the endoplasmic reticulum. After synthesis is terminated, it is transported to the Golgi apparatus to be used and passed into the plasma of the liver. The entire synthesis occurs in approximately 15 to 20 min. Normal albumin synthesis varies between 150 and 200 mg/kg/day. Table 10-1 details normal human metabolism (modified from Ref. 17).

Albumin synthesis is dependent on adequate protein and nitrogen intake.[19,22] Starvation will decrease albumin synthesis, and a decreased amino acid supply will increase RNA breakdown. In all probability, the level of amino acids is the basic regulator of RNA metabolism and consequently of albumin synthesis. Adequate protein intake is therefore essential for adequate albumin synthesis.

Another regulating factor of albumin synthesis is osmotic pressure. When exogenous albumin or serum expanders are infused *in vivo,* there is a marked decrease in albumin synthesis.[17] Studies of the ability of the cirrhotic to synthesize albumin reveal that patients with ascites have elevated exchangeable albumin pools. In severely ill subjects, there is no apparent reduction in albumin metabolism.[17,23] Of note also is the fact that both normal thyroid and cortisol metabolism are essential in maintaining RNA metabolism.[17] Hormones and insulin appear to have a lesser effect.

The exact site and mechanism of albumin degradation are poorly understood. It probably occurs somewhere in the intra- and extravascular spaces,

Table 10-1. Normal Albumin Metabolism

Subject of study	Serum albumin level (g/100 ml)	Albumin synthesis (mg/kg/day)	Albumin degradation (% plasma albumin pool/day)	Exchangeable albumin pool (g/kg)	Intravascular albumin (%)
Adults					
Male	3.5–4.5	120–200	6–10	4.5–5.0	38–45
Female	3.5–4.5	120–150	6–10	3.5–4.5	38–45
Children					
13 days–14 months	3.3–4.3	180–300	10–11	6.0–8.0	33–43
3–8 yr	4.2–5.0	130–170	6–9	3.0–4.0	46–51

Table 10-2. Protein Requirement for Maintenance of Nitrogen Balance in Cirrhosis[a]

Protein intake (g/day)	Days of study per patient	Nitrogen balance (g/day)	Number of patients	
			Total	Negative
0	6	−4.7 to −2.2	6	6
10–25	6–9	−4.2 to −3.1	2	2
50	6–12	−1.6 to +3.8	11	2
75	6–12	+1.9 to +6.1	10	0
75	15–30	+1.6 to +6.8	17[b]	0
85–100	6–12	+4.3 to +9.5	6	0

[a]Modified from Gabuzda and Shear.[15]
[b]Patients with ascites.

but there is no proof as to exactly where degradation occurs and in which organ. Although large amounts can be lost in disease of the intestine or kidney, the mechanism and site of normal degradation are poorly understood.

The misuse of intravenous albumin replacement therapy in chronic liver disease is common. It may be indicated when a transient rise in oncotic pressure is desired, or pre- or postoperatively, when it may be urgent. However, most often low albumin levels should be corrected by encouraging endogenous synthesis[24,25] through correct nutrition.

10.7.2. Pathophysiology

In patients with severe liver disease, such as cirrhosis, studies reveal that protein requirements are normal for maintenance of nitrogen balance. Cirrhotic subjects were fed varying amounts of protein in diets providing adequate calories (1600 to 3500) daily.[15] Patients fed 75 g had a positive balance and the overwhelming majority fed 50 g maintained a positive nitrogen balance. This study is of marked significance when one considers the diet treatment of the severely ill patient with liver disease. The data of the study (Table 10-2) show that patients with an intake below 25 g will definitely be in negative balance even though there is sufficient caloric intake.

10.7.3. Protein Diet Therapy

Treatment of severe liver disease with diet protein varies widely—from increased intake to marked restriction to selective supplementation. From our present understanding of the pathophysiologic processes, it would seem that protein restriction is indicated in hepatic encephalopathy; however, it is essential to maintain positive nitrogen balance in order to maintain normal hepatocyte RNA activity and its resultant maintenance of the hepatocyte and production of essential serum proteins. The protein minimum required by a patient of average body build to achieve positive nitrogen balance ranges from 35 to 50 g/day and would be guaranteed with 1 g/kg/day. In a patient without hepatic coma and showing no signs of deterioration, adequate caloric and protein intakes should be maintained in order to treat any form of severe liver disease. Specific variations of severe liver disease diets in relation to other factors such as sodium intake will be discussed.

Concerning specific amino acid therapy, there is too little information available. However, it does appear that branched-chain amino acids, when infused or eaten, will ameliorate encephalopathy.[26-28] The ratio of branched-chain to aromatic amino acids is markedly decreased in chronic liver disease.[29] Although this theory is promising, it needs more confirmation.[27]

Some foods have a higher ammonia content than others and are therefore detrimental to the patient with hepatic encephalopathy.[30] The foods highest in ammonia content are primarily cheeses, meats, and some vegetables such as: domestic blue cheese, salami, cheddar cheese, lima beans, egg yolk, and bacon. They were noted out of a spectrum of 64 foods tested.

There has also been controversy in the literature concerning the differential hyperammonemia levels and clinical effects produced by foods.[31,32] Some work showed milk and casein less harmful in hepatic encephalopathy than certain meats.[32] Another detailed study in these patients revealed that vegetable-protein diets were less encephalopathic than animal protein.[32] The authors could not identify the exact reason but felt that the explanation resided in the lower amounts of such amino acids as methionine and other such materials in vegetable proteins. Much more work is needed to ascertain which foods, containing which amino acids, are more beneficial or less harmful to the patient with severe liver disease. (Low-protein diets and the ammonia content of foods are listed in Chapter 20.)

10.8. VITAMINS

The normal human physiology of all water- and fat-soluble vitamins has been described in Chapter 3. In early studies of liver disease, several vitamin deficiencies were incriminated as possible etiologic agents of cirrhosis. However, these deficiencies are now understood to be part of the chronic liver disease process. Deficiencies occur because of poor intake,[33,34] malabsorption,[35-37] abnormal conversion in the liver,[38,39] decreased hepatic storage, and concurrent diseases, such as chronic pancreatitis.[40,41]

Because of the importance of coagulation in chronic liver disease, deficiencies of fat-soluble vitamins are often most worrisome. In the study of Leevy et al.[42] of alcoholics, fat-soluble vitamin deficiencies were actually less common than water-soluble vitamin deficiencies, whereas a study of nonalcoholic subjects by British scientists revealed just the opposite.[43] Poor absorption of fat-soluble vitamins may be due to a decreased bile acid pool size. When steatorrhea is present, it may also cause decreased absorptive efficiency of fat-soluble vitamins.

Measurements of plasma vitamin A levels show a decrease that is partially due to a defect in the synthesis of transport protein.[44]

Vitamin K deficiency results in prolongation of the prothrombin time. The deficiency may occur due to poor intake or poor absorption; the latter is most common in obstructive jaundice. Decreased prothrombin production in the liver also results in prolongation of the prothrombin time. Decreased production of prothrombin may be differentiated from decreased vitamin K absorption by a simple vitamin K parenteral administration. One or two doses will correct a deficiency of vitamin K, whereas it will be useless in a damaged liver that cannot produce prothrombin.

It is generally agreed that water-soluble vitamin deficiencies are common in chronic liver disease. Ascorbic acid, cyanocobalamin, folic acid, nicotinic acid, pantothenic acid, pyridoxine, riboflavin, and thiamine are reported to be significantly decreased in both alcoholic and nonalcoholic patients with chronic liver disease.[33,42,43] Thiamine has often been incriminated in chronic alcoholic liver disease as the cause of many symptoms that are correctable by replacement therapy.

Folic acid deficiency is manifested often in chronic liver disease as megaloblastic anemia. Patients may become deficient because of poor intake, malabsorption, or concurrent disease.[34,36,37] Alcoholics may absorb some synthetic folate polyglutamates, but may be unable to digest and absorb some polyglutamates that are in foods.[39] In the latter study by Baker et al., water-soluble vitamin absorption of 37 alcoholic patients with liver disease was compared with normal nonalcoholic subjects. All alcoholics absorbed riboflavin and pantothenic acid, but had decreased absorption of thiamine and pyridoxine. However, they were able to absorb synthetic pyridoxine and synthetic folylmonoglutamate. It was clearly suggested from this work that folic acid, pyridoxine, and thiamine deficiencies in alcoholic liver disease were due to poor absorption of these substances from dietary products. Although most evidence indicates deficiencies in folic acid and cyanocobalamin in alcoholics, there are some studies that indicate these substances may be increased in selected alcoholic subjects.[45,46]

10.9. COAGULATION FACTORS

Factors are decreased in subjects with severe liver disease.[47,48] When these deficiencies become clinically apparent, hemorrhage is often present. The acute situation is best treated by intravenous fresh frozen plasma, which corrects most of the deficiencies. We have found that at least 3 units are needed to have any effect on coagulation tests.

Platelet counts may be low and often are due to the effect of alcohol. Platelet transfusion is not usually necessary.

If the hemorrhaging liver disease patient survives an acute insult, all coagulation factors usually improve with appropriate protein nutrition. If not, the liver disease is usually terminal.

10.10. MINERALS ASSOCIATED WITH LIVER DISEASE

10.10.1. Magnesium

Magnesium deficiency is often noted in patients with portal cirrhosis.[49,50] Although approximately one-half of all magnesium is present in bone, severe liver disease appears to be associated with deficiency. Clinical features of the deficiency are signs of increased neuromuscular excitability, including hyperesthesia, carpopedal spasm, muscular cramps, tetany, and even convulsions.[51] The exact reason for the deficiency in hepatic disease is not clear. It may be due to poor intake or increased excretion.

10.10.2. Zinc

Vallee and associates[52,53] have clearly demonstrated low serum zinc levels in patients with cirrhosis of the liver, which are corrected by increased oral zinc intake. One study shows that zinc deficiency is associated with chronic alcoholism more often than with chronic liver disease.[110] Plasma zinc levels have a significant diurnal variation, but they remain low in patients with cirrhosis of the liver, as well as for those with malabsorption.[54] Fasting levels were 71 and 76 μg/100 ml and postmeal levels 60 and 64 μg/100 ml in cirrhotic and malabsorptive subjects, respectively, compared with fasting levels of 97 μg/100 ml and postmeal levels of 81 μg/100 ml in control subjects. There is a close correlation of plasma zinc and plasma albumin levels, suggesting that the low zinc levels are related to albumin deficiency in the cirrhotic.[54] Urinary zinc excretion appears to be increased in cirrhotics but not in patients with malabsorption.[54]

10.10.3. Copper

Normal humans ingest approximately 2 to 5 mg of copper daily and absorb approximately 1 to 2.5 mg. Balance is maintained by excretion of an equal amount in the bile and feces. In Wilson disease there is a defect of the mechanism of copper excretion through the biliary system. Associated with this retention of copper is a deficiency in the liver's synthesis of serum copper-binding protein, ceruloplasmin. Patients have serum ceruloplasmin levels less than 20 mg/100 ml (normal: 20 to 40 mg/100 ml). Urinary copper excretion, however, is greatly increased (normal: less than 50 μg/24 hr). Increased deposition of copper occurs due to decreased transport in the liver, brain, kidney, and other organs.[55,56] There is limited success in treating the disease by chelation with D-penicillamine. The removal of copper from tissues is slow, but this can delay organ damage and in some cases reverse the course of the disease.[56,57]

Copper may be increased in the liver of biliary cirrhosis or chronic active hepatitis and correlates in these diseases with the duration of cholestasis. Chronic active hepatitis may then be confused with Wilson disease.[55,57]

10.10.4. Iron

Some patients with chronic liver disease develop gastrointestinal lesions that bleed, such as hemorrhagic gastritis or esophageal and gastric varices. Chronic blood loss leads to iron deficiency and resultant microcytic, hypochromic anemia. It should not be forgotten that most often patients with chronic alcoholism and cirrhosis have macrocytic indices due to low folic acid and/or B_{12}.[58] The cause of bleeding should be corrected and the anemia, if chronic, treated by oral or parenteral iron replacement.

More dramatic in liver disease is the association of *iron storage disorders.* Grace and Powell[59] have developed the following classification:

1. Idiopathic (familial hemochromatosis).
   a. Latent or precirrhotic stage.
   b. Cirrhotic stage.
2. Cirrhosis of the liver with secondary iron overload.
3. Liver disease associated with certain anemias (thalassemia, pyridoxine-responsive anemias, hereditary spherocytosis).
4. Liver disease associated with dietary iron overload (as seen in South African blacks using iron cooking utensils).
5. Liver disease associated with increased transfusion.
6. Congenital transferrin deficiency.

Increased iron storage results in hemochromatosis. With increased iron deposition in parenchymal cells, fibrosis of the liver occurs. Other organs of the body, such as the pancreas, endocrine glands, and heart, are similarly affected. The increased iron in the body may be due to increased absorption or, more commonly, increased intake associated with increased absorption. Transfusions are obvious reasons for increased intakes.

In most cases, it is related to increased intake often associated with alcoholism. South African blacks brew their alcoholic beverages and beers in iron-containing pots, resulting in an extremely high intake of iron.[60,61] Other significant factors are that alcohol by itself can increase iron absorption[62] and a low-protein diet enhances iron overload.[63]

Idiopathic familial hemochromatosis is a rare autosomal-recessive disorder.[64] Serum ferritin may be helpful in early identification of susceptible members of a family.[65,66] This becomes increasingly important with the Scandinavian observation that iron-fortified foods have increased the occurrence of high serum iron levels.[67] Certainly, the patient susceptible to hemochromatosis should receive a low-iron diet.

10.11. WATER AND ELECTROLYTES

Water and electrolyte balance becomes a problem occasionally in acute hepatic failure, and more frequently in chronic hepatic insufficiency, when the body begins to hold on to water and sodium. In cirrhosis of the liver, total body water and Na are usually increased. The clinician most often does not have a hint of this until peripheral edema or ascites develop,[68,69] at which point there is usually peripheral hyponatremia,[70] though total body Na is high or normal. In the chronic ascites patient the serum sodium will usually be recorded at around 120 meq/liter and is regularly below 130 meq/liter. The serum potassium will be normal or low. Measurements of total body potassium are usually low in the cirrhotic.[71,72] When azotemia develops, serum potassium may be high and replacement therapy is no longer indicated. The mechanisms for potassium disturbances will be discussed but hypochloremia is less clearly understood. It may occur but is often complex due to alkalosis.[73]

In the patient who develops ascites due to acute and chronic liver disease, several factors are working at the same time, any one of which may be more to blame than the others; all must be treated, however, because in a given patient it is virtually impossible to determine which factor is paramount. The following are all pathologic mechanisms contributory to the development of ascites:

1. A rise in hydrostatic venous pressure.
2. A fall in plasma osmotic pressure (decreased serum albumin).[74]
3. Increased hepatic lymph formation with decreased reabsorption of lymph.[75]
4. Retention of sodium and water by:
   a. Hyperaldosteronism due to liver inability to inactivate aldosterone.
   b. Possible intrarenal vasoconstriction with poor renal perfusion.[76]
   c. Possible renal tubular defects.[77]
   d. Possible other humoral influences.[78]

The *dietary treatment* of the liver disease patient retaining water and sodium is essential in the successful control of either peripheral edema or ascites. Most often, it is also necessary to add an aldosterone antagonist or intermittent diuretic therapy.

The first step is restriction of sodium intake. The diet should contain between 500 and 1000 mg of sodium (see Chapter 20, diet 12). The second step, which is more severe and difficult to maintain, is a restriction of total fluid intake to less than 1 liter/day (see Chapter 20, diet 15). The third step requires the use of increasing doses of aldosterone antagonists to offset the effect of the hyperaldosteronism. The fourth step of therapy is the use of sodium-excreting diuretics. A mild case treated early may be controlled by steps one and two, whereas a severe case, dyspneic and distended, may require all four. The clinician must be judicious in the use of these four steps to control ascites.

If the severely ill and hypoalbuminemic patient has required previous paracentesis or has been malnourished, one may have to add a fifth step to commence diuresis. This would be either administration of albumin if the patient is hypoalbuminemic, in order to increase plasma osmotic pressure, or initial removal of some ascitic fluid to increase renal perfusion.

The patient on strict fluid oral restruction often becomes extremely thirsty. This can be counteracted by the addition of small sugar candies to the diet. It is extremely difficult to manage a restricted protein intake as well as restricted sodium and water intakes. This will be discussed under hepatic failure.

In azotemia there may be marked ascites and yet restriction of water may not be indicated. In this situation, the clinician is often faced with the possibility of the hepatorenal syndrome, which is poorly understood. However, when azotemia develops, and is increasing, with a rising BUN or creatinine, it is most helpful to perfuse some water. A prerenal azotemia may be present in these cases. Increased perfusion of water permits better renal function. Once diuresis begins, BUN and creatinine begin to fall. As they return toward normal, one can restrict fluid intake in order to further continue diuresis. However, this sequence of clinical events may take days to weeks and requires the greatest sophistication in fluid balance therapy.

Most often, ascites and peripheral edema can be controlled by these measures. However, at times a shunt procedure is necessary. Many institutions have now found the peritoneojugular (LeVeen) shunt a safe and effective method to reduce intractable ascites.[79,80] A tube is inserted into the peritoneal cavity. A one-way valve is placed extraperitoneally and deep to the abdominal muscles and is connected to a venous tube placed subcutaneously that is inserted into the jugular artery. This is reserved for the patient in whom the above fails, and it can be most successful. A last resort is to proceed to a portocaval vascular shunt procedure. This is a drastic step, for if the patient is cooperative, ascites usually can be controlled by diet and diuretic management on the LeVeen shunt.

10.12. DIET THERAPY OF LIVER DISEASE

10.12.1. Acute Liver Disease

There are three common clinical acute liver disease states seen in the adult: Acute viral hepatitis, toxic hepatitis, and alcoholic hepatitis.

At the present state of knowledge, there is no specific diet therapy for any of these conditions. The principles of diet therapy that hold for any acute illness, or for chronic liver disease, are used in acute liver disease. During any acute illness, spicy or so-called "rich" (very fatty) foods may be difficult to tolerate. Therefore, a bland diet is most easily digested and absorbed. The only concern is avoidance of high-fat intake (see discussion on alcoholic fatty liver). In most situations, if the patient's appetite is not affected, he should have an ad lib diet. Patients with acute and chronic liver disease do have an increased incidence of dysosmia, dysgeusia, and disordered gustatory acuity.[81] As a consequence, the patient with acute disease may have a loss of appetite, which decreases proper intake. The *ad lib diet* should be encouraged so that the patient will eat food he enjoys. There is absolutely no concrete evidence to indicate that high intake of carbohydrate, fat, or protein increases morbidity. There is some evidence suggesting that a high-fat diet may increase fat deposition in the liver. Therefore, it may be wise to avoid high-fat intake. The patient should receive at least 4 to 5 g of carbohydrates per kilogram body weight to maintain adequate energy. This type of intake can be maintained early in the course of the disease with hard candy, carbonated beverages, fruit juices, and sweets that are easily tolerated. If the acutely ill patient cannot tolerate this level of carbohydrates after a few days of illness, he should be placed on intravenous therapy to maintain adequate cellular physiology. It is also important to avoid excessive water loading with carbohydrates if intravenous therapy is started; consequently, it may be necessary to use 10% glucose rather than 5% to maintain adequate carbohydrate intake.

Some evidence indicates that some alcoholic hepatitis patients do better if there is adequate essential amino acid intake.[82] Intravenous amino acids are available for peripheral vein administration without the use of central vein hyperalimentation (see Chapter 19). If the patient is without adequate protein intake for more than 5 to 7 days, it should be considered for reversal of nitrogen loss, which may be important in the final outcome. Some studies of chronic liver disease stress the importance of adequate protein intake,[14,15] while others suggest too much may be harmful.[16] Studies of American soldiers in Asia and Asians in Singapore showed no benefit from maintaining protein intake early in the course of acute hepatitis.[84,85] Therefore, protein intake for acute liver disease is still controversial. There is little in the literature on the effects of specific amino acids and other food products in toxic and other hepatitides in humans, but a large literature on animals, which has been reviewed in detail by Schiff and Klatskin.[86]

10.12.2. Chronic Liver Disease

The major chronic liver disease problem that presents for treatment is cirrhosis of the liver. Varying with societies, the etiology may be infectious, congenital, idiopathic, or alcoholic-associated cirrhosis. The first step is to remove all toxins, alcohol or others, from the diet. The next step presents a real challenge to the clinician. Because cirrhotic liver complications are so varied, *the diet recommendation must depend on the clinical status.* Judicious juggling of the various factors is important in assisting the patient through the various complications. Dietary intake of water, sodium, protein, energy, and vitamins will depend on the complication and status of the disease. At times, each of these factors should be increased, decreased, or maintained constant. The following are recommended.

10.12.2.1. The Stable Cirrhotic

This patient requires no specific dietary management, feels well, has no gastrointestinal complaints, no evidence of peripheral edema or ascites, a normal white blood count, and only minor abnormalities in liver chemical tests. He should be encouraged to maintain an adequate diet, with maintenance nutrition as outlined in Chapter 2. However, I do encourage this type of patient to take a general vitamin tablet daily. The stable cirrhotic is cautioned about total sodium intake, as are all persons.

10.12.2.2. The Development of Peripheral Edema

This clinical finding heralds the increase of total body water and sodium. Consequently, the diet must become restrictive in salt intake (see Chapter 20, diet 12). It is often difficult to restrict the patient to 200 to 500 mg of sodium; therefore, if there is only minimal peripheral edema, a 1-g sodium diet should be tried at the onset. If it is not successful, the total sodium intake should be dropped to 10 to 20 meq (200 to 500 mg). With the restriction of sodium intake, caution, and instruction on fluid intake, restriction should begin. However, the patient need not be as strict as the ascitic.

10.12.2.3. The Ascitic

This patient is most difficult to manage and requires not only rigid restriction of sodium intake (200 to 500 mg), but an absolute limit of total water intake (see Chapter 20, diet 15). The patient who is compulsive about his diet most often can control the ascites. However, there are some who develop intractable fluid retention and aldosterone antagonists must then be added to the regimen; if these are unsuccessful, more vigorous diuretics follow. Often, it is impossible to fully assess whether the patient is restricting sodium and fluid. An excellent method is evaluation of urine water and sodium excretion. However, the patient may not be cooperative and outpatient facts are difficult to establish. The ascitic often suffers other complications of the cirrhosis, such as hypersplenism or encephalopathy. If there are no significant complications and sodium and fluid restriction in addition to diuretics fail, shunting procedure should be considered for control of the ascites.

10.12.2.4. Hepatic Encephalopathy

The earliest sign of encephalopathy is a minor personality change, followed by weakness, somnolence, or slight confusion. Physical findings, electroencephalography, and serum arterial ammonia levels confirm the clinical impression. It may or may not be associated with fluid retention and hence the diet will vary. However, in our experience, once encephalopathy develops, the patient does require at a minimum some sodium and fluid restriction. The degree of encephalopathy varies and the tolerance to protein varies accordingly. Some patients are extremely sensitive to small amounts of protein. Although they may tolerate 20 to 40 g, they can go into significant clinical encephalopathy with 50 or 60 g. Once the diagnosis of hepatic encephalopathy is established, the clinician must establish the amount of optimal dietary protein the patient can tolerate. Rudman et al.[30,87] have established that different amino acids and foods have different hyperammoniacal effects. Diet 14 of Chapter 20 lists increased ammonia content and relative ammonia-producing effects of certain foods. Arginine, aspartic acid, glutamic acid, and tryptophan are the least ammoniagenic amino acids.[87] This has great potential in the diet of the encephalopathic patient. Milk contains fewer ammoniagenic amino acids than eggs or meat and therefore appears to be the best basic food for the patient with severe liver failure (see Chapter 20, diet 10). Berlyne and Epstein[89] have demonstrated that dialyzed milk may be employed in patients requiring strict sodium management.

Fischer's theory of "false neurotransmitters" as a primary cause of encephalopathy includes the thesis that aromatic amino acids (AAA, phenylalanine and tyrosine) are increased and compete with branched-chain amino acids (BCAA, valine, leucine, and isoleucine) in crossing the blood–brain barrier. The BCAA are supposed to be protective. Whether diets increased in BCAA or IV BCAA therapy prove to be effective remains to be determined, but they now are recommended.[26-29] (See Chapter 20, diet 13.)

10.12.2.5. Hypoalbuminemia

The patient with ascites, or hepatic encephalopathy and coma, or following gastrointestinal hemorrhage, will often have a serum albumin level below 3.5 mg/100 ml. This may aggravate fluid retention by lowering osmotic pressure. Adequate protein caloric intake is necessary to reverse the low serum albumin level caused by nutritional deficiency. At least 50 g/day are needed to keep a cirrhotic in positive balance and at least as much is needed to synthesize albumin.

During hepatic encephalopathy the patient may not tolerate protein; consequently, the vicious cycle of poor intake and resultant low serum albumin ensues. In these instances parenteral feeding of calories or of salt-poor albumin becomes necessary. One should not use intravenous albumin unless the low serum protein level appears to be definitely contributing to uncontrolled fluid retention.[24,25] When this occurs, at least 2 to 4 units of albumin is needed. Use of i.v. amino acids and TPN has gained wide acceptance. However, there is no experience yet in treating the severely cirrhotic patient. But it is clear in other stress situations that amino acid infusion can improve albumin synthesis.[90]

10.12.2.6. The Bleeding Patient

The cirrhotic subject may begin to bleed and consequently will require either simple iron replacement, when the bleeding does not require blood transfusion, or blood replacement, when the bleeding is massive. This patient often has abnormal coagulation. If the condition is one of hypersplenism and low serum platelets, fresh blood or component platelet transfusion is necessary. If it is one of all coagulation factors, then fresh frozen plasma may be adequate. However, at least 3 to 4 units of fresh frozen plasma is often necessary to correct coagulation defects. Theoretically, the patient with obstructive jaundice can be corrected with vitamin K therapy; however, this does not work in the usual chronic cirrhotic patient.

10.12.2.7. Mental Confusion

On occasion it is difficult to establish whether the patient is in mild hepatic encephalopathy or there are vitamin and mineral deficiencies. Therefore, in the confused state, the patient should have adequate thiamine replacement, as well as a check for serum magnesium and other electrolyte replacement.

The following checklist is used for the chronic liver disease patient:

1. Establish fluid and electrolyte status.
   a. Possibly restrict sodium.
   b. Possibly restrict water intake.
2. Establish the need for vitamins.
   a. Vitamin K, thiamine.
   b. General replacement.
3. Treat hepatic encephalopathy if present.
   a. Control gastrointestinal bleeding.
   b. Limit all sedation.
   c. Correct any electrolytic imbalance, hypokalemia, or alkalosis created by diuretics.
   d. Control any infection.
   e. Manage protein restriction.
   f. Use antibiotics or lactulose for bowel cleansing.
4. Establish adequate caloric and protein intake.

10.12.2.8. Lactulose

Lactulose[91] is a synthetic disaccharide that is not metabolized in the upper gastrointestinal tract of man. It is degraded by the action of bacteria in the lower gut to its component monosaccharides, which are hydrolyzed to lactic and acetic acids. These may produce an osmotic diarrhea and concomitantly lower the pH of the stool. It is reported as effective as neomycin. The suggested dose is anywhere from 100 to 200 ml/day.

10.13. ALCOHOL

One gram of alcohol contains approximately 7 cal. Despite a high energy yield, alcohol is one of our greatest social hazards. It is at least the fourth leading cause of death; nobody knows how many deaths are caused because of accidents related to the neurologic phenomena caused by the drug. Its chronic use is associated with the development of cirrhosis of the liver and all of its fatal complications. No book on nutrition could be published without some discussion of alcohol, although it is not a recommended nutrient.

Alcohol or ethanol is produced by the fermentation of glucose by yeast enzymes. It requires no modification in the gut and can be absorbed passively. In the gastrointestinal tract, it is a direct irritant to the mucosa, causes erythema, relaxes the cardioesophageal sphincter, and sends the ampulla of Vater into spasm. Once absorbed, it enters the circulation and is primarily metabolized by the liver, where alcohol dehydrogenase converts it to acid aldehyde. Nicotinamide adenine dinucleotide is also required for the reaction. The interested reader is referred to reviews by Lieber[16,92] for all of the hepatic and metabolic effects of alcohol.

A continuous excess of alcohol (more than 5 to 6 oz of whiskey per day for several weeks) causes fatty liver.[92] Why one subject will develop alcoholic hepatitis or cirrhosis of the liver and another subject will merely be plagued by increased deposition of fat in the liver is still not clear. Animal studies and some human studies reveal that injury is associated with malnutrition and particularly protein deficiency. Cirrhosis caused by alcohol has been attributed to choline deficiency in animals, but that relationship, although present, is not the cause in humans.[93,94]

Alcohol may have a deleterious effect on absorption.[95] It may impair absorption of D-xylose, thiamine,[96] B_{12},[97] folic acid,[98] and ascorbic acid.[99]

The effect on the small bowel is related to dose and duration of intake and may decrease carrier-mediated absorption, alter the physical properties of enterocytic membranes, or induce intestinal secretion of water and electrolytes by its effect on the cyclic AMP system.[100] Nutritional deficiencies in alcoholics may occur not only due to poor intake but simultaneous poor absorption as well.[100]

Metabolism of the fatty liver is discussed above. Alcohol is one of the main causes of fatty liver. Diet fat is known to be deposited in the liver of alcoholics, and a high-fat diet can increase the amount deposited. However, fat composed of medium-chain triglycerides is more beneficial than long-chain triglycerides: Alcoholic subjects consuming medium-chain rather than long-chain triglycerides have a less fatty liver. Naturally, this is relative and it is difficult to suggest that someone imbibing harmful amounts of alcohol assist their liver by simultaneously taking medium-chain triglycerides. Nevertheless, it is of some help to note that the patient with fatty liver can decrease the amount of fat in the liver by substituting medium-chain triglycerides for long-chain triglycerides.[101]

Chronic alcoholism is also associated with mineral deficiencies such as magnesium and zinc. Those have been described above in the dietary treatment of patients with chronic hepatic disease.

10.14. MYCOTOXINS

Aflatoxin is a toxin produced by the fungus *Aspergillus flavus*. Oettle[10] has reported epidemiologic evidence that aflatoxins are associated with hepatoma of the liver in Africa. His thesis is now fairly well accepted. Several episodes of hepatitis have also been demonstrated after ingestion of products contaminated with toxins produced by *A. flavus*. Study of a West Indian epidemic caused by contaminated maize revealed that consumption between 2 and 6 mg of aflatoxin daily over a period of months produced hepatitis. The hepatitis was severe enough to cause portal hypertension and many deaths.[102]

10.15. CONGENITAL DISEASES OF THE LIVER

10.15.1. Galactosemia

Galactosemia, a deficiency of galactose-1-phosphate uridyl transferase, is life threatening for the infant. Symptoms occur in the first few days of life, and unless galactose is totally removed from the diet, extensive liver damage causes death.[104,105]

10.15.2. Fructose Intolerance

Fructose intolerance is somewhat similar to galactosemia. Fructose-1-phosphate aldolase is deficient due to an autosomal recessive trait. A similar condition involving fructose-1, 6-diphosphatase deficiency has also been described. Serious damage can occur to the liver with resultant cirrhosis. The disease can be controlled by strict elimination of both fructose and sucrose from the diet.[106,107] (See Chapter 20, diet 18.)

Detailed lists of genetic factors in liver disease and other congenital abnormalities, such as tyrisonemia, can be found in recent monographs.[106,108,109]

10.15.3. Congenital Hyperammonemic Syndromes

Congenital hyperammonemic syndromes are caused by deficiencies of urea cycle enzymes, by some disorders of ornithine and lysine metabolism, and by some inborn errors of branched-chain amino acid metabolism. Ammonia toxicity produces a pathophysiology similar to that of hepatic encephalopathy. Early effects of the disorder are reversible. The exact biochemical diagnosis is made with esoteric diagnostic methods. Treatment is the restriction of protein intake and prevention of constipation. However, treatment may fail even with the additional use of agents such as lactulose, and hyperammonemia in the newborn can be fatal. Genetic counseling is wise once a diagnosis is made in a newborn.[109]

10.15.4. Fasting and Dietary Effects on Serum Bilirubin

Fasting increases the plasma bilirubin concentration. The effect is greater in patients with chronic nonhemolytic unconjugated hyperbilirubinemia than in normal subjects; it is also more obvious in patients with liver disease and hemolytic anemia. Dietary factors can also increase the level of unconjugated bilirubin. There are complex interrelationships that affect the level of serum bilirubin. Carbohydrate feeding tends to increase bilirubin levels and appears to be associated with lipid withdrawal. Infusion of both glucose and amino acids in normal males causes a twofold increase in plasma bilirubin; in other experiments, however, an increase of dietary lipid to 9% of the energy intake results in a stable serum bilirubin level.

REFERENCES

1. Newsholme TA: Role of the liver in integration of fat and carbohydrate metabolism and clinical implications in patients with liver disease, in Popper H, Schaffner F (eds): *Progress in Liver Disease.* New York: Grune & Stratton, 1976, vol V.

2. Unger RH, Orco L: The essential role of glucagon in the pathogenesis of diabetes mellitus. *Lancet* 1:14, 1975.

3. Tranquada R: Lactic acidosis. *Calif Med* 101:450, 1964.

4. Owen OE, Morgan AP, Kemp HG, *et al:* Brain metabolism during fasting. *J Clin Invest* 46:1589, 1967.

5. Record CO, Alberti KGMM, Williamson DH, *et al:* Glucose tolerance and metabolic changes in human viral hepatitis. *Clin Sci* 45:677, 1973.

6. Alberti KGMM, Record CO, Williamson DH, *et al:* Metabolic changes in active chronic hepatitis. *Clin Sci* 42:591, 1972.

7. Rehfeld JF, Juhl E, Hilden M: Carbohydrate metabolism in alcohol-induced fatty liver. *Gastroenterology* 64:445, 1973.

8. Berkowitz D: Glucose tolerance, free fatty acid and serum insulin responses in patients with cirrhosis. *Am J Dig Dis* 14:691, 1969.

9. Holdsworth CD, Nye L, King E: The effect of portacaval anastomosis on oral carbohydrate tolerance and on plasma insulin levels. *Gut* 13:58, 1972.

10. Megyesi C, Samols E, Marks V: Glucose tolerance and diabetes in chronic liver disease. *Lancet* 2:1051, 1967.

11. Johnson DG, Alberti KGMM, Faber OK, *et al:* Hyperinsulinism of hepatic cirrhosis. Diminished degradation or hypersecretion. *Lancet* 1:10, 1977.

12. Hoyumpa AM Jr, Greene HL, Dunn GD, *et al:* Fatty liver: Biochemical and clinical considerations. *Am J Dig Dis* 20:1142, 1975.

13. Jenkins DJA, Leeds AR, Gassull MA, *et al:* Decrease in postprandial insulin and glucose concentration by guar and pectin. *Ann Intern Med* 86:20, 1977.

14. Miller LL, Whipple GH: Liver injury, liver protection, and sulfur metabolism: Methionine protects against chloroform liver injury even after anesthesia. *J Exp Med* 76:421, 1942.

15. Gabuzda GJ, Shear L: Metabolism of dietary protein in hepatic cirrhosis. *Am J Clin Nutr* 23:479, 1970.

16. Lieber CS: Pathogenesis and early diagnosis of alcoholic liver injury. *N Engl J Med* 298:888, 1978.

17. Rothschild MA, Oratz M, Schreiber SS: Regulation of albumin metabolism. *Annu Rev Med* 26:91, 1975.

18. Peters T Jr: Serum albumin. *Adv Clin Chem* 13:37, 1970.

19. Rothschild MA, Oratz M, Schreiber SS: Serum albumin. *Am J Dig Dis* 14:711, 1969.

20. Schultze HE, Heremans JE: *Molecular Biology of Human Proteins.* New York: Elsevier, 1966.

21. Rothschild MA, Oratz M, Schreiber SS: Albumin metabolism. *Gastroenterology* 64:324, 1973.

22. Waterlow JC: Observations on the mechanism of adaptation to low protein intake. *Lancet* 2:1091, 1968.

23. Wilkinson P, Mendenhall CL: Serum albumin turnover in normal subjects and patients with cirrhosis measured by I-labeled human albumin. *Clin Sci* 25:281, 1963.

24. Tullis JL: Albumin. 1. Background and use. *JAMA* 237:355, 1977.

25. Tullis JL: Albumin. 2. Guidelines for clinical use. *JAMA* 237:460, 1977.

26. Soeters PB, Fischer JE: Insulin, glucagon, amino acid imbalance and hepatic encephalopathy. *Lancet* 2:880, 1976.

27. Baker AL: Amino acids in liver disease: A cause of hepatic encephalopathy? *JAMA* 242:355, 1979.

28. Freund H, Yoshimura N, Fischer JE: Chronic hepatic encephalopathy. Long-term therapy with a branched-chain amino acid-enriched elemental diet. *JAMA* 242:347, 1979.

29. Morgan MY, Milson JP, Sherlock S: Plasma ratio of valine, leucine and isoleucine to phenylalanine and tyrosine in liver disease. *Gut* 19:1068, 1978.

30. Rudman MD, Smith RB III, Salam AA, *et al:* Ammonia content of food. *Am J Clin Nutr* 26:487, 1973.
31. Fenton JCB, Knight EJ, Humpherson PL: Milk and cheese diet in portal-systemic encephalopathy. *Lancet* 1:164, 1966.
32. Greenberger NJ, Carley J, Schenker S, *et al:* Effect of vegetable and animal protein diets in chronic hepatic encephalopathy. *Am J Dig Dis* 22:845, 1977.
33. Leevy CM, Baker H, Tenhove W, *et al:* B-complex vitamins in liver disease of the alcoholic. *Am J Clin Nutr* 16:339, 1965.
34. Klipstein FA, Lindenbaum J: Folate deficiency in chronic liver disease. *Blood* 25:443, 1965.
35. Lindcheer WG: Malabsorption in cirrhosis. *Am J Clin Nutr* 23:448, 1970.
36. Halsted CH, Robles EA, Mezey E: Decreased jejunal uptake of labeled folic acid (^{3}H-PGA) in alcoholic patients: Roles of alcohol and nutrition. *N Engl J Med* 285:701, 1971.
37. Halsted CH, Robles EA, Mezey E: Intestinal malabsorption in folate-deficient alcoholics. *Gastroenterology* 64:526, 1973.
38. Leevy CM, Tamburro C, Smith F: Alcoholism, drug addiction, and nutrition. *Med Clin North Am* 54:1567, 1970.
39. Baker H, Frank O, Zetterman RK, *et al:* Inability of chronic alcoholics with liver disease to use food as a source of folates, thiamin, and vitamin B_6. *Am J Clin Nutr* 28:1377, 1975.
40. Sun DCH, Albacete RA, Chen JK: Malabsorption studies in cirrhosis of the liver. *Arch Intern Med* 119:567, 1967.
41. Marin GA, Clark ML, Senior JR: Studies of malabsorption occurring in patients with Laennec's cirrhosis. *Gastroenterology* 56:727, 1969.
42. Leevy CM, Thompson A, Baker H: Vitamins and liver injury. *Am J Clin Nutr* 23:4, 1970.
43. Morgan AG, Kelleher J, Walker BE, *et al:* Nutrition in cryptogenic cirrhosis and chronic aggressive hepatitis. *Gut* 17:113, 1976.
44. Smith FR, Goodman DS: The effects of diseases of the liver, thyroid, and kidneys on the transport of vitamin A in human plasma. *J Clin Invest* 50:2426, 1971.
45. Carney MWP: Serum folate and cyanocobalamin in alcoholics. *Q J Stud Alcohol* 31:816, 1970.
46. Cowling DC, Mackay IR: Serum vitamin B_{12} levels in liver disease. *Med J Aust* 2:558, 1959.
47. Shapiro SS: Disorders of the vitamin K-dependent coagulation factors, in Williams WJ, et al (eds): *Hematology,* ed 2. New York: McGraw–Hill Book Co, 1977.
48. Sherlock S: *Diseases of the Liver and Biliary System,* ed 5. Oxford: Blackwell Scientific Publications, 1975.
49. Wacker WEC, Parisi AF: Magnesium metabolism. *N Engl J Med* 278:712, 1968.
50. Lim P, Jacob E: Magnesium deficiency in liver cirrhosis. *Q J Med* 41:291, 1972.
51. Hersh T, Siddiqui DA: Magnesium and the pancreas. *Am J Clin Nutr* 26:362, 1973.
52. Vallee BL, Wacker WEC, Bartholomay AF, *et al:* Zinc metabolism in hepatic dysfunction. Serum zinc concentrations in Laennec's cirrhosis and their validation by sequential analysis. *N Engl J Med* 256:403, 1956.
53. Vallee BL, Wacker WEC, Bartholomay AF, *et al:* Zinc metabolism in hepatic dysfunction. II. Correlation of metabolic patterns with biochemical findings. *N Engl J Med* 257:1055, 1957.
54. Walker BE, Dawson JB, Kelleher J, *et al:* Plasma and urinary zinc in patients with malabsorption syndromes or hepatic cirrhosis. *Gut* 14:943, 1973.
55. Sternlieb I, Scheinberg IH: Chronic hepatitis as a first manifestation of Wilson's disease. *Ann Intern Med* 76:59, 1972.
56. Deiss A, Lynch RE, Lee GR, *et al:* Long-term therapy of Wilson's disease. *Ann Intern Med* 75:57, 1971.
57. Strickland TG, Leu ML: Wilson's disease. Clinical and laboratory findings in 40 patients. *Medicine* 54:113, 1975.
58. Crosby WH: Certain things physicians do. Red cell indices. *Arch Intern Med* 139:23, 1979.
59. Grace ND, Powell LW: Iron storage disorders of the liver. *Gastroenterology* 64:1257, 1974.
60. Bothwell TH, *et al:* Iron overload in Bantu subjects. Studies on the availability of iron in Bantu beer. *Am J Clin Nutr* 14:47, 1964.
61. Bothwell TH, Bradlow BA: Siderosis in the Bantu. A combined histopathological and chemical study. *Arch Pathol* 70:279, 1960.

62. Charlton RW, *et al:* Effect of alcohol on iron absorption. *Br Med J* 2:1427, 1964.

63. Buchanan WM: The importance of protein malnutrition in the genesis of Bantu siderosis. *S Afr J Med Sci* 36:99, 1971.

64. Finch SC, Finch CA: Idiopathic hemochromatosis: Au vin storage disease. *Medicine* 34:381, 1955.

65. Kidd KK: Genetic linkage and hemochromatosis. *N Engl J Med* 301:209, 1979.

66. Beaumont C, Simon M, Fauchet R, *et al:* Serum ferritin as a possible marker of the hemochromatosis allele. *N Engl J Med* 301:169, 1979.

67. Olssen KS, Heedman PA, Staugard F: Preclinical hemochromatosis in a population on a high-iron-fortified diet. *JAMA* 239:1999, 1978.

68. Clowdus BF II, Summerskill WHJ, Casey TH, *et al:* Isotope studies of development of water and electrolyte disorders and azotemia during the treatment of ascites. *Gastroenterology* 41:360, 1961.

69. Birkenfeld LW, Leibman J, O'Meara MP, *et al:* Total exchangeable sodium, total exchangeable potassium, and total body water in edematous patients with cirrhosis of the liver and congestive heart failure. *J Clin Invest* 37:687, 1958.

70. Hecker R, Sherlock S: Electrolyte and circulatory changes in terminal liver failure. *Lancet* 2:1121, 1956.

71. Casey TH, Summerskill WHJ, Orvis AL: Body and serum potassium in liver disease. I. Relationship to hepatic function and associated factors. *Gastroenterology* 48:198, 1965.

72. Casey TH, Summerskill WHJ, Bickford RH, *et al:* Body and serum potassium in liver disease. II. Relationships to arterial ammonia, blood pH, and hepatic coma. *Gastroenterology* 48:208, 1965.

73. Atkins EL, Schwartz WB: Factors governing correction of the alkalosis associated with potassium deficiency: The critical role of chloride in the recovery process. *J Clin Invest* 41:218, 1962.

74. Atkinson M, Losowsky M: The mechanism of ascites formation in chronic liver disease. *Q J Med* 30:153, 1961.

75. Witte M, Witte C, Dumont A: Progress in liver disease: Physiological factors involved in the causation of cirrhotic ascites. *Gastroenterology* 61:742, 1971.

76. Epskin M, Berk D, Hollenberg N, *et al:* Renal failure in the patient with cirrhosis: The role of active vasoconstriction. *Am J Med* 49:175, 1970.

77. Chaimovitz C, Szylman P, Alroy G, *et al:* Mechanism of increased renal tubular sodium reabsorption in cirrhosis. *Am J Med* 52:198, 1959.

78. Denison E, Lieberman F, Reynolds T: 9-alpha-fluorohydrocortisone induced ascites in alcoholic liver disease. *Gastroenterology* 61:497, 1971.

79. LeVeen HH, Christoudias G, Ip M, *et al:* Peritoneo-venous shunting for ascites. *Ann Surg* 180:580, 1974.

80. Berkowitz HD, Mullen JL, Miller LD, *et al:* Improved renal function and inhibition of renin and aldosterone secretion following peritoneovenous (LeVeen) shunt. *Surgery* 84:120, 1978.

81. Smith FR, Henkin RI, Dell RB: Disordered gustatory acuity in liver disease. *Gastroenterology* 70:568, 1976.

82. Eckhardt RD, Falcon WW, Davidson CS, *et al:* Improvement of active liver cirrhosis in patients maintained with amino acids intravenously as the source of protein and lipotropic substances. *J Clin Invest* 28:603, 1949.

83. Freeman JB, Stegink LD, Meyer PD, *et al:* Metabolic effects of amino acids vs. dextrose infusion in surgical patients. *Arch Surg* 110:916, 1975.

84. Chalmers TC, *et al:* The treatment of acute infectious hepatitis. Controlled studies of the effects of diet, rest, and physical reconditioning on the acute course of the disease and on the incidence of relapses and residual abnormalities. *J Clin Invest* 34:1163, 1955.

85. Fung WP, Tan KK, Tye CY: The effect of diet on the severity of liver damage in acute alcoholic hepatitis. *Singapore Med J* L3:224, 1972.

86. Schiff L, Klatskin G: Toxic and drug-induced hepatitis, in *Diseases of the Liver,* ed 4. Philadelphia: JB Lippincott Co, 1975.

87. Rudman D, Galambos JT, Smith RB III, *et al:* Comparison of the effect of various amino

acids upon the blood ammonia concentration of patients with liver disease. *Am J Clin Nutr* 26:916, 1973.

88. Mitchell HS, Rymberger HJ, Anderson L, *et al: Nutrition in Health and Disease.* Philadelphia: JB Lippincott Co, 1976.

89. Berlyne GM, Epstein N: Dialysed milk in low-sodium and low-potassium diets for renal, hepatic, and cardiac disease. *Lancet* July 3, 1971.

90. Skillman JJ, Rosenoer VM, Smith PC, *et al:* Improved albumin synthesis in postoperative patients by amino acid infusion. *N Engl J Med* 295:1037, 1976.

91. Fessel JM, Conn HO: Lactulose in the treatment of acute hepatic encephalopathy. *Am J Med Sci* 226:103, 1973.

92. Lieber CS: Hepatic and metabolic effects of alcohol. *Gastroenterology* 65:821, 1973.

93. Best CH, Hartcroft WS, Lucas CC: Liver damage produced by feeding alcohol or sugar and its prevention by choline. *Br Med J* 2:1001, 1949.

94. Barak AJ, Tuma DJ, Sorrell MF: Relationship of ethanol to choline metabolism in the liver: A review. *Am J Clin Nutr* 26:1234, 1973.

95. Krasner N, Cochran KM, Russell RI, *et al:* Alcohol and absorption from the small intestine. *Gut* 17:245, 1976.

96. Tomasulo PA, Kater RM, Iber FL: Impairment of thiamin absorption in alcoholism. *Am J Clin Nutr* 21:1341, 1968.

97. Lindenbaum J, Lieber CS: Alcohol-induced malabsorption of vitamin B_{12} in man. *Nature (London)* 224:806, 1969.

98. Halsted CH, Robles EA, Mezey E: Decreased jejunal uptake of labelled folic acid (H_3-PGA) in alcoholic patients: Roles of alcohol nutrition. *N Engl J Med* 285:701, 1971.

99. O'Keane M, Russell RI, Goldberg A: Ascorbic acid status of alcoholics. *J Alcohol* 7:6, 1972.

100. Wilson FA, Hoyumpa AM Jr: Ethanol and small intestine transport. *Gastroenterology* 76:388, 1979.

101. Malagelada JR, Linscheek WG, Houtsmuller UMT, *et al:* Effect of medium-chain triglycerides on liver fatty acid composition in alcoholics with or without cirrhosis. *Am J Clin Nutr* 26:738, 1973.

102. Oettle AG: The aetiology of primary carcinoma of the liver in Africa: A critical appraisal of previous ideas with an outline of mycotoxin hypothesis. *S Afr Med J* 39:817, 1965.

103. Krishnamachari KAVR, Bhat RV, Nagarajan V, *et al:* Hepatitis due to aflatoxicosis. An outbreak in Western India. *Lancet* 1:1061, 1975.

104. Applebaum MN, Thaler MM: Reversibility of extensive liver damage in galactosemia. *Gastroenterology* 69:496, 1975.

105. Isselbacher KJ, Anderson EP, Kurahashi K, *et al:* Congenital galactosemia, a single enzymatic block in galactose metabolism. *Science* 123:635, 1956.

106. Silverberg M, Davidson M: Nutritional requirements of infants and children with liver disease. *Am J Clin Nutr* 23:604, 1970.

107. Baker L, Winegrad AI: Fasting hypoglycemia and metabolic acidosis associated with deficiency of hepatic fructose-1-6-diphosphate activity. *Lancet* 2:13, 1970.

108. Brunt PW: Genetics of liver disease. *Clin Gastroenterol* 2:615, 1973.

109. Hsia YE: Inherited hyperammonemic syndromes. *Gastroenterology* 67:347, 1974.

110. Kiilerich S, Dietrichson O, Loud FB, *et al:* Zinc depletion in alcoholic liver disease. *Scand J Gastroenterol* 15:363, 1980.

Gallbladder Diseases

11.1. NORMAL FUNCTION OF THE GALLBLADDER

The gallbladder is the site of concentration of biliary secretions so that hepatic bile is converted to gallbladder bile. Due to the action of cholecystokinin–pancreozymin (CCK-PZ) and complex neurogenic reflexes, it contracts during feeding. The major stimuli of CCK-PZ are fat-containing foods. When they enter the duodenum, the hormone is secreted and the gallbladder contracts, secreting its contents through the common duct into the duodenal sweep, where it mixes with the bolus of food to help form micelles for absorption of fats and cholesterol. Postcholecystectomy patients secrete bile at a continuous rate, but they adapt so that there is adequate secretion of bile during feeding, preventing malabsorption. The gallbladder and its contents and functions are described in more detail in Chapter 1.

11.2. ACUTE CHOLECYSTITIS

11.2.1. Etiology

The exact incidence is not clearly understood, but one would suspect that it is probably the same as cholelithiasis. The cause of acute cholecystitis in highly industrialized societies most often is cholelithiasis. However, in less developed areas of the world, it might be more commonly of infectious origin, and in rare cases it is idiopathic. In either event, the gallbladder becomes inflamed, which results in the following clinical picture.

11.2.2. Clinical Picture

The patient experiences acute right upper quadrant pain, which is often associated with nausea and vomiting. The pain is persistent and unrelenting. The stoic patient may weather the attack without medical aid, but most often, in the acute episode, the physician's help is sought.

The right upper quadrant is often tender and there may be guarding. The temperature is usually low grade but may be elevated markedly. There is usually no sign of chronic disease, except when the attack occurs in the post-operative period, where it is more common than usual, or in association with some chronic disorder, as cirrhosis.

The diagnosis is usually confirmed by the clinical picture, elevated white blood count, normal pancreatic enzymes, nonvisualization of the gallbladder on oral or intravenous opaque dye X-ray study or HIDA scan, and by ultrason-ography to visualize either a distended gallbladder or one filled with stones.

11.2.3. Treatment

The treatment in optimal conditions is surgical.[1] However, where the diagnosis is delayed, or where the patient's condition indicates that surgery may be treacherous, a medical approach may be adopted. In these cases and in preparation for surgery, the patient is given adequate analgesia and anticholinergic medication. The dietary management should be no feeding, or if vomiting is prevalent, nasogastric suction during the acute onset. Once the patient can tolerate some liquids, fat-free foods should be fed. Although controversy exists concerning the significance of a low-fat diet in chronic disease, there is no question that significant amounts of fats do induce contraction of the gallbladder and, consequently, should be eliminated in the acute stage. (See Chapter 20, diets 20 and 21.)

11.3. CHRONIC CHOLECYSTITIS

11.3.1. Etiology

As with acute cholecystitis, this disorder is most common in the United States and Europe, where the incidence of cholelithiasis is high. The role of diet in the etiology of this disease will be discussed under cholelithiasis. Chronic cholecystitis is most often associated with chronic inflammation of the gallbladder and relative degrees of scarring. Scarring and chronic inflammation may be so severe that the gallbladder actually can become nonfunctioning and self-destructive.

11.3.2. Clinical Picture

The clinical picture presents a varied spectrum. Classically, it is recurrent right upper quadrant pain following meals. However, the pain, in cases, may be atypical and refer to the epigastrium, or left upper quadrant, or to the subscapular areas. One would suspect that the pain would be most common after fatty meals, but that is not substantiated by clinical studies. Actually, most clinical studies indicate the pain is acute, and chronic cholecystitis is not clearly associated with fatty foods.[2] In a study of 1000 patients who were referred for oral cholecystography because of symptoms, it was found that bloating, belching, and so-called dyspepsia were no more common in patients with abnormal radiographs than in those with normal studies.[3] Another study evaluated 146 women, employing careful symptoms and cholecystogram correlation. The results revealed that food intolerance occurred in 53% of those with normal oral cholecystograms and in 50% of those with abnormal studies.[4] Furthermore, the groups with normal gallbladders had a 28% incidence of fried and fatty food intolerance, whereas those with the abnormal cholecystograms had only a 4% incidence. The picture is further confused by the study of Taggart and Billington,[5] who fed 12 subjects complaining of fatty food intolerance prepared foods containing disguised fats unrecognizable by the subjects. The study revealed that 92% of the subjects could not correlate their symptoms with the fatty foods, although they had given such an initial history. From all of these observations it must be concluded that fried or fatty food intolerance is not clearly associated with symptoms in patients with cholelithiasis or chronic cholecystitis.[2-5]

The diagnosis of chronic cholecystitis is confirmed by the presence of an abnormal oral or intravenous cholecystogram or abnormal gallbladder ultrasonography.

11.3.3. Treatment

Preferred treatment is removal of the gallbladder. However, research is developing rapidly in the treatment of so-called lithogenic bile and dissolution of gallstones by the administration of bile salts (specifically, chenic acid) or bile salts and dietary factors. As of this time, it is not clear whether this is a preferred or acceptable long-term treatment. However, studies are actively progressing and it is assumed that this mode of therapy may become readily acceptable, especially for the patient who is not a surgical candidate. The medical and dietary treatments of chronic cholecystitis are now controversial. Most clinicians and authors of long-established texts, such as Bockus,[6] remove all high-fat intake from the diet. As mentioned above, this is controversial in the literature. Nevertheless, it is a well-accepted form of therapy, and both clinicians and patients attest to its merit.[6] However, others do not profess the use of a low-fat diet.[7] I still prefer to keep the patient who experiences repeated episodes of acute classic clinical colic on a low-fat diet. However, the patient who merely experiences so-called dyspeptic symptoms of bloating and upper abdominal discomfort is advised to eat what he or she can tolerate. A low-fat diet is included in Chapter 20.

11.4. GALLSTONES (CHOLELITHIASIS)

11.4.1. Pathophysiology

Cholesterol and mixed gallstones are different from pigment stones. Pigment gallstones are associated primarily with other diseases, such as chronic hemolytic states, alcoholic cirrhosis, chronic infection of the biliary tree, and are more common in the Orient.[8] The cholesterol gallstone is commonly associated with cholecystitis and symptomatic disease in Western societies. Analysis of the composition of stones from eight countries reveals the cholesterol composition varies from 60 to 94% with varying amounts of calcium and other compounds. The composition of the stones varies with sex and age. Female patients have more cholesterol in their stones, while males have much more calcium.[9]

There can be no question that gallstones are related to diet. The question that does arise is, how significant is the diet in preventing the formation or assisting in the dissolution of gallstones? Approximately one-third of a million gallbladders are removed each year in the United States, the overwhelming majority containing gallstones. Because cholecystectomy is such a common procedure in the Western world, and because it is almost invariably due to cholelithiasis that has caused symptoms, it behooves everyone to carefully understand the future role of diet in the treatment of this disorder. The discussion concerning diet is related primarily to the cholesterol gallstone.

Burkitt and Trowell[11] have stressed that cholelithiasis is one of the degenerative diseases of Western society.[8] They have emphasized that the disease is little seen in underdeveloped countries and that the cause is related to the differences in diet between underdeveloped and developed societies.

There are ethnic and racial differences in gallstone formation. Blacks living in Western societies have as high an incidence of cholelithiasis formation as whites,[12] but they do not have a high incidence when living in Africa.[11] Groups such as the Pima Indians of Arizona have an incidence of 73% in females between the ages of 25 and 34.[13] Physiologic bile acid balance studies in these subjects revealed a difference in their ability to maintain cholesterol solubility that may not be due to diet, but rather to genetic variation.

Autopsy studies in Uganda and other underdeveloped areas have revealed an almost miniscule incidence of cholelithiasis. In one study of 4395 autopsies done between 1923 and 1955, there was not one documented gallstone, and in another survey there were only 27 small stones noted in a 5-yr period.[14,15] Similar observations have been reported by many surgeons and clinicians working in Africa. In rural areas in the Far East, there also appears to be a very low incidence of cholesterol gallstones.[16]

Of further interest is the fact that populations from underdeveloped areas that undergo so-called Westernization, which includes many social and economic factors, but also definite changes in diet, appear to show a markedly increased incidence of cholesterol-containing gallstones. This can be noted from observations in the southern United States, where it was rare to see a gallstone in the black population,[17] but now the incidence in blacks in northern cities is the same as that in whites.[12] Similarly, the previously very low incidence of cholesterol stones in Japanese subjects living in rural areas has evolved so that the incidence of cholesterol stones in the people of Japanese cities has increased, while pigment stones still remain high in rural areas.[16] All

of these facts suggest that cholesterol gallstones are as much due to environmental factors, such as dietary intake, as they are due to genetic factors. There
is no question that certain groups of peoples do have a greater chance of
developing cholesterol gallstones, such as the Pima Indians, although the epidemiologic evidence implicating the diet is very strong.

A diet survey that compared 91 cholelithiasis subjects and 86 control subjects revealed that women with gallstones had a significantly decreased intake
of crude fiber from bakery products.[18] This study was based on diet recall. A
more sophisticated research effort is needed to relate specific diet factors to
stone formation. The following discussion on lithogenic factors details what is
known about bile and the diet.

11.4.1.2. Lithogenic Bile and Gallstone Formation

The sequence of cholesterol gallstone formation is saturation of bile with
cholesterol, precipitation or crystallization of cholesterol in the bile, and
growth of a stone.

The saturation of bile with cholesterol depends on the *lithogenicity* of the
bile. Three factors are important: cholesterol, lecithin (phospholipid), and bile
salt content.[8,21]

The solubility of cholesterol in bile depends on the ratio of bile acids and
phospholipids to cholesterol. The lithogenic index refers to the saturation of
cholesterol in bile. When bile is supersaturated with cholesterol, the index is
greater than one, and when cholesterol is fully soluble in the bile, it is less than
one.[19,20] Figure 11-1 shows the triangular coordinates that graphically plot the
ratios of the three factors responsible for cholesterol saturation. Bile that
forms in the so-called metastable zone (shaded area) is subject to other specific
factors causing slow stone formation; bile that forms above the metastable
zone precipitates cholesterol rapidly.[21] Although stone formation occurs primarily in the gallbladder, it is readily apparent that hepatic bile can have as
much cholesterol as gallbladder bile. The liver has a primary role because
cholesterol, bile acid, or lecithin can be increased or decreased by liver production. However, the liver is influenced greatly by systemic metabolic factors as
well as by dietary intake.

In persons with a high percentage of cholesterol in the bile, resulting in a
lithogenic index greater than one, the bile becomes supersaturated with cholesterol, which precipitates and forms cholesterol stones. So-called cholesterol
stones are most often mixed, and stone formation will relate to other factors in
the bile such as calcium and pigments. When the bile is supersaturated with
cholesterol, cholesterol crystals form, and these crystals may precipitate on
bile pigments, protein matter, and foreign bodies such as bacteria or intestinal
contents that reflux into the biliary tree. Once they precipitate, the nidus is
formed. It may continue to grow as the factors that permitted the original
precipitation of cholesterol continue.[22,23]

Bile saturated with cholesterol of high lithogenic index occurs in many
subjects living in developed societies and is associated with risk factors such as

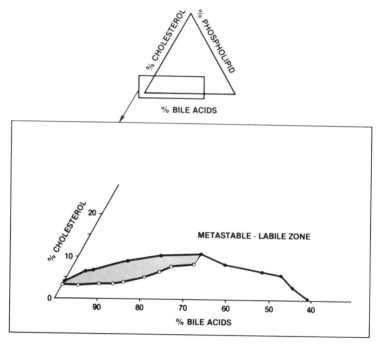

Figure 11-1. Cholesterol, phospholipid, and bile acid relative molar proportions plotted on a triangular coordinate. (●———●) Admirand and Small line; (o———o) Holzbach line.[20] Cross-hatching indicates metastable-labile zone (see text). Cholesterol crystallization occurs above top line.

obesity, high caloric intake, clofibrate therapy, chronic diseases causing bile acid malabsorption, and female sex hormones.[8] Of note is that lithogenic bile is not seen in subjects living in countries with low incidences of gallstones.[24,25]

11.4.1.3. Bile Acids and Foods

Bile acids are produced in the liver, as described in Chapter 1. Bile acid synthesis is controlled by a negative feedback mechanism. The rate of production by the liver varies inversely with the rate of return of bile acids to the liver in the enterohepatic circulation.

The size of the bile acid pool is important. Measurement of bile acid pools in subjects with gallstones has shown that most have a decreased total bile acid pool compared to subjects without stones.[25-28] The fact that the bile acid pool is small in female Indians, where the incidence is so high,[25] leads some observers to believe this represents a defect in the synthesis by the liver. Decreased pool size is noted in other systemic disease states where bile acids may be lost from the gut. The evidence clearly indicates that a decreased bile acid pool is associated with increased cholesterol gallstone formation. Of even more importance may be the bile acid flow rate.[8] At low flow rates, such as during fasting, hepatic bile may become saturated. Prolonged fasting is now clearly associated with increased attacks of acute gallbladder disease.

The diet and bile acid metabolism are interrelated, even if the significance of the relationship is not clear.[29] The definite relationships that exist between the bile acids, cholesterol, and phospholipids complicate the study of the effect of dietary factors on bile acid synthesis and pool size turnover. All three factors are so interrelated that it is difficult to study any one in relation to foods without the others.[29]

Animal experiments reveal diets low in fiber and high in simple sugars or starch cause a significant reduction in bile salt synthesis.[30,31] In studies conducted on Rhesus monkeys with chronic exteriorized enterohepatic circulations, safflower oil and triolein supplements increase bile salt synthesis and pool size but do not alter the relative proportions of all three factors.[12] The same workers found that medium-chain triglycerides actually decrease bile salt synthesis and pool size but do not alter the saturation of cholesterol, whereas fasting and a fat-free diet tend to decrease bile acid synthesis and pool size and increase the lithogenicity of the bile. These experiments clearly point out the importance of studying all three components of bile when evaluating dietary factors. The bile acid pool size may change, but its relationship to cholesterol and phospholipids is most important when studying gallstones. This has been further emphasized by studying the synthesis rates of bile acid and cholesterol in Rhesus monkeys.[33] Interruption of the flow of either factor to the liver has a marked effect on the synthesis rate. The circulation to the liver from absorbed dietary and enterohepatic circulating factors must exist and, consequently, play a primary role in keeping cholesterol soluble.[33] Different bile acids appear to have different inhibitory feedback effects. Deoxycholic acid inhibits cholesterol synthesis as well as absorption, while chenic does not.[34] Much more experimental data are needed in humans to fully understand these complex relationships.

Increased intake of fiber is associated in animal experiments and in humans with increased fecal bile salt excretion.[35-37] Increased excretion has been shown to occur with cellulose,[35] bengal gram,[38] maize and wheat,[39] and bagasse.[40] However, some of the evidence is controversial,[41] and more controlled studies on this subject are clearly needed.

Of interest are the experiments[42,43] of feeding bran to volunteers that reveal a reduction in the amount of deoxycholate in the bile. The circulating pool of deoxycholic acid was reduced and there was a concomitant increase in the chenic pool, which was probably the result of increased hepatic production of chenic acid to compensate for the decreased deoxycholic acid. This would have the same effect as feeding chenic acid. These studies suggest the interesting possibility that a high-fiber diet that would bind deoxycholic acid in the colon would have the same beneficial effect as chenic acid administration.

Eastwood and Boyd[44] have shown that significant quantities of bile acids are bound to nonabsorbable foods in the small intestine of the rat. Eastwood and Hamilton[45] found lignin to be a very potent binder, and other vegetable fibers can perform a similar function.[45] The greatest binding occurs in the colon, where deoxycholic acid is most abundant. Other investigators have shown that pectin,[46] alfalfa, sugar cane pulp, sugar beet pulp, bran, and a variety of foodstuffs are capable of binding bile salts under various physical conditions.[47-49] From these works it is apparent that a wide variety of fiber substances can bind bile salts within the intestine. It is postulated that the bile salt binding effect is related to the hypocholesterolemic effect and one could then postulate that this also is a mechanism for increasing cholesterol solubility in the bile.

11.4.1.4. Phospholipids

The phospholipid concentration in the bile is a second factor of great importance in maintaining cholesterol solubility. Bile acids appear to increase the synthesis and secretion of biliary phospholipids.[50,51] Lecithin, the major phospholipid in human bile secretion, is actually stimulated by bile salts. It is not increased by lecithin feeding.[51] However, experiments indicate that choline feeding may stimulate biliary lecithin synthesis and have a greater effect on phospholipid secretion than bile salts.[52]

11.4.1.5. Cholesterol

Bile acids have been shown to inhibit the hepatic synthesis of cholesterol by directly acting on hepatocytic cholesterol synthesis and indirectly by affecting intestinal absorption of cholesterol.[8,34,53] There can be no question that cholesterol metabolism is closely related to bile salt physiology. Anything that will affect bile salt physiology may well have an effect on cholesterol metabolism. It also appears that the reverse may be true—that anything that will affect cholesterol metabolism may have an effect on bile salt metabolism.

Although many studies have been performed on the effect of diet on serum cholesterol, little has been reported on the effect of diet on the cholesterol content of the bile. One study has shown increasing cholesterol intake results in saturation of the bile with cholesterol, but subjects tested were also eating a diet rich in sucrose and very low in fiber.[54] Conversely, it has also been shown that a low-cholesterol diet, which is primarily unsaturated fat, has seemed to increase gallstone formation.[55] However, all of the components of a high- or low-cholesterol diet are important in evaluating the effect on gallstone formation and cholesterol saturation of bile. As pointed out by Hofmann and associates,[56] in the patients who followed a low-saturated-fat diet that increased cholesterol saturation, the unsaturated fat did contain increased amounts of plant sterols. Further observations of the effect of chenic acid on gallstone patients placed on low-cholesterol diets, with or without high plant sterols, revealed that the low-cholesterol diet potentiated the effect of chenic acid in reducing the lithogenicity of bile, but a high-plant-sterol diet did not potentiate this effect.[57]

Studies correlating the effect of the general diet with cholesterol formation reveal that there is increased gallstone formation and lithogenesis of bile when there is increased caloric intake.[8,58,59] Patients with gallstones have a statistically significant increased intake of calories irrespective of dietary composition compared to patients who do not have gallstone formation.[58] These observations have further been emphasized by the fact that overnutrition causes a hypersecretion of cholesterol, which is probably explained by the fact that obese subjects have increased production of cholesterol by the liver.[60]

Although there are limited studies correlating cholesterol level in bile with dietary intake, there have been numerous studies correlating serum cholesterol levels with dietary intake. Keys and associates were the first to demonstrate that serum cholesterol levels were significantly lowered by increased ingestion of pectin.[61] This work has been confirmed by numerous investigators.[62-65] Other fiber substances have also been shown to decrease serum cholesterol. This is true for guar,[65] legumes,[67,68] oats,[67] beans,[62] and bran.[69] The results of bran are controversial, and many authors have reported that bran had no effect on serum cholesterol level.[70-73] The studies with bran and wheat vary tremendously in the amount of bran fed. Some researchers believe large amounts of bran are required to have an effect. Other substances such as bagasse, maize, and cellulose have also been shown not to have a lowering effect on serum cholesterol.[40,41,62,74] Munoz *et al.*[75] fed ten healthy men, aged 19 to 54, 26 g of soft white wheat bran, corn bran, soybean hulls, textured vegetable protein, hard red spring wheat, apple fiber, or carrot fiber for 28 to 30 days in a well-controlled study. They found red spring wheat lowered serum cholesterol and low-density lipoprotein cholesterol and raised high-density lipoprotein cholesterol. Corn bran also raised the percentage of high-density lipoprotein cholesterol. The other substances had little effect, except a lowering of plasma cholesterol by soybean hulls.[75]

All of these data indicate that there is an effect of diet on serum cholesterol, but the effect on the bile cholesterol is not yet clear. The diet may well have an effect on it, and the effect might be either positive or negative, depending on the dietary intake. Studies employing chenic acid to dissolve gallstones[76] in association with dietary factors have also indicated that the effect of chenic might be enhanced by dietary factors such as a low-cholesterol diet[57] or the addition of lecithin when cholic acid is used.[77]

11.4.2. Treatment

The accepted treatment is surgical removal of gallstones. However, there are situations in which the patient is either not a good candidate for surgery or refuses surgery. In those instances, medical treatment is indicated.

At the present time, diet therapy for gallstones is a very active research field. Although there are definite implications that gallbladder bile can be altered by the diet, there have been no controlled studies or protocols to demonstrate that diet has a significant effect on gallstones (as already discussed). However, these should be forthcoming as initial evaluations do reveal that dietary factors have an influence on the three components of bile associated with maintaining cholesterol solubility, that is, cholesterol itself, bile acids, and phospholipids.

The most exciting advance has been the use of oral administration of chenodeoxycholic acid (chenic) to dissolve gallstones.[76] From most evaluations it is apparent that a dose of chenic acid of about 15 mg/kg/day or more has an excellent chance of causing dissolution of gallstones if it can be tolerated. Diarrhea or side effects may be a problem, necessitating a decrease in dosage. There appears to be no definite liver toxicity from long-term use of this bile acid, as was seen with the use of other bile acids.[21,76-80] Use of the drug decreases the concentration of cholesterol in the bile, while there is a concomitant increase in biliary chenic acid with no evidence of any unusual bile acid formation. The major factor to influence the response, providing the patient can take an adequate dose, is the type of gallstone being treated. Common duct stones are as responsive to dissolution as gallbladder stones.[78,79] An epimer of chenic acid, ursodeoxycholic acid, also appears capable of causing dissolution of cholesterol gallstones when administered orally. This new drug needs further testing to determine its long-range effects.[81]

The Food and Drug Administration has still not cleared chenic acid for use in the dissolution of gallstones, although it has been cleared in 25 countries throughout the world. The results of a national cooperative gallstone study in the United States are awaited.

Of interest are reports that dietary factors may moderate oral bile acid (chenic) administration. In one study dietary fiber, in conjunction with cholestyramine, decreased the lithogenicity of bile.[82] In another, lecithin, when added to cholic acid, helped the dissolution of gallstones.[77] Other investigators have also found that addition of β-sitosterol to chenic acid may be effective in the treatment of cholesterol gallstones. Three grams of -sitosterol was added to the dosage of chenic acid in a double-blind study, and complete dissolution of stones was higher in the group that took the sitosterol.[83] Begemann and colleagues[84] have also shown that β-sitosterol can decrease biliary cholesterol saturation, but some of their experimental subjects were postcholecystectomy. These studies are somewhat in conflict with a previous study using plant sterols.[57] This field requires more research so that one can be sure of the effect of plant sterols alone as well as in conjunction with chenic acid therapy.

It is important to stress to patients either prone to forming cholesterol stones or who have cholesterol stones that both fasting[84] and certain drugs such as clofibrate[86] increase the lithogenicity of bile.[85] It is also important to remember that patients on a strict low-cholesterol dietary regimen for the treatment of complications of atherosclerosis had a higher incidence of gallstones.[55]

In summary, one can state that the medical treatment of gallstones is not clear. Some clinicians feel strongly that there is no role, as yet, for diet, and that the clinician should merely advise avoidance of indiscretion. There are others who feel strongly that a low-cholesterol, high-fiber diet will help reduce gallstone formation. And, finally, early results of treatment with chenic acid indicate that the lithogenicity of bile can be decreased with the administration of this naturally occurring bile acid (see Chapter 20, diet 26, for low-cholesterol, low-saturated-fat, high-fiber diets).

REFERENCES

1. Editorial: Treatment of acute cholecystitis. *Lancet* 1:182, 1976.
2. Koch JP, Donaldson RM: A survey of food intolerances in hospital patients. *N Engl J Med* 271:657, 1964.
3. Hinkle CL, Moller GA: Correlation of symptoms, age, sex and habitus with cholecystographic findings in 1000 consecutive examinations. *Gastroenterology* 32:807, 1957.
4. Price WH: Gallbladder dyspepsia. *Br Med J* 2:138, 1963.
5. Taggert D, Billington BP: Fatty foods and dyspepsia. *Lancet* 2:464, 1964.
6. Bockus HL: *Gastroenterology,* ed 3. Philadelphia: WB Saunders Co, 1976, Vol III, p 777.
7. Spiro HM: *Clinical Gastroenterology,* ed 2. New York: Macmillan, 1977.
8. Bennion LJ, Grundy SM: Risk factors for the development of cholelithiasis in man. *N Engl J Med* 299:1161, 1221, 1978.
9. Sutor DJ, Wooley SE: A statistical survey of the composition of gallstones in eight countries. *Gut* 12:55, 1971.
10. Glenn F: Calculous biliary tract disease: An increasing burden on surgical facilities. *Ann Surg* 171:163, 1970.
11. Burkitt DP, Trowell HC: *Refined Carbohydrate Foods and Disease.* New York: Academic Press, 1975.
12. Trotman BW, Soloway RD: Influence of age, race and sex on pigment and cholesterol gallstone incidences. *Gastroenterology* 65:573, 1973.
13. Sampliner RE, Bennett PH, Comess LJ, *et al:* Gallbladder disease in Pima Indians. *N Engl J Med* 283:1358, 1970.
14. Parnis RO: Gallbladder disease in Nigeria. A five-year review. *Trans R Soc Trop Med Hyg* 58:437, 1964.
15. Edington GM: Observations on hepatic disease in the Gold Coast with special references to cirrhosis. *Trans R Soc Trop Med Hyg* 51:48, 1957.
16. Maki T: Cholelithiasis in the Japanese. *Arch Surg* 82:599.
17. Cunningham JA, Hardenbergh FE: Comparative incidence of cholelithiasis in the Negro and white races. *Arch Intern Med* 97:78, 1956.
18. Smith DA, Gee MI: A dietary survey to determine the relationship between diet and cholelithiasis. *Am J Clin Nutr* 32:1519, 1979.
19. Metzger AL, Heymsfield S, Grundy SM: The lithogenic index—a numerical expression for the relative lithogenicity of bile. *Gastroenterology* 62:499, 1972.
20. Holzbach RT, Marsh M, Olszewski M, *et al:* Cholesterol solubility in bile: Evidence that supersaturated bile is frequent in healthy man. *J Clin Invest* 52:1467, 1973.
21. Goldstein LI, Schonfeld LJ: Gallstones: Pathogenesis and medical treatment, in Stollerman E (ed): *Advances in Intestinal Medicine.* Chicago: Year Book Medical Publishers, 1975, p 89.
22. Redinger R, Small DM: Bile composition of bile salt metabolism and gallstones. *Arch Intern Med* 130:618, 1972.
23. Boochier IAD, Copperband SR, el Kodsi BM: Mucous substances and viscosity of normal and pathological bile. *Gastroenterology* 49:343, 1965.
24. Biss K, Ho KJ, Mikkelson B, *et al:* Some unique biologic characteristics of the Masi East Africa. *N Engl J Med* 284:694, 1971.
25. Northfield TC, Hofmann AF: Biliary lipid secretion in gallstone patients. *Lancet* 1:747, 1973.
26. Vlahcevic ZR, Bell CC, Gregory DH, *et al:* Relationship of bile acid pool size to the formation of lithogenic bile in female Indians of the southwest. *Gastroenterology* 62:73, 1972.
27. Swell L, Bell CC, Bohac I, *et al:* Relationship of bile acid pool size to biliary lipid excretion and the function of lithogenic bile in man. *Gastroenterology* 61:716, 1971.
28. Vlahcevic ZR, Bell CC, Bohac I, *et al:* Diminished bile acid pool size in patients with gallstones. *Gastroenterology* 59:165, 1970.
29. Low-Beer TS: Diet and bile acid metabolism. *Clin Gastroenterol* 6:165, 1977.
30. Portman OW, Murphy D: Excretion of bile acids and β-hydroxysterols by rats. *Arch Biochem Biophys* 76:367, 1958.

31. Hellstrom K, Sjovall J, Wigand G: Influence of semisynthetic diet on type of fat, on the turnover of deoxycholic acid in the rabbit. *J Lipid Res* 3:405, 1962.
32. Redinger RN, Herman AH, Small DM: Primate biliary physiology. X. Effects of diet and fasting on biliary lipid secretion and relative composition and bile salt metabolism in the rhesus monkey. *Gastroenterology* 64:610, 1973.
33. Strasberg SM, Petrunka CN, Ilson RD: Effect of bile acid synthesis rate on cholesterol secretion rate in the steady state. *Gastroenterology* 71:1067, 1976.
34. Gallo-Torres HE, Miller ON, Hamilton JG: Some effects of deoxycholate administration on the metabolism of cholesterol in man. *Am J Clin Nutr* 32:1363, 1979.
35. Shurpalekar KS, Doraiswamy TR, Sundarvalli OE, *et al:* Effect of inclusion of cellulose in an "atherogenic" diet on the blood lipids of children. *Nature (London)* 232:554, 1971.
36. Stanley MM, Paul D, Gacke D, *et al:* Effects of cholestyramine, Metamucil, and cellulose on fecal bile acid excretion in man. *Gastroenterology* 65:889, 1973.
37. Pomare EW, Heaton KW, Low-Beer TS, *et al:* The effect of wheat bran upon bile salt metabolism and upon the lipid composition of bile in gallstone patients. *Am J Dig Dis* 21:251, 1976.
38. Mathur KS, Khan MA, Sharma RD: Hypocholesterolaemic effect of bengal gram: A long-term study in man. *Br Med J* ii:30, 1968.
39. Antonis A, Bersohn I: The influence of diet on fecal lipids in South African white and Bantu prisoners. *Am J Clin Nutr* 11:142, 1962.
40. Walters RL, Baird RM, Davies PS, *et al:* Effects of two types of dietary fiber on fecal steroid and lipid excretion. *Br Med J* iii:536, 1975.
41. Eastwood MA, Kirkpatrick JR, Mitchell WD, *et al:* Effects of dietary supplements of wheat bran and cellulose on feces and bowel function. *Br Med J* iv:392, 1973.
42. Pomare EW, Heaton KW: Bile salt metabolism in patients with gallstones in functioning gallbladders. *Gut* 14:885, 1973.
43. Pomare EW, Heaton KW: Alteration of bile salt metabolism by dietary fiber (bran). *Br Med J* 4:262, 1973.
44. Eastwood MA, Boyd GS: The distribution of bile salts along the small intestine of rats. *Biochim Biophys Acta* 137:393, 1967.
45. Eastwood MA, Hamilton D: Studies on the absorption of bile salts to nonabsorbed components of the diet. *Biochim Biophys Acta* 152:165, 1968.
46. Leveille GA, Sauberlich HE: Mechanism of the cholesterol-depressing effect of pectin in the cholesterol-fed rat. *J Nutr* 88:209, 1966.
47. Story JA, Kritchevsky D: Binding of bile salts by nonnutritive fiber. *Fed Proc Fed Am Soc Exp Biol* 33:663, 1974.
48. Kritchevsky D, Story JA: Binding of bile salts *in vitro* by nonnutritive fiber. *J Nutr* 104:458, 1974.
49. Story JA, Kritchevsky D: Binding of sodium taurocholate by various foodstuffs. *Nutr Rep Int* 11:161, 1975.
50. Entenman C, Holloway RJ, Albright ML, *et al:* Bile acids and lipid metabolism. I. Stimulation of bile lipid excretion by various bile acids. *Proc Soc Exp Biol Med* 127:1003, 1968.
51. Nilsson S, Schersten T: Importance of bile acids for phospholipid secretion into human hepatic bile. *Gastroenterol* 57:525, 1969.
52. Robins SJ, Armstrong MJ: Biliary lecithin secretion. II. Effects of dietary choline and biliary lecithin synthesis. *Gastroenterology* 70:397, 1976.
53. Webster KH, Lancaster MC, Hofmann AF, *et al:* Influence of primary bile feeding on cholesterol metabolism and hepatic function in the rhesus monkey. *Mayo Clin Proc* 50:134, 1975.
54. DenBesten L, Connor WE, Bell S: The effect of dietary cholesterol on the composition of human bile. *Surgery* 73:266, 1973.
55. Sturdevant RAL, Pearce ML, Dayton S: Increased prevalence of cholelithiasis in men ingesting a serum-cholesterol-lowering diet. *N Engl J Med* 288:24, 1973.
56. Hofmann AF, Northfield TC, Thistle JL: Can a cholesterol-lowering diet cause gallstones? *N Engl J Med* 288:46, 1973.
57. Maudgal DP, Bird R, Enyobi VO, *et al:* Chenic acid in gallstone patients: Effect of low cholesterol and of high plant sterol diets. *Gut* 18:419, 1977.

58. Sarles H, Pommeau CCY, Save E, *et al:* Diet and cholesterol gallstones. *Am J Dig Dis* 14:531, 1969.
59. Sarles H, Hauton J, Plance NE, *et al:* Diet, cholesterol gallstones, and composition of the bile. *Am J Dig Dis* 15:251, 1970.
60. Grundy SM, Duane WC, Adler RD, *et al:* Biliary lipid outputs in young women with cholesterol gallstones. *Metabolism* 23:67, 1974.
61. Keys A, Grande F, Anderson JT: Fiber and pectin in the diet and serum cholesterol in man. *Proc Soc Exp Biol Med* 106:555, 1961.
62. Grande F, Anderson JT, Keys A: Effect of carbohydrates of leguminous seeds, wheat and potatoes on serum cholesterol concentration in man. *J Nutr* 86:313, 1965.
63. Palmer GH, Dixon DG: Effect of pectin dose on serum cholesterol levels. *Am J Clin Nutr* 18:437, 1966.
64. Jenkins DJA, Leeds AR, Newton C, *et al:* Effect of pectin, guar gum and wheat fiber on serum cholesterol. *Lancet* 1:1116, 1975.
65. Durrington PN, Manning AP, Bolton CH, *et al:* Effect of pectin on serum lipids and lipoproteins, whole-gut transit-time, and stool weight. *Lancet* 2:394, 1976.
66. Degroot AP, Luyken R, Pikaar A: Cholesterol-lowering effect of rolled oats. *Lancet* 2:303, 1963.
67. Keys A, Anderson JT, Grande F: Diet-type (fats constant) and blood lipids in man. *J Nutr* 70:257, 1960.
68. Luyken R, Pikaar A, Polman H, *et al:* The influence of legumes on the serum cholesterol level. *Voeding* 23:447, 1962.
69. Persson I, Raby K, Fonns-Bech P, *et al:* Bran and blood-lipids. *Lancet* 2:1208, 1975.
70. Durrington P, Wicks ACB, Heaton KH: Effect of bran on blood-lipids. *Lancet* 2:133, 1975.
71. Connell AM, Smith CL, Somsel M: Absence of effect of bran on blood-lipids. *Lancet* 1:496, 1975.
72. Jenkins DJA, Hill S, Cummings JH: Effect of wheat fiber on blood lipids, fecal steroid excretion and serum iron. *Am J Clin Nutr* 28:1408, 1975.
73. Truswell AS, Kay AM: Absence of effect of bran on blood-lipids. *Lancet* 1:922, 1975.
74. Antoni A, Bersohn I: The influence of diet on serum lipids in South African white and Bantu prisoners. *Am J Clin Nutr* 10:484, 1962.
75. Munoz JM, Sanstead HH, Jacob RA: Effects of dietary fiber on plasma lipids of normal men. *Clin Res* 26:584, 1978.
76. Editorial: Chenodeoxycholic acid. *Lancet* 1:805, 1978.
77. Toouli J, Jablonski P, Watts MM: Gallstone dissolution in man using cholic acid and lecithin. *Lancet* 1:1124, 1975.
78. Thistle JL, Hofmann AF, Ott BJ, *et al:* Chemotherapy for gallstone dissolution. I. Efficacy and safety. *JAMA* 239:1041, 1978.
79. Hofmann AF, Thistle JL, Klein PD, *et al:* Chemotherapy for gallstone dissolution. II. Induced changes in bile composition and gallstone response. *JAMA* 239:1138, 1978.
80. Iser JH, Dowling RH, Mok HYI, *et al:* Chenodeoxycholic acid treatment of gallstones. *N Engl J Med* 293:378, 1975.
81. Nakagawa S, Makino I, Ishizaki T, *et al:* Dissolution of cholesterol gallstones by ursodeoxycholic acid. *Lancet* 1:367, 1977.
82. Hall RC, Klauda HC, Waite VM, *et al:* Improved lithogenicity of fasting hepatic bile by dietary fiber and cholestyramine: Studies in mice and humans. *Surg Forum* 26:435, 1975.
83. Gerolami A, Sarles H: β-Sitosterol and chenodeoxycholic acid in the treatment of cholesterol gallstones. *Lancet* 1:721, 1975.
84. Begemann F, Bandomer G, Herget HJ: The influence of β-sitosterol on biliary cholesterol saturation and bile acid kinetics in man. *Scand J Gastroenterol* 13:57, 1978.
85. Soloway RD, Schoenfield LJ: Effects of meals and interruption of enterohepatic circulation of flow, lipid composition, and cholesterol saturation of bile in man after cholecystectomy. *Am J Dig Dis* 20:99, 1975.
86. Thistle JL, Schoenfield LJ: Induced alterations in composition of bile for persons having cholelithiasis. *Gastroenterology* 61:488, 1971.

Pancreatic Diseases

Although the pancreas is the major site of production of enzymes necessary for digestion of fats and proteins, information concerning its role in human nutrition has gathered slowly. The nutritionist, physiologist, and gastroenterologist have all been intrigued by its role, but because of its anatomic position it has remained an organ difficult to study.

Specific social and economic problems weigh heavily in pancreatic disease. Examples are the excessive use of alcohol in Europe and the United States and malnutrition in Asia. These factors are related to the incidence of acute and chronic pancreatic disease and their resultant morbidity and mortality.[1]

12.1. ACUTE PANCREATITIS

12.1.1. Etiology

The most common predisposing factors in highly developed societies are biliary tract disease and alcoholism. Biliary tract disease is claimed as an antecedent factor in anywhere from 10 to 70% of the cases and alcoholism anywhere from 0 to 75%. Gallstones are the most commonly associated biliary tract disease. Their presence always raises the question of transient pancreatic duct obstruction. Depending on the area of the world, approximately one-third of the pancreatitis cases are associated with the following factors: trauma; drugs, such as corticosteroids, furosemide, thiazides, azothiaprine, and estrogens; viral infections; vasculitis and ischemia; pregnancy; types I and V hyperlipoproteinemias; heredity factors; and a variety of endocrine and metabolic disorders.[2]

12.1.2. Pathophysiology

The exact mechanism by which these associated factors initiate acute inflammation of the pancreas is not clear. However, the pathology is thought to occur because of the activation of proteolytic enzymes of the pancreas that escape into surrounding pancreatic tissues and the pancreas itself. Freed trypsin activates other proenzymes. Lecithinase (phospholipase A) is capable of converting lecithin to lysolecithin, which is extremely toxic to tissues. Another enzyme, elastase, is capable of digesting connecting tissue. Many of the substances increased in pancreatitis, such as histamine, kinins, kallikrein, and a myocardial depressant factor, have all been demonstrated in experimental studies to cause systemic cardiovascular effects.[3] Large amounts of fluid, particularly plasma, may escape into the peritoneal and retroperitoneal tissues. There is resultant systemic shock and fat necrosis of tissues within and around the pancreas. This explains the severe morbidity and the possible mortality. Associated medical complications may be either renal or respiratory failure. The most severe cases of pancreatitis are often labeled hemorrhagic or necrotic. Of prognostic significance is the rapid fall in serum calcium, which is due to the formation of calcium salts as well as to a metabolic disturbance of calcium metabolism. A rapid fall is associated with a poorer prognosis.

12.1.3. Diagnosis

The diagnosis of pancreatitis is usually easy to establish when there is severe abdominal pain associated with elevation of pancreatic enzymes. When shock is present, the prognosis is guarded. When only pain is present, it is helpful to have early elevation of serum amylase, followed by a later elevation of serum lipase. Mild pancreatic pain can be epigastric and usually radiates under the ribs or through to the back. The urine amylase may be of help when the blood levels are not elevated. Amylase may be cleared more rapidly through the kidney during disease.[4,5] Often noted during the acute disease are hyperglycemia and/or hyperlipemia. These are usually transient phenomena that are resolved once the disease subsides.

12.1.4. Treatment

In the mild form of the disease, characterized by pain without severe vomiting and no evidence of a shocklike state, the patient can be treated with analgesics, bed rest, a fat-free diet, and antacids.

However, when there is any evidence of vomiting and pain is severe, with the threat of shock, the following therapy outline should be considered.

1. There must be absolute relief of pain by the use of narcotics and, if necessary, nerve block.

2. If shock is at all present, there must be adequate replacement of fluids intravenously. This may require large amounts of fluids. If it is not readily controlled by fluid replacement, there must be consideration for peritoneal dialysis to remove free enzymes and vasoactive substances.

3. The use of nasogastric suction is controversial.[6] *However,* most clinicians use it with effectiveness. This author particularly believes in nasogastric suction because it removes all gastric and acid secretions from the stomach, which are the primary stimulants of pancreatic secretion in the fasting patient.

4. Administration of drugs, e.g., anticholinergics, cimetidine (H_2 inhibitor), should be considered to reduce gastric and pancreatic secretions.

5. Controversy exists concerning the use of antibiotics, corticosteroids, trasylol (aprotinin), and glucagon.[7-9] In scattered reports, all of these yielded varied results. Some authors approve of their use, and others disapprove.

6. In our own experience, peritoneal dialysis has been extremely helpful in the acute case. This therapy is now available in most hospitals with excellent expertise. The rationale for its use is that it removes the vasoactive substances and enzymes that are liberated in the peritoneal cavity.[10] Of course, peritoneal dialysis does not get into the retroperitoneal space and, consequently, is of no definite help in that area. In our limited experience, this has been of some assistance.

7. Finally, there have been scattered reports that surgical intervention can be of help, but this entire field still remains controversial.[11,12]

Dietary Therapy. The patient without vomiting should first be observed; if there is no sign of significant temperature elevation, no recurrent pain, and no significant abdominal abnormalities, then fat-free food intake may be permitted. (Fat-free diets are outlined in Chapter 20, diets 20 and 21.) A liquid fat-free diet is prescribed for the first day, followed by progression to more semisolid foods.

If the patient is vomiting, nasogastric suction may be started and no food intake allowed until signs of the active disease have subsided.

A new approach to the management of severe pancreatitis has been nutritional support via oral or intravenous amino acid and glucose feeding. When the disease is protracted, Blackburn and colleagues[13] have found this form of nutritional support helpful in maintaining and bringing the patient through the critical stages of the illness. This is a more aggressive approach to sustaining nutrition in acutely ill patients who may have a protracted course of the disease. These authors found that by sustaining the patient on intravenous protein and glucose (Chapter 19), they could carry the patient through the protracted stage of the illness, resulting in a better candidate for surgical correction of such complications as abscess or pseudocyst. Controlled studies are needed on the effectiveness of this form of therapy.

12.2. CHRONIC PANCREATITIS AND PANCREATIC INSUFFICIENCY

12.2.1. Etiology

As with acute pancreatitis, the etiologies of the chronic disease are multiple. Debate exists concerning the correlation of the variety of pathologic abnormalities[14] and the variety of clinical forms of the chronic disease.[15] There may be repeated episodes of the disease that simulate repeated acute attacks. Unless a pathology is available, some feel this is merely acute recurrent pancreatitis. When pathology specimens are available, it is possible to demonstrate chronic disease of the organ, which may or may not result in calcification.[14] There are forms of painless pancreatitis as well.[15]

The most common etiologic association in Europe and America is with alcoholic disease. Gallstone-related pancreatitis usually tends to display normal histology of the pancreas and, consequently, no true chronic form of the disease.[15] The overwhelming number of cases are associated with chronic alcohol intake and nutritional and metabolic diseases.

12.2.2. Clinical Picture

The presentation of chronic pancreatitis is most often either recurrent abdominal pain or the consequences of pancreatic endocrine or exocrine insufficiency.[14] The type with recurrent pain may be debilitating and cause addiction to narcotics, or result in surgery for the destruction of nerves leading to the pancreas. Pancreatic insufficiency may present as diabetes or steatorrhea, or both. Less frequently, chronic pancreatitis may present with jaundice or bleeding.

The diagnosis is often difficult. Endoscopic retrograde cannulation may demonstrate abnormal ducts with saccular formation. In a patient with the clinical symptoms, this is excellent evidence of the disease. The diagnosis can also be readily established when this cannot be demonstrated if there is evidence of pancreatic insufficiency and a normal small-bowel mucosa. The gastroenterologist and clinician usually have no difficulty in making the diagnosis in a serious case. The patient with chronic pain and no signs of insufficiency or debilitation can be a diagnostic dilemma that taxes the clinician's acumen. However, the nutritional aspects of the disease are most important in chronic pancreatitis that results in pancreatic insufficiency, and this usually is easily diagnosed.

Other tests employed with varying success in different institutions are secretin stimulation with collection of duodenal materials and the so-called Lundh test meal. I have found both to be of relatively little use, but many institutions profess their accuracy.[15] Naturally, once there is evidence of calcification of the pancreas, this is absolute evidence that the pancreas has undergone damage.

12.2.3. Pathophysiology

Because the pancreas is the major site of production of enzymes for the digestion of fat, protein, and carbohydrate, injury to the pancreas results in decreased production of enzymes, diminished digestion and absorption, and finally malabsorption.

The endocrine functions of the pancreas are also decreased, and glucose intolerance ensues with resultant mild diabetes. The diabetes is often very mild and does not require supplementation with insulin. The functions of the exocrine pancreas are discussed in detail in Chapter 1.

Diabetes occurs in approximately 25% of all patients with pancreatitis and in 75% when calcification is evident.[16] The degree of fat and nitrogen loss with chronic pancreatitis is less clear. However, once it develops, it is often dramatic. The steatorrhea in these patients may be mild or massive. As much as 30 to 50 g of fat is readily seen in a 24-hr stool.

Vitamin B_{12} malabsorption is also recorded in chronic pancreatic insufficiency.[17-19] Studies have indicated that a factor elaborated by the pancreas assists in normal vitamin B_{12} absorption. However, it is rare to see severe B_{12} deficiency or megaloblastic anemia associated with pancreatic disease.[20,21]

12.2.4. Treatment

It is essential that any known cause of pancreatitis be removed or else the disease will progress and therapy will become impossible. Of particular note is the use of alcohol. Some patients are susceptible to small amounts, inducing progression of the disease. However, with others, the severity of the disease is related to the amount of intake.[22] The best approach is to attack the cause, and the removal of alcohol will be a major factor in permitting adequate treatment and nutrition.

The treatment of pancreatic *diabetes* is usually simple. Insulin therapy is most often not required, for most patients show no evidence of ketosis or angiopathy. However, if severe hyperglycemia persists, oral hypoglycemic agents may be of help. The rare patient will require insulin replacement. The use of dietary management for diabetes will be discussed separately under that topic below.

12.2.4.1. Enzyme Replacement

Table 12-1 lists the various pancreatic enzyme products available and the manufacturer's recommended dose. We have found it most useful to administer either the capsules, granules, or tablets with the meal and with all snacks. At least two tablets are usually required with each meal and with snacks. It is not unusual for patients to require three or four tablets with moderate-size meals in order to increase their efficiency of absorption. This is a direct addition of enzymes to those secreted into the duodenum, increasing the level of active digestion in order to assist in absorption. Success in the use of replacement enzyme therapy depends on the cooperation of the patient. Several authors have proposed hourly replacement therapy; others readily find success when the enzymes are given with meals and snacks.[20]

Table 12-2 shows the results of a careful analysis of all of the pancreatic extracts on the market. It is seen that some supplements have far greater activity than others, and that two have virtually no amylase. I have found Cotazym® most reliable, and apparently it does have twice as much enzyme activity as the other preparations.[21]

There has been a report that the addition of cimetidine to the regimen increases the effectiveness of enzyme replacement, and consequently de-

Table 12-1. Some Common Pancreatic Preparations[a]

Product and manufacturer	Composition	Preparation	Manufacturer's recommended dose
Pancreatin N.F.	Whole hog pancreatic extract		
Pancreatin (Lilly)	Triple strength, Pancreatin N.F.	Tablet, 325 mg[b]	1–2 with meals
Panteric (Parke Davis)	Triple strength, Pancreatin N.F.	Capsule, 325 mg[b] Tablet, 325 mg[b] Granules[b]	1–2 after each meal 1–2 after each meal 1 tsp after each meal
Ilozyme (Warren Teed)	Quadruple strength, Pancreatin N.F.	Tablet, 400 mg[c]	1–3 with meals
Viokase (Viobin)	Quadruple strength, Pancreatin N.F.	Tablets, 325 mg[c] Powder, 1/2 tsp=750 mg	1–3 with meals 1/2 tsp with meals
Pancrealipase	Whole hog pancreatic extract, lipase enriched		
Cotazyme (Organon)	CaCO₃, 50 mg added to preparation	Capsule, 300 mg Packet, 800 mg	1–3 prior to each meal 1–2 packets prior to each meal

[a] Adapted from Taubin and Spiro.[20]
[b] Enteric coated.
[c] Not enteric coated.

creases the amount of steatorrhea.[23] The rationale for this therapy is that a decrease in acid production by the stomach will permit better pancreatic enzyme replacement without destruction of the enzymes.

12.2.4.2. Diet Therapy

When diabetes is a problem, simple sugars should be restricted, as in any other diabetic regimen. However, in the absence of diabetes, simple sugars can be used readily in the diet as a good source of calories. Weight loss is prevalent in most patients with chronic pancreatitis; consequently, good sources of calories are urgently needed. The problem of carbohydrate intake also exists with fat intake. Fats are a good source of calories, but large amounts can increase pancreatic activity in an already sick gland and potentially increase any active inflammation. Furthermore, increased fat intake with marginal fat absorption will increase steatorrhea. The increased amounts of undigested fats in the stool may result in diarrhea and perhaps aggravate absorption of calcium and other electrolytes.

Proteins appear to be handled more easily by the patient with pancreatic insufficiency. It is also important to note that when there is poor protein intake and protein malnutrition, the mucosa of the small intestine may be negatively

Table 12-2. Enzyme Analysis of Pancreatic Supplements[a]

Drug	Manufacturer	Form	Tablet or capsule weight (mg)	Protease[b]	Trypsin[c]	Activity per tablet or capsule		
						Trypsinogen[d]	Amylase[e]	Lipase[f]
Accelerase	Oraganon Inc.	Capsule	372±20.0	70	10,300	2,530	19,850	2,600
Convertin	B.F. Ascher & Co. Inc.	Tablet	777±46.4	18	2,210	1,090	3	78
Cotazym	Organon Inc.	Capsule	487±30.0	214	63,800	14,300	60,470	4,630
Dactilase	Lakeside Laboratories Inc.	Tablet	583±14.1	39	17,000	0	5,810	501
Enzypan	Norgine Laboratories, Inc.	Tablet	402±10.9	179	4,630	887	6,640	830
Festal	Hoechst Pharmaceuticals, Inc.	Tablet	578±28.2	36	289	16,600	15,290	1,270
Kanulase	Dorsey Laboratories	Tablet	801± 6.0	19	96	32	10	36
KU-zyme	Kremers–Urban Co.	Capsule	554± 8.9	40	11	349	1,055	94
Lipan	Spirt & Co., Inc.	Tablet	540±30.2	120	7,400	324	27,510	2,380
Pancreatin	Eli Lilly & Co.	Tablet	352± 8.4	26	3,560	197	5,140	507
Panteric	Parke Davis & Co.	Tablet	769±19.4	85	2,460	1,770	17,190	1,620
Pro-gestive	Nutrition Control Products	Tablet	339±10.1	46			15,900	1,220
Uticon	Layton Laboratories	Tablet	872±20.0	70	14,200	2,440	14,720	680
Viokase	Viobin Corp.	Tablet	397± 7.5	112	2,440	3,140	25,590	2,230

[a] Adapted from Pairent and Howard.[21]
[b] Micromoles of (5%) TCA-soluble tyrosine equivalents per minute, pH 7.5, 37°C (hemoglobin substrate).
[c] Increase in 410-nm absorbance per 15 minutes, pH 8.1, 37°C (benzoylarginyl-p-nitroanilide substrate).
[d] Additional increase in 410-nm absorbance per 15 minutes, pH 8.1, 37°C (benzoylarginyl-p-nitroanilide substrate), after 30 min, 37°C incubation of sample with enteropeptidase.
[e] Grams of glucose equivalents per 30 minutes, pH 7.2, 37°C (Somogyi units × 1000).
[f] Milliliters of 0.050 N sodium hydroxide required to neutralize acid formed during 18-hr incubation at 37°C (Cherry–Crandall method).

affected and further absorptive inefficiency may develop.[24,25] Therefore, it is important to replace protein stores as early as possible.

It is necessary to deliver as many calories as possible to increase weight and ultimately to maintain adequate energy and weight. At the present state of our knowledge, the diet indicated is one of high protein and high carbohydrate if there is no evidence of diabetes, and one of high complex carbohydrate if there is evidence of diabetes, and fat maintained at a moderate level (see Chapter 20, diets 23 and 26).

Note should be taken of the use of oral elemental and parenteral alimentation. Patients with pancreatic insufficiency have benefited from the addition of such therapy to the diet, particularly patients with complications of pancreatitis, such as infected pseudocysts or fistula (see Chapters 18 and 19 on elemental diets and hyperalimentation).[26-29] Elemental diets can deliver large amounts of protein to the gut. These may be particularly helpful early in the treatment of chronic pancreatic insufficiency. Further studies are needed in this area.

The use of medium-chain triglycerides is also of importance. These triglycerides are easily absorbed, will not add to the steatorrhea or help cause diarrhea, as occurs with ordinary triglycerides, and can be an added source of energy and calories for the patient with pancreatic insufficiency (see Chapter 1).[30] Long-term use of medium-chain triglycerides is sometimes difficult for patients. In a cooperative patient, it can be accomplished when they are made palatable (see Chapter 20, diet 7).

12.3. CARCINOMA OF THE PANCREAS

Carcinoma of the pancreas has become the second most common gastrointestinal cancer of Western societies, but it remains rare in underdeveloped countries. The cause for its rapid increase in highly industrialized countries is not clear.

The disease may progress rapidly after onset and only in rare cases is there a surgical cure. Chemotherapy to date has offered little prolongation of life. Relief of pain and malnutrition is most important to make the patient comfortable.

Diet treatment of carcinoma of the pancreas is limited to pancreatic insufficiency and is the same as described in Section 12.2.4.2. Proper nutrition with home hyperalimentation[29] is now undergoing trials in patients with incurable cancer. It is hoped that this will make their limited time more comfortable and perhaps enable them to function at an acceptable level.

12.4. CYSTIC FIBROSIS

Cystic fibrosis is one of the most common gastrointestinal disorders of infants. With our increased ability to treat the disorder, children now survive and live to adulthood. It is an autosomal recessive disorder that results in abnormal functioning of exocrine glands. The disturbance in glandular secretion may affect a multiplicity of organs including the pancreas, lungs, and sweat glands. Its incidence in whites is approximately one in 1500 live births. It is less frequent in blacks.[31]

12.4.1. Pathophysiology

The cause of the disorder is still unknown despite much research on the effect of abnormal biochemical synthesis and resultant disturbances of glandular secretions. It is obvious that multiple glands are affected. Of interest are the observations by several groups of a disarrangement of fatty acid metabolism and related altered prostaglandin production.[33,34] This mechanism of the disease process is suggested by the fact that growth patterns of children significantly improved with intravenous administration of essential fatty acids, in the form of Intralipid®.[33,35]

Patients with cystic fibrosis have very low levels of linoleic and arachidonic acid in their blood lipids. Linoleic acid is metabolized to a by-product, which is then oxidized to a prostaglandin. It is suggested that the deficiency in prostaglandins could be related to an increased sensitivity to cyclic AMP, which is seen in cystic fibrosis patients. Similarly, it is suggested that prostaglandin deficiency could explain alterations of skin electrolyte secretion. What is not clearly understood is whether essential fatty acid deficiency is the primary defect of the disease, or whether it is merely due to the malabsorption that occurs with the disease.

12.4.2. Diagnosis

The clinical picture is one of an infant ill with respiratory disease or severely malnourished. One of the classic presentations in infancy is meconium ileus, which occurs in approximately 10% of the cases. In these infants, an intestinal obstruction due to inspissated meconium occurs because of the disturbed secretion of small-bowel glands. When the disease is mild, or progresses into childhood and adolescence, the most common presentations are those of repeated upper respiratory infections and huge stools due to malabsorption. Less frequent presentations in adulthood are failure of reproductive organs,[36] chronic liver disease, and even portal hypertension.[37]

The diagnosis can be established by the abnormal excretion of sodium in the sweat in conjunction with the clinical picture. A confirmatory sweat test using 9α-fluorohydrocortisone can be employed. Other tests that have been used are rectal biopsy, which reveals dilated rectal glands, stool for fecal fat, and stool for protein digestion such as can be done on a gelatin strip. However, most of these tests are superfluous if a significantly abnormal sweat electrolyte can be demonstrated[38] in conjunction with the above clinical picture.

12.4.3. Treatment

1. It is important that there be careful management of all respiratory infections.

2. It is essential that any evidence of any organ failure be treated vigorously.

3. Nutritional supplementation and pancreatic enzyme replacement are essential for maintaining adequate growth and development. (a) Pancreatic enzyme replacement is employed as described for pancreatic insufficiency. (b) Nutritional support may be important at various degrees and stages of the disease. Allan and associates[39] have improved the growth of severely malnourished children by using a nutritional supplement of protein hydrolysate, a glucose polymer, and medium-chain triglycerides. Berry and co-workers,[40] using the same supplementation, were also able to demonstrate a significant weight gain and an increase in serum albumin in their patients. Any oral-supplementation diet should ensure that essential fatty acids are present, as essential fatty acid deficiency can occur when there is prolonged use of elemental diets deficient in essential fatty acids.[41]

There have been suggested complications of treatments. Some researchers have noted hyperuricosuria[42] with the increased use of pancreatic enzymes. It has also been reported that patients with cystic fibrosis tend to have lithogenic bile, which accounts for an increased incidence of gallstones, but in patients treated with pancreatic enzymes, bile became less lithogenic.[43]

In summary, it is of utmost importance to ensure that proper nutrition is maintained. Simple enzyme replacement may accomplish this goal; however, when there is a recorded decreased growth rate, supplementation with some form of diet, as recommended by Allen et al.[39] and Berry et al.,[40] is essential (see Chapter 20, diet 22).

12.5. DIABETES MELLITUS

12.5.1. Definition and Pathophysiology

Diabetes mellitus is the loss of homeostatic control of blood glucose levels associated with a deficiency of insulin secretion and a probable abnormality in glucagon secretion, which result in abnormal metabolism of carbohydrates, fats, and proteins. When the disorder is severe, multiple organ systems are affected pathologically. Blood glucose is regulated by the liver, which stores excess glucose as glycogen and releases it when the blood glucose level is too low. Organs such as the adrenal and thyroid glands also affect blood glucose. The organ that principally maintains glucose stability is the pancreas, via its insulin secretion. The purpose of this text is to discuss the relationship of diabetes to the gut and the role of the diet. The interested reader is referred to the latest text on diabetes mellitus for more detailed information on the pathophysiology.[44]

12.5.2. Etiology

The etiology of mild diabetes mellitus remains uncertain in many cases. In severe diabetes mellitus, which requires insulin, there is an inherited insulin secretory deficiency.[44] Similarly, there is decreased insulin production, with severe damage to the pancreas such as may occur with pancreatitis or violent infections. There may be, as yet, undescribed disturbances in glucagon production, either inherited or acquired.

Epidemiologists have begun to accumulate a large amount of data, which is substantiated by some experimentation, that suggests glucose intolerance and the mild form of diabetes, which requires minimum dietary management or insulin coverage, may be due to environmental factors. Several reviews[45-47] of epidemiologic trends reveal that certain native populations in underdeveloped societies have much lower incidences of diabetes mellitus and its complications. Trowell[45] and Cleave,[46] working in Great Britain, have postulated that this is clearly due to dietary factors.[45,46] Their evidence incriminates the high use of refined carbohydrates concomitant with the decreased use of complex carbohydrates, primarily fiber. Review of the literature reveals there had been no experimental design to test this hypothesis until Jenkins and associates[48,49] and Kiehm and associates[50] showed that the addition of guar flour and/or pectin to a test meal or a high-polysaccharide diet can decrease postprandial glycemia and raise serum insulin. Jenkins and associates[48] added 10 g of pectin to breakfast marmalade and/or fed guar-rich bread and demonstrated significant drops in postprandial blood glucose and insulin levels. In a more clinical application to diabetic patients, they were able to show that the addition of 16 g of guar and 10 g of pectin to a controlled meal containing 106 g of carbohydrate significantly decreased the rise in blood glucose and insulin levels in three insulin-dependent diabetic subjects.[49] This hypothesis has further been tested in eight insulin-dependent patients who were given alternating diets containing either 3 g or 20 g of fiber for a 10-day period while total caloric intake was maintained.[51] The mean plasma glucose level on the high-fiber diet was 121 mg/ml, while on the low-fiber diet it was 169 mg/ml. In addition, glucagon levels fell in the high-fiber-diet period, and patients maintained on the same insulin had more hypoglycemic episodes.[51] High-fiber diets, which are high-polysaccharide carbohydrate, also have been shown by others to decrease insulin requirements in diabetics,[50] stabilize insulin responses to carbohydrate,[52] and ameliorate carbohydrate hyperlipidemia.[52] Studies of the effect of fiber on normal subjects revealed definitely improved carbohydrate tolerance, but no change in insulin receptor numbers and affinity,[53] and also that the dietary form in which carbohydrate is fed rather than glucose chain length is important in changing the insulin response.[54] These initial experiments clearly implicate high-fiber intake as an important dietary mechanism for controlling glucose intolerance. It further emphasizes the points made by Trowell concerning the role of low-fiber intake in the cause of mild diabetes mellitus. Epidemiologic evidence, such as demonstrated in Table 12-3, reveals changes in diet among ethnic groups correlate with changes in incidence of diabetes. These observa-

Table 12-3. Incidence of Diabetes in Relation to Diet in South Africans[a]

Group	Number	Known (%)	Diabetes discovered (%)	Total (%)	Crude fiber mean (g/day)	Sugar mean (g/day)	Fat mean (g/day)
Transvaal Bantu	2015	—	1 case	—	24.0	30–60	65
Cape Bantu	1027	0.9	2.7	3.3	6.5	55–85	85
Cape White	1650	0.8	2.4	3.2	4.3	80–140	110
Cape Indian	1520	4.3	4.2	8.5	4.0	70–90	80

[a]Modified from Trowell.[45]

tions are important and should be stressed in future research and probably in the control and treatment of the disease in our society.

12.5.3. Gastrointestinal Manifestations[55,56]

Most patients with severe diabetes mellitus have some form of gastrointestinal manifestation, as described below.

12.5.3.1. Gallbladder

There is an increased incidence of cholelithiasis in subjects suffering from diabetes mellitus.[55] Concomitant with the cholelithiasis is an increased incidence of acute and chronic cholelithiasis and all of the complications that occur from these diseases. Diabetic subjects will have a greater problem with infections in general, and, consequently, any site that becomes infected, such as the gallbladder, will result in more serious complications. There also appears to be a correlation between obesity and cholelithiasis in these subjects.

12.5.3.2. Esophagus

It has been reported that the esophagus may also be involved in the disease process.[56] Manometric studies[57] reveal generalized atony of the esophagus in patients with associated neuropathy. It may be a decreased force of contractions and an increased number of nonperistaltic contractions. This could be associated with symptoms of esophageal disease, although there is no study, as yet, that demonstrates an increased incidence of ordinary esophagitis or its complications in diabetic subjects.

12.5.3.3. Stomach

Although various studies indicate that disturbances of the mucosa, such as gastritis, occur, it is not clear that this is a specifically associated disorder. However, so-called *gastroparesis diabeticorum* is a definite clinical entity. It may occur in as many as 20% of all diabetics. In this condition, the stomach empties poorly and at times may become acutely dilated. The condition may cause vomiting or severe epigastric pain. The diagnosis can be made by barium study of the stomach. It will empty poorly and appear dilated.[55,56,58] It is best managed by adequate diabetic control and symptomatically.

12.5.3.4. Small Intestine

More common gastrointestinal manifestations of diabetes mellitus occur in patients who have some form of involvement of the small intestine. Minor to severe degrees may be present and are manifested by either diarrhea, nocturnal or daytime, or some form of malabsorption. The symptoms are often referred to as diabetic enteropathy.

The involvement of the small intestine may be due to neuropathy, microangiopathy, episodes of stress associated with hypoglycemia or electrolyte imbalance, or disturbances in hormone production such as occur with insulin and glucagon.[55,56,59] A certain percentage of diabetic subjects will have adult onset of gluten enteropathy.[60] Therefore, it is essential to conduct a thorough evaluation of any diarrhea malabsorption in these subjects and obtain a peroral small-bowel biopsy to rule out gluten enteropathy.[60]

Probably the most common cause of diabetic diarrhea is the neuropathy associated with the interneuronal pathways in the pre- and paravertebral sympathetic ganglia. Whalen and associates[61] carefully evaluated 13 diabetic patients suffering from diarrhea and found 5 had a mild degree of steatorrhea, but intestinal biopsy specimens and pancreatic exocrine function studies were normal. The studies left no other explanation than a disturbance in sympathetic and parasympathetic pathways, as has been noted in autopsy studies on such patients.[61]

When steatorrhea is associated with diarrhea in diabetes mellitus, many etiologic factors have to be considered. It is generally agreed that a deficiency in pancreatic secretion is not responsible, and most cases, as mentioned above, are associated with an autonomic neuropathy.[61] The usual symptomatology in neuropathy is associated with decreased sweating, postural hypotension, impotence, and bladder dysfunction. Radiographic studies of the small bowel reveal abnormalities of gastrointestinal motility such as rapid transit or a prolonged transit time with disturbed motility. When this occurs, the causes of malabsorption syndromes must be ruled out. Celiac disease should be considered, and in its absence, bacterial overgrowth becomes a possibility. Increased growth of colonic bacteria has been recorded in diabetics and attributed to abnormal bowel motility.[62-64] Some of these patients benefit from antibiotic therapy,[64] and some do not.[60] For all of these reasons, the diabetic who presents with diarrhea, with or without steatorrhea and malabsorption, is a challenge to the clinician and must be thoroughly evaluated and managed correctly according to findings, whether they be abnormal small-bowel mucosa (celiac, necessitating a gluten-free diet), bacterial overgrowth (necessitating antibiotic therapy), or neuropathy, requiring rigid glucose metabolic control.

12.5.4. Diagnosis

The diagnosis of diabetes mellitus is established by the demonstration of significant glucose intolerance in relationship to the clinical picture. The glucose intolerance may be related to many clinical entities. All possible associations should be evaluated and the severity of the disease then established. Glucose tolerance, symptoms of polyuria, polydipsia, and weight (maintenance, loss, or obesity) are extremely important in establishing not only the diagnosis but also its severity. When an estimate is accurately made, therapy can be approached.

12.5.5. Treatment

Treatment of the diabetic revolves about three factors: (1) The diet; (2) when necessary, the addition of glucose stabilization drugs such as the sulfonylureas or insulin; and (3) careful education so that factors (1) and (2) succeed.

Most diabetic patients are of two types: The young patient with a severe case that definitely requires long-term insulin management, and the older patient who has only recently developed diabetes. The older patients usually comprise two groups: those who are obese and can be controlled by management of their diet and obesity, and those who require the addition of drugs to manage glucose homeostasis.

It is not the purpose of this text to discuss the use of sulfonylureas or insulin. The interested reader is referred to the latest diabetes textbooks on the subject.[65-67]

Diet management and the appropriate education to maintain that management are essential factors in controlling the disease. The objective of the diet should be to maintain *optimal weight* as well as to maximize utilization of nutrition.

It is important to assess the appropriate necessary total caloric intake for each subject. The recommended caloric intake according to age, sex, body build, and activity are recorded in tables in Chapter 2. A very sedentary older patient will require far fewer calories than a very vigorous young athlete, such as a baseball pitcher. The tables in Chapter 2 should be helpful in establishing optimal caloric intake. The maintenance of the appropriate weight level is the final common denominator to what the caloric intake should be. For this reason, the patient must be followed carefully during an extended period of time to evaluate the caloric intake in conjunction with weekly or monthly weight. In this way, the clinician or dietician can establish the correct caloric intake for a given individual, taking into account that individual's activity and metabolism.

It is also of great importance to distribute the caloric intake over the waking hours. Diabetic patients tend to easily become hypoglycemic; this occurs even more so while they are on insulin coverage. Therefore, it is recommended that the caloric intake be divided evenly over the day. Such division is demonstrated in Table 12-4.[67] Various diets will alter carbohydrate availability during the day. Of specific note has been the fact that although high-carbohy-

Table 12–4. Examples of Recommended Daily Distribution of Caloric Content of Energy Intake[a]

| Type of diabetic patient | Time of day and fraction of total calories | | | | | |
	Break-fast	Mid-morning	Noon	After-noon	Supper	Bed-time
Stable insulin-requiring	2/7		2/7		2/7	1/7
Labile insulin-requiring	2/10	1/10	2/10	1/10	3/10	1/10
Stable, non-insulin-requiring	2/7		2/7		3/7	

[a]Modified from Skillman and Tzagournis.[67]

drate polysaccharide diets are of help in decreasing insulin requirement, as well as lowering fasting glucose levels, they tend to cause more hypoglycemia, especially in patients who are on insulin.[51,68] Therefore, caution should be used in a given individual when changing the pattern of caloric intake, both in the quantity and in the type.

12.5.5.1. Carbohydrates

It has been traditional in many clinics to carefully limit the amount of carbohydrate intake. The main concern was the hyperglycemic effect of simple sugars. As described above, investigators have found that when high carbohydrate, composed primarily of polysaccharides and oligosaccharides, is used, fasting blood glucose levels can be controlled and insulin requirements decreased.[48-52]

In several institutions, high-carbohydrate diets are the treatment of choice for diabetic patients who require sulfonylureas, or less than 30 units of insulin per day. One series of experiments documented a statistically significant decrease of insulin requirement, as well as lack of need of sulfonylureas in subjects requiring these drugs after they had been on a 2-week diet consisting of 75% carbohydrates in which the crude dietary fiber intake averaged 14.2 g/day.[50] Diabetologists working with children have also reported that they do much better on a high-oligosaccharide diet.[69] Consequently, review of the most recent experimentation indicates that a high-fiber, high-polysaccharide diet (approximately 75%) is of benefit to the patient with diabetes mellitus (see Chapter 20, diet 23). Although this is relatively new information, it seems significant. However, the traditional low-carbohydrate diet is still the choice of some clinicians and may still be tolerated by a significant number of patients (see low-carbohydrate diabetic diet, Chapter 20, diet 24).

12.5.5.2. Fats and Proteins

If one recommends a 65 to 75% caloric intake of high carbohydrates, then the remainder of calories should be derived from fat and protein. The necessary daily requirement of protein, as outlined in Chapter 2, should be maintained. This leaves little for higher-fat intake. The limitation of saturated fat and cholesterol intake in a diabetic seems extremely important considering the

Table 12–5. Strategies for the Two Main Types of Diabetic Patients

Dietary strategy	Non-insulin-dependent obese patients	Insulin-dependent nonobese patients
Decrease calories	Yes	No
Protect or improve β-cell function	Very urgent priority	Seldom important because β-cells are usually extinct
Increase frequency and number of feedings	Usually no	Yes
Day-to-day consistency of intake of calories, carbohydrate, protein, and fat	Not crucial if average caloric intake remains in low range	Very important
Day-to-day consistency of the ratios of carbohydrates, protein, and fat for each of the feedings	Not crucial	Desirable
Consistency of timing of meals	Not crucial	Very important
Extra food for unusual exercise	Not usually appropriate	Usually appropriate
Use of food to treat, abort, or prevent hypoglycemia	Not necessary	Important

high incidence of atherosclerosis in this group of patients. It is particularly of importance in the obese and the diabetic with hyperlipoproteinemia. These patients should be on restricted saturated-fat and cholesterol intake (see diet for type 4 hyperlipoproteinemia, Chapter 20, diet 25).

The maintenance of optimal weight in these patients cannot be overstressed. The undernourished diabetic needs calories, while the obese diabetic will improve greatly merely by loss of weight. To accomplish the goal of optimal weight, patient education is essential. Table 12–5, as adopted from West,[65] lists important strategies in handling the two types of diabetic patients. The education must be accomplished either through the physician, the diabetic counselor, or the dietician. However, it is a waste of time to hand a diet to a diabetic patient without a careful educational program and observation through the treatment period to ensure that the diet plan is being followed.

REFERENCES

1. Sarles H: An international survey on nutrition and pancreatitis. *Digestion* 9:389, 1973.
2. Schmidt H, Creufeldt W: Etiology and pathogenesis of pancreatitis, in Bockus HC (ed): *Gastroenterology*, ed 3. Philadelphia: WB Saunders Co, 1976, Vol III, p 1005.

3. Takada Y, Appert HE, Howard JM: Vascular permeability induced by pancreatic exudate formed during acute pancreatitis in dogs. *Surg Gynecol Obstet* 143:779, 1976.

4. Warshaw AL, Fuller AF Jr: Specificity of increased renal clearance of amylase in diagnosis of acute pancreatitis. *N Engl J Med* 292:325, 1975.

5. Johnston SG, Ellis CJ, Levitt MD: Mechanism of increased renal clearance of amylase in acute pancreatitis. *N Engl J Med* 295:1214, 1976.

6. Levant JA, Secrist DH, Resin H, *et al:* Nasogastric suction in the treatment of alcoholic pancreatitis: A controlled study. *JAMA* 229:51, 1974.

7. Finch WT, Sawyers JL, Schenker S: A prospective study to determine the efficacy of antibiotics in acute pancreatitis. *Ann Surg* 183:667, 1976.

8. Trapnell JE, Rigby CC, Talbot CH, *et al:* A controlled trial of Trasylol in the treatment of acute pancreatitis. *Br J Surg* 61:177, 1974.

9. Knight JM, Condon JR, Smith R: Possible use of glucagon in the treatment of pancreatitis. *Br Med J* 1:440, 1971.

10. Bolooki G, Gliedman ML: Peritoneal dialysis in treatment of acute pancreatitis. *Surgery* 64:466, 1968.

11. Wise L, Grana W: Role of emergency laparatomy in acute pancreatitis. *Am Surg* 42:128, 1976.

12. Warshaw AL, Imbembo AL, Civetta JM, *et al:* Surgical intervention in acute necrotizing pancreatitis. *Am J Surg* 127:484, 1974.

13. Blackburn GL, Williams LF, Bistrian BR, *et al:* New approaches to the management of severe acute pancreatitis. *Am J Surg* 131:114, 1976.

14. Sarles H, Dayan H, Tasso F, *et al:* Chronic pancreatitis, relapsing pancreatitis, calcifications of the pancreas, in Bockus HC (ed): *Gastroenterology.* Philadelphia: WB Saunders Co, 1976, Vol III, p 1040.

15. Marks IN, Banks S: Clinical aspects, in Bockus HC (ed): *Gastroenterology.* Philadelphia: WB Saunders Co, 1976, Vol III, p 1052.

16. Marks IN, Bank S: Treatment of steatorrhea due to pancreatic insufficiency. *Mod Treat* 2:326, 1965.

17. Toskes PP, Hansell J, Certa J, *et al:* Vitamin B_{12} malabsorption in chronic pancreatic insufficiency. *N Engl J Med* 284:627, 1971.

18. Von der Lippe G: The absorption of vitamin B_{12} in chronic pancreatic insufficiency. *Scand J Gastroenterol* 12:257, 1977.

19. LeBauer EK, Smith K, Greenberger NT: Pancreatic insufficiency and vitamin B_{12} malabsorption. *Arch Intern Med* 122:423, 1968.

20. Taubin HL, Spiro HM: Nutritional aspects of chronic pancreatitis. *Am J Clin Nutr* 26:367, 1973.

21. Pairent FW, Howard JM: Pancreatic exocrine insufficiency. *Arch Surg* 110:739, 1975.

22. Sarles H, Cross RC, Nicholas R, *et al:* Quantity of alcohol consumption and risk of chronic pancreatitis. *Gut* 17:826, 1976.

23. Regan PT, Malagelada JR, DiMagno EP, *et al:* Cimetidine as an adjunct to oral enzymes in the treatment of malabsorption due to pancreatic insufficiency. *Gastroenterology* 74:468, 1978.

24. Tandon BN, George PK, Sama SK, *et al:* Exocrine pancreatic function in protein–calorie malnutrition disease of adults. *Am J Clin Nutr* 22:1476, 1969.

25. Brunner O, Castillo C, Araya M: Fine structure of the small intestinal mucosa in infantile marasmic malnutrition. *Gastroenterology* 70:495, 1976.

26. Dudrick SJ, Ruberg RL: Principles and practices of parenteral nutrition. *Gastroenterology* 61:901, 1971.

27. Adham N, Delman M, Cado R: Effect of parenteral hyperalimentation on pancreatic function. *Gastroenterology* 62:717, 1972.

28. Bury KD, Stephens RV, Randall HT: Use of a chemically defined, liquid, elemental diet for nutritional management of fistulas of the alimentary tract. *Am J Surg* 121:174, 1971.

29. Jeejeebhoy KN, Langer B, Tsallas G, *et al:* Total parenteral nutrition at home: Studies in patients surviving 4 months to 5 years. *Gastroenterology* 71:943, 1976.

30. Hashin SA, Robalt HB, van Itallie TB: Pancreatogenous steatorrhea treated with medium chain triglyceride. *Clin Res* 10:394, 1962.

31. Rosenlund ML, Lustig HS: Young adults with cystic fibrosis. *Ann Intern Med* 78:959, 1973.
32. diSant'Agnese PA, Davis PB: Research in cystic fibrosis. *N Engl J Med* 295:467, 534, 597, 1976.
33. Elliott RB: A therapeutic trial of fatty acid supplementation in cystic fibrosis. *Pediatrics* 57:474, 1976.
34. Campbell IM, Crozier DN, Caton RB: Abnormal fatty acid composition and impaired oxygen supply in cystic fibrosis patients. *Pediatrics* 57:480, 1976.
35. Rivers JPW, Hassam AG: Defective essential-fatty-acid metabolism in cystic fibrosis. *Lancet* 2:642, 1975.
36. Kaplan E, Shwachman H, Perlmutter AD, *et al:* Reproductive failure in males with cystic fibrosis. *N Engl J Med* 279:65, 1968.
37. Danielson GK, Kyle GC, Denton R: Portal hypertension in cystic fibrosis. *JAMA* 195:217, 1966.
38. Jones JD, Steige H, Logan GB: Variations of sweat sodium values in children and adults with cystic fibrosis and other diseases. *Mayo Clin Proc* 45:768, 1970.
39. Allan JD, Masson A, Moss AD: Nutritional supplementation in treatment of cystic fibrosis of the pancreas. *Am J Dis Child* 126:22, 1973.
40. Berry HK, Kellogg FW, Hunt MM, *et al:* Dietary supplement and nutrition in children with cystic fibrosis. *Am J Dis Child* 129:165, 1975.
41. Dodge JA, Salter DG, Yassa JG: Essential fatty acid deficiency due to artificial diet in cystic fibrosis. *Br Med J* 1:192, 1975.
42. Stapleton FB, Kennedy J, Nousia-Arvanitakis S, *et al:* Hyperuricosuria due to high-dose pancreatic extract therapy in cystic fibrosis. *N Engl J Med* 295:246, 1976.
43. Roy CC, Weber AM, Morin CL, *et al:* Abnormal biliary lipid composition in cystic fibrosis. *N Engl J Med* 297:1301, 1977.
44. Felig P: Pathophysiology of diabetes, in Sussman EK, Metz RJS (eds): *Diabetes Mellitus,* ed 4. New York: American Diabetes Association, 1975.
45. Trowell H: Diabetes mellitus and obesity, in Burkitt DP, Trowell H (eds): *Refined Carbohydrate Foods and Disease.* New York: Academic Press, 1975.
46. Cleave TL, Cambell GD, Panter NS: *Diabetes, Coronary Thrombosis, and the Saccharine Disease,* ed 2. Bristol: John Wright, 1969.
47. West KM: Diabetes in American Indians and other native populations of the new world. *Diabetes* 23:841, 1974.
48. Jenkins DJA, Leeds AR, Gassull MA, *et al:* Decrease in postprandial insulin and glucose concentrations by guar and pectin. *Ann Intern Med* 86:20, 1977.
49. Jenkins DJA, Leeds AR, Wolever TMS, *et al:* Unabsorbable carbohydrates and diabetes: Decreased post-prandial hyperglycemia. *Lancet* 2:172, 1976.
50. Kiehm TG, Anderson JW, Ward K: Beneficial effect of a high carbohydrate, high fiber diet on hyperglycemic diabetic men. *Am J Clin Nutr* 29:895, 1976.
51. Miranda PM, Horwitz DL: High-fiber diets in the treatment of diabetes mellitus. *Ann Intern Med* 88:482, 1978.
52. Albrink MJ, Newman T, Davidson PC: Effect of high- and low-fiber diets on plasma lipids and insulin. *Am J Clin Nutr* 32:1486, 1979.
53. Wigand JP, Anderson JH Jr, Jennings SS, *et al:* Effect of dietary composition on insulin receptors in normal subjects. *Am J Clin Nutr* 32:6, 1979.
54. Wahlquist ML, Wilmshurst EG, Murton CR, *et al:* The effect of chain length on glucose absorption and the related metabolic response. *Am J Clin Nutr* 31:1998, 1978.
55. Katz LA, Spiro HM: Gastrointestinal manifestations of diabetes. *N Engl J Med* 275:1350, 1966.
56. Scarpello JHB, Sladen GE: Diabetes and the gut. *Gut* 19:1153, 1978.
57. Mandelstam P, Siegel CI, Lieber A, *et al:* The swallowing in patients with diabetic neuropathy-gastroenteropathy. *Gastroenterology* 56:1, 1969.
58. Goyal RK, Spiro HM: Gastrointestinal manifestations of diabetes mellitus. *Med Clin North Am* 55:1031, 1971.
59. Casdary WF, Creutzfeld W: Gastrointestinal manifestation of diabetes, in Sussman EK, Metz RJS (eds): *Diabetes Mellitus,* ed 4. New York: American Diabetes Association, 1975.

60. Vinnik IE, Kern F, Struthers JE Jr: Malabsorption and the diarrhea of diabetes mellitus. *Gastroenterology* 43:507, 1962.

61. Whalen GE, Soergel KH, Greenen JE: Diabetic diarrhea. A clinical and pathophysiologic study. *Gastroenterology* 56:1021, 1969.

62. Sume SM, Finlay JM: On pathogenesis of diabetic steatorrhea. *Ann Intern Med* 55:994, 1961.

63. Malins JM, French JM: Diabetic diarrhea. *Q J Med* 36:467, 1957.

64. Green PA, Bergg KG, Sprague RG, *et al:* Control of diabetic diarrhea with antibiotic therapy. *Diabetes* 17:385, 1968.

65. West KH: Dietary therapy of diabetes: Principles and applications, in Sussman EK, Metz RJS (eds): *Diabetes Mellitus,* ed 4. New York: American Diabetes Association, 1975.

66. West KW : Prevention and therapy of diabetes mellitus, in *Present Knowledge in Nutrition,* ed 4. New York: The Nutrition Foundation, Inc, 1976.

67. Skillman TJ, Tzagournis M: *Diabetes Mellitus.* Kalamazoo, Mich: The Upjohn Co, 1975.

68. Lenner RA: Studies of glycemia and glucosuria in diabetics after breakfast meals of different composition. *Am J Clin Nutr* 29:716, 1976.

69. Drash A: Diabetes mellitus in childhood: A review. *J Pediatr* 78:919, 1971.

13

Malabsorption and Small-Intestine Diseases

The small intestine is the most important organ relating to nutrition. Humans need at least the duodenum and part of the proximal jejunum to survive by eating. With the advent of intravenous hyperalimentation, life can be maintained by intravenous feedings. However, to lead a normal life, the small intestine must function adequately. Until 1956 the small intestine had occupied a minor, but interesting, role in medicine. In that year, Drs. Margot Shiner[1] and William H. Crosby and H. W. Kugler[2] independently developed instruments to obtain peroral biopsy specimens from human small bowel, which made the study of intestinal mucosal pathology on fresh specimens possible. It led to the discovery of mucosal lesions in tropical and nontropical sprue, exudated enteropathy, and many other diseases. There occurred a reawakening of interest in the role of the small intestine in metabolism. The addition of the triple lumen tube to the peroral biopsy technique permitted careful correlation of histology and chemistry and opened new vistas for study that has resulted in a vast amount of information on man.

13.1. ANATOMY AND HISTOLOGY

The small intestine is a convoluted tube ranging from 16 to 30 ft in length and extending from the gastric pylorus to the ileocecal valve. It is contained in the central and lower part of the abdominal cavity, but a portion extends below the superior opening of the pelvis and the front of the rectum. It has three divisions, the duodenum, jejunum, and ileum.

The *duodenum* is the shortest and most fixed part of the intestine. It has a "C" shape extending from the pylorus of the stomach and is divided into four segments, the superior, descending, horizontal, and ascending portions, often referred to as the bulb, second, third, and fourth parts of the duodenum. Its nervous supply is from the celiac plexus. Its very rich blood supply is from both branches of the hepatic and superior mesenteric arteries. It joins the jejunum at the ligament of Treitz, at which point the intestine is fixed. The *jejunum* is 8 to 10 ft long. It is wider, thicker, more vascular, and deeper in color than the ileum, although there is no definite landmark that separates the two. The *ileum* is smaller in diameter than the jejunum and its walls are thinner and less vascular. Its aggregate lymph nodules (Peyer's patches) are more numerous and larger than those found in the jejunum. The terminal portion of the ileum usually lies in the pelvis, from which it ascends to end in the medial side of the large intestine. Both the jejunum and ileum are attached by a fold of peritoneum, the mesentery, to the posterior abdominal wall. The junction with the posterior abdominal wall is fan-shaped and about 15 cm long, but the width from base to intestinal wall is longer and allows the freest motion so that each coil can accommodate itself in changes of form and position.

The very rich arterial blood supply of the small intestine is derived mainly from the superior mesenteric artery, which leaves the aorta at almost the level of the first lumbar vertebra, courses downward and forward, crosses the third portion of the duodenum to enter the root of the mesentery, and then curves downward toward the right into the right iliac fossa. Numerous branches to the small intestine, 10 to 16 in number, arise from the convex left side of the curving vessel. These mesenteric or intestinal arteries run obliquely downward and forward within the mesentery and divide and redivide to form a series of arches. A series of straight arteries arise from the final arches (see Fig. 13-1).

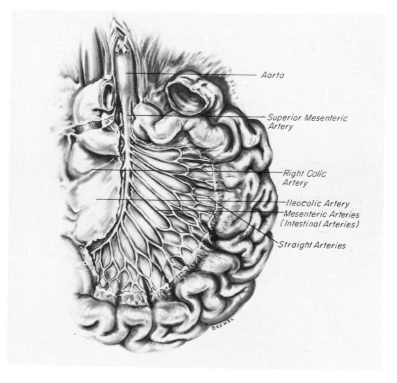

Figure 13-1. Arterial supply of the small intestine. Note the arched arteries arising from the mesenteric arteries. These arches give off straight arteries that encircle the wall. All this forms a very rapid and rich flow of blood to the villi.

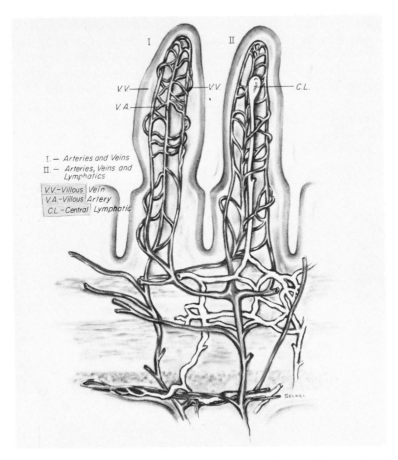

Figure 13-2. Villus arterial, venous, and lymphatic supply and drainage, which runs through the lamina propria, to offer rapid circulation to the epithelium.

The straight arteries then branch to either side of the gut and encircle it; hence, the small bowel is encircled throughout its length by arterial rings. In turn, the arterial circles give off branches of diminishing size that penetrate the muscular layer of the intestinal wall and supply the costs of the bowel. Inside the intestinal wall the penetrating artery branches in anastomosis. Small arterial twigs shunt into the center of every villus and capillaries project outward toward the periphery of the villus (see Fig. 13-2). This very rich blood supply permits the small bowel to perform its function of rapid absorption and adaptability for fluid exchange from the gut lumen to the intravascular circulation. A reverse network of venous drainage accompanies the arterial structure. Also associated is a rich lymphatic drainage that begins in the villus as a central lymph vessel. Solitary lymph nodules may be found in the jejunum. Aggregate lymph nodules, known as Peyer's patches, are found in the mucosa and submucosa of the ileum. Larger lymph vessels emerge from the intestinal wall and enter the mesentery. In this area, the lacteals are often visible as whitish strands. They pass through various nodes and finally drain into a cluster of nodes known as the superior mesenteric glands. This group of nodes is situated in the origin of the superior mesenteric artery. Large lymph channels emerge from here and from the gastrointestinal lymph trunk that empties into the cisterna chyli.

Both sympathetic and parasympathetic fibers innervate the small bowel. The sympathetic fibers derive from the fifth to ninth segments of the spinal cord. The fibers from these segments pass through corresponding ganglia of the sympathetic chain and emerge to form the greater splanchnic nerves to the celiac ganglia. Parasympathetic innervations stem from the vagus nerve. Branches from the celiac plexus continue to the small bowel in conjunction with sympathetic fibers to the superior mesenteric plexus and hence into the small bowel.

The wall of the small intestine is subdivided into four layers: serosal, muscular, submucosal, and mucosal (Fig. 13-3). The serosa is derived from the peritoneum and comprises the outer coat. The muscularis consists of an inner circular layer and an outer longitudinal layer. The submucosal coat connects the mucosa and muscularis layers and contains the larger blood vessels, lymphatics, and nerves in a network of loose areolar tissue.

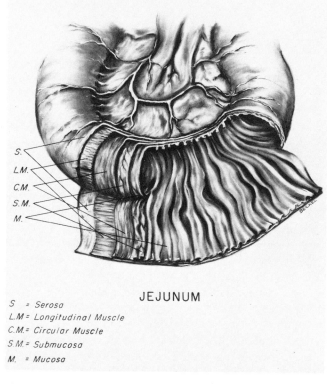

JEJUNUM

S = Serosa
L.M = Longitudinal Muscle
C.M.= Circular Muscle
S.M.= Submucosa
M. = Mucosa

Figure 13-3. The four layers of the wall of the small intestine. The circular folds (valvulae conniventes) are evident in the mucosal layer.

Figure 13-4. Dissecting microscope (× 10) photograph of human intestinal villi. The fingerlike villi and their vascular network can be seen.

The mucosa constitutes the inner lining of the small intestine. It is a thick layer separated from the submucosa by the muscularis mucosae, which is composed of an outer longitudinal and an inner circular layer of muscle fibers. Internal to the muscularis mucosae is the lamina propria, which forms the core of the villi. The intestinal mucosa is distinguished by the presence of the structures described below.

The mucosal surface is thrown into a series of folds that project into the lumen of the bowel. These circular folds are often referred to as the valvulae conniventis (plicae circulares or valves of Kerckring). These folds are composed of reduplications of the mucous membrane, the two layers of the folds being bound together by submucosal tissue. Unlike the gastric folds, these are prominent and do not disappear when the lumen is distended. They do not occur in the ileum and are most prominent in the duodenum and jejunum. Usually, they run transversely across the bowel surface, while others form complete circles. Larger folds often alternate with smaller ones. They are longer in the jejunum, but midway through the segment of the small bowel they begin to diminish in size and almost disappear from the lower ileum. As a result, the ileum is thinner than the proximal small bowel. The circular folds, like the villi, add to the absorptive area and they *serve to retard the passage of food.*

The intestinal villus, like the nephron of the kidney, constitutes the functional unit of the small bowel (Fig. 13-4 to 13-7). In the jejunum, the individual villus measures between 320 and 575 μm in height and between 110 and 135 μm

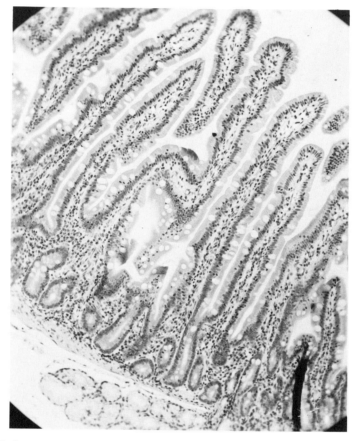

Figure 13-5. Low-power photomicrograph (× 100) of normal human duodenal mucosa. Note the long, thin villi. Brunner's glands are evident in submucosa.

in thickness. The average thickness of the glandular portion is 250 μm. The villi are highly vascular structures that project from the mucosal surface into the lumen, usually rising in height from one to two-and-a-half times the total width of the mucosa. They are largest and most numerous in the duodenum and jejunum and undergo rhythmic contractions that serve to pump the contents of the central lacteal into the lymph channels of the submucosa. The appearance and shape of the villi may vary in different patients and at different levels of the small intestine. Figure 13-4 is a low-magnification photograph of villi and their fingerlike projections. Figure 13-5 reveals a cross section of the villus structure and Fig. 13-6 the appearance of the tips of the villi and their columnar epithelial cells. The villi are usually long and slender. However, a certain proportion (up to 20%) may be leaf or tongue shaped. Essential parts of the villus are the lacteal vessel, the blood vessels, the surface epithelium, the basement membrane, the striated border, and the muscular tissue. All of these structures are supported and held together by connective tissue in which are scattered various types of cells—lymphocytes, eosinophils, histiocytes, fibroblasts, plasma cells, and endothelial cells.[3,4] All of these structures are contained within the lamina propria, which forms the core of the villus and which is surrounded by the basement membrane. This membrane is a stratum of endothelial cells upon which the epithelial cell rests.

The epithelial cell, or enterocyte, has been the subject of intensive investigation. It is the site of absorption, digestion, and maintenance of major fluid and electrolyte balance. A normal functioning epithelium is essential for maintenance of human nutritional and fluid homeostasis. Each cell possesses a clear oval nucleus situated toward the base of the cell. On the luminal surface of the cell, there is a highly refractive striated or brush border made up of microvilli (Figs. 13-6 and 13-7). The enterocyte (Fig. 13-8) is contained within a continuous membrane that behaves as if it were fenestrated by tiny water-filled channels or pores, too small to be seen, but yet large-enough to permit free permeability of water, monovalent ions, and hydrophilic solutes of small molecular size, such as urea.[5] The electron microscope reveals the "brush" border con-

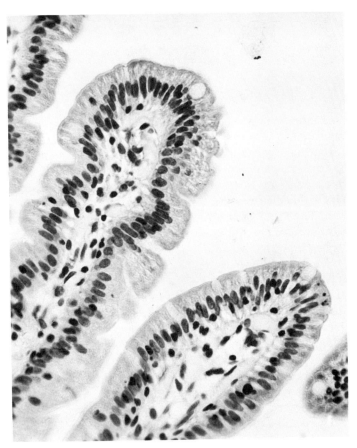

Figure 13-6. High-power photomicrograph (× 400) of normal jejunal villi. Note the brush border along lumen side of epithelial cells and orderly arrangement of nuclei along basement membrane side of epithelium. Few round cells can be seen in central lamina propria.

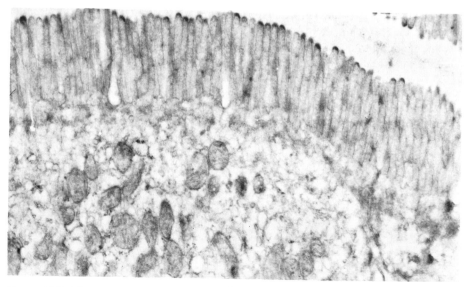

Figure 13-7. Electron micrograph of brush border seen vaguely in Fig. 13-6. Note the thin "matchstick-like" microvilli that line the epithelial cells on their lumen side.

sists of rodlike structures projecting off the cell into the lumen. They contain receptors important in absorption. The inside of the cell contains fine organelles, mitochondria, Golgi apparatus, etc. The interested reader is referred to the literature for a more detailed evaluation of these intensely studied structures.[6-8] Goblet cells are occasionally interspersed between enterocytes, are filled with mucus that distends the upper half of the cell, and secrete the mucus into the lumen. The bases of the villi form crypts lined by epithelial cells that resemble those seen in the middle and tips of the villi, but are less differentiated. The epithelial cells originate in the crypts and migrate to the tips of the villi, where they are finally extruded as dead cells. The cell turnover ranges from 3 to 5 days.[9,10]

Situated within the crypt areas are other cells, such as the enterochromaffin cells of Kultschitsky (argentaffin or endocrine cells) and the Paneth cells. The Kultschitsky cells have been shown to contain serotonin. Carcinoid tumors are assumed to arise from this cell. The Paneth cell contains very large eosinophil, periodic acid–Schiff staining granules that are secretory in nature.

The lamina propria contains the entire spectrum of peripheral circulating leukocytes, lymphocytes being the most common. There may be selective increase of lymphocytes, plasma cells, polymorphonolean cells, or eosinophils when pathology occurs. Those will be described under individual disease entities.

Lymphocytes and plasma cells of the intestine play a very specific role in normal immunology. Histologic and immunofluorescent studies reveal that both T and B lymphocytes are present in the small intestine. The T-cells are more predominant in young animals, and B-cells are more predominant in adult animals. The sources of these T and B lymphocytes appear to be Peyer's patches and the appendix. The lymphocytes are mobile and easily extend into the lamina propria. Very immature lymphocytes are converted to plasma cells within the lamina propria,[11,12] making the intestine a lymphoid organ that produces a great amount of immunoglobulins. IgA makes up approximately 80 to 85% of the globulin produced within the gut, with IgM next, and then IgG. Very small amounts of immunoglobulin E and IgD are present in the normal intestine, less than IgG.[12] The role of the gut in immunology remains to be fully explained, although selective diseases associated with both deficiency and increase of immunoglobulins are reported.

Peroral small-bowel biopsy has enabled the clinician to verify and clarify his diagnostic impressions concerning the function and the presence of diseases in the small bowel. The usefulness of the technique as it correlates with disease will be discussed below.

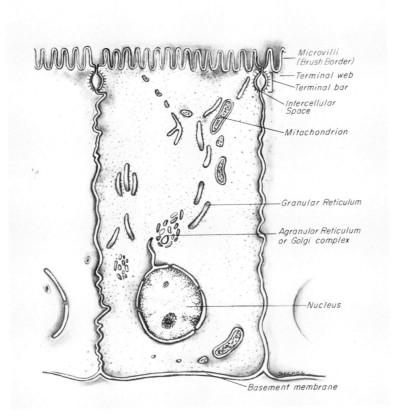

Figure 13-8. Schematic representation of the enterocyte demonstrating relationships of brush border, nucleus, and cytoplasmic structures.

13.2. ABSORPTION AND DIGESTION[13,14]

Fats are absorbed primarily in the duodenum and proximal jejunum, but they may be absorbed throughout the small bowel. The absorption of simple triglycerides depends on intraluminal micellar formation. Pancreatic secretion of lipase into the duodenum is an essential factor for partial digestion of triglycerides to glycerol, monoglycerides, and diglycerides before normal micelle formation and absorption. The micelle is a physiochemical complex of lipids, cholesterol, and bile salts whose formation is dependent on the presence of normal intraluminal amounts of conjugated bile acids and phospholipids. Once micelles are formed, the anatomic and functional integrity of the intestinal enterocyte are essential to complete absorption. Triglyceride components are absorbed, reformed within the enterocyte, become coated with protein to form chylomicrons, and pass primarily into the lymphatics. The integrity of the lymphatic system is important in assisting with this process. Medium-chain and short-chain fatty acids are similarly reformed to triglycerides, but they pass primarily into the portal blood. They are handled more expeditiously than long-chain fatty acids. Absorption of cholesterol occurs through the enterocyte. More details of the absorption of fats are discussed in Chapter 1.

Simple and complex carbohydrates (mono-, di-, and polysaccharides) and starch are absorbed primarily in the proximal small bowel. The digestion of starch requires salivary and pancreatic amylases to break chemical bonds to release maltose, maltotriose, and dextrose. The major action of disaccharides is on the brush border of the enterocytes, with peak activity probably in the distal jejunum. Disaccharides are broken down into monosaccharides in the lumen and along the microvilli. Sucrose, lactose, and maltose are the primary disaccharides in our diet. A deficiency in any one of the corresponding disaccharidase enzymes results in malabsorption of the corresponding sugar. The one most seen in our society is lactose malabsorption (see Chapter 1 for further discussion of the absorption of simple carbohydrates). The digestion and absorption of more complex carbohydrates, such as storage and algyl polysaccharides, fiber, etc., are less well understood. These are discussed in Chapter 4.

Protein digestion occurs primarily in the lumen of the proximal small bowel. Some whole undigested protein molecules may be absorbed intact. Pepsin may produce minimal protein hydrolytic digestion in the stomach, but the majority of protein breakdown occurs in the proximal small bowel by the action of pancreatic trypsin, chymotrypsin, elastase, and carboxypeptidase, which digest complex proteins into polypeptides, peptides, and amino acids. Proteins that are not broken down prior to absorption may be digested in the brush border. Active transport occurs through the enterocyte into the bloodstream (see Chapter 1 for more details on protein absorption).

The absorption of vitamins and the relationship of gastrointestinal hormones produced in the small intestine are discussed in detail in Chapter 1.

One of the major functions of the small intestine is the maintenance of *fluid and electrolyte balance.*[13-16] It is integral to the digestive and absorptive processes as well as maintenance of systemic cellular and intravascular homeostasis. Anywhere from 5 to 10 liters of water enter the small intestine during a given day of a healthy eating adult. Approximately 3 liters are derived from the diet, and 1 liter from saliva, 2 from gastric juice, 1 from bile, 2 from pancreatic juice, and a certain amount from the small bowel itself. It is estimated that absorption of 30 to 50% of the total fluid load occurs in the jejunum, 20 to 40% in the ileum, and 10% in the colon.

Enterocyte membranes are lipid and therefore water, electrolytes, and all water-soluble substances must pass through the intestinal membrane by carrier- mediated transport mechanisms or through so-called "pores," which are revealed by electron microscopy to be water-filled spaces.

Carrier transport can simply be thought of as the acceptance of the solute by some part of the cell membrane, which then transports it across to the interior of the cell. This mechanism can be active—when electrolytes or other substances, such as glucose and amino acids, are transported against an electrochemical gradient or concentration gradient. There is also a carrier exchange mechanism for electrolytes. This active process can occur down an energy gradient and is then referred to as facilitated diffusion.[14]

Noncarrier transport occurs through the intercellular junctions where water can pass back and forth depending on osmolality in order to maintain isotonicity. Electrolytes can also pass with the water through these "pores." It is a passive diffusion in response to a concentration gradient. Electrical and pH gradients also cause movement of electrolytes and weak acids and bases. These movements of simple compounds are summarized in Table 13-1. The average output of fluid leaving the intestine is approximately 600 ml/day, the concentration of sodium being 125 meq/liter, potassium 9 meq/liter, chloride 60 meq/liter, and HCO_3 44 meq/liter.[14] This can be measured in subjects with ileostomies.

Although there is a net absorption of water and electrolytes from the small bowel, it is important to remember that there is a concomitant great flux of water and electrolytes across the cell wall membrane during digestion and absorption. The system has tremendous adaptability in accommodating all sorts of concentrations of foods and chemicals presented to it. Lymph flow also increases as water is absorbed. The greatest flow occurs during the absorption of isotonic saline.[13]

The net absorption of *sodium* results from the difference between the two opposing fluxes. The major net absorption occurs in the jejunum and decreases down the gut. Some absorption is active through a carrier mechanism, and passive via diffusion when the electrochemical gradient is in the direction from the intestinal lumen to the cytoplasm of the cell.[13,15,16] Sodium and *potassium* absorption also occurs during active glucose and galactose absorption by solvent drag. However, active absorption of sodium readily occurs in the ileum against a concentration gradient (see Table 13-1). Potassium absorption also is a result of the difference between two opposing fluxes. Potassium absorption in the jejunum occurs primarily by electrochemical gradient to maintain equilibrium.

Table 13-1. Na and H_2O Transport Characteristics of Different Regions of the Intestine

	Duodenum and jejunum	Ileum
Permeable to		
Water	Permeable	Permeable
NaCl, small solutes (xylose, urea)	Permeable	Relatively impermeable
Carrier-mediated transport		
Absorption	Na^+, glucose, galactose, fructose, amino acids	Na^+, Cl^-, glucose, galactose, fructose, amino acids
Secretion	H^+	H^+, HCO_3
Major mechanism of NaCl absorption	Passive (solvent drag)	Active transport via "carrier"

[a]Modified from Breberdorf.[14]

Chloride and *bicarbonate* absorption and secretion follow sodium exchange, but they also have very separate and active processes. The efforts to understand the pathologic fluid phenomenon, such as occurs in cholera, led to the understanding that the active absorption of anions occurs in the ileum, which is stimulated by cyclic AMP. When chloride is absorbed, bicarbonate is secreted. Chloride absorption is against a concentration gradient; to maintain neutrality of solutions, it will carry cations along with it.

Bicarbonate is readily absorbed in the jejunum. The ileum maintains a luminal bicarbonate concentration of approximately 45 mM and will absorb or secrete it in order to maintain that concentration. Usually, there is an exchange of chloride for bicarbonate, but either may be absorbed or secreted as a primary function.[13]

13.3. MALABSORPTION

Malabsorption may be due to primary small-bowel or pancreatic disease of idiopathic, congenital, inflammatory, or infectious origin, or it may occur secondary to many systemic diseases. It may be very selective, as occurs in B_{12} or trehalose malabsorption,[17] or it may be very broad, with multiple absorption defects such as occur in pancreatic enzyme deficiencies and celiac sprue. Malabsorption is not a disease, but the result of the failure of nutrient or mineral absorption due to a disease or disorder.

In order to correctly treat the afflicted patient, one must understand the clinical and pathophysiologic presentations of malabsorption and identify the disease causing it by laboratory tests. In mammals, the small intestine is the only organ that absorbs nutrients. However, malabsorption is not synonymous with small-intestine disease and may occur readily because of systemic disease that secondarily affects the small bowel. The following broad pathophysiologic working classification may be employed:

1. *Defects of gastric function*
   Due to mucosal deficiency such as secretion of intrinsic factor
   Due to surgical resection
2. *Defects of digestion*
   Due to biliary or hepatic deficiencies
   Due to pancreatic secretory deficiencies
3. *Defects of small-intestine absorption*
   Due to primary mucosal disease
   Due to surgical disease
   Due to vascular lesions, both small or large vessels
   Due to neoplastic lesions
   Due to infectious or parasitic diseases
   Due to drugs or chemotherapeutic agents
   Due to selective enzyme deficiencies
4. *Systemic diseases* causing defects of gastric function or small-intestine absorption.

The historical and physical presentation of a patient with malabsorption will vary greatly depending on the disease at fault and its severity. The physician's approach to evaluation and differential diagnosis will also be influenced by the area of the world in which he is working or from which the patient comes. For instance, a physician seeing an adult patient in cold temperate climates would first think of idiopathic steatorrhea, pancreatic disease, or malignancy, but when evaluating a patient with malabsorption in semitropical climates, early consideration would be given to parasitosis, infection, and tropical sprue.

Diseases causing malabsorption can occur at any age. Statistics of a differential diagnosis on the basis of age are not available. The classic syndrome of gluten enteropathy or celiac disease, for example, can be seen in children or in adults. Between 20 and 30% of all individuals with celiac disease will have symptoms during childhood.[18] Infectious diseases are seen at all ages. Neoplastic disease and chronic pancreatitis are usually seen in older patients and rarely in childhood; lymphoma is an exception.

A careful history of exposure to chemicals, drugs, and travel is essential. Chronic use of neomycin and exposure to chemotherapeutic agents may cause mucosal defects in the small intestine. *Giardia lamblia* and tropical sprue are classic examples of diseases contracted during travel.

A family history of celiac disease is reportedly present in only 8 to 24% of celiac patients.[19] Consequently, the diagnosis of gluten enteropathy is frequently made without any family history.

The incidence of signs and symptoms in gluten enteropathy and tropical sprue is listed in Table 13-2.

Weight loss is one of the cardinal symptoms of malabsorption and occurs in 50 to 100% of patients, depending on the groups of patients and diseases analyzed. It results from the loss of nutrients as well as anorexia. Some patients with selective malabsorption disorders, such as lactose intolerance (lac-

Table 13-2. Comparison of Symptoms and Signs in 100 Patients with Idiopathic
Steatorrhea and 100 Patients with Tropical Sprue[a]

Symptoms and signs	Idiopathic steatorrhea (%)	Tropical sprue (%)
Lassitude	96	97
Weight loss	97	96
Diarrhea	80	95
Flatulence	59	98
Glossitis	90	90
Nausea and vomiting	37	15
Abdominal discomfort	31	90
Skin lesions	20	60
Tetany	50	0
Purpura	10	—

[a]From Floch.[3]

tase deficiency) or trehalose intolerance (diarrhea from mushrooms), may not have weight loss. Certain subjects may have a moderate degree of malabsorption without weight loss because of compensating adequate caloric intake. Others may present with the symptoms of selective absorption defects, such as osteomalacia related to calcium and vitamin D deficiencies or anemia from a selective iron or B_{12} deficiency before weight loss. The very astute clinician should suspect a malabsorption syndrome without the presence of weight loss. Keeping these facts in mind, it is still the rare patient who does not eventually show weight loss.

Diarrhea is the other major symptom of malabsorption. It was present in 80–97% of all patients with celiac disease in three different studies.[18] In our experience with tropical sprue, diarrhea occurred in 97% of all patients.[3] Most physicians have great difficulty in making the diagnosis of a malabsorption syndrome without the presence of diarrhea. However, diarrhea may be totally absent even in the presence of severe steatorrhea.

The character of the diarrhea is often the first hint that a malabsorption syndrome is developing. The patient will complain of a change in bowel habits. The stool becomes softer, movements become more frequent, with or without a foul odor, and the stool has a greasy or foamy appearance. Because of increased amounts of gas within the stool, the fecal material floats. Some patients may complain of a greatly increased amount of stool in addition to increased frequency. Others may simply complain of increased frequency and not increased amount. The character of the stool may vary with the disease causing the malabsorption. The stool seen in pancreatic insufficiency is often grayish and pasty, whereas the stool in sprue is often yellowish brown. Patients with malabsorption will rarely complain of being awakened at night to have a bowel movement or of perianal complications of hemorrhoidal or fistulous disease, as is seen with inflammatory bowel disease.

Abdominal discomfort is often a part of the symptomatology. Patients note a bloating sensation after eating. This can be correlated with dilated loops of the small bowel and disturbed motor activity from radiographic study.[14] Abdominal distension may limit food intake and aggravate weight loss. This is often manifest by the male who unbuttons his belt and the female who loosens her tight garments. Nausea is much more common than vomiting, which is not copious when it occurs.

Aphthous ulcers and burning tongue (glossitis) occur in up to 50% of all patients. The aphthous ulcer is a characteristic "punched out" ulcer of the mouth mucosa. The glossitis ranges from painful erythema, to swelling of the tongue, to atrophy of the papillae.

The physician must consider malabsorption diseases when secondary manifestations cause atypical presentations. Bone pain as a result of *osteomalacia* can be the presenting picture of idiopathic steatorrhea. *Skin manifestations* are very common in all types of malabsorption disorders, from the very small lesion of dermatitis herpetiformis,[20-22] to the flushing noted with the carcinoid syndrome, to the deep ridging or spooning of nails noted with iron deficiency anemia of hookworm disease.

The *physical findings* will vary with the disease at fault and its severity. Often the patient will present with obvious weight loss and muscle wasting, and occasionally with emaciation, while some may lose weight and look relatively well. Others lose weight precipitously. Depending on their weight before the malabsorption developed, they may literally look starved.

Few physical abnormalities may be present, or the total syndrome may be seen. The hair may be thinned, from a loss of texture as well as actual loss of hair follicles. The skin of the face may become pigmented in some disorders. In specific disorders such as carcinoid syndrome, flushing might be seen; in mast cell disease, pigmentation may also occur. Aphthous ulcers of the mouth may be seen in their characteristic form. The tongue may be reddened and "beefy" or actually white with total loss of papillae. Numerous types of skin lesions may be associated and vary from those seen in dermatitis herpetiformis to that of psoriasis and pityriasis rosacea.[20-22] The nails may be ridged or spooned, depending on the severity of iron deficiency anemia. Clubbing of the fingers may also be noted. Peripheral edema may occur due to nutritional or renal failure.[22] The abdomen may be distended, with mild tenderness. Depending on the etiology of the disease, osteomalacia or osteoporosis with associated bone tenderness may be present. With chronic poor absorption of vitamin D and calcium, there may be deformities of the extremities and evidence of pathologic fractures. If the malabsorption results from cirrhosis, the complications of that disease—ascites, increased abdominal peripheral circulation, and the splenomegaly of portal hypertension—may be seen. Peripheral edema is found occasionally in mild malabsorptive states and frequently in severe malabsorption associated with hypoproteinemia. When there is neurologic involvement, peripheral neuropathies may be evident on physical examination.

The literature has become replete with simple and complex tests to diagnose malabsorption of materials. Almost all of these have remained research tools employed by selective institutions, and only a few have been evaluated by many. The actual diagnosis of a disease that produces malabsorption may require one or many laboratory tests, the number depending on the disorder and *not* a random "battery."

All studies of absorption are based on establishing either the biochemical or the anatomic integrity of the small bowel.

Normal absorption of fats, carbohydrates, proteins, and hematologic factors is dependent on the specific functions within the bowel. Absorption studies are based on these functions.[13,23] An understanding of the basic processes of absorption, as described above, is necessary in employing the correct tests. Of selective clinical importance are the facts that (a) absorption of calcium occurs primarily in the upper small bowel, but may occur throughout the bowel if necessary, and (b) magnesium, whose depletion is being recognized regularly of late, is absorbed primarily in the ileum. Deficiency of either of these minerals leads to marked symptomatology.

Hematologic factors are important in malabsorption studies as malabsorption is often first manifested by anemia.[20,24,25] Iron is absorbed throughout the intestinal tract and requires selective processes of uptake and then transfer through the enterocyte. There is a rate-limiting process of absorption through the small-bowel mucosa that is dependent on body iron store levels.

Folic acid is absorbed primarily in the jejunum,[26] and the efficiency of absorption depends on the enterocyte. The absorption of vitamin B_{12} which occurs in the distal ileum, depends first on binding by the intrinsic factor produced in the stomach and then on receptors on the enterocyte.[25] The vitamin B_{12} intrinsic factor complex is absorbed into the enterocyte and converted to transcobalamin. Defects in gastric function, the integrity of the ileal cell, bacterial competition for vitamin B_{12} and autoimmune mechanisms may all interfere with normal absorption, resulting in anemia. Naturally, iron deficiency produces a microcytic anemia, whereas folic acid and vitamin B_{12} deficiencies produce megaloblastic anemias. A combined deficiency of iron and one of the other factors may produce a dimorphic anemia with individual characteristics not apparent until one factor is corrected.

Depending on the clinical presentation, the physician may progress rapidly to perform confirmatory tests, or may be inclined to merely screen for some parameter of malabsorption in order to decide if he should go on to a more expensive or complicated examination. Table 13-3 lists common, useful screening tests and more detailed confirmatory tests.

13.3.1. Screening for Malabsorption

13.3.1.1. Hematologic

The total white blood count may be normal. However, after folic acid or vitamin B_{12} deficiency develops, white blood cell maturation becomes affected and more mature white blood cells may be noted in the bone marrow or on the peripheral blood smears. When anemia develops, the total red blood cell count

Table 13-3. Tests Used in Screening for and Confirming the Presence of Malabsorption

Screening tests
  Hematologic
    White cell, red cell, hemoglobin, hematocrit, serum vitamin B_{12}, serum folic acid
  Biochemical
    Fats: serum carotene, serum cholesterol
    Carbohydrates: xylose tolerance test (urine), lactose tolerance
  Radiographic
    Small-bowel barium X-ray examination

Confirmatory tests
  Hematologic
    Schilling test with and without intrinsic factor
  Biochemical
    72-hr stool chemical fat
  Breath tests
    H_2 excretion, lactose intolerance
    Bile salt malabsorption
  Parasitic
    Stool and duodenal aspiration analysis
  Peroral biopsy
    Pathology
    Enzyme analysis (lactase)

may fall. Depending on whether the anemia is due to a deficiency of iron, folic acid, or vitamin B_{12}, the type of peripheral red blood cell will vary. Iron deficiency will produce microcytic hypochromic cells, whereas the other types produce macrocytic hyperpigmented cells.[25,26]

Serum vitamin B_{12}[25] *and serum folic acid levels*[26] are easily determined and can be of great value in assessing the deficiency. Similarly, serum iron levels may be of some help, but one must always be cautious in interpreting serum iron levels as they may reflect a blood loss rather than nutritional deficiency or malabsorption.

13.3.1.2. Biochemical

Determination of *serum carotene* and serum cholesterol can be of value in screening for malabsorption.[20–23] Because both of these blood levels reflect intake as well as malabsorption, they are merely screening tests. The serum carotene level has been of value when markedly depressed. Specifically, a serum level of less than 20 μg is rarely seen except in gluten enteropathy and severe tropical sprue. The serum carotene level is spuriously high in diabetes even if malabsorption is present. Low serum cholesterol levels are seen in hyperthyroid patients, and while rarely of diagnostic value in malabsorption syndromes, they may arouse clinical suspicion.

The *serum albumin* level reflects stores, and a drop to abnormal levels indicates malabsorption, loss, or liver disease (see Chapters 5 and 10).

Carbohydrate tolerance tests are of value. The *xylose tolerance* test,[20,23] which is performed by oral ingestion of 5 or 25 g followed by a 5-hr urine collection, has gained worldwide acceptance. In Asian countries it is used for massive screening. It correlates well with small-bowel malabsorption despite the facts that false-negative tests occur and false-positive results are occasionally seen in pancreatic diseases. It is useful as a screening test when small-bowel biopsy is premature in the diagnostic work-up.

Xylose is absorbed completely from the upper small bowel and only small amounts are metabolized. The majority is excreted in the urine. Any patient unable to excrete at least 20% of the ingested dose within 5 hr is suspect of having a malabsorption syndrome. Technical factors are important in performing the test. There should be adequate fluid intake and adequate urine output. Renal function must be normal, as poor clearance will result in false-negative results.

The *lactose tolerance* test [27-29] is of importance in diagnosing both primary and secondary lactose malabsorption (also see Chapter 17). The primary syndrome is hereditary, while the secondary form is associated with all types of small-bowel disease. Therefore, once lactose intolerance is demonstrated, at least a small-bowel barium study should be done. The easily performed test requires the ingestion of 50 g of lactose and serial blood samples. It is important that a blood sample be taken within 15 to 30 min after ingestion, as the peak rise occurs most often in that period. It should be at least 20 mg higher than the fasting sample. The rise may occur early and consequently should not be missed. A 2-hr test is more than adequate to diagnose lactose intolerance. As some patients will show almost normal rises in blood glucose but still have symptoms from the disaccharide, most observers feel that this test's greatest value is in its correlation with symptoms. The majority of patients who do have an adequate rise in blood sugar will not have symptoms, and those who have lactose malabsorption will have abdominal distension, discomfort, or diarrhea from the lactose. However, the clinician must keep in mind that there is a group of patients in whom no symptoms develop even though they have flat blood levels, and a group in whom symptoms are produced even though there is an adequate rise in blood sugar. If I find no symptoms after a lactose load, I do not make a diagnosis of lactose intolerance or obtain blood levels, for it would be meaningless clinically. The purist will perform a small-bowel biopsy and have it assayed for lactase, but this is hardly necessary in routine diagnostic work, for it is the symptom complex induced by lactose that is important.

The hydrogen breath analysis test has been described in the assessment of lactase deficiency and lactose intolerance.[30,31] After the administration of only 12.5 g of lactose, which is the amount present in a glass of milk, six hourly breath analyses determine whether hydrogen has been absorbed from the intestine in significant amounts and excreted in the breath. With a negative baseline this would occur only after bacterial production of hydrogen after the ingestion of the lactose. The same principle is employed in the diagnosis of sucrase deficiency[32] and intestinal bacterial overgrowth with resultant malabsorption.[33]

The *glucose tolerance test* is essentially useless in a differential diagnosis of the malabsorption syndrome. On occasion, a flat test may raise suspicion, but false-negative results are so frequent that it is useless alone.

13.3.1.3.   Radiologic (Figs. 13-9 and 13-10)

The only reliable radiologic technique for examining the small bowel is the oral ingestion of barium. Details of this examination and its interpretations are described elsewhere.[3,34] Several simple facts are important in the interpretation of the barium study. There should not be any significant obstruction at the gastric outlet, and a significant amount of barium should be ingested orally. There has been controversy over whether 8 oz or up to 24 oz is important, but interpretations can readily be made with at least an 8-oz barium meal.

The important findings are a loss of change of the folds of the small bowel, flocculation of the barium, puddling and segmentation of the barium medium, and dilatation of the lumen in the bowel. The exact causes of these phenomena in small-bowel disease are not clear. Many observers feel that they result from a change in the luminal contents of the small bowel. Regardless of the cause, the findings are present when there is significant malabsorption and absent except when there are selective singular absorption defects. The use of very refined barium media may create false-negative tests. I have seen complete masking of the malabsorption pattern. Consequently, a negative small-bowel finding is no longer reassuring if refined micronized barium is used. If standard barium mixtures are used, it is very helpful.

The classic pattern of malabsorption does not often differentiate the cause of the malabsorption; any disease causing malabsorption may give the same appearance on barium evaluation. Filling defects or stenosis of the small bowel noted from the barium study point to malignancy or inflammatory bowel disease. Carcinoma, lymphoma, or inflammatory disease may merely give a diffuse malabsorption pattern, but when filling defects or stenotic areas are noted, then these disorders should be suspected. This is of importance because the physician may often be forced to employ an open surgical pathological diagnosis when those findings are present.

13.3.2.   Confirmatory Tests

13.3.2.1.   Hematologic

The Schilling test,[25] or radioactive vitamin B_{12} absorption study, is of importance in identifying diseases causing vitamin B_{12} malabsorption. In classic pernicious anemia, intrinsic factor (IF) is absent. When vitamin B_{12} is ingested in the absence of IF, the test is abnormal, but when IF is added, the test result is normal. However, when diffuse small-bowel disease is present, there is abnormal absorption with or without IF. The exact diagnostic figures will vary in different isotope laboratories.

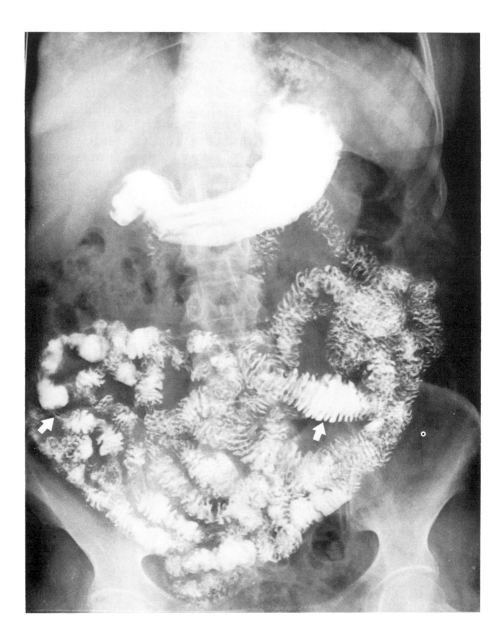

Figure 13-9. Photograph of barium-filled normal small intestine. Stomach is outlined at top. **Note** the feathery filling of the jejunum on the right with one loop filled (arrow) outlining the valvulae conniventes. Note the numerous contractions in the normal partially filled ileum at the head of the barium meal on the left (arrow).

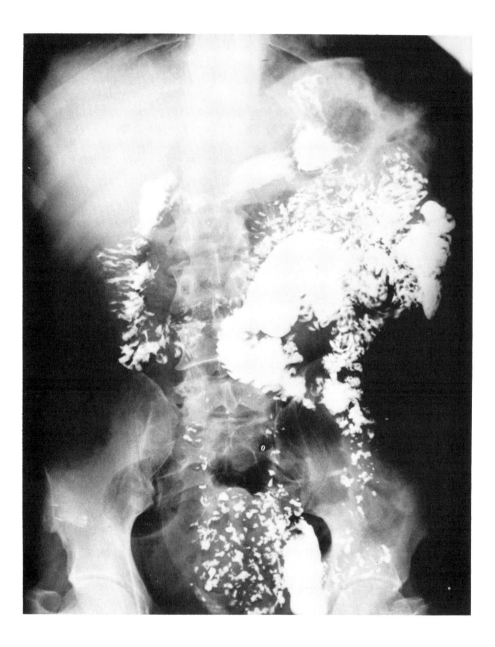

Figure 13-10. Photograph of barium-meal X-ray of a patient with tropical sprue taken 15 min after ingestion of the barium. Compared to Fig. 13-9, note the loss of normal markings, the puddling and segmentation of the barium, and the dilated loop of jejunum in the center with loss of the normal folds.

13.3.2.2. Biochemical

Determination of steatorrhea by the chemical measurement of *fecal fat* is the only universally accepted test.[20,23] It is confirmatory of nonspecific malabsorption and essential in making a diagnosis. In a normal individual at least 94% of ingested fat is absorbed. Therefore, a subject eating 100 g of fat per day will have less than 6 g of fat in the stool per day. Because bowel patterns and habits are so varied, the test is most reliable when a 3-day stool excretion is measured. In some laboratories a 1-day test has been found effective; others have stressed the need for 5-day collections. However, it is well accepted that a 72-hr collection of stool correlates best with disease. Excretion of between 6 and 8 g/day is considered a marginal value, but over 8 g/day is abnormal. The finding of abnormal excretion of fat indicates steatorrhea and is clinically synonymous with a malabsorption syndrome. In selective malabsorption entities such as carbohydrate or vitamin malabsorption syndromes, steatorrhea is not present, but in all diffuse malabsorption disorders the chemical stool fat is elevated.

Stool nitrogen may also be an indicator of malabsorption, but it is a difficult test and rarely employed in a differential diagnosis of malabsorption disorders.

Breath tests are the most recent addition to the armamentarium of absorption studies. Isotopically labeled compounds can be measured in expired gas as a result of malabsorption due to bacterial breakdown in the intestine, or as the result of normal absorption due to normal enzymatic digestion. Abnormal gases also can be measured.

The appearance of hydrogen in expired air is abnormal. It can be used as a measure of lactase deficiency and lactose malabsorption, as was first reported by Levitt and Donaldson[35] and Calloway and colleagues[36] and advocated by Metz and colleagues.[30-33] It is particularly useful in the study of carbohydrate malabsorption. When the particular carbohydrate being ingested is metabolized by bacteria, hydrogen is produced, which is absorbed and then expired in air. Ordinarily, hydrogen is not present in expired air in any significant amount. The ability to measure it after the ingestion of any carbohydrate such as lactose indicates that the carbohydrate has been metabolized by bacteria rather than broken down by the gut enzymes.

Also of clinical application has been the use of [^{14}C]cholyglycine in analyzing the malabsorption of bile acid.[37-39] Although it appears to be slightly less sensitive than other parameters of distal ileal function, it is still useful when available because of its simplicity. When there is abnormal absorption and metabolism of bile acids, there is an increased excretion of $^{14}CO_2$. This occurs because of more rapid release of glycine, which is absorbed and rapidly metabolized, resulting in increased amounts of labeled CO_2 excretion. Although this test is useful, the sensitivity must be interpreted carefully in measuring increased bacterial contamination of the small bowel.[30,40]

The use of labeled Triolein appears to be the most specific and sensitive screening test for steatorrhea.[41] It correlates very well with fecal fat. In this test, lack of recovery of labeled carbon in expired air means malabsorption of the ingested labeled Triolein.[41]

A study for *infectious and parasitic agents* is extremely important. In acute diarrheal states the bowel should be studied for infectious agents. Transient malabsorption has been well documented with giardiasis, strongyloidiasis, and hookworm.[42,43] Helminths are readily found in stool, but *Giardia lamblia* can be missed and it may be necessary to perform duodenal drainage to find the organisms. By applying both stool and duodenal drainage studies for *Giardia* the incidence of recovery can be increased.[42]

Rare entities such as trehalose intolerance, as suggested by a careful history of mushroom ingestion, can be diagnosed only by clinical suspicion and specific testing.[17]

The greatest use of *peroral small-bowel biopsy* is to obtain histologic evidence for the diagnosis of the cause of a malabsorption disorder.[45] Once malabsorption is suspected and confirmed by chemical stool fat, or the suspicion is great enough from radiographic study, then a peroral small-bowel biopsy is essential. The techniques most easily performed today use the simple Carey or Crosby capsules or a suction tube such as the Brandborg–Rubin tube. Research centers may use the multiple biopsy retrieval tube, but this is rarely needed for diagnostic purposes and should be employed only in research centers. *Peroral biopsy is essential in the differential diagnosis of most diseases causing malabsorption and no patient should be relegated to long-term treatment without a biopsy specimen confirmation of the disorder.* It is essential for a diagnosis of either gluten enteropathy or tropical sprue, and a diagnosis of eosinophilic gastroenteritis, lymphoma, Whipple's disease, lymphangiectasia, or acanthocytosis may be confirmed in the biopsy specimen. In most malabsorption disorders of other origin, however, the specimen may be normal or merely demonstrate some shortening of villi, increased round cell infiltration in the lamina propria, and mild disturbance of the epithelial cells. The classic findings in gluten enteropathy are loss of all villi, but no thinning of the mucosa, for there is a concurrent increase in the crypt area; marked chronic round cell infiltration of the lamina propria; and round cell infiltration of the epithelium with a change from normal columnar to cuboidal cellular epithelial structure. The brush border containing decreased microvilli becomes relatively ineffective.

It is not possible to review all of the histologic findings in all disorders, and the interested reader is referred to Table 13-4 and a monograph on the subject.[45]

Small-bowel mucosa may be needed for biochemical analysis in identifying selective enzyme deficiencies.[29] Specifically, this has been used in determining lactase, sucrase, and maltase deficiency in disaccharidase malabsorption. However, most clinical centers and almost all clinicians not working in academic centers will rarely use it for this purpose.

13.3.3. Differential Diagnosis and Dietary Management

When a diagnosis of steatorrhea is established by the presence of abnormal fecal fat excretion, a protocol for the differential diagnosis of gross fat malabsorption can be carried out, such as that shown in Fig. 13-11.

Evidence of pancreatic insufficiency must be obtained to make a diagnosis of *pancreatic disease* as the cause of malabsorption. The patient with cystic fibrosis will demonstrate severe malabsorption but also have an abnormal sweat electrolyte test, decreased stool trypsin activity, and a relatively normal small-bowel biopsy. The other biochemical tests or roentgen findings may be of no help in the differential diagnosis.

The patient with chronic pancreatic insufficiency caused by chronic pancreatitis demonstrates abnormal glucose tolerance, a history of elevation of amylase or lipase, and possibly pancreatic calcification. Endoscopic retrograde pancreatolography is of some help in the diagnosis of chronic pancreatitis. Usually the clinical history and one or more of these findings will make the diagnosis easy. The small-bowel biopsy is relatively normal, although the roentgen series and chemical stool fat are markedly abnormal.

The establishment of the disease or disorder causing the malabsorption may tax the acumen of the internist and gastroenterologist. Those caused by pancreatic and gastric diseases and their correct nutritional management are discussed in Chapters 8 and 12. In this chapter, we will concern ourselves with the nutritional treatment of small-bowel diseases.

Table 13-4. Information Provided by Small-Intestine Biopsy[a]

Disorders in which biopsy is diagnostic	

Diffuse lesions

Whipple's disease	Abetalipoproteinemia
Immunodeficiency syndromes	(Gluten enteropathy)[b]

Patchy lesions

Lymphoma	Parasitic infestations:
Lymphangiectasia	Giardiasis
Eosinophilic enteritis	Coccoidiosis
Systemic mastocytosis	Strongyloidiasis
Amyloidosis	Capillariasis
Crohn's disease	Cryptosporidiosis

Disorders in which biopsy is abnormal but not diagnostic	
Gluten enteropathy (celiac)[b]	Severe folate and B_{12} deficiency
(may be classic)	Radiation enteritis
Tropical sprue	Radiomimetic drugs
Unclassified sprue	Malnutrition
Infectious gastroenteritis	
Intraluminal bacterial overgrowth	

Disorders in which biopsy is usually normal	
Postgastrectomy malabsorption	Cirrhosis
Pancreatic exocrine insufficiency	Hepatitis
Ulcerative colitis	Iron deficiency anemia

[a]Modified from Trier.[44]
[b]Some gastroenterologists feel this is a classic lesion and may be diagnostic when other diseases are ruled out.

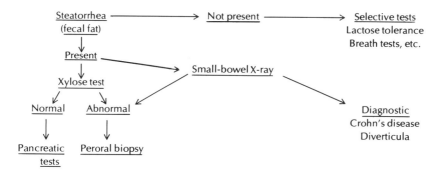

Figure 13-11. Protocol for malabsorption work-up.

The following classification of malabsorption disorders was first presented by Wilson and Dietschy[47] and modified by Gray.[23] The diseases and their recommended nutritional therapy are listed. The classic small-bowel malabsorption disorder, gluten enteropathy, is discussed in detail as well as nutritional therapy for those diseases that require nonspecific support rather than specific therapy.

Disease	Recommended Diet Therapy
1. Intramural small-bowel disease	
Gluten enteropathy	Gluten-free diet (Chapter 20, diet 16)*
Disaccharidase deficiency	Lactose-, maltose-, or sucrose-free diets (Chapter 20, diets 17–19)
Tropical sprue	Supportive
Dermatitis herpetiformis	Plus or minus gluten free
Nongranulomatous jejunitis	Supportive
Whipple's disease	Supportive
Amyloidosis—primary or secondary	Supportive
Eosinophilic gastroenteritis	Variable
Food allergy	Variable
Small-bowel ischemia	Supportive
Small-bowel resection—minor or massive	Supportive
Intestinal lymphagiectasia	Supportive
Abetalipoproteinemia	Supportive
Lymphoma	Supportive

*See discussion for treatment failures.

Disease *(cont.)*	Recommended Diet Therapy *(cont.)*
2. *Parasitosis*	
Hookworm	Supportive
Stronglyoidiasis	Supportive
Coccidioidomycosis	Supportive
Schistosomiasis	Supportive
Capillariasis	Supportive
Giardiasis	Supportive
Radiation enteritis	Supportive
3. *Malabsorption caused by multiple defects*	
Ileal dysfunction due either to resection or to Crohn's disease of the ileum	Supportive
Zollinger–Ellison syndrome	Supportive
Scleroderma	Supportive
Postgastrectomy	Supportive
Mast cell disease	Supportive
Diabetes mellitus	Supportive
Endocrinopathies	Supportive
4. *Insufficient intraluminal pancreatic enzyme activity*	
Chronic pancreatitis	Supportive and pancreatic replacement as needed
Pancreatic carcinoma	Supportive and pancreatic replacement as needed
Pancreatic resection	Supportive and pancreatic replacement as needed
Cystic fibrosis	Supportive and pancreatic replacement as needed
5. *Insufficient intraluminal bile acid activity*	
Biliary obstruction with and without jaundice	Supportive
Bacterial overgrowth syndrome due to stasis or diverticula	Supportive

Disease *(cont.)*	Recommended Diet Therapy *(cont.)*
Cholecystocolonic fistula	Supportive
(Biliary resection and Crohn's disease)	Supportive

Diet therapy can be divided into
1. Specific nutritional treatment
   a. Gluten enteropathy
   b. Disaccharidase deficiencies
2. Nonspecific supportive nutritional therapy

Whenever there is a specific treatment that will cure the malabsorption, there is no need for supportive treatment following successful treatment. The need prior to completion of specific therapy varies. Gluten enteropathy is treated by removal of gluten from the diet. The patients usually respond rapidly and little else is needed. However, should the patient be markedly anemic, then iron or immunologic factors may be needed initially in the therapy. Direct replacement of gross deficiencies are needed during acute illness. In a disorder such as Whipple's disease, following successful treatment with antibiotics, the cause of the malabsorption is eradicated and the patient absorbs normally. However, should there have been some gross deficiency due to long-term malabsorption, then an added nutritional supplement may be needed. After a specific treatable malabsorption disorder is cured, and after early nutritional supplementation, the disease should not recur if the diagnosis and treatment are correct and followed by the patient.

13.4.   GLUTEN ENTEROPATHY

If one understands this classical disease entity, then one understands small-bowel function and all possible diseases of the organ. Almost every possible manifestation of abnormal digestion and absorption has been noted. Almost every possible systemic manifestation, from a variety of skin disorders to malignancies, is associated with the disease. It was first clearly described in 1888 by Samuel Gee[48] as the celiac affliction and later came to be known as celiac disease. Herter[49] publicized the disorder in 1908 and recognized increased loss of fat in the stool resulting from impaired intestinal absorption. It was not until the 1950s that Dicke, Weijers, and van de Kamer,[50–53] in a series of classical experiments, found that removal of wheat from the diet caused the signs and symptoms of celiac disease to disappear, whereas reintroduction of the offending agent caused a recrudescence.

Gee had first recognized the similarity between idiopathic steatorrhea and celiac disease, but it was Cooke and co-workers[54] who showed that 43 of 100 patients with idiopathic steatorrhea had a definite history of childhood celiac disease. Later, Shiner[2] demonstrated the similarity of mucosal biopsy specimens between patients with childhood celiac disease and those with adult idiopathic steatorrhea. It was then demonstrated that mucosal changes could be reversed on a gluten-free diet and reproduced locally by instilling wheat through a tube onto the mucosa in a susceptible patient.[55,56]

Most physicians agree that the definition of celiac disease should include only those cases that are caused by gluten and can be reversed on a gluten-free diet.[59] However, there are some who still feel that the disease should be described as a celiac disorder and include such conditions as collagenous sprue in which the mucosal atrophy cannot be reversed by a gluten-free diet.

The exact incidence of the disease is not known. Cases from Africa and China have not been described, although they have been reported from India. The disease occurs primarily in Europe, North America, and Australia. The incidence is highest in Switzerland, where it occurs in one of 890 births, and lowest in Sweden, where it is seen in one of 6500 births, the frequency varying between these two extremes in the other temperate-climate countries of the world.[57]

The basic defect in celiac disease remains controversial. One group believes that there is a specific lack of an intestinal peptidase necessary for the ultimate detoxification of gluten, while the other group believes that patients with gluten-sensitive enteropathy are genetically prone to the development of localized immune reactions when exposed to dietary gluten.[58–65] However, it is clear the α-gliadin fraction of gluten is responsible for the damage to the small-bowel mucosa.[59] Gluten is the protein of flour and consists of four peptides: Gliadins, glutenins, albumins, and globulins. The latter two are only present in small amounts. Glutenins are very high in molecular weight and relatively insoluble. The gliadins consist of four amino acid groupings that can be demonstrated by starch-gel electrophoresis to consist of rapid α, β, and γ fractions, and slow "w" fraction. It is in the α fraction that the toxic properties exist. Three components have been demonstrated in the α fraction.[60]

The toxic α-gliadins are primarily found in wheat gluten, but occur in rye and barley as well.[61,62] Analysis of gluten by other workers reveals a definite histologic toxicity from wheat, barley, and rye, but no apparent toxicity is produced by oats.[63,64] The work of Dissanayake and colleagues[64] is impressive, but the effect of oat protein is controversial and the clinician must keep this in mind. All agree that maize and rice flour are harmless to all gluten-sensitive patients.[64]

Cornell and Townley[62] suggest that the major defect in gluten-sensitive individuals is their inability to digest α-gliadin. This inability occurs because of a lack of a specific peptidase in the intestinal mucosa. Their evidence as yet is not accepted by all observers.[62]

There is no question that the ingestion of gluten in a toxic subject activates a local mucosal immune system. When this happens, there is a demonstrable antigen–antibody reaction in the mucosa.[65] Further evidence that the immunologic picture is important in this disease is the appearance of antireticular antibodies,[66] the appearance of circulating immune complexes,[67] the reactivity of lymphocytes,[68] and the most recent correlation of HLA-DW3[69] and HLA-B8 antigens.[69,70] This association of HLA antigens suggests that the fundamental abnormality of gluten enteropathy is a binding reaction between gluten protein and intestinal cells that is determined by a histocompatible gene. This binding causes a local mucosal immune reaction. This explanation would further the theory that the disease is a genetically inherited condition. Of particular interest is the association of dermatitis herpetiformis and gluten enteropathy. These patients also seem to have increased frequency of HLA-B8 antigens and their intestinal disease is always reversed, and in some patients skin lesions as well, by gluten withdrawal.[71,72]

The clinical picture can have great variation. The initial descriptions of the classic celiac child and adult were emaciated subjects with signs of avitaminosis and the symptoms of marked steatorrhea. The classic celiac patient presents with weight loss, diarrhea consisting of large, bulky stools often foul smelling and buoyant, loss of appetite, distension after meals, gaunt appearance, perhaps a diffuse variety of skin lesions. However, the advent of the peroral biopsy and awareness of the selectivity of malabsorption reveal patients may present early with mild anemia, osteoporosis, or mild chronic infections and first appear on orthopedic, neurologic, immunologic, gynecologic (infertility), psychiatric, or derminologic services.[73] Reports have even demonstrated that they may present as diffuse pulmonary disease.[74,75] The symptoms and signs are described above (Table 13-2).

Once suspicion is aroused, the diagnosis for that particular patient should be pursued. As noted above, classic gluten enteropathy is characterized by decreased xylose absorption, very low serum carotene, moderately to markedly increased fecal fat excretion, abnormal breath test, decreased vitamin B_{12} absorption, and a diffusely abnormal peroral small-bowel biopsy specimen that reveals sub-total villus atrophy with complete loss of villi and markedly abnormal epithelium. The lamina propria will be heavily infiltrated with round cells. The total mucosal width may be normal because of increased infiltration and decreased maturation in the crypt areas (see Figs. 13-12 and 13-13). Although the biopsy specimen findings are not specific, once the clinical picture is matched with the abnormal chemistry and the peroral small-bowel biopsy findings, the clinician can feel safe in making the diagnosis of gluten enteropathy. It should be kept in mind that these patients have an increased incidence of lymphoma that may well exist in association with the abnormal mucosa. The barium radiographic study of the small bowel is of help in this differential diagnosis. It may reveal the classic malabsorption pattern and also yield information on selective infiltrating lesions in the bowel. It also appears that other neoplasms may be more common in these patients, but ischemic heart disease is less common.[76]

A select group of these patients have IgA deficiency. This may lead to impaired resistance to certain diseases such as poliovirus poliomyelitis.[77,78] Although there are scattered reports of IgA deficiency in gluten enteropathy patients, the entire spectrum of serum and intestinal immunoglobulins is much more varied. In any given celiac patient there may be increases and decreases both in the serum and in the excretions into the gut. IgA is often raised, while IgM and IgC are often decreased.[59]

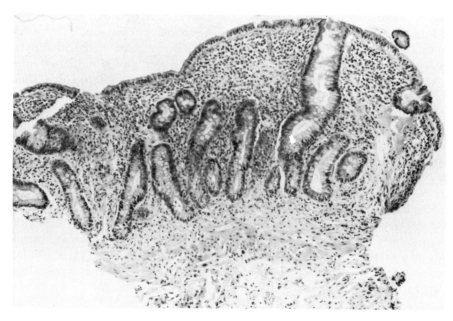

Figure 13-12. Low-power photomicrograph (× 100) of a peroral small-bowel biopsy specimen from a patient with gluten enteropathy. Compared to Fig. 13-5, note the complete loss of villi and the heavy infiltrate of white blood cells in the lamina propria.

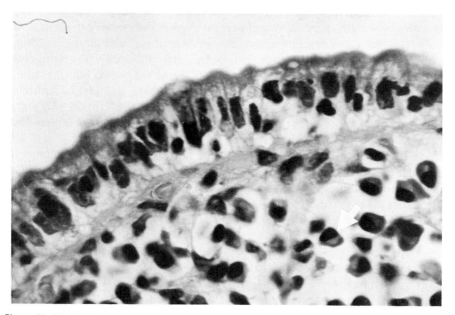

Figure 13-13. High-power photomicrograph (× 400) of epithelium on villus of gluten enteropathy small-bowel specimen shown in Fig. 13-12. Compared to Fig. 13-6, note the heavy infiltrate of lymphocytes in the epithelial cells, the loss of basement orientation of nuclei, and the large number of plasma cells in the lamina propria (arrow).

The *treatment* is total removal of gluten from the diet. The experiments of Rubin *et al.*[56] and Shriner and Ballard[65] demonstrate small amounts of gluten can cause a marked reaction in the intestinal mucosa. Therefore, the sick gluten enteropathic patient must completely eliminate gluten from the diet. The patient and his family must be carefully instructed about the diet and the reactions caused by the protein. The patient with classic gluten enteropathy should respond within weeks and the dramatic improvement will be rewarding to both physician and patient.

The patient who has few symptoms and merely a mild anemia may be less motivated to undertake the rigorous dietary regimen. Furthermore, once patients are improved and several years pass, there is evidence that indicates they tend to return to some form of gluten intake. An evaluation of 130 patients who were lost to follow-up and did not continue their diets revealed 13% were anemic and only one patient had died of a small-bowel lymphoma.[79] It appears that the gluten-free diet may be important in preventing malignancy in these patients.[80] However, there is no question that once these patients are relieved of the main thrust of the disease, they appear to be able to tolerate some gluten. The question then arises, is it absolutely necessary to insist on a lifelong gluten-free diet? Only the patient, in consultation with the physician, can make this decision. It is essential that gluten be eliminated from the diet to relieve the patient of severe signs and symptoms of malabsorption. However, once these are relieved, the above question can then be discussed. Those who accept the statistics of associated malignancy will insist that the restriction be lifelong. Those who wish to gamble will weigh the odds.

A complete list of recommended intake and diets for the gluten-free patient is included in Chapter 20, and there are several texts on the diet that should also be consulted.[81,82]

It is also important to remember that the patient suffering from active gluten enteropathy will often have marked intolerance to lactose because of his disaccharidase deficiencies.[83] It is wise to avoid milk and other lactose-containing foods during the acute phase until there is regrowth of the intestinal mucosa and redevelopment of the disaccharidase enzymes (Chapter 20, diet 17). Once the gluten enteropathic patient returns to normal, he should be able to tolerate lactose and disaccharidases.

Treatment failure may be due to concomitant pancreatic insufficiency.[83a]

13.5. INTESTINAL BRUSH-BORDER PROTEIN DEFICIENCY DISEASES (CARBOHYDRATE INTOLERANCES)

These diseases were described as carbohydrate intolerance. However, this is less accurate than the term suggested by Alpers and Seetharam.[84] The intolerances occur because of enzyme deficiencies in the brushborder of the epithelial cells of the intestinal mucosa. When an enzyme is deficient, the intestine is unable to digest the particular substance on which the enzyme acts, which results in an intolerance or malabsorption of that substance.

The protein enzymes are situated within the microvilli and include disaccharidases (maltase, sucrase, lactase, trehalase), peptidases (oligoaminopeptidase, γ-glutamyltranspeptidase, enterokinase), and phosphatases (alkaline, calcium, magnesium ATPase). They hydrolyze the particular substrate in the lumen or on the outside of the cell surface. The digested monosaccharides are then absorbed. Apparently, the protein enzymes can be easily digested by proteolytic enzymes within the lumen; consequently there is a dynamic homeostatic process of production and destruction of the disaccharidase enzymes.

Table 13-5, modified from Alpers and Seetharam,[84] is a classification of enzyme disturbances and proposed mechanisms associated with each.

The dietary treatment for each of these disorders is included in Chapter 20. Lactase deficiency is most common and will be described in detail. The interested reader is referred to Gryboski's[85] text on infants for more detailed descriptions of the less common sucrase–isomaltase deficiency[84-88] and even less common glucose–galactose malabsorption.[89-93]

13.5.1. Lactase Deficiency–Lactose Intolerance

Although lactose is only one of the three main disaccharides found in foods, the others being sucrose and maltose, it is the only carbohydrate associated with a significant amount of clinical intolerance. In 1910, von Reuss[93] first noted an abnormality in disaccharide absorption and excretion. Not until 1959 did Holzel and co-workers[94] carefully describe defective lactose absorp-

Table 13-5. Abnormal Brush-Border Enzymes in Man[a]

Disease	Alteration in protein	Proposed mechanism
Sucrase–isomaltase deficiency	Sucrase absent	Decreased or altered synthesis
Enterokinase deficiency	Enterokinase absent	Decreased or altered synthesis
Lactase deficiency	Decreased lactase	Decreased or altered synthesis
Vitamin D deficiency	Decreased ATPase	Decreased or altered synthesis
Familial vitamin B_{12} absorption	Altered receptor	Decreased or altered synthesis
Glucose–galactose malabsorption	?	Decreased or altered synthesis
Neutral amino acid malabsorption	?	Decreased or altered synthesis
Stagnant-loop syndrome	Decreased disaccharidases	Increased degradation
Gastroenteritis	Decreased lactase	Increased degradation
Alcoholic pancreatic insufficiency	Increased disaccharidases	Decreased degradation
Cystic fibrosis	Increased disaccharidases	Decreased degradation
Diabetes mellitus	Increased disaccharidases	Unknown
Trehalose malabsorption	Decreased trehalase	Unknown
Primary bile-salt malabsorption	?	Unknown

[a]Adapted from Alpers and Seetharam.[84]

tion that resulted in infant disease. Since that time, lactose malabsorption and intolerance in adults have been carefully studied, clinically defined, and become treatable. The incidence of the disorder varies throughout the world. It occurs in 85 to 100% of Asian subjects. Some observers report a 100% incidence in the Chinese.[96] In Negro subjects the incidence is anywhere from 70 to 90%, in Mediterranean subjects anywhere from 40 to 90%, and in English, Nordic, and Caucasian subjects from 5 to 20%.[27,95–100] The wide range of incidence among Asians, Negroes, Mediterraneans, and Caucasians indicates certain genetic and developmental lines from which several controversial theories have evolved. Simons[102] carefully reviewed the published reports on lactose intolerance throughout the world and concluded that there is no simple explanation for the differences among groups and that "one must acknowledge the role of a series of interrelated influences deriving not only from human physiology but from human ecology and culture as well." Some have felt that lactase sufficiency, resulting in the ability to handle lactose, is the result of a genetic ability to develop enough lactase, whereas others have felt that this has been strictly adaptation, not genetic.[101,102] Regardless, there are clear differences among populations in the world today. Lactose intolerance is also clearly associated with disease of the small bowel.[97] In these instances, the enzymes are depleted due to damage to the epithelium.

13.5.2. Pathophysiology

Lactase, as other disaccharidases, appears in the intestine in adequate amounts in infancy. It catalyzes the hydrolysis of lactose to its two component hexoses, glucose and galactose. It has been detected in 15- to 20-week-old fetuses. In healthy humans able to tolerate lactose, the lactase level ranges from 39 to 258 units/g intestinal protein.[85] There are three different types of lactase.[88] The first hydrolyzes lactose and cellobiose optimally at pH 6; the second hydrolyzes lactose and synthetic substrate at pH 4.5; the third hydrolyzes only synthetic substrate at pH 6. Both the first and the third forms have been shown to be deficient in lactose intolerance, whereas the second type increases in lactase insufficiency.[88] In general, the levels of intestinal lactase are greatest in the perinatal period and lowest in adulthood. In subjects able to tolerate lactose, the lactase levels remain high throughout life, whereas lactase-insufficient subjects have lower levels in adulthood. The lactase level of lactase-insufficient adults appears to fall to about 3 units/g intestinal protein.[103] Studies have shown no significant differences in the activities of human galactosidase.

Symptoms may occur when there is insufficient lactase, lactose being neither hydrolyzed nor absorbed. When unabsorbed lactose remains within the lumen of the bowel and moves intact down the small intestine to the areas where the intestinal microflora is abundant, bacteria ferment the carbohydrate to produce organic acids, lactic acid, carbon dioxide, and hydrogen. In addition, the increased amount of disaccharides increases the osmotic pressure, which causes water to enter the intestinal lumen. The presence of the organic acids and the increased water load results in symptoms of either discomfort due to intestinal distension or diarrhea.

13.5.3. Diagnosis

Lactase insufficiency is either *primary*, as occurs in those patients without intestinal disease, or *secondary*, as occurs when there is inflammation in the intestine or disease of the epithelial cells. In either type, the diagnosis depends on the demonstration of symptoms and malabsorption after the administration or an oral dose of lactose.

In the standard *lactose-tolerance test*, 50 g of lactose is given in water or a fruit drink. Blood glucose levels are determined at 0, 15, 30, 60, and 120 min after the oral dose. The blood glucose level should rise at least 20% over the basal level if lactose is hydrolyzed and absorbed. The 50-g dose usually discriminates between lactase-sufficient and -insufficient intestine for it correlates well with decreased lactase levels in peroral intestinal mucosal biopsy specimens. However, for practical reasons, the most important observation is whether the subject develops symptoms during the first few hours after ingestion. Results fall into four categories. (1) Of the patients who are lactase insufficient, over 80% will develop symptoms. However, (2) 20% are insufficient and will not develop symptoms. Furthermore, (3) there are lactase-sufficient subjects who develop symptoms, and (4) there are lactase-sufficient subjects who can handle this large dose of oral lactose without symptoms.

Because of the overlap in responses to the oral test, the clinician may have to repeat the test or continue his observations with lactose-containing foods such as milk.

A test that is still in the developmental phase may be very helpful and simpler to perform than repeated blood analyses. In the *breath test*, oral lactose is given and serial samples of breath collected and measured for hydrogen.[30,31,36] The test is more difficult in that the laboratory must be able to measure small amounts of hydrogen; nevertheless the test is very simple to perform. Radioactive carbon in carbon dioxide can also be used. When bacteria ferment nonabsorbed lactose, they free the labeled carbon so that $^{14}CO_2$ is produced, which can be measured in expired air.[104,105] Although these tests are sophisticated and helpful, they are not ordinarily necessary to make the diagnosis.

Barium studies are also performed in which lactose is added to the barium mixture. When patients are intolerant, the small-bowel pattern will appear abnormal. This test is also not required to make a diagnosis, but it can be used as an added test or substituted for the oral or breath test.

Measurement of the stool pH is particularly helpful in children when lactose intolerance is suspected. Ordinarily the stool pH is close to neutral; when there is increased fermentation due to disaccharides in the stool, the pH may fall to 5, and certainly lower than 6.

During the ordinary lactose-tolerance test, the stool can be screened for the presence of disaccharides with clini-test tablets.[106] Any amount of sugar in the stool resulting in greater than a 1+ test is suspicious. This test is particularly helpful with children, where it is difficult to obtain blood.

13.5.4. Correlation of Lactose Load and Development of Symptoms

Bond and Levitt,[105] using isotope and breath tests, studied the amount of lactose malabsorbed in lactase-deficient subjects. After a 15-g dose they found that from 42 to 75% of the lactose was not absorbed, compared to 0 to 8% in lactase-normal subjects. Their study suggested that individual differences in susceptibility to diarrhea after milk ingestion was probably due to differences in the quantity of lactose not absorbed. Lebenthal and associates[107] studied patients aged 6 weeks to 50 years. Performing over 1000 jejunal biopsies and correlating lactase activity with tolerance and milk consumption, they found that after the age of five, two distinct groups of subjects emerged. One group, approximately 25% of the total population, was characterized by low lactase activity; the second group maintained the same level of lactase activity as that of the first 3 yr of life. The low-lactase group developed clinical lactose intolerance, consumed relatively small amounts of milk, and usually had a flat curve; the second group consumed more milk, averaging approximately 1 quart/day, and usually had normal oral lactose tolerance test results. Bond and Levitt concluded that lactase insufficiency manifests itself in the form of lactose intolerance usually after the age of five.[107]

Bedine and Bayless[108] published a definitive study that quantitated the amount of milk that could be tolerated by low-lactase subjects. Their study pointed out that although a subject may or may not be able to tolerate 50 g of lactose, it is only the subject's ability to tolerate a specific amount that really matters to that subject. Twenty low-lactase subjects were given from 3 to 96 g of lactose to determine the threshold level for symptoms. Seventy-five percent of these subjects had symptoms of either flatulence, distension, cramps, or loose bowel movements with 12 g of lactose (which is the amount in 8 oz of milk) or less. Two of the subjects were symptomatic with as little as 3 g of lactose, three developed symptoms with 6 g, ten with 12 g, and two with 24 g of lactose. Two of the twenty subjects needed 48 g to become symptomatic, and one developed symptoms with 96 g. When these same subjects were tested with polyethylene glycol, a nonabsorbable marker, while being given varying doses of lactose, fluid accumulation in the gut resulted, thereby giving objective evidence of fluid and sodium accumulation in the small bowel due to small lactose loads.[108]

Since lactase has been shown to be a non-adaptable enzyme,[109,110] the fact that patients with lactase insufficiency have a varying ability to tolerate doses of lactose has tremendous clinical significance. A very careful dietary history should be obtained and the patient then tested to determine whether he or she can tolerate small amounts. In this way, Lactase-insufficient patients who do not develop symptoms from small amounts of lactose will be able to ingest some products containing milk, and in our society milk is an important calcium food. Jones and associates[111] fed low-lactose milk to lactose-intolerant subjects and found that the milk (at least 50% lactose reduced) was tolerated well by most of their whole-milk-intolerant subjects. The lactose of low-lactose milk frequently has been digested into glucose and galactose, which makes the carbohydrate available as well as reduces the symptoms. They also found that low-lactose milks are palatable and acceptable to the majority of lactose-intolerant subjects. In line with this, Lee and Lillibridge[112] have developed a semi-quantitative determination of lactose that is suitable for a variety of foods and have published rough contents of common foods, which is of some help in preparing diets for lactose intolerance. Chapter 20 lists lactose-containing foods and the quantity wherever such information is available. The clinician must carefully educate his patient about the quantities of lactose present in foods so that the patient can manage his own intake. (See also Chapter 20 for diets.) There is available a lactase powder that can be added to milk or foods to render it more digestable.[111] Reports reveal that the hydrolyzed milk can be used effectively.[113,114]

13.6. DIETARY MANAGEMENT OF MALABSORPTION DISORDERS NOT CORRECTABLE BY SPECIFIC TREATMENT

Malabsorption caused by a disease that is chronic and not correctable by specific treatment requires long-term nutritional supplementation. Examples of this can be pancreatic insufficiency, bile acid disturbances, short-bowel syndrome, etc. These diseases are noted in the classification of malabsorption disorders. When a specific treatment is available, then the malabsorption caused by a disease is corrected and only specific nutritional defects need be corrected while the disease is initially being treated. However, when there is no specific treatment, chronic malabsorption ensues and long-term supplementation is necessary.

The following nutritional factors must be added to the diet in varying amounts, depending on the clinical situation. The physician must evaluate the patient carefully to determine which deficiencies must be more vigorously treated.

1. In most chronic malabsorption disorders there should be daily vitamin replacement. The vitamins are taken orally on rare occasions by intermittent intramuscular injection. In rare instances it may be necessary to replace vitamin B_{12} by monthly injections. This occurs occasionally in iatrogenic malabsorption due to vegetarian diets, in massive resections of the ileum, and in intrinsic factor deficiencies.

Deficiencies of other water-soluble and fat-soluble vitamins can usually be avoided by long-term oral supplementation. However, when there is severe malabsorption, initial replacement by intravenous or intramuscular therapy may be necessary.

2. Mineral deficiencies may be apparent, such as iron deficiency. Here, oral replacement may be too slow, and intravenous replacement will be rewarding. Evidence of reticulocytosis and correction of the iron deficiency anemia may be dramatic after intravenous therapy.

Zinc deficiency has recently been reported in chronic inflammatory disease, and replacement by intravenous therapy may prove beneficial to some patients.

Specific mineral requirements for calcium, magnesium, etc., are discussed in Chapter 1.

3. Replacement of calories, essential fatty acids, essential amino acids, and minimal carbohydrate requirements may be necessary early in the management of a newly discovered malabsorption disorder. They also may be necessary in the long-term management of a debilitated patient. Treatment by intravenous methods is now possible and is discussed in Chapters 18 and 19.

On occasion, the patient with poor absorption can be aided by medium-chain triglycerides, which do not require chylomicron formation and pass through the liver rather than the lymphatic circulation. These are available in liquid form. Patients suffering from selected chronic malabsorption disorders can add fat calories orally by using medium-chain triglycerides. They can be added to the diet by mixing with salads, used in cooking, or, if tolerated, by adding them to drinks. Directions for their use are found in Chapter 20.

Amino acid supplementation is also readily available through the use of oral solutions. This is discussed in Chapter 18. It is rare that these have to be used in most common malabsorption disorders. They are most effective in severe malnutrition resulting from chronic inflammatory diseases of the gut.

The patient with chronic malabsorption seems to tolerate spicy foods poorly; therefore it always seems prudent to add the bland regimen to their diet recommendation, especially when they are most symptomatic. These judgements should be individualized. It is silly to make a rule that all diets be restrictive if it destroys appetite and enjoyment, and yet to some the restrictive diet may remove all symptoms. These considerations hold for lactose restriction as well. Early in most malabsorption diseases lactose is poorly tolerated, but later the restriction can be moderated or removed.

Occasionally a case with malabsorption will fall into a category of questionable response to treatment. The most difficult of these are patients who fill the diagnostic criteria for gluten enteropathy but do not respond to a gluten-free diet, or those who seem lactose intolerant as part of another disease process. It behooves the physician to be diligent in ascertaining that both the prescribed diet is followed and that all diagnostic studies are fully evaluated. A flat small-bowel biopsy is not always gluten enteropathy.[3, 115] When there is a failure, the patient's cooperativeness and the diet intake must be checked thoroughly while reconsidering the diagnosis. Other food proteins can cause a similar picture.[115] Conversely, gluten can get into foods least suspect; therefore care must be given to the evaluation. At times the reason for a sugar intolerance also may not be clear. It may be transient due to gut bacterial enzyme deactivation,[117] and a patient should not be relegated to lifelong fears of lactose without complete understanding of the cause. Successful dietary management of the problem case requires cooperation and patient education. To succeed the facts must be carefully taught to the patient and food provider.

REFERENCES

1. Shiner M: Duodenal biopsy. *Lancet* 1:17, 1956.
2. Crosby WH, Kugler HW: Intraluminal biopsy of the small intestine: The intestinal biopsy capsule. *Am J Dig Dis* 2:236, 1957.
3. Sheehy TW, Floch MH: *The Small Intestine—Its Function and Diseases.* New York: Hoeber Medical Division, Harper & Row Publishers Inc, 1964.
4. Astaldi G, Stroselli E: Biopsy of normal intestine. *Am J Dig Dis* 5:175, 1960.
5. Ingelfinger FJ: Gastrointestinal absorption. *Nutr Today* 2:2, 1967.
6. Shiner M: Electron microscopy of jejunal mucosa. *Clin Gastroenterol* 3:33, 1974.
7. Haubrich WS, Watson JHL, O'Driscoll W, *et al:* Electron microscopy of the free border of human intestinal epithelial cell. *Henry Ford Hosp Med Bull* 7:113, 1959.
8. Trier JS: Morphology of the epithelium of the small intestine, in Code C (ed.): *Handbook of Physiology. Alimentary Canal.* Washington, DC: American Physiology Society, 1968, vol. 3.
9. Leblond CP, Stevens CE: Constant renewal of the intestinal epithelium in the albino rat. *Anat Rec* 100:357, 1948.
10. Williamson RCN: Intestinal adaptation structural, functional, cytokinetic changes. *N Eng J Med* 298:1393, 1444, 1978.
11. Parrot DMV: The gut as a lymphoid organ. *Clin Gastroenterol* 5:211, 1976.
12. Brandzaeg P, Baklien K: Immunoglobulin-producing cells in the intestine in health and disease. *Clin Gastroenterol* 5:251, 1976.
13. Davenport HW: *Physiology of the Digestive Tract*, ed 4. Chicago: Year Book Medical Publishers, chap 14.
14. Breberdorf F: Normal mechanisms of water and electrolyte absorption by the gastrointestinal tract, in Dietschy JM (ed): *Disorders of the Gastrointestinal Tract.* New York: Grune & Stratton, 1976, p 33.
15. Fordtran JS, Dietschy JM: Water and electrolyte movement in the intestine. *Gastroenterology* 50:263, 1966.
16. Curan PF: Water absorption from the intestine. *Am J Clin Nutr* 21:781, 1968.
17. Bergoz R: Trehalose malabsorption causing intolerance to mushrooms. *Gastroenterology* 60:909, 1971.
18. Cooke WT, Asquith P: Coeliac disease. *Clin Gastroenterol* 3:1, 1974.
19. Stokes PL, Asqeroluith P, Cooke WT: Genetics of coeliac disease. *Clin Gastroenterol* 2:547, 1973.
20. Brow JR, Parker F, Weinstein WM, *et al:* The small intestinal mucosa in dermatitis hepetiformis. Severity and distribution of the small intestinal lesion and associated malabsorption. *Gastroenterology* 60:355, 1971.
21. Marks J, Shuster S: Intestinal malabsorption and the skin. *Gut* 12:938, 1971.
22. Moorthy AV, Zimmerman SW, Maxim PE: Dermatitic hepetiformis and coeliac disease. Association with glomerulonephritis, hypocomplementia, and circulating immune complexes. *JAMA* 239:2019, 1978.
23. Gray GM: Maldigestion and malabsorption: Clinical manifestations and specific diagnosis, in Sleisinger M, Fordtran J (ed): *Gastrointestinal Diseases*, ed 2. Philadelphia: WB Saunders Co, 1978, p 272.
24. Hines JD, Hoffbrand AV, Molleir DL: The hematologic complications following partial gastrectomy. *Clin J Med* 43:555, 1967.
25. Toskes PP, Deren JJ: Vitamin B_{12} absorption and malabsorption. *Gastroenterology* 65:662, 1973.

26. Halsted CH, Robles EA, Mezey E: Intestinal malabsorption in folate-deficient alcoholics. *Gastroenterology* 64:526, 1973.

27. Bayless TM, Rosenweig NS: A racial difference in incidence of lactose deficiency. A survey of milk intolerance and lactose deficiency in healthy adult males. *JAMA* 197:968, 1966.

28. Sahi T: The inheritance of selective adult-type lactose malabsorption. *Scand J Gastroenterol* 9(suppl 30), 1974.

29. Welsch JD, Robrer V, Knudsen KB, et al: Isolated lactose deficiency. Correlation of laboratory studies and clinical data. *Arch Intern Med* 120:261, 1967.

30. Metz G, Jenkins DJA, Peters TJ, et al: Breath hydrogen as a diagnostic method for hypolactasia. *Lancet* 1:1155, 1975.

31. Maffei HVL, Metz GL, Jenkins DJA: Hydrogen breath test: Adaptation of a simple technique to infants and children. *Lancet* 1:1110, 1976.

32. Metz G, Jenkins DJA, Newman A, et al: Breath hydrogen in hyposucrasia. *Lancet* 1:1119, 1976.

33. Metz G, Drasar BS, Gassull MA, et al: Breath hydrogen test for small intestinal bacteria colonization. *Lancet* 1:668, 1976.

34. Marshak RH, Lindner AE: *Radiology of the Small Intestine*, ed 2. Philadelphia: WB Saunders Co, 1976.

35. Levitt MD, Donaldson RM: Use of respiratory hydrogen (H_2)) excretion to detect carbohydrate malabsorption. *J Lab Clin Med* 75:937, 1970.

36. Calloway DH, Murphy EL, Bauer D: Determination of lactose intolerance by breath analysis. *Am J Dig Dis* 14:811, 1969.

37. Fromm H. Hofmann AF: Breath test for altered bile-acid metabolism. *Lancet* 2:621, 1971.

38. Sherr HP, Sasaki Y, Newman A, et al: Detection of bacterial deconjugation of bile salts by a convenient breath-analysis technique. *N Eng J Med* 285:656, 1971.

39. Fromm H, Thomas PJ, Hofmann AF: Sensitivity and specificity in tests of distal ileal function: Prospective comparison of bile acid and vitamin B_{12} absorption in ileal resection patients. *Gastroenterology* 64:1077, 1973.

40. Lauterberg BH, Newcomer AD, Hofmann AF: Clinical value of the bile acid breath test. *Mayo Clin Proc* 53:227, 1978.

41. Newcomer AD, Hofmann AF, DiMagno EP, et al: Triolein breath test. A sensitive and specific test for fat malabsorption. *Gastroenterology* 76:6, 1979.

42. Kamath MD, Murugasu R: A comparative study of four methods for detecting *Giardia lamblia* in children with diarrheal disease and malabsorption. *Gastroenterology* 66:16, 1974.

43. Peterson H: Giardiasis (lambliasis). *Scand J Gastroenterol* 7(suppl 14), 1972.

44. Trier J: Diagnostic value of peroral biopsy of the proximal small intestine. *N Engl J Med* 285:1470, 1971.

45. Whitehead R: *Mucosal Biopsy of the Gastrointestinal Tract*. Philadelphia: WB Saunders Co, 1973.

46. Trier JS: Diagnostic usefulness of small intestinal biopsy. *Viewpoints Dig Dis* 9:1, 1977.

47. Wilson FA, Dietschy JM: Differential diagnostic approach to clinical problems of malabsorption. *Gastroenterology* 61:911, 1971.

48. Gee S: On the coeliac affection. *St. Bartholomew's Hosp Rep* 24:17, 1888.

49. Herter CA: *On Infantalism from Chronic Intestinal Infection*. New York: Macmillan, 1908, p 14.

50. Dicke WK: Celiac disease: A study of the damaging effect of some cereals, especially wheat, caused by a factor outside of their starch, on the fat absorption of children with celiac disease. Presented at the International Congress of Paediatrics, Zurich, 1950.

51. Dicke WK, Weijers HA, van de Kamer JH: Celiac disease: Presence in wheat of factor having deleterious effect in cases of celiac disease. *Acta Paediatrica* 42:34, 1953.

52. Weijers HA, van de Kamer JH, Dicke WK: Celiac disease. *Adv Pediatr* 9:277, 1957.

53. van de Kamer JH, Weijers HA: Coeliac disease. V. Some experiments on the cause of the harmful effect of wheat gliadin. *Acta Paediatrica* 44:465, 1955.

54. Cooke WT, Peeney ALP, Hawkins CF: Symptoms, signs and diagnostic features of idiopathic steatorrhea. *Q J Med* 22:59, 1953.

55. Anderson CM: Histological changes in the duodenal mucosa in coeliac disease: Reversibility during treatment with a wheat gluten free diet. *Arch Dis Child* 35:419, 1960.

56. Rubin CE, Brandborg LL, Phelps PC, Taylor HC: The apparent identification and specific nature of the duodenal and proximal jejunal lesion in celiac disease and idiopathic spine. *Gastroenterology* 38:28, 1960.

57. McNeish AS, Anderson CM: Celiac disease. The disorder in childhood. *Clin Gastroenterol* 3:127, 1974.

58. Asquith P: Celiac disease immunology. *Clin Gastroenterol* 3:213, 1974.

59. Patey AL: Gliadin: The protein mixture toxic to coeliac patients. *Lancet* 1:722, 1974.

60. Evans DJ, Patey AL: Chemistry of wheat proteins and the nature of the damaging substances. *Clin Gastroenterol* 3:199, 1974.

61. Cornell HJ, Townley RRW: The toxicity of certain cereal proteins in coeliac disease. *Gut* 15:862, 1974.

62. Cornell HJ, Townley RRW: Investigation of possible intestinal peptidase deficiency in coeliac disease. *Clin Chim Acta* 43:113, 1973.

63. Dissanayake AS, Jerrome DW, Offord RE, *et al:* Identifying toxic fractions of wheat gluten and their effect on the jejunal mucosa in coeliac disease. *Gut* 15:931, 1974.

64. Dissanayake AS, Truelove SC, Whitehead R: Lack of harmful effect of oats on small-intestinal mucosa in coeliac disease. *Br Med J* 4:189, 1974.

65. Shiner M, Ballard J: Antigen–antibody reactions in jejunal mucosa in childhood coeliac disease after gluten challenge. *Lancet* 1:1202, 1972.

66. Seah PP, Fry L, Rossiter MA, *et al:* Anti-reticulin antibodies in childhood coeliac disease. *Lancet* 2:681, 1971.

67. Mowbray JF, Holborow EJ, Hoffbrand AV, *et al:* Circulating immune complexes in dermatitis herpetiformis. *Lancet* 1:400, 1973.

68. Sikora K, Anand BS, Truelove SC, *et al:* Stimulation of lymphocytes from patients with coeliac disease by a subfraction of gluten. *Lancet* 2:389, 1976.

69. Keuning JJ, Pena AS, van Leeuwen A, *et al:* HLA-DW3 associated with coeliac disease. *Lancet* 1:506, 1976.

70. Scott BB, Swinburne ML, Rajah SM, *et al:* HL-A8 and the immune response to gluten. *Lancet* 2:374, 1974.

71. Mann DL, Katz SI, Nelson DL, *et al:* Specific B-cell antigens associated with gluten-sensitive enteropathy and dermatitis herpetiformis. *Lancet* 1:110, 1976.

72. Fry L, Seah PP, Riches DJ, *et al:* Clearance of skin lesions in dermatitis herpetiformis after gluten withdrawal. *Lancet* 1:288, 1973.

73. Barry RE, Baker P, Read AE: Celiac disease. The clinical presentation. *Clin Gastroenterol* 3:55, 1974.

74. Hood J, Mason AMS: Diffuse pulmonary disease with transfer defect occurring with coeliac disease. *Lancet* 1:445, 1970.

75. Smith MJL, Benson MK, Strickland ID: Coeliac disease and diffuse interstitial lung disease. *Lancet* 1:473, 1971.

76. Whorwell PJ, Foster KJ, Alderson MR, *et al:* Death from ischemic heart-disease and malignancy in adult patients with coeliac disease. *Lancet* 2:113, 1976.

77. Mawhinney H, Tomkin GH: Gluten enteropathy associated with selective IgA deficiency. *Lancet* 2:121, 1971.

78. Beale AJ, Parish WE, Douglas AP, *et al:* Impaired IgA responses in coeliac disease. *Lancet* 1:1198, 1971.

79. McCrea WM, Eastwood M, Martin M, *et al:* Neglected celiac disease. *Lancet* 1:187, 1975.

80. Stokes PL, Holmes GKT: Malignancy. *Clin Gastroenterol* 3:159, 1974.

81. Sheedy CB, Keifetz N: *Cooking for the Celiac Child.* New York: The Dial Press Inc, 1969.

82. Wood MN: *Gourmet Food on a Wheat Free Diet.* Springfield: Charles C Thomas Publishers, 1967.

83. Malis F, Lojda Z, Fric P, *et al:* Disaccharides in celiac disease and mucoviscidosis. *Digestion* 5:40, 1972.

83a. Regan PT, Dimagno EP: Exocrine pancreatic insufficiency in celiac sprue. *Gastroenterology* 78:484, 1980.

84. Alpers DH, Seetharam B: Pathophysiology of diseases involving intestinal brush-border proteins. *N Engl J Med* 296:1047, 1977.

85. Gryboski J: *Gastrointestinal Problems for Infants.* Philadelphia: WB Saunders Co, 1975.

86. Antonowicz I, Lloyd-Still JD, Khaw KT, *et al:* Congenital sucrase–isomaltase deficiency. *Pediatrics* 49:847, 1972.

87. Ament M, Perera D, Esther L: Sucrase isomaltase deficiency—A frequently misdiagnosed disease. *J Pediatr* 83:721, 1973.

88. Gray GM, Conklin KA, Townley RRW: Sucrase–isomaltase deficiency: Absence of an inactive enzyme variant. *N Engl J Med* 294:750, 1976.

89. Lindquist B, Meenwise GW: Chronic diarrhea caused by monosaccharide malabsorption. *Acta Paediatr Scand* 51:674, 1962.

90. Anderson CM, Kerry KR, Townley RR: An inborn defect of intestinal absorption of certain monosaccharides. *Arch Dis Child* 40:1, 1965.

91. Lindquist B, Meeuwisse GW: Diets in disaccharidase deficiency and defective monosaccharide absorption. *J Am Diet Assoc* 48:307, 1966.

92. Meeuwisse GW, Lindquist B: Glucose–galactose malabsorption. *Acta Paediatr Scand* 59:74, 1970.

93. von Reuss A: Über alimentare Saccharosurie bei darmkranken Sauglingsalter. *Wien Klin Wochenschr* 23:123, 1910.

94. Holzel A, Schwarz V, Sutcliff KW: Defective lactose absorption causing malnutrition in infancy. *Lancet* 1:1126, 1959.

95. Gilat T, Kuhn R, Gelman E, *et al:* Lactase deficiency in Jewish communities in Israel. *Am J Dig Dis* 15:895, 1970.

96. Huang SS, Bayless TM: Milk and lactose intolerance in healthy orientals. *Science* 160:83, 1968.

97. Sahi T: Dietary lactose and the aetiology of human small-intestinal hypolactosia. *Gut* 19:1074, 1978.

98. Bayless TM, Rosensweig NS: A racial difference in the incidence of lactase deficiency. *JAMA* 197:968, 1966.

99. Gray M: Congenital and adult intestinal lactose deficiency. *N Engl J Med* 294:1057, 1976.

100. Paige DM, Bayless TM, Mellits ED: Lactose malabsorption in preschool black children. *Am J Clin Nutr* 30:1018, 1977.

101. Dahlquist A: Specificity of the human small intestinal disaccharidases and indicative for heredity disaccharide intolerance. *J Clin Invest* 41:463, 1962.

102. Simons FJ: New light on ethic differences in adult lactose intolerance. *Am J Dig Dis* 18:595, 1973.

103. Lebenthal E, Tsuboi K, Kretchmer N: Characterization of human intestinal lactose and hetero-β-galctosidases of infants and adults. *Gastroenterology* 67:1107, 1974.

104. Newcomer AD, McGill DB, Thomas PJ, *et al:* Progressive comparison of indirect methods for detecting lactose deficiency. *N Engl J Med* 293:1232, 1975.

105. Bond JH, Levitt MD: Fate of soluble carbohydrate in the colon of rats and man. *J Clin Invest* 57:1158, 1976.

106. Kerry KR, Anderson M· A ward test for sugar in feces. *Lancet* 1:981, 1964.

107. Lebenthal E, Antonowicz I, Shwachman H: Correlation of lactose activity, lactose tolerance and milk consumption in different age groups. *Am J Clin Nutr* 28:595, 1975.

108. Bedine MS, Bayless TM: Intolerance of small amounts of lactose by individuals with low lactase levels. *Gastroenterology* 65:735, 1973.

109. Gilat T: Lactase in man: A non-adaptable enzyme. *Gastroenterology* 62:1125, 1972.

110. Rosensweig NS, Herman RH: Diet and disaccharidase. *Am J Clin Nutr* 22:99, 1969.

111. Jones DV, Latham MC, Kosikowski FV, *et al:* Symptom response to lactose-reduced milk in lactose-intolerant adults. *Am J Clin Nutr* 29:633, 1976.

112. Lee DE, Lillibridge CB: A method for qualitative identification of sugars and semiquantitative determination of lactose content suitable for a variety of foods. *Am J Clin Nutr* 29:428, 1976.

113. Paywe-Bose D, Welsh JD, Gearhardt HL, *et al:* Milk and lactose-hydrolyzed milk. *Am J Clin Nutr* 30:695, 1977.

114. Paige DM, Bayless TM, Huang S, *et al:* Lactose hydrolysed milk. *Am J Clin Nutr* 28:818, 1975.
115. Barer AL, Rosenberg IH: Refractory sprue: Recovery after removal of nongluten dietary proteins. *Ann Intern Med* 89:505, 1978.
116. Katz AJ, Grand RJ: All that flattens is not "sprue." *Gastroenterology* 76:376, 1979.
117. Bampoe V, Avigad S, Sapsford RJ, *et al:* Lactose degradation by human enteric bacteria. *Lancet* 2:125, 1979.

Constipation, Diarrhea, Gas, and the Intestine

The topics of this chapter are grouped together because they often result in abdominal pain or cramps, increased flatus or eructation, or diarrhea or constipation. At times an etiology is found, but most often the patient is plagued and a nutritional approach to the symptoms is sought. The nutritional advice given such patients can be of enormous help or useless, but should always be on a sound physiologic basis. The following, I hope, is a rational approach to using the diet as the treatment of these patients.

14.1. CONSTIPATION

Constipation is more easily felt than defined. It is one of the major complaints of Western societies. Witness the tremendous use of laxatives and the recent success of adding fiber foods to the diet. Constipation was defined by Osler[1] as merely a retention of feces for any reason. It is more clearly defined as "defecation that occurs with insufficient frequency so that there is insufficient stool or it may be abnormally hard or dry."[2] The definition implies that there are satisfactory frequencies and consistencies of the stool for every individual. It is important to remember that colonic function can vary greatly from individual to individual, some individuals comfortable with only a few bowel movements per week, others feeling the necessity to evacuate at least daily in relatively large amounts.

The classification of constipation is difficult because there are so many related causes, but the one proposed by Bockus[3] for simple constipation and modified to add systemic diseases can be useful when a subject is first evaluated.

14.1.1. Causes

    1. Social and economic:
        a. Dietary factors.

    b. Upright position.

    c. Improper training and neglect.

    d. Lack of toilet facilities.

    e. Iatrogenic: abuse of drugs, laxatives, narcotics, anticholinergics, tranquilizers, enemas.

2. Decreased expulsive power:

    a. Weakness of the diaphragm, abdominal wall, or pelvic floor.

    b. Chronic constipation with or without megacolon (this may occur due to severe or chronic malnutrition or aging, or may be associated with decreased neuromuscular function.[4]

3. Deficient peristaltic activity of the intestine:

    a. Decreased gastrocolic reflex, which may be correctable.[4]

    b. Increased cholinergic activity, which may be associated with diseases such as Hirschsprung's disease or actual spinal cord damage.

4. Obstruction of movement:

    a. So-called spastic constipation.

    b. Obstructive organic lesions, such as congenital, adhesions, or true tumors.

    c. Abnormalities of the anal sphincter that limit outflow.

5. Neurologic diseases (although some of these are included above, the paraplegic and severe spinal cord cases must be classified).

6. Metabolic and electrolyte disorders (hypothyroid, hypercalcemia).

7. Psychogenic causes.

Understanding the pathophysiology of constipation is essential for the determination of the proper treatment.

14.1.2. Pathophysiology

Ordinarily, the colon receives approximately 1.5 liters of fluid per day with an electrolyte content as discussed below in Section 14.2. The diet of Western societies has been shown to result in the absorbance of 95% of the water and almost all of the sodium chloride, and the excretion of significant amounts of potassium.[5] What is not known is the effect on the colon of a high-fiber diet, where there appears to be less constipation. Figure 14-1 is a graphic representation of the classic data collected by Burkitt and Trowell,[6] revealing the marked increases in stool weight and decreased transit times resulting from high-fiber diets compared to the so-called "normal stool" resulting from the diet in which most colon physiologic studies have been done to date.

If the definition of constipation as the insufficient passage of stool, or too hard or dry stool, is accepted, then the observations of Burkitt and Trowell[6]—that constipation does not exist in societies whose diets are very high in fiber—must also be accepted. If this is true, then most simple constipation can be attributed to low-fiber intake. The greatly increased use of bran products is an observation that tends to bear this out, as well as the frequent testimony of patients that their constipation is cured by increased fiber intake. Fiber affects

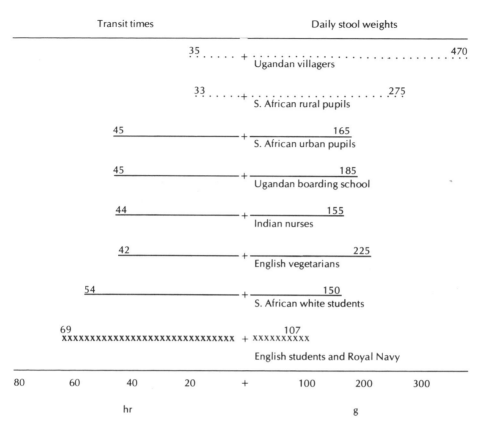

Figure 1. Average intestinal transit times and stool weights in communities on different diets. (.......) Unrefined carbohydrate foods; (———) intermediate carbohydrate foods; (xxxxx) refined carbohydrate foods. Adapted from Burkitt and Trowell.[6]

the physiology of the colon by retaining water. The "normal stool" output per day is approximately 100 to 125 g. The nonconstipated high-fiber-diet stool ranges from 200 to 400 g/day.[6] The increase in stool size is largely water.[7] Careful stool weight water analyses of subjects on high-bran intake reveal that both percentages of fecal dry weight and water increase equally; therefore, although there is increased water, there is also increased dry matter.[8] It can be concluded that a decreased intake of fiber can be one of the causes of functional or physiologic constipation. The colon's function is to absorb free water, and the healthy colon will do this efficiently. If enough fiber is present, it will bind more water and yield more bulk, which will result in decreased transit time and softer stools (see Chapter 4).

Another function of the colon is to propel the fecal mass. Contractility may be either decreased or increased.[4] Motility of the large intestine has been difficult to study in man. It is important to understand the movements of the colon and the function of the sphincters for an understanding of constipation and defecation.[9] The main movements are segmental contractions and peristaltic actions. Segmental contractions are seen as changes in haustrae and best explain maintenance of the contents within the large intestine. Peristaltic motion is less common and best defined as movement of the contents to the rectum and external sphincters. The number of peristaltic waves vary, account for 6 to 8% of contractions[9] and in some subjects occur only once a day (when measurements can be recorded).

A series of important neuromuscular activities occur for normal defecation once a significant amount of feces is sent into the rectum. A disturbance in defecation that results in constipation or diarrhea can occur when there is a loss of neuromuscular function, either from aging or from discrete lesions. Ordinarily, the rectum is distended by stool, which creates a sensation of pressure in the rectum. The impulses pass through the spinal cord. The internal anal sphincter is not under voluntary control and is not paralyzed when the spinal cord is damaged. The external anal sphincter is under voluntary control and is usually paralyzed after destruction of the lower spinal cord. Once distension of the rectum occurs from approximately 100 cm^3 of feces and a pressure of 18 mm Hg, the brain receives an impulse and the urge to defecate occurs, but the sphincter remains closed. When the volume is increased to approximately 200 cm^3 and the pressure to 55 mm Hg, both the internal and external sphincters relax and the contents of the rectum are expelled. The rectum is greatly distensible and large amounts can be held, but when the pressure rises to approximately 55 mm Hg, the sphincters will open. At somewhere between 18 and 55 mm Hg, voluntary action of the normal bowel movement is best accomplished in the squatting position. There is a necessary tightening of the abdominal wall muscles followed by a Valsalva-type maneuver and the opening of the sphincters, expelling the stool.

Relaxation of the internal sphincter does not occur in Hirschsprung's disease or where there is local destruction of the pathways for transmission of somatic impulses. Defecation can also be impaired where pathways for the external sphincter or weakening of the abdominal wall muscles occur. Voluntary closure of the external sphincter may be strong enough to retain large amounts of feces. Consequently, a functional megacolon can occur.[4] Therefore, psychologic as well as organic causes of constipation can be related to voluntary disturbances of the normal defecatory mechanisms.

In summary, the normal functions of the colon are to absorb water and electrolytes and to pass fecal material in a regular fashion. When there is a disturbance in the normal dietary intake of substances necessary to form adequate waste, or when there is a disturbance in the motility or sphincter function, constipation can result.

14.1.3. Diagnosis and Clinical Findings

When a disease such as hypothyroidism or a neurologic disorder is the cause of constipation, proper treatment must await the diagnosis of that disease. Occasionally, constipation is treated for an extended period of time before an identifiable disease is uncovered. The patient complaining of constipation should have a careful history, physical examination, and clinical screening to determine if an organic disease process is causing the disorder. If organic disease is absent, then constipation per se can be treated.

It is important to establish that constipation does in fact exist before treatment is started. Many patients will complain of constipation when they are having regular bowel movements but only incomplete evacuation. In these cases, the possibility of a physiologic disorder must be considered. A feeling of incomplete evacuation must be thoroughly investigated for organic causes. The sensation of incomplete evacuation with regular bowel movements indicates some form of psychiatric or organic disease and must be properly treated.

Constipation due to organic disorders falls into two broad categories.

14.1.4. Functional Megacolon and Encopresis

This condition is most often seen in children who have no evidence of organic disease. The cause of the disorder is unknown, but psychologic factors often have an overwhelming influence. Functional megacolon may lead to an encopresis: The colon becomes dilated and feces-filled. Children and some adults may then develop an overflow phenomenon that causes staining of underwear. Both the inability to defecate and the social horror of incontinence become a gross problem to all concerned.

Most adults presenting with this problem have organic disease. Most children progressing to adulthood have been treated. Treatment varies greatly with the associated psychiatric factors. Often they are paramount, and handled correctly the problem will be solved. At other times, the approach is physiologic.

Hirschsprung's disease should be ruled out by adequate radiographic and, if necessary, biopsy studies. If that diagnosis is made, the treatment is surgical.

In a limited number of cases, I have found the following therapeutic regimen successful:

1. A careful psychiatric evaluation must be done by the physician. At times, a social worker and a psychiatrist must be included in the evaluation. If it is determined that the family is stable and well adjusted, then pharmacologic and diet therapy may be tried. However, if there are severe social problems at home—either a mother–child problem or a marital problem that affects the child—then psychiatric intervention is invariably needed. The presence of an encopresis usually signals an emotional/psychiatric problem at home.

2. A simple physiologic approach is retraining the child. I have found it most successful to induce a chronic diarrhea with laxatives and then gradually withdraw the laxatives. During this period of retraining, the child may be undergoing psychiatric treatment and reassurance. The period of retraining can be as long as several months. I have found it more successful to use harsh laxatives, such as strong neurogenic stimulants, rather than mineral oil, which has been employed by others. Mineral oil causes a constant ooze, which is most uncomfortable and embarrassing. A stimulatory evacuation is certainly preferred over a constant leak. As a consequence, such stimulants as

bisacodyl, casanthranol, danthron, and the milder senna concentrate are much more effective.

3. At the same time, a correct dietary regimen should be encouraged, such as a high-fiber diet (see Chapter 20). However, these patients are not cured by fiber alone. Its use will not be helpful unless steps (1) and (2) are followed.

14.1.5. Chronic Idiopathic Constipation

This is the most common form of simple constipation. Without an obvious psychiatric or organic cause, it presents most often in the young female or elderly patient. The prognosis has not been good; however, the recent use of high-fiber substances has improved the prognosis. Previously, small amounts of fiber were employed. The recognition that large amounts are needed to significantly increase bulk as well as decrease transit time has resulted in much more success.

Successful treatment of constipation can be accomplished with foods if they are included in sufficient quantity. The classic work of Hoppert and Clark[10] revealed that the addition of bran to muffins had a definite effect on stool volume and transit time.[10] However, it is a quantitative effect—enough bran must be added to affect stool bulk. The manner in which the food is used in the diet depends entirely on the individual.

Bran has been the most popular fiber agent employed. A sudden increase in the intake of fiber can cause bloating and symptoms. It is best to begin the patient on 1 teaspoon of whole bran daily with an increase of a teaspoon every 2 to 3 days until soft, regular bowel movements are induced. Most patients will find between 1 and 2 tablespoons a successful regimen. It takes approximately 1 week to begin to regulate the bowel movement. As noted in Section 15.1 (Diverticular Disease), this has been successful in at least 50% of all patients (see references on diverticular disease). It has been this author's experience that, barring psychologic bowel fixation, the cooperative patient is able to reverse a constipated bowel pattern by the use of bran or high fiber.

All fiber substances do not necessarily absorb water. It is well to remember that some fiber substances, particularly bran, have a laxative effect, whereas other substances, such as lignin and pectin, may have a constipating effect.[16–19] This is discussed in more detail in Chapter 4. Many of my patients who had suffered constipation will not part with their bran. A whole spectrum of patients has been completely relieved by a high fiber (e.g., bran) supplement. Initial reports are very positive, but only time will tell the extent to which our society has been relieved by this nutritional addition to the armamentarium.[11–15]

Although bran is the most popular, other foods can be used with success. An adequate intake of fruits and vegetables can increase stool bulk.[20] A diet including such water-binding foods as berries, cabbage, corn, dates, etc., decreases transit time as well as increases bulk.[10] It is best to accomplish a softer stool by a balanced diet, but individual needs must be considered. Some patients will succeed only with bran, while others will need a variety of foods.

Future dietary goals should include high-fiber intake for all, but the attainment of this will depend on ethnic, economic, and political influences. At this time, bran is therapeutic for constipation and the broad use of a high-fiber diet should be encouraged, but its use will depend on the individual and society's influence (see Chapter 20, diets 26 and 27.)

14.2.   DIARRHEA

Diarrhea can be defined as a change from usual bowel habits to those of increased fluid content, often associated with increased frequency of passage. There may be increased fluid and volume without increased frequency. Stool may or may not contain blood, mucus, or pus.

Such a definition of diarrhea implies tremendous variation from individual to individual. In Western societies, the stool volume ranges from 100 to 150 ml/day. In Asian and African societies, the volume of a normal stool is 200 to 400 ml/day. What is considered normal for African people would be diarrhea for Western people. As the Western diet increases in fiber, the stool size may increase. What is most important when considering clinical diarrhea is that there is a change in bowel habit to either a more fluid or a more frequent movement that may or may not be associated with blood or mucus. The diet will alter the consistency of the stool: Increasing fiber increases the stool size, but usually concomitantly increases both the fluid and the solid intake, with the percentage of water remaining the same (approximately 70 to 75% of the stool is water). The character of the stool will also change, depending on the cause of the diarrhea. Secretory and osmotic diarrhea are more fluid, whereas exudative diarrhea causes greater frequency with more evidence of blood and cellular elements.

14.2.1.   Pathophysiology of Water and Electrolyte Secretion and Excretion [21-24]

A knowledge of the exchange of fluids within the gut is most important in understanding diarrhea. These are discussed in detail in the chapters on the physiology of eating and the small intestine. The following are important to remember when relating normal fluid and electrolyte exchange to diarrhea.

14.2.1.1.   Fluid Exchange in the Gut

The following approximate amounts are secreted and absorbed along the course of the gut so that a final fluid volume is left in the rectum to be excreted. (See Table 14-1.)

The final normal volume of fluid excreted in the rectum may be 100 to 200 ml, and as high as 300 ml, depending upon the type of fiber intake. This will represent approximately 70 to 75% of the fecal output, the rest being waste products and substances bound to fiber and bacteria.

Table 14-1. Approximate Gut Fluid Exchange (in Liters)

	Secreted	Absorbed	Excreted
Diet	2[a]	—	—
Saliva	1	—	—
Stomach	2	—	—
Liver	1	—	—
Pancreas	2	—	—
Small bowel	1		
Jejunum		4–5	—
Ileum		3–4	—
Colon		0.5–1.5	—
Rectum			0.1–0.2[a]

[a] Varies with type of diet intake.

14.2.1.2. Electrolyte Exchange

A continuous flux of solutes, such as electrolytes, occurs, in addition to the approximately 9 liters/day of fluid flux in humans (Table 14-1). Furthermore, as discussed in the chapter on the small intestine, active processes of absorption and secretion of electrolytes also affect water exchange. This understanding has come about because of a better understanding of secretory diarrhea, wherein the secretion of electrolytes drags water along, causing diarrhea. The classic example of cholera diarrhea best demonstrates this mechanism. The normal concentrations of the predominant electrolytes present in the intestine are listed below.

What happens in the gut is that the chyme containing hydrogen ions and electrolytes is poured into the duodenum, where the digestive and absorptive processes take place. Large amounts of water and electrolytes are exchanged by mechanisms described in Chapters 1 and 2. Water, sodium, chloride, and potassium largely are reabsorbed by so-called solvent drag along with nutrients and products of digestion that enter the intestinal epithelial cells as simple solutes due to osmotic gradients and passive absorption. This process is most active in the duodenum and jejunum and slows down in the ileum, where active processes (so-called sodium pump) take over and there is active absorption of sodium with secretion of bicarbonate. As a consequence, in the distal ileum the average concentration of sodium is approximately 125 meq/liter and potassium 9 meq/liter, with chloride and bicarbonate being 60 and 74 meq/liter, respectively. Because approximately 600 ml is passed into the right side of the colon, the normal amounts of excreted electrolytes (Table 14-2) are respectively 75, 5, 36, and 44 for sodium, potassium, chloride, and bicarbonate. The right side of the colon then continues the active reabsorption of water and electrolytes so that sodium is absorbed and potassium excreted. This mechanism of potassium excretion is a most important exchange mechanism because it is coupled to sodium absorption. Concomitantly, there is a chloride for bicarbonate exchange, whereby carbonate is actively secreted while chloride is absorbed. As noted in Table 14-2, the actual total excretion of cations (Cl^- + HCO_3^-) does not balance that of the anions. This is because large amounts of organic acids produced by the colonic bacteria reduce the bicarbonate concentration. As a result, in the normal colon, in which anywhere from 100 to 200 ml of fluid is present, there would be a total output of anywhere from 4 to 8 eq of sodium and 9 to 18 eq of potassium. This is extremely important in diarrhea situations because the actively working colon can excrete large amounts of potassium and bicarbonate that can deplete the body of both and result in *hypokalemic acidosis.*

The normal colon absorbs approximately 500 ml of water per day in conjunction with the exchanges of electrolytes. When a diarrhea occurs, large amounts of fluid are poured into the right side of the colon. The average colon can approximately quadruple its effectiveness of active reabsorption of fluids and electrolytes. However, when they cannot be accommodated, the colon is overcome and large amounts of fluid containing sodium, chloride, potassium, and neutralized bicarbonate are lost. The body can become depleted not only of potassium and bicarbonate but also of sodium and chloride. This is again exemplified by the classic cholera diarrhea.

Classified according to cause, diarrheas are simple, osmotic, secretory, motility and transit, exudative, drug induced, and complex, as are those of malabsorption and metabolic disorders. They are discussed with appropriate dietary therapies below.

14.2.2. Osmotic Diarrhea

Lactose intolerance is the classic example of an osmotic diarrhea. Lactose is a disaccharide of glucose and galactose that is absorbed in the upper small intestine after degradation of the disaccharide in the presence of the enzyme lactase. When lactase is insufficient, lactose is not absorbed and remains within the gastrointestinal lumen. In large amounts, it can create an osmotic gradient in which fluid goes from the intestinal wall into the intestinal lumen, resulting in an increased fluid load and diarrhea. The unabsorbed disaccharide can also become fertile ground for bacterial fermentation with resultant increased production of carbon dioxide, hydrogen, and organic anions. The short-chain organic acids can decrease the pH, which can have a negative effect on colonic absorption and also act as a cathartic. The syndromes associated with disaccharide deficiencies are discussed and referenced in Chapter 13.

Table 14-2. Gut Electrolyte Exchange and Excretion

	Na^+	K^+	Cl^-	HCO_3^-	H_2O
Ileum concn meq/liter	125	9	60	74	
Normal excretion (meq)	75	5	36	44	600
Stool concn meq/liter	40	90	15	30	
Normal excretion (meq)	4	9	2	3	100

Treatment of the most common lactose intolerance is either the removal of lactose from the diet or the addition of lactase, which is now readily available. Most lactose-intolerant patients learn to tolerate small amounts and are careful to avoid large amounts that might occur in prepared foods or milk products. For details of treatment and low-disaccharide diets, see Chapters 13 and 20, diets 17 to 19.

Other causes of osmotic diarrheas are saline laxatives. Most of these contain phosphates and magnesium sulfate. The addition of large amounts to the intestinal fluid causes retention of fluid within the small bowel, resultant fluid overload, and a massive cathartic effect. In a well-hydrated patient, this is probably the most physiologic of laxatives, accomplished simply by causing an osmotic overload.

14.2.3. Secretory Diarrhea

Increased intestinal secretion is the cause of many diarrheas and may be due to the following.

1. Bacterial enterotoxins, such as produced by *Vibrio cholerae, Staphylococcus, Clostridia perfringence,* toxigenic *Escherichia coli* (traveler's diarrhea), and occasionally by some forms of *Shigella* and *Pseudomonas.* The classic mechanism has been elucidated for cholera. The enterotoxin enters the crypt area of the intestinal villus, where it stimulates the production of adenyl cyclase. Adenyl cyclase stimulates production of cyclic AMP, which in turn stimulates chloride secretion, and decreased sodium absorption and a flux of water into the bowel result.[25,26]

2. Neoplasms, such as islet-cell tumors, that produce pancreatic cholera or the watery diarrhea hypochlorhydric syndrome. Intestinal secretion is stimulated in these diseases by gastrointestinal hormones, e.g., VIP.[27] Other pancreatic tumors, such as the hypergastrinoma of the Zollinger–Ellison syndrome, cause gastric hypersecretion that in turn impairs small-bowel absorption. Medullary thyroid cancer may also cause a secretory diarrhea.

3. Bile acids and the long-chain fatty acids that are associated with steatorrhea.[28] Dihydroxybile acids can be increased in the lower bowel due to ileal resection and result in a secretory stimulus.

4. Stimulatory laxatives, such as the anthraquinone cathartics.

14.2.4. Motility and Transit Diarrhea

The gastrocolic reflex occurs most often after eating. In many subjects this may be hyperactive.[22] It may result in intestinal hurry or an irritable bowel syndrome. Some patients often have diarrhea or an explosive bowel movement of sudden onset following eating. When it does not occur after eating, the syndrome often perplexes the patient and the clinician. It is assumed that the diarrhea occurs because of strong propulsive waves that move the large small-bowel volume through to the colon, which cannot accommodate rapidly. The diarrhea of rapid transit may also occur due to a strong peristaltic wave through the right side of the colon, which would empty the colon. In addition to the rapid peristalsis, there may be inadequate mixing of intestinal contents, which can result in a cathartic osmotic effect in the colon (see Chapter 16).

14.2.5. Exudative Diarrhea

Exudative diarrheas are caused by any pathologic process that produces inflammation and disturbs the normal histology of the intestine. The classic lesions are produced by infectious agents.[26] However, inflammatory bowel disease, such as idiopathic ulcerative colitis, and Crohn's disease cause similar diarrheas. Parasitic organisms, antibiotic lesions, and neoplasms can also cause an exudative process, "exudative" because the diarrhea often consists of serum proteins, blood, and mucus produced by the bowel wall. The volume of fluid in these diarrheas is less than seen with osmotic and secretory diarrheas, but often much more painful and associated with more toxic symptoms.

The invading bacteria belong to the shigella, typhoid, paratyphoid, salmonella, and invasive E. coli groups. The classic pathology sequence is (1) invasion of the colonic mucosa, (2) inflammatory reaction consisting of hemorrhage, necrosis, and ulceration, and (3) diarrhea consisting of blood, pus, and mucus. Bacteria such as Shigella often secrete associated toxins that can simultaneously cause spasm and secretory diarrhea as well.

14.2.6. Diarrhea and Malabsorption

Malabsorption is commonly associated with diarrhea but may be absent in 10% of cases (see Chapter 13). The processes associated with malabsorption are discussed in more detail under specific gastric, pancreatic, and small-bowel disorders. In summary, the diarrhea of malabsorption is complex. It is associated with steatorrhea, fatty acids, bile acids, and carbohydrates, all of which can decrease colonic absorption of water and often stimulate intestinal secretion or cause an osmotic diarrhea. The malabsorption associated with either a damaged mucosa or a short-bowel syndrome is associated with multiple defective absorptive processes resulting in diarrhea. The malabsorption of bacterial overgrowth is complex and associated with the osmotic phenomenon of increased organic acids produced by bacteria, as well as the phenomena caused by steatorrhea and the malabsorption of bile acids.

14.2.7. Diarrhea and Metabolic Disorders

Diarrheas associated with metabolic disorders are poorly understood. Those caused by hyperthyroidism are probably due to hyperactivity of the bowel and those associated with adrenal insufficiency and hypoparathyroidism are secondary to stress induced on the bowel.

The diarrhea associated with diabetes mellitus is the result of either a diabetic enteropathy due to nerve damage, an associated gluten enteropathy, or bacterial overgrowth in the bowel. A thorough evaluation for malabsorption is needed in the patient with diabetes mellitus who develops diarrhea before the mechanism causing the diarrhea can be established (see Chapter 12).

14.2.8.   Drug-Induced Diarrhea

Once the association to the drug is made, treatment is obvious. The difficulty in diagnosis is often denial of drug use by the patient or an inability to identify the drug by the physician.

14.2.9.   Treatment

Most diarrheas are treated symptomatically. When it is a simple diarrhea, the use of a current antispasmodic or antidiarrhea agent usually suffices. However, the complex metabolic and malabsorption diarrheas require specific management as described in the chapters on gastric, pancreatic, and small-bowel disease. The most common osmotic diarrhea is treated as described under lactose intolerance with lactose-free or low-lactose diets.

The secretory diarrheas require diagnostic evaluation in order to institute the appropriate therapy. Hypergastrin-secreting tumors can be treated by removal of the tumor or with H_2 inhibitors; villus adenoma-secreting tumors require removal of the tumor; but medullary thyroid tumors have most recently been treated with nutmeg.[29-31] One teaspoon of ground nutmeg four times daily has appeared to control this diarrhea, which is associated with increased prostaglandinlike activity.

The secretory diarrhea associated with exotoxin stimulation of the bowel, due either to mild traveler's diarrhea or severe cholera, requires clinical evaluation for appropriate therapy. If the diarrhea is mild and the patient is not dehydrated or toxic, then it can be treated by oral therapy. If the patient is becoming dehydrated, then it will require intravenous therapy.

Oral diet therapy can be employed to treat toxigenic diarrhea. The glucose absorption mechanism is not damaged in the small intestine, and consequently the intake of glucose will activate the pump to absorb sodium and water with the glucose, resulting in a net absorption of water to counteract the secretory effect of the toxin.

Glucose is effective in the treatment of cholera. The coupled absorption of glucose and sodium takes place despite massive secretion of fluid by the intestine in this disease. If the patient is not vomiting and dehydration has not caused collapse, then oral glucose electrolyte solutions can reverse the net water secretion and correct the diarrhea. It is amazing that adults can require as much as 750 to 1000 ml/hr for several hours in order to stop the diarrhea.[32] Approximately 5% of all patients cannot tolerate oral glucose solutions. The glucose solution should contain electrolytes. Various investigators have used a variety of electrolyte–glucose combinations. Naalin and colleagues[33] have added glycine to the oral glucose–electrolyte solution and demonstrated that it can decrease the amount of diarrhea more effectively than either glucose or glycine alone (see Chapter 20). Because of a whole host of economic and social problems, controversy still exists as to the best oral solution to use in underdeveloped countries. It is surprising that in the United States or in countries where traveler's and tourist's diarrhea is prevalent, oral therapy is not readily used by physicians. Hirschhorn and Denny[34] compared the clinical responses of two treatment regimens in acute diarrhea in Apache children. They found that treatment with oral glucose-electrolyte solutions could restore well-being and appetite effectively. The solution they used orally (GE-SOL; composition: 81 mM sodium, 18 mM potassium, 71 mM chloride, 28 mM bicarbonate, and 139 mM glucose) was capable of restoring moderate degrees of volume depletion in children. Furthermore, it was effective in supplementing intravenous fluids and in restoring positive gut water balance.

Because glucose is difficult to acquire, it has been suggested that sucrose can be substituted in oral solutions and have the same effect.[35] But sucrose solutions can only be used in lesser concentrations, because in higher concentrations it will add to the diarrhea.[36] Controversy exists as to the effectiveness of sucrose. Nalin[36] reports 1% failure with glucose solutions, but 18% with sucrose solutions.

Mann[37] has recommended what he terms "Phillip's fluid" that can be prepared at home using Caro Syrup and cream of tartar to give a concentration of 92 mM sodium, 18 mM potassium, 24 mM bicarbonate, 64 mM chloride, and 125 mM glucose, which is effective. The World Health Organization recommends a formula that can be prepared at home and yields similar concentrations.[38] (See Chapter 20 for formulas.) Whether one uses Gelsol, Phillip's fluid, or a low-concentration sucrose fluid, it is quite evident that all glucose–electrolyte solutions can correctively treat toxigenic diarrhea until the infection is brought under control or reversed.[38,39]

Treatment of other forms of diarrhea are discussed in the sections where those specific titles are listed.

14.3. GAS

Humans are both blessed and troubled by gases produced in the gastrointestinal tract. From the first time a mother burps her baby, through the endless obsessions and jokes that children have about belching and flatus, to the social implications in Asian societies of burping after a meal, to the social implications of passing wind at a society dinner, or having halitosis at a business meeting, human beings are constantly reminded of gas within the gastrointestinal tract. An understanding of normal gas production can be of help in enjoying its production or in attempting to control its volatility.

14.3.1. Belching

Belching occurs when the lower esophageal sphincter opens to permit air to arise from the stomach in a forceful manner. The details of the physiology of belching are described in Chapters 1 and 7. Most often, belching is a voluntary event, but it can occur involuntarily.

Aerophagia occurs when there is frequent release of air from the stomach into the esophagus. Most patients who complain of frequent belching are aerophagics who swallow a great deal of air either due to frequent eating or as a result of emotional strain. They are usually unaware of the frequent swallowing of air and can be helped by intensive education or control of their emotional unrest. Air that accumulates in the stomach from swallowing air is not due to fermentation. There is no evidence of a clinical syndrome in which either food or fermentation is the cause of increased production of gas within the stomach that results in increased belching. Increased eructations due to increased amounts of air in the stomach are due to the swallowing of air. Pyloric obstruction or extensive gastric carcinoma occasionally may cause mild belching or abnormal tastes due to overflow and retention in the stomach, but they rarely ever cause the belching seen from large amounts of swallowed air.

14.3.2. Halitosis and Flatus

Halitosis and intestinal gas can be partially due to swallowed air that goes from the stomach into the intestine, but most often they are due to bacterial fermentation of foods. Gas that accumulates within the intestinal tract is passed out through the anus, or absorbed into the bloodstream and passed out through expired air. Halitosis results when odiferous gases are produced, absorbed, and expired.

Humans produce anywhere from 500 to 1000 cm^3 of gas per day, as measured by rectal tube collections.[40-42] Ninety-nine percent of intestinal gas is composed of nitrogen, oxygen, hydrogen, methane, and carbon dioxide, with varying and small amounts of odiferous gases.[43] Nitrogen is usually the predominant gas. Oxygen occurs in very low concentrations because it is rapidly used up by intestinal bacteria. Hydrogen, methane, and carbon dioxide make up approximately 50% of intestinal gas. They are produced by bacterial action on intestinal contents, as they are not normal constituents of the atmosphere. They are produced in the gut rather than passed on down from the stomach. All gases are produced primarily in the colon but can be produced in the small bowel, depending on luminal contents and the distribution of responsible bacteria.

Methane (CH_4) is the best studied of the gases. Approximately one-third of all human beings are capable of producing it by colonic bacterial action.[44] Unlike other gases, CH_4 is related to some inherited or developmental familial characteristics. When both parents are producers of methane, their children will have at least a 95% chance of being methane producers. If only one parent is a producer, then their children have approximately a 50% chance of being methane producers, and if neither parent is blessed with methanogenic bacteria, then the chance of a child developing such a flora is approximately less than 10%.[44]

The production of hydrogen and carbon dioxide is directly related. Increased amounts of hydrogen are associated with increasing percentages of carbon dioxide. Carbon dioxide is rapidly absorbed from the gut and is therefore difficult to measure in flatus. However, intestinal washout techniques and isotope-labeled expired air studies on subjects fed beans or onions reveal hydrogen production is related to carbohydrate and protein intake, whereas methane production is related to familial and bacterial activity.[45-48]

14.3.3. Relationship of Foods to Intestinal Gas Production

Beans contain large amounts of oligosaccharides such as stachyose. Stachyose cannot be split by human enzymes and requires intestinal bacterial action, which occurs in the colon. The fermentation results in a marked increase in expired hydrogen and a fivefold increase in flatus.[42] Earlier workers also showed a general high-carbohydrate diet increases flatus 50%, and soy protein increases the volume, but to a lesser degree.[40] A careful study on the comparison of a fiber-free diet to a high-fiber diet consisting of brussels sprouts indicated the sprouts increase flatus 44%. Davies[41] studied ten flatulent healthy subjects on three different diets. The first, a high-carbohydrate bean diet containing approximately 9.4 g of crude fiber and 148 g of carbohydrate, resulted in a mean flatus production of approximately 50 ml of gas per hour. The second, a very low crude fiber diet of approximately 2.4 g of fiber (the average American intake) and 100 g of carbohydrate, resulted in approximately 25 ml of gas per hour. The third diet, a liquid chemical elemental diet containing no crude fiber and 407 g of carbohydrate, produced approximately 10 ml of gas per hour. This study clearly indicates that the amounts of oligosaccharides and long-chain polysaccharides contained in crude fiber have a significant effect on the volume of gas produced.[41]

Employing both breath gas and rectal flatus analysis, Hickey and associates[49] carefully studied ingested fruits and fruit juices in a group of healthy young male volunteers free of any apparent food hypersensitivity or gastrointestinal disease. Orange juice and apricot nectar produced no increase in intestinal gas production. Apple juice markedly increased hydrogen production and flatus. Grape juice, raisins, and bananas significantly elevated hydrogen production, but to a lesser degree than apple juice. Prune juice caused a marked increase in breath hydrogen but caused diarrhea, which did not enable careful rectal hydrogen collection. It is interesting to note that apple juice will increase hydrogen production but yet have a lesser effect on the bulk size of stool than formed apples.[50] It is suggested that the sorbitol present in apples may cause increased hydrogen excretion, but equal amounts are present in apricot, which does not cause hydrogen ion secretion.[49] Analysis of both citric and malic acid contents of fruit juices reveals no significant relationship to the production of hydrogen.[49] Much more work is needed on the correlation of specific compounds within fruits and all foods that are related to gas production.

Subjects who do not break down lactose leave it for the fermentation of bacteria. Consequently, lactose malabsorbers frequently complain of increased flatus (see Chapter 13).

Although there have been extensive studies on the relationship of the intestinal microflora to diseases in human subjects, there has been little actual correlation of specific bacterial species and the production of gases. What is known is that certain humans harbor methanogenic bacteria and that a wide variety of the flora are capable of fermenting carbohydrates and proteins that human enzymes are not capable of digesting. Although much is known about the bacteria involved in rumen fermentation, and in many animal fermentation models, very little is known concerning the fecal flora as it directly relates to clinical gas production in humans.[51,52]

When halitosis becomes a clinical problem and oral pathology is ruled out, the cause is intestinal-produced trace gas, absorbed and excreted through the lungs. Trace gases make up less than 1% of flatus and include ammonia, hydrogen sulfide, and volatile amino acids and short-chain fatty acids. By means of sensitive gas chromatography, they can be detected in expired air. Much more work is needed to understand these phenomena, which may prove to be therapeutically helpful in the future.[53-55]

14.3.4. Treatment

Therapy of the patient complaining of increased flatus or halitosis depends upon a careful dietary history and then dietary recommendations. It should be stressed that therapy will vary in each individual because of the uniqueness of the individual's bacterial flora.

The following clinical and dietary factors must be evaluated and manipulated or eliminated to determine their effect:

1. The presence of any malabsorption.
2. Lactose.
3. Excessive high-fiber intake in the form of beans or fruits, as discussed above.

Most often, the patient complaining of halitosis and increased flatus cannot be treated effectively. However, in light of recent advances in diet and food research, new information should become available that will be helpful in the treatment of these patients. The interested physician should attempt manipulation of the diet by the three steps suggested above, ruling out lactose malabsorption, ruling out any malabsorption disorder that will result in increased availability of substances for bacterial fermentation, and finally, in dietary manipulations of oligosaccharides and fruit substances that produce hydrogen and carbohydrate increases. The methane producers are extremely difficult to treat and there is no readily available information to offer these subjects on a scientific basis.

REFERENCES

1. Osler W: *Principles and Practice of Medicine*, ed 6. Sidney, London: Appleton, 1906, p 527.
2. Hurst AF: *Constipation and Allied Intestinal Disorders*, ed 2. London: Oxford Press, 1921.
3. Bockus HL: *Gastroenterology*, ed 3. Philadelphia: WB Saunders, 1976, vol 2, p 936.
4. Schuster MM: Constipation and anorectal disorders. *Clin Gastroenterol* 6:643, 1977.
5. Phillips SF, Giller J: The contribution of the colon to electrolyte and water conservation in man. *J Lab Clin Med* 81:733, 1973.
6. Burkitt DP, Trowell HC: *Refined Carbohydrate Foods and Disease*. New York: Academic Press, 1975, p 76.
7. Floch MH, Fuchs HM: Modification of stool content by increased bran intake. *Am J Clin Nutr* 31:185, 1978.
8. Fuchs HM, Dorfman S, Floch MH: The effect of dietary fiber supplementation in man. II. Alteration in fecal physiology and bacterial flora. *Am J Clin Nutr* 29:1443, 1976.
9. Davenport HW: *Physiology of the Digestive Tract*, ed 4. Chicago: Year Book Medical Publishers, 1977, p 72.
10. Hoppert CA, Clark AJ: Digestibility and effect on laxation of crude fiber and cellulose in certain common foods. *J Am Diet Assoc* 21:157, 1945.
11. Olmstead WH, Williams RD, Buaerlein T: Constipation: The laxative value of bulky foods. *Med Clin North Am* 20:449, 1936.
12. Cleave TL: Natural bran in the treatment of constipation. *Br Med J* 1:461, 1941.
13. Walker ARP: Crude fibre, bowel motility and pattern of diet. *S Afr Med J* 35:114, 1961.
14. Kramer P: The meaning of high and low residue diets. *Gastroenterology* 47:649, 1964.
15. Eastwood MA, Hamilton T, Kirkpatrick JR, *et al*: The effects of dietary supplements of wheat bran and cellulose on faeces. *Proc Nutr Soc* 32:22A, 1973.
16. Heaton KW, Heaton ST, Barry RE: An *in vivo* comparison of two bile salt binding agents, cholestyramine and lignin. *Scand J Gastroenterol* 6:281, 1971.
17. Eastwood MA, Girdwood RH: Lignin: A bile salt sequestrating agent. *Lancet* 2:1170, 1968.
18. Williams RD, Olmstead WH: The effect of cellulose, hemicellulose and lignin on the weight of stool. A contribution to the study of laxation in men. *J Nutr* 11:433, 1936.
19. Bender WA, Whistler RL, Bemiller JN: *Industrial Gum*. New York: Academic Press, 1959, p 377.
20. Kelsay JL, Behall KM, Prather ES: Effect of fiber from fruits and vegetables on metabolic responses of human subjects. I. Bowel transit time, number of defecations, fecal weight, urinary excretions of energy and nitrogen and apparent digestibilities of energy, nitrogen, and fat. *Am J Clin Nutr* 31:1149, 1978.
21. Binden HJ: Net fluid and electrolyte secretion: The pathophysiological basis of diarrhea. *Viewpoints Dig Dis* 12:2, 1980.
22. Davenport W: *Physiology of the Digestive Tract*, ed 4. Chicago: Year Book Medical Publishers, 1977.
23. Phillips SF: Diarrhea—A broad perspective. *Viewpoints Dig Dis* 7:5, 1975.
24. Phillips SF: Diarrhea: A current view of the pathophysiology. *Gastroenterology* 63:495, 1972.
25. Sharp GWG: Action of cholera toxin on fluid and electrolyte movement in the small intestine. *Ann Rev Med* 24:19, 1973.
26. DuPont HL: Etiologic diagnosis of acute diarrhea. *Ann Intern Med* 88:707, 1978.
27. Schwartz CJ, Kimberg DV, Sheerin HE, *et al*: Vasoactive intestinal peptide stimulation of

adenylate cyclase and active electrolyte secretion in intestinal mucosa. *J Clin Invest* 54:536, 1974.

28. Ammon E, Phillips S: Inhibition of ileal water absorption by intraluminal fatty acids. *J Clin Invest* 53:205, 1974.

29. Fawell WN, Thompson G: Nutmeg for diarrhea of medullary carcinoma of thyroid. *N Engl J Med* 289:108, 1974.

30. Bennett A, Gradidge CF, Stamford IF: Prostaglandins, nutmeg and diarrhea. *N Engl J Med* 289:110, 1974.

31. Barrowman JA, Hillenbrand P, Rolles K, *et al:* Nutmeg for diarrhea. *N Engl J Med* 289:810, 1974.

32. Editorial: Oral glucose/electrolyte therapy for acute diarrhea. *Lancet* 1:79, 1975.

33. Nalin DR, Cash RA, Rahman M, *et al:* Effect of glycine and glucose on sodium and water absorption in patients with cholera. *Gut* 11:768, 1970.

34. Hirschhorn N, Denny KM: Oral glucose–electrolyte therapy for diarrhea: A means to maintain or improve nutrition? *Am J Clin Nutr* 28:189, 1975.

35. Moenginah PA, Suprapto, Soenarto J, *et al:* Oral sucrose therapy for diarrhea. *Lancet* 1:323, 1975.

36. Nalin DR: Oral sucrose therapy for diarrhea. *Lancet* 1:1308, 1975.

37. Mann GV: Diarrhea. *Am J Clin Nutr* 28:804, 1975.

38. Sack RB, Pierce NF, Hirschhorn N: The current status of oral therapy in the treatment of acute diarrheal illness. *Am J Clin Nutr* 31:2252, 1978.

39. Hirschhorn N, McCarthy BJ, Ranney B, *et al:* Ad libitum oral glucose–electrolyte therapy for acute diarrhea in Apache children. *J Pediatr* 83:562, 1973.

40. Blair HA, Derm RJ, Bates PL: The measurement of volume of gas in the digestive tract. *Am J Physiol* 149:688, 1947.

41. Davies PJ: Influence of diet on flatus volume in human subjects. *Gut* 12:713, 1971.

42. Steggerda FR, Dimmick JF: Effect of bean diets on concentration of carbon dioxide in flatus. *Am J Clin Nutr* 19:120, 1966.

43. Bond JH, Levitt MD: A rational approach to intestinal gas problems. *Viewpoints Dig Dis* 9:2, 1977.

44. Bond JH, Engel RR, Levitt MD: Factors influencing pulmonary methane excretion in man. *J Exp Med* 133:572, 1971.

45. Levitt MD: Production and excretion of hydrogen gas in man. *N Engl J Med* 281:922, 1969.

46. Levitt MD: Volume and composition of human intestinal gas determined by means of an intestinal washout technique. *N Engl J Med* 284:1394, 1971.

47. Kirk E: The quantity and composition of human colonic flatus. *Gastroenterology* 12:782, 1949.

48. Calloway DH: Respiratory hydrogen and methane as affected by consumption of gas-forming foods. *Gastroenterology* 51:383, 1966.

49. Hickey CA, Calloway DH, Murphy EL: Intestinal gas production following ingestion of fruits and fruit juices. *Am J Dig Dis* 17:383, 1972.

50. Haber GB, Heaton KW, Murphy D, *et al:* Depletion and disruption of dietary fiber—Effects on satiety, plasma-glucose, and serum-insulin. *Lancet* 2:679, 1977.

51. Hungate RE: Microbial activities related to mammalian digestion and absorption of food, in Spiller GA, Amen RA (eds): *Fiber in Human Nutrition.* New York: Plenum Press, 1976.

52. Drasar BS, Hill MJ: *Human Intestinal Flora.* New York: Academic Press, 1974.

53. Pauling L, Robinson AB, Teranishi R, *et al:* Quantitative analysis of urine vapor and breath by gas–liquid partition chromatography. *Proc Nat Acad Sci USA* 68:2374, 1971.

54. Chen S, Mahadevan V, Zieve L: Volatile fatty acids in the breath of patients with cirrhosis of the liver. *J Lab Clin Med* 75:622, 1970.

55. Chen S, Zieve L, Mahadevan: Mercaptans and dimethyl sulfide in the breath of patients with cirrhosis of the liver. Effect of feeding methionine. *J Lab Clin Med* 75:628, 1970.

Diverticular and Inflammatory Bowel Diseases

This chapter deals with three colon diseases that have been recognized with increased frequency during the 20th century: diverticulosis of the colon, nonspecific ulcerative colitis, and Crohn's disease (granulomatous ileitis and/or colitis). Although diverticular disease is probably of different etiology than nonspecific ulcerative colitis and Crohn's ileitis, all three have in common an unknown etiology, inflammation of the colon, an increasing incidence during this century, and suggested environmental factors in their etiology.

15.1. DIVERTICULAR DISEASE OF THE COLON

15.1.1. Incidence

Incidences are determined by either barium enema or autopsy studies. The first cases were reported early in the century and since then the diagnosis has been made with increasing frequency.[1] Reports from the United States, United Kingdom, Sweden, France, and Australia reveal the general incidence varies from 2 to 45%, and also varies with age. A careful autopsy study by Hughes[2] in Australia revealed an overall incidence of 43%. The female to male ratio was almost equal, the incidence increased with age, being 9% between 31 and 50 yr, 36% between 51 and 70 yr, and 56% over the age of 71. The overall incidence past the age of 50 was 48%. The distribution in the colon is important and the study showed 15% of the subjects had few diverticula, 26% had many in the sigmoid, 30% had them through the descending colon and sigmoid, 4% in the sigmoid, descending colon, and transverse colon, 16% in the entire colon, and only 5% in the cecum.

15.1.2. Etiology

The cause of diverticular formation in the colon remains uncertain. Most diverticula develop in areas of the muscle wall where the muscle is weakest.

Because the disease increases in frequency with age, it is logical to suspect that a weakness throughout the colon develops that permits diverticular formation. However, diverticula do occur in young people, and autopsy studies reveal multiple diverticula without significant atrophy of the muscle wall. Fleschner[3] attempted to classify diverticulosis into two forms. One is a simple massed diverticulosis that involves primarily the descending sigmoid colon, results in a barium-enema appearance of studded diverticula, and in a gross distortion of the descending colon. The other type is a "spastic colon diverticulosis" that appears on barium enema to have less diverticula but is associated with more spasm. It is usually seen in the sigmoid and descending colon but may involve the entire colon and often is associated with deformed haustral markings and occasionally with the appearance of thumb prints.

The appearance of muscle spasm with diverticula reinforces the theory of hyperactive muscle activity as an associated cause. Painter and Burkitt[1] presented evidence that indicates there is increased muscle activity in subjects with diverticular disease.[1] However, Eastwood and associates[4] found that subjects who do not have diverticular disease exhibit variation in colonic function similar to those who do have diverticular disease. Nevertheless, their data do confirm that the general population that presently is subject to formation of diverticula does have a significant variation and potential abnormal colonic motility.[4] This raises the possibility that a single environmental factor may be related to the high incidence of diverticular disease in present-day Western societies. That smooth muscle spasm is associated with an increased refined-carbohydrate diet and concomitant decrease in high-fiber or bulk-forming diet is now generally accepted.[1,5] Diverticula are virtually unreported in societies with high-fiber diets. The relationship to low-fiber intake is fortified by evidence that the incidence is much lower in vegetarians living in Western societies[5] and that stool–dietary analysis of diverticular disease subjects reveals they excrete less β-sitosterol and, hence, must eat less plant sterol as a group.[6] The emergence of diverticular disease in the urban South African black further emphasizes the epidemiologic dietary theories.[8]

The fiber deficiency correlates well with the physiology of disturbed motility and a weak muscle wall. Very simply, the colon that has less bulk is caused to contract more, according to LaPaz' law. Having less bulk makes less of a colon content and, as a consequence, the contracting bowel has less to distribute the force. The force of contraction results in a greater distribution of pressure on the wall and a resultant diverticular formation. When circular muscles, which cause segmental delineation, and longitudinal muscles, which increase the central force, simultaneously contract, something has to give. When there is a limited luminal content to which pressure can be distributed, the weakest point in the muscle wall permits the out-pocketing to occur. This theory then relates increasing diverticular formation with age because of increasing weakness of the muscle wall, the decreased bulk due to fiber deficiency that is prominent in Western societies, and the increase in spasm of the colon of the subjects with diverticular formation.

15.1.3. Pathology

The taeniae of the colon are three longitudinal muscle bands approximately one-third equidistant from the mesentery, and they circle the colon. Blood vessels reach the colon from the mesentery, then enter through the wall near the two antimesenteric taeniae. It is at this point that diverticula most commonly develop.[9] The uncomplicated diverticula on histologic examination consists of a mucosa, submucosa, and peritoneal wall, but no muscularis. They are most common in the sigmoid colon.[9] Muscular abnormalities are the most consistent and important feature after inflammation occurs in the bowel wall. The taeniae coli and circular muscle become thick and may become corrugated in appearance.[9,10] At times the muscular thickening forms cliffs so that there are actually no diverticula, but out-bulgings.. Morson[10] has described these changes in careful detail and correlated them with barium enema findings. It is not clear as to whether muscular abnormalities occur without inflammation. However, diverticula do become inflamed and the lymphoid tissues associated with them become hyperplastic. The inflammation often spreads longitudinally from the apex of the diverticulum parallel to the outer aspect of the deep muscle layers to form a small dissecting abscess. Free perforation rarely occurs, but small localized abscesses occur and give symptoms. The pericolic abscesses become filled with pus and may be necrotic. In most cases, they are walled off, but may spread into adjacent organs such as the urinary bladder.

A high incidence of bleeding also is associated with diverticula. It is estimated approximately 10% of all people suffering from diverticular disease will develop either serious inflammation or bleeding. Surprisingly, the bleeding that occurs is more often from the right side of the colon than the left. This is important in diagnostic evaluation of the acutely bleeding patient who may require surgery.

The *diagnosis* of diverticular disease is usually made in the uncomplicated case by barium-enema examination. However, once there is an acute episode, the diagnosis can be made clinically by the finding of lower abdominal pain in association with tenderness, low-grade temperature, an elevated white blood count, and, on barium-enema examination, the findings of incomplete filling or spasticity in the area of the inflamed bowel. Sigmoidoscopic examination is usually negative or reveals marked spasm at 15 to 20 cm. During the acute stage, colonoscopy should be postponed. After barium enema, and if there is any concern about differentiating a spastic segment of bowel from carcinoma, colonoscopy can be done. Briefly, the mild case is diagnosed by barium enema, the acute, painful, or febrile by clinical signs and barium enema, and the bleeding patient by endoscopy and angiography followed by a barium contrast study, if needed.

15.1.4. Treatment

The treatment of the disease depends entirely on the symptoms and stage. The asymptomatic patient is treated entirely different than the patient with pain or evidence of active inflammation.

Once serious complications such as obstruction, severe abscess formation, or severe bleeding have occurred, surgical intervention may be necessary.[11]

15.1.4.1. The Asymptomatic Patient

When there is little pain and the diagnosis has been established by barium-enema examination, the recommended treatment is a high-fiber diet to increase stool bulk. Considering the present theories of the etiology of diverticular disease, one must advise a high-fiber diet. It is extremely important that the disease be "staged" when the high-fiber diet is employed. The patient with signs of obstruction and inflammation certainly should not be treated with increased bulk. It is the patient with minimal symptoms, such as mild pain and persistent constipation, or alternating bowel pattern, who should be treated with a high-bulk diet. Painter and colleagues[12] treated 70 patients with diverticular disease with a high-fiber diet consisting primarily of supplemented unprocessed bran. An approximately 2-yr follow-up revealed a relief of major symptoms in approximately 85% of the subjects.[12] Plumley and Francis[13] supplemented a regular diet with a high-residue, crisp bread made of bran and studied 42 subjects.[13] They found 71% had controlled symptoms being followed for 6 to 36 months. Findlay and associates[14] fed 20 g of unprocessed bran daily to both normal and colonic diverticular patients. They found increased stool weight, shortened transit time, a modified fecal flow pattern, and reduction of the increased intraluminal pressure in response to stimuli in the distal colon in subjects with diverticular disease. There was no change in baseline motility studies, but a definite change in the response to stimuli. Another study on 40 subjects fed 23 g of wheat bran per day showed an excellent clinical response in 33. However, most studies are poorly controlled and there is no double blind or control group. Nevertheless, simultaneous motility studies do reveal a decrease in high colonic pressure in diverticular disease.[14] Comparison of the effect of bran on motility to the effect of a bulk-forming agent, sterculia, and anticholinergic drugs reveals bran is more comparable to a combination of bulk-forming plus anticholinergic agents rather than to merely bulk-forming agents alone.[15,16] The effect of long-term bran use or high-fiber diet seems to be related to a change in motility that is characterized by a decrease in intraluminal pressure.

The treatment of the asymptomatic patient or the patient with mild symptoms should include a high-fiber diet. To date the only studies performed have used bran. Diets 26 and 27 of Chapter 20, include the high-fiber diets recommended to increase dietary fiber to at least 40 g/day (equal to crude fiber of approximately 10 to 15 g). Each individual must be assessed and advised individually. Some subjects will not be able to tolerate bran, while others will not be able to tolerate large amounts of fruits. The diet lists substitutes for each high-fiber food. Each subject should vary the fiber intake in order to build up his own high-fiber diet. There is no significant clinical research to date to indicate which fiber substances may be preventative, or best tolerated by the subject with diverticular formation. The best practice for now is to increase the total fiber intake, and this may be accomplished by bran in 20-g doses as suggested by Painter and others or by a broad high-fiber diet as listed in Chapter 20, diets 26 or 27.

15.1.4.2. The Symptomatic Patient

The patient with mild abdominal pain but having bowel movements, either constipation, alterating irregular bowel habits, or loose bowel movements can still be treated with a high-fiber diet, but it is recommended that an anticholinergic agent be added to the regimen.

The patient who develops significant pain that is cramplike or colicky in nature and causes an alteration in the usual daily bowel activity must be observed carefully. If there is no sign of obstructive disease or focal perforation, a high-fiber diet can be continued and anticholinergics employed to relieve pain. When pain relief requires medication, narcotics that cause spasm should be *absolutely* avoided, e.g., morphine, codeine, as they increase segmental contractions. Mild pain-relieving medication may be employed to complement the anticholinergics.

When the bowel becomes *acutely inflamed* or there is evidence of *focal abscess formation*, the management must include a fiber-free or liquid diet. The severity of the dietary restriction should be correlated with the severity of the symptoms and findings. When there is evidence of marked tenderness, a mass, an elevated white blood count, temperature, or definite obstruction, a liquid diet should be instituted. If obstruction is suspected and confirmed by radiographic findings, the patient should be placed in a hospital, on intravenous feeding, nasogastric long-tube drainage, and appropriate antibiotic therapy with the hope of relieving a temporary obstruction. Surgical intervention is necessary if the obstruction cannot be relieved with this therapy.

During the acute attack, patients can be treated at home with the following regimen:

1. Fiber-free liquid diet (Chapter 20, diet 29)
2. Anticholinergic therapy
3. Appropriate antibiotic therapy

If the attack is severe and there is obstruction, the patient should be treated with the following regimen:

1. Hospitalization
2. Intravenous therapy
3. *Non per os* and long-tube nasogastric suction

If surgery is required, or fistular or abscess formation has occurred in a patient who cannot tolerate surgery, nutritional supplementation should be considered. These regimens are discussed in detail in Chapters 18 and 19.

As the attack subsides, the patient can gradually be returned to foods containing more fiber. It is also recommended that the diet be bland (Chapter 20, diet 31).

Once the acute attack has completely subsided, the patient can be returned to a high-fiber regimen. However, this should be done gradually by first increasing the number of vegetables and fruits and then bran and cereals. The increase of fiber should not be undertaken until one establishes the absence of any obstruction in the bowel and a full free flow of fecal material.

15.2. INFLAMMATORY BOWEL DISEASE

1. *Ulcerative proctitis* can be defined as an inflammatory disease of the epithelium of the rectum of unknown cause that invariably remains limited to the rectum.

2. *Ulcerative colitis* can be defined as an inflammatory disease of the mucosa of the colon of unknown cause that usually involves the rectum and may involve the entire colon.

3. *Crohn's colitis* (granulomatous colitis) can be defined as an inflammatory disease that involves the entire wall of the colon and is most often characterized by segmental involvement, but may upon occasion involve the entire colon.

4. *Regional ileitis* or *Crohn's ileitis* (terminal ileitis) can be defined as an inflammatory bowel disease that involves the entire wall of the small intestine, is most characteristic when it involves the distal ileum, but may involve skip areas through the entire gut.

5. *Crohn's ileocolitis* can be defined as an inflammatory disease that involves the entire gut wall and is characterized by skip areas involving both the ileum and the colon.

Although selected cases can be easily classified as one of the five above disorders, all too frequently a clinician or pathologist cannot clearly separate the disorders. What is often thought to be simple ulcerative colitis turns out, upon pathologic surgical study, to be Crohn's colitis. On occasion, what is believed to be a simple ulcerative colitis extends to the ileum following ileostomy. What is on occasion thought to be a simple limited ulcerative proctitis extends to a full-blown ulcerative colitis or even, more rarely, a Crohn's colitis. With increasing incidence of the diseases and increasing experience of clinicians and pathologists, the clear separation of disorders, now classified as inflammatory bowel disease, often is less apparent.

The purpose of this chapter is to elaborate on the nutrition of patients with inflammatory bowel diseases rather than give details on these diseases. The following discussion will group all off the disorders together because nutrition therapy is nonspecific but varies with the symptoms and complications of the disease. For detailed discussion of other aspects of each one of these diseases, the interested reader is referred to the gastroenterologic texts.[17-19]

It is shocking that so little is found in the medical literature concerning the dietary management of patients with these disorders. The following discussion will attempt to give a rationale for the dietary management of these patients with appropriate reference to accepted scientific fact and specific diets.

The *cause* of these diseases remains unknown. The incidence of both ulcerative colitis and Crohn's colitis and ileitis has increased during the past century.[17] The reason for the increase was first thought to be a better diagnostic approach, but this does not appear to be true from statistical evaluation. Many theories have been proposed as to the cause, but none have been substantiated or sustained in animal models, and hence not accepted on a universal basis. The present theories most accepted are that the cause of the diseases lies either within some immunologic deficiency or reactivity, or is of infectious origin. There has been no demonstration that any inflammatory bowel disease is transmittable from person to person. Studies reveal approximately 10 to 15% of affected families will have either one of the diseases. Multiple reports also are present in which multiple members of a family have one or more of the other diseases. However, no clear genetic pattern has yet been determined.

Theories on the immunologic mechanisms as a cause of the inflammation began to develop early in the clinical discovery of the disease. First, Anderson proposed that colitis was due to an allergic reaction to foods and subsequently proposed that milk was the most common inciting agent.[20] Truelove[20a] then presented a group of patients with ulcerative colitis who were markedly improved when cow's milk and milk products were removed from the diet, and had frank attacks of ulcerative colitis when it was reintroduced into the diet. He further attempted to substantiate the theory that milk was an etiologic factor when he demonstrated an increased frequency of the antibodies to the protein fractions, casein, lactalbumin, and lactoglobulin, in the serum of patients with ulcerative colitis.[21] However, the later work was not confirmed and clinical experience revealed that patients with inflammatory bowel disease have a high incidence of lactose intolerance that results in an increase in diarrhea and symptoms (see Chapter 13). With the realization that lactose does cause symptoms and the lack of substantiation that milk protein is an inciting agent, theories on foods as a direct cause through an allergic mechanism gained little support. However, acute inflammatory lesions in the colon can be produced by classic hypersensitivity reactions such as the Shwartzman phenomenon.[22,23] Further interest developed in immunologic theories when it became possible, through immunofluorescent techniques, to demonstrate antigens on the surface of colonic mucosa and antibody in the serum that react with extracts from colonic mucosa of patient with ulcerative colitis.[24,25] But no clear patholigic mechanism has evolved from these studies. The development of a better understanding of lymphocyte and plasma cell function in the mucosa of the gut has produced many important observations such as the importance of immunologic body competency, the production of immunoglobulins, and the reactivity of this tissue to luminal contents. Even though the immunologic capacity of the gut is better known, evidence continues to accrue that the cause of the disease is related to immunologic mechanisms, but not the basic mechanism. Carlsson and associates[26] have shown that antibodies to colon antigen occur in patients with many other diseases as well. Falchuk and Isselbacher[27] have shown that circulating antibodies to bovine albumin are present in 93% of patients with ulcerative colitis, 86% with Crohn's disease, in 5 with untreated celiac disease, and in 43% of normal subjects. These findings suggest that the absorption of antigenic material and stimulation of antibody production occurs in association with intestinal mucosal damage as well as in normal subjects.[27] This evidence shows that many of the immunologic findings are secondary to normal immunologic reactivity of the gut to food antigens that can pour through a porous thick mucosa.

Initial studies evaluating infectious agents reveal that an acute colitis indistinguishable from ulcerative colitis could be caused by the classical Shigella infections. Subsequent attempts to evaluate bacterial or viral infectious agents still reveal no conclusive evidence and the cause of these diseases remain unknown.[28-34]

15.2.1. Pathology

The intestine reacts to disease in a limited fashion. Inflammation is nonspecific and may be characterized by acute or chronic round cell infiltration in the mucosa, submucosa, or serosa. When this occurs, epithelial cell elements can react by decreased growth with resultant decreased development of glandular elements. Selective responses may occur such as the formation of granulomas, neuromas, or neurofibromas. Specific reactions may be seen, such as pyloric gland metaplasia in Crohn's disease.

Once inflammation becomes chronic, pathologic complications may occur as atrophy of mucosal elements causing loss of function, erosion into blood vessels causing bleeding, thinning of the tensile strength of the wall causing perforation, or a chronic fibrotic reaction causing scarring and stenosis.

Each of the inflammatory bowel diseases has a more characteristic pathologic response. However, there is an overlap, and at times the most astute pathologist cannot differentiate between simple ulcerative colitis, which is supposed to classically involve only the mucosa, and Crohn's disease, which is supposed to be a transmural disease.

Table 15-1, as modified from Morson,[10] may be helpful in differentiating the two diseases.

Classical *signs and symptoms* of Crohn's disease involving the small intestine include either right lower quadrant pain with or without a mass, diarrhea or change in bowel habit, or anemia. Crohn's disease is associated with numerous extraintestinal manifestations and may present with oral lesions, gastritis or duodenitis, anal disease, or skin lesions such as erythema nodosum or pyoderma gangrenosum.[17-19]

The presentation in ulcerative colitis is most often that of fulminating diarrhea associated with rectal bleeding. However, once again, extraintestinal manifestations may be the presenting symptom, such as pyoderma gangrenosum, erythema nodosum, or spondylitis.

A physical examination may be helpful in diagnosis when the classic mass is present in the right lower quadrant, or the sigmoidoscopic findings of ulcerative mucosa are limited to the rectal vault in ulcerative proctitis. The physician must weigh all the findings in order to place the patient in the correct category.

Table 15-1. Differential Characteristics of Crohn's Disease & Ulcerative Colitis

	Crohn's disease	Ulcerative colitis
Course and gross findings		
Distribution	Discontinuous; segments of disease with normal intervening mucosa	In continuity from rectum proximally; not segmental
Rectum	Normal, 50%	Almost always involved
Anal lesions	Chronic fissure, anal fistula, chronic abscess; edematous anal tags in 75%	Acute fissure; anal excoriation; acute abscess in 10%
Mucosa	Serpiginous ulceration; cobblestone appearance; fissuring	Diffusely granular and hemorrhagic; no fissuring
Vascularity	Rarely pronounced (except in acute cases)	Often intense
Histologic findings		
Depth of inflammation	Transmural	Mucosal and submucosal
Submucosa	Widened	Normal width or reduced
Focal aggregates of lymphocytes	Usual (often transmural)	Sometimes (restricted to mucosa and submucosa)
Crypt abscesses	May occur	Very common
Goblet cell population of mucosa	Normal or slight reduction	Reduced in active disease
Granulomas	Present about 60%	Absent
Fissuring	Common	Absent
Precancerous epithelial changes	No	Yes
Intravascular platelet thrombi	No	No
Submucosal edema	Yes	Usually absent
Fibrosis	Yes	No
Pyloric gland metaplasia	Yes	No
Muscle necrosis	Never	Only in toxic megacolon
Clinical findings		
Ethnic distribution	More frequent in Jews	More frequent in Jews
Positive family history	Approximately 10%	Approximately 10%
Incidence	2–4/100,000	3–6/100,000
Site	50%+ small bowel diseased	Colon only, backwash ileitis rare
Hematochezia	Less or absent	Almost universal
Diarrhea	Present, but less	Early and frequent
Fever	30–50% if small bowel involved	Rare if uncomplicated
Steatorrhea	Frequent if small bowel involved	Rare
Skin lesions: Erythema nodosum Pyderma gangrenosum, etc.	Less common	17% of hospitalized patients
Arthritic manifestations Ankylosing spondylitis	Less common	20–25%

(continued)

Table 15-1 (Continued)

	Crohn's disease	Ulcerative colitis
Clinical findings (Continued)		
Liver involvement: Fatty, focal necrosis; cirrhosis	Somewhat less	50% at autopsy
Diagnostic aids		

Sigmoidoscopy	Normal or small discrete
	    aphthoid ulcers occasionally
		Principal diagnostic aid—
		    diffuse pinpoint ulcers

Abdominal mass	Quite frequent in RLQ	Rare
Roentgen study	Right colon frequent; segmental-skip areas	Rectum and up—continuous
	Eccentric asymmetrical	Complete circumferential involvement
	Cobblestone appearance	Pseudopolyposis
	Terminal ileum involved 50%	Terminal ileum normal; backwash occasional

	Deep ragged ulcers; fissuring	Minute diffuse marginal ulcers;
		    occasional large diffuse
		    marginal ulcers

Biopsy	(See *Histologic findings*)	
Complications		
Toxic megacolon	Rare	5% hospital cases
Free perforation	Rare	1.6–0.3%

The *diagnosis* of Crohn's ileitis involving the distal ileum with skip areas in the small intestine is made in the presence of a normal barium enema, normal sigmoidoscopic findings, and a small-bowel series that reveals a characteristic lesion in the terminal ileum or in the small intestine.

Differential diagnosis is between ulcerative colitis and Crohn's colitis in which the disease is limited to the colon and does not involve the small intestine (approximately 50%). The factors listed in Table 15-1 are often helpful and enable the clinician to differentiate the two diseases. However, as stated above, at times it is virtually impossible and even the pathologist cannot make a definite decision between the two forms of inflammatory bowel disease (IBD).

The complications of IBD are (1) nutritional wasting, (2) perforation, (3) stricture and obstruction, and (4) hemorrhage. The management of (2) through (4) is described elsewhere,[16–19] and only diet and drug treatment will be discussed here.

15.2.2. Prognosis

A simple dictum to follow is that less disease means a better prognosis whereas more disease means a worse prognosis. Hospitalization usually indicates the disease will be worse. The patient who never requires hospitalization usually has fewer complications and a better prognosis. Simple ulcerative proctitis is not a life-threatening disease as long as it remains confined to the rectum. Approximately one-third extend to the more proximal colon and then the prognosis becomes that of ulcerative colitis. The prognosis of ulcerative colitis is worse if the first attack is severe and requires hospitalization, or the disease starts very early in life, or the disease is extensive. Approximately one-third of hospitalized patients will usually require some form of operation.[16]

The results of long-term follow-up studies after surgery indicate at least a 50% recurrence rate with Crohn's disease, and a significant number with simple ulcerative colitis require revision of their ileostomy site after total colectomy.[16-19,35,36] The overall prognosis is varied depending on the complications, both systemic and intestinal, as listed in Table 15-1. However, the overall picture in IBD improves yearly with the increased efficacious use of corticosteroids and nutritional support.

15.2.3. Treatment

At the present time the medical treatment of inflammatory bowel disease includes the following.

1. *Use of Sulfasalazine.* Most clinicians feel it is very helpful in the acute stage of all inflammatory bowel disease if the patient can tolerate the medication. Controversy still exists as to whether the drug should be used in all subjects for long-term maintenance. The controversy is greatest in regional enteritis, and most observers feel that it is of benefit in ulcerative colitis.[17-19]

2. *Corticosteroids.* The use of oral, rectal, or intravenous corticosteroid (including ACTH) therapy has been instrumental in prolonging life and decreasing complications in all IBD.[17-19] Often a horrifying attack can be ameliorated by the use of oral or intravenous medication and the appropriate nutritional support in order to maintain body function until remission can be induced. For details of use of corticosteroids and ACTH, once again the reader is referred to gastroenterologic texts.[17-19]

3. *Symptomatic Treatment.* The judicious use of anticholinergics and opiates is extremely helpful in the acute or chronic management. I have found it most helpful to use 15 or 30 mg of codeine sulfate every 6 to 8 hr to help manage mild pain or diarrhea. When the patient cannot tolerate codeine, the use of one of the myriad anticholinergics or antidiarrheal agents, such as Lomotil or Immodium, is helpful. These drugs are much more expensive than codeine and, although they may be very helpful in the acute state, run into high costs when used long-term.

4. *Rest.* Often the patient will become fatigued during the acute stage of the disease and rest at home. Hospitalization may be necessary; this depends on the individual patient's tolerance and the management course set by the physician.

5. *Psychological Support and Patient Education.* This is one of the most important aspects of managing the patient with IBD. The acute case requires

family and patient education. This often has to be supplemented by some form of psychological support in order to tolerate the trials and tribulations and complications of the chronic course. The physician may be able to handle this himself or it may require the more skilled management of a trained counselor or psychiatrist.

6. *Nutritional Therapy.* It is important to stress certain specific aspects of absorption and malabsorption in IBD in order to approach nutritional therapy.

The degree of malabsorption correlates with the degree of small-bowel involvement or resection. A differential analysis of absorption studies in mild cases shows fecal fat and nitrogen excretion is increased while xylose absorption remains relatively normal. Massive small-bowel resection has a greater degree of fecal fat excretion and associated xylose malabsorption.[37] A study of 65 patients with regional enteritis by Gerson and associates[38] reveals a close correlation between the length of bowel resected or diseased ileum for both vitamin B_{12} and fat malabsorption. When the length of ileal dysfunction exceeds 90 cm, it is common to find an abnormal Schilling test. The D-xylose test is usually a poor indicator of the amount of disease. These facts should be kept in mind when the clinician approaches the replacement of energy and caloric intake in the patient with Crohn's disease.

Selected deficiencies are also a distinct problem in IBD. Lactose intolerance due to lactase deficiency is more common in inflammatory bowel disease. Consequently, many patients cannot tolerate large amounts of lactose while they may be able to tolerate small amounts. In order to get enough energy into diets in patients who like milk products, it may be necessary to add lactase to the milk.

Vitamin B_{12} deficiency may occur due to B_{12} malabsorption.

Both calcium and magnesium deficiencies occur in patients with severe small-bowel inflammatory disease. The incidence again is greatest when there is the greatest involvement or resection of the small intestine.[39,40] Krawitt and associates[40] noted that calcium malabsorption in Crohn's disease was only present in 13% of their patients. Furthermore, they were able to correlate calcium malabsorption best with enteric protein loss,[41] but not with steatorrhea or xylose absorption, bacterial colonization, or corticosteroid therapy. Others have observed a rapid serum-zinc depletion associated with corticosteroid therapy. The patients studied were burn victims and not those with IBD.[42] Nevertheless, zinc deficiency may become clinically apparent. Sturniolo et al.[43] demonstrated decreased zinc absorption and reduced zinc-serum concentrations ($65 \pm 14.1 \mu$g) in 23 patients with Crohn's disease. Poor growth and zinc deficiency in children with celiac disease have been reversed with the addition of zinc to the diet.[44]

Growth failure occurs in anywhere from 20 to 30% of patients under the age of 21 with Crohns' disease, and to a lesser degree in children with ulcerative colitis. Grand and Homer[45] demonstrated malabsorption defects and bacterial overgrowth in children, both of which result in nutritional deficiencies. These children benefit from nutritional supplementation and, if necessary, treatment with corticosteroids.[45–47] The course of their disease is interrupted sufficiently in order for them to increase their linear growth rate.

Many cases are reported in which normal growth never occurs. The reason for this is a suspected pituitary or metabolic genetic deficiency, as total-colectomy subjects with ulcerative colitis have continued to have decreased growth and actually remain dwarfs. When surgical amelioration of the inflammatory bowel plus nutritional supplementation does not correct decreased growth, the clinician is stumped and the cause is often never delineated.

One of the frequent complications of chronic inflammatory bowel disease is nephrolithiasis. The stones most often contain calcium and oxylate. It is now known that increased urinary oxylate excretion occurs because of increased absorption of oxylates from the intestine,[48] with the colon being one of the major sites of hyperabsorption of oxylates.[49]

At the present state of our knowledge, nutritional therapy falls into three categories: (1) symptomatic treatment, (2) replacement therapy, and (3) preventive therapy. There is all too little information to tell exactly which foods may and which may not be of benefit. Research in this area is developing rapidly, and consequently the future holds great promise for more preventive and corrective diet therapy. However, the majority of the present treatments are symptomatic or replacement.

15.2.3.1. Acute Ulcerative Colitis

During this phase, diarrhea, tenesmus, and frequent bowel movements are a problem. In order to control the symptoms, drugs usually must be used. However, it is essential to control the diet as well. The extent of diet control is also dependent on the severity of the symptoms. A patient rapidly dehydrating, having more than a dozen bowel movements per day, and suffering from severe abdominal cramps will often need hospitalization with peripheral or central venous feeding and absolute bowel rest. This type of patient can be supported for a few days on peripheral glucose and amino acid solutions, and vitamin therapy. However, if the course is fulminating and it appears that the patient will require longer than 1 week of bowel rest, the clinician must consider total parenteral nutrition, and proceed with a several week commitment to TPN. The methods and use of this are described in Chapter 19. This is usually only needed in a severely ill ulcerative colitis patient undergoing a massive acute exacerbation.

Should TPN not be required, but some form of bowel rest necessary, a diet that requires little effort for digestion and absorptive processes, and produces no residue material that reaches the inflamed colon, can be chosen. This type of diet can be chemical or elemental, as described in Chapter 18. It is also possible to feed nonchemical, palatable foods that are relatively basic and relatively free of fiber, as described in the Chapter 20, diet 29. This form of diet will produce energy and protein and permit small-bowel function, but can totally limit the amount of roughage that reaches the colon. However, the clinician should be aware that increased activity in the small intestine may well increase the amount of liquid that is presented to the right side of the colon, and consequently fiber-free feeding can aggravate diarrhea. When absolute rest is needed, removal of all significant oral intake is necessary.

As the patient improves, and the number of bowel movements subsides, dietary intake can be increased from an essential fiber-free diet to a low-fiber diet and finally, as symptoms totally subside, to a regular diet that includes fiber to a degree that the patient feels is optimal.

Because lactose intolerance is common in inflammatory bowel disease, it is wise to limit milk intake. The question of milk allergy is always present in these patients, and consequently milk intake may be a selective problem (see Chapter 17). However, all too often it is helpful and necessary to use milk or milk products such as ice cream to help add calories to the diet. It is best to first establish whether there is a lactose intolerance by appropriate testing, and if this is so, lactase can be added to the milk or milk products. A diet free of lactose may be helpful to the intolerant patient (see Chapter 20, diet 17). The exact effect of spices on the colon is not known. The only studies done have been on gastric mucosa (see Chapter 9), and no work is available on the effect on the intestine after eaten spices have traversed the gut. Although it seems prudent to limit the intake of harsh spices such as garlic and pepper, there is no scientific evidence to substantiate the harm or danger of these substances in the colon.

Because nutrition is often poor for long periods, *all* vitamins and minerals should be replaced and maintained in accordance with requirements (see Chapter 3).

15.2.3.2. Chronic Ulcerative Colitis

The dietary treatment of chronic ulcerative colitis depends on the degree of symptoms. If there is diarrhea, a low-fiber diet should be encouraged. However, the clinician should realize that this preference is largely on the basis of clinical anecdotal experience. There are no good studies to demonstrate that high fiber has a deleterious effect. As a matter of fact, epidemiologic studies indicate that it may be preventive. Furthermore, there is no information to indicate whether certain foods could help decrease diarrhea. Studies are badly needed in order to determine whether selected fiber diets can be of benefit. Some clinicians do add psyllium seed as a bulk former in mild colitis. At the present state of the art, however, most clinicians feel high-fiber substances should be eliminated from the patient suffering from any form of diarrhea. Consequently, low-fiber diets listed in Chapter 20, diet 29, are recommended. They largely eliminate fruits and vegetables, as well as bulk cereals. Many dieticians confuse the term "bland" with roughage. Many high-fiber substances, which in reality are high-roughage substances, are bland, whereas many bland foods contain no roughage.

Besides a low-fiber diet, variation in the quantity of food eaten at any meal may be helpful. Often patients can tolerate small feedings whereas a very large meal will induce more peristaltic activity. Furthermore, very cold foods may induce peristalsis as opposed to room-temperature or slightly warm foods.

The patient with chronic ulcerative colitis who has no diarrhea and no abdominal pain can often tolerate roughage. If increased amounts of fiber do not induce symptoms, there is no known reason why these patients cannot eat as much fiber as they desire. As a matter of fact, the epidemiologic evidence indicates that fiber may be preventive for inflammatory bowel disease; consequently, the clinician should not limit the intake of fiber if it does not induce symptoms.

15.2.3.3. Acute Ileitis

The dietary treatment of an acute fulminating attack of ileitis, or jejunal ileitis, is similar to the acute episode of ulcerative colitis, with some exception. In the patient having diarrhea, the situation is extremely similar and if the diarrhea, pain, and toxicity are severe, then everything said for bowel rest and the use of elemental, or easily absorbed, diets holds for small-bowel inflammatory disease.

The acute ileitis subject with evidence of complete obstruction must be kept off all feeding on nasogastric suction, and on peripheral or central vein feeding. As in severe fulminating colitis, if the treatment is anticipated to be longer than 1 week, TPN is indicated. As obstruction relieves, a gradual increase in intake is attempted.

In acute ileitus with evidence of intermittent or partial obstruction, liquid or elemental diets can be tried. Most often, attacks of ileitis will cause partial obstruction. These patients can be tried on liquids for a few days and then progressed to low-fiber diets.

If either the complete or partial obstructive case does not respond, surgical resection of the stenotic bowel is indicated. Because the disease recurs and small-bowel tissue is not replaceable, attempts at preventing resection are vigorous. Strobel and associates[50] were able to induce a remission on home TPN in 12 of 17 severely ill children after all other medical treatment failed. Other reports of successful elemental and TPN therapy are described with the techniques in Chapters 18 and 19.

Because the small bowel is the site of vitamin and mineral absorption, it is often necessary to replace these more vigorously than in simple colitis.

15.2.3.4. Chronic Ileitis

Chronic ileitis with or without diarrhea is treated the same as chronic colitis. Often these patients will have selective vitamin or mineral deficiencies because of impaired absorption. When this occurs, zinc, magnesium, calcium, and vitamin B_{12} deficiency should be suspected and treated by replacement. As noted previously, zinc can reverse some growth problems.[43,44] It is safest to advise routine vitamin and mineral supplementation in any ileitis subject with evidence of significant small-bowel involvement.

The nutritional management of enterocutaneous or enteral fistulas is controversial, but many report success with hyperalimentation.[55,56]

Table 15-2. Normal Ileostomy Effluent [51,52,54]

	Daily ranges	
	Total output (av)	Concentration (av)
Wet weight	200–600 g	—
Dry weight	28–48 g	—
Water	88–94%	—
pH	—	6.0–6.5
Sodium	30–80 meq (55)	100–130 meq/liter (115)
Potassium	3–6 meq (4)	5–11 meq/liter (8)
Chloride	15–30 meq (20)	15–140 meq/liter (45)
Calcium	15–40 meq (18)	10–64 meq/liter (25)
Magnesium	7–9 meq (8)	10–28 meq/liter (15)
Phosphorus	122–202 meq (150)	
Nitrogen	0.6–2.4 g (1)	
Fat	1.5–3.8 g (2.2)	

15.3. Ileostomy

Ileostomy is necessitated when the colon is removed because of severe colitis, infarction, cancer, or traumatic injury. The body rapidly adjusts to colectomy and fluid balance is maintained although the ileostomate is more prone to electrolyte and fluid imbalance.

Table 15-2 lists the average daily outputs from ileostomy subjects.[51,52] Although more water, sodium, and other minerals are lost, the average small bowel compensates by increased absorption if intake is adequate. Several texts describe the problems of ileostomy subjects and the general, psychological, and local care needed.[53,54]

Although the ileostomy subject cured of illness can eat a regular diet, some modifications are recommended.

1. A regular diet containing low- to moderate-fiber intake. It seems reasonable that the subject with an ileostomy maintain a low-fiber diet in order to maintain the comfort of a low ileostomy output. Because the role of the high-fiber diet still remains controversial, it certainly seems that the ileostomy subjects should not be put through increased output until the preventive role of a high-fiber diet is clearly understood.

2. Limitation of high-electrolyte-content fluids several hours prior to retiring because they increase output during sleep.

3. Maintaining adequate vitamin B_{12} replacement if there is significant ileum resection or any suspicion of B_{12} malabsorption.

Ileostomy obstruction has been reported due to large food particles.[54] When this occurs the particle causing the obstruction can usually be removed through a sigmoidoscope or with manipulation. Because this can occur, and because there may be some natural narrowing at the stoma, the ileostomy subject is cautioned not to eat large particles of food without chewing carefully. The rapid swallowing of food that is not chewed carefully should be avoided.

Whenever the ileostomy subject has more than a few days of increased ileostomy fluid, the physician should be alerted to the potential need for intravenous fluid and salt replacement. Studies done varying sodium intake in ileostomy subjects reveal they can deplete sodium if intake is curtailed below 10 meq/day and that high-sodium intake can increase ostomy output.[52]

And finally, except for individual idiosyncracies, there is no known necessary dietary restriction for the healthy ileostomy patient.

15.4. Colostomy

Dietary management in the subject with a colostomy will vary depending on the amount and reason for colon resection. The subject with a left-sided colostomy needs no precautions in dietary intake as compared to a normal person. Indiscretion in the left-sided colostomy subject usually results in the same phenomenon that occurs in the patient with normal anal function. When indiscretion may cause diarrhea, the left-sided colostomy subject suffers the problem of increased cleansing, but no deleterious or dangerous effects while the right side of the colon is functioning.

The colostomy subject who has an ascending colostomy or cecostomy can suffer from problems mentioned for the ileostomy subject above. The subject with a transverse colostomy may occasionally run into water and electrolyte depletion, but this is unlikely if the right side of the colon is functioning well. It is only when the right side of the colon function is removed that water and electrolyte replacement will be needed.

As opposed to ileostomy, colostomy is done more commonly for partial resection of the colon, for carcinoma of the colon, and for other inflammatory or obstructing lesions of the colon, in addition to inflammatory disease. It is less commonly done for IBD than ileostomy, but when it is, the distinct risk of recurrence is present in the remaining colon tissue. When and if IBD recurs, all of the dietary treatments mentioned under IBD apply to the patient with active inflammatory bowel disease, plus the colostomy. If the disease is acute, limitation of intake is necessary. If the disease is chronic, moderation is necessary as outlined above. When the colostomy is done for carcinoma of the colon, there is little specific dietary advice needed.

As always, patients develop their own patterns of gastrointestinal function. The subject with a very rapid gastrocolic reflex and an ostomy must learn to live with that function. Subjects with colostomies can often control their colonic function so that they have relatively no excretion, and control their flow by a colostomy irrigation. For these effects and management, the interested reader is referred to the appropriate text.[53,54]

REFERENCES

1. Painter NW, Burkitt DP: Diverticular disease of the colon, a 20th century problem. *Clin Gastroenterol* 4:3, 1975.
2. Hughes LE: Postmortem survey of diverticular disease of the colon. *Gut* 10:336, 1969.
3. Fleschner FG: Diverticular disease of the colon. *Gastroenterology* 60:316, 1971.
4. Eastwood MD, Brydon WG, Smith AN, *et al*: Colonic function in patients with diverticular disease. *Lancet* 1:1181, 1978.
5. Grimes DS: Refined carbohydrate, smooth-muscle spasm and disease of the colon. *Lancet* 1:395, 1976.
6. Gear JSS, Ware A, Fursdon P, *et al*: Symptomless diverticular disease and intake of dietary fiber. *Lancet* 1:511, 1979.
7. Miettinen TA, Tarpila S: Fecal β-sitosterol in patients with diverticular disease of the colon and in vegetarians. *Scand J Gastroenterol* 13:573, 1978.
8. Segal M, Solomon A, Hunt JA: Emergence of diverticular disease in the urban South African Black. *Gastroenterology* 72:215, 1977.
9. Slack WW: The anatomy, pathology and some clinical features of diverticulosis of the colon. *Br J Surg* 50:185, 1962.
10. Morson BC: Pathology of diverticular disease of the colon. *Clin Gastroenterol* 4:37, 1975.
11. Colcock BP: Diverticular disease: Proven surgical management. *Clin Gastroenterol* 4:99, 1975.
12. Painter NS, Almeida AZ, Colebourne KW: Unprocessed bran in treatment of diverticular disease of the colon. *Br Med J* ii:137, 1972.
13. Plumley PF, Francis B: Dietary management of diverticular disease. *J Am Diet Assoc* 63:527, 1973.
14. Findlay JM, Mitchell WD, Smith AN, *et al*: Effects of unprocessed bran on colon function in normal subjects and in diverticular disease. *Lancet* 1:146, 1974.
15. Brodribb AJM, Humphreys DM: Diverticular disease: Three studies. *Br Med J* 1:424, 1976.
16. Srivastava GS, Smith AN, Painter NS: Sterculia bulk-forming agent with smooth-muscle relaxant versus bran in diverticular disease. *Br Med J* 1:315, 1976.
17. Spiro HM: *Clinical Gastroenterology*, ed 2. New York: MacMillan, 1977, unit 6.
18. Sleisenger MD, Fordtran JS: *Gastrointestinal Disease*, ed 2. Philadelphia: Saunders, 1978.
19. Bockus HL: *Gastroenterology*, ed 3. Philadelphia: Saunders, 1976, vol 2, sect 5.
20. Andresen AFR: Gastrointestinal manifestations of food allergy. *Med J Record* 122:271, 1925.
20a. Truelove SC: Ulcerative colitis provoked by milk. *Br Med J* i:154, 1961.
21. Taylor KB, Truelove SC: Circulating antibodies to milk proteins in ulcerative colitis. *Br Med J* ii:924, 1961.
22. Rider JA, Moeller HC: Hypersensitivity factors in ulcerative colitis. *JAMA* 183:545, 1963.
23. Patterson M, Terrell JC, Waldron RL, *et al*: The Shwartzman phenomenon in the colon of rabbits. *New Series* 8:213, 1963.
24. Klavins JV: Cytoplasm of colonic mucosal cells as site of antigen in ulcerative colitis. *JAMA* 183:547, 1963.
25. Broberger O, Perlmann P: Demonstration of epithelial antigen in colon by means of fluorescent antibodies from children with ulcerative colitis. *J Exp Med* 115:13, 1962.
26. Carlsson HE, Lagercrantz R, Perlmann P: Immunological studies in ulcerative colitis. *Scan J Gastroenterol* 12:707, 1977.
27. Falchuk KR, Isselbacher KJ: Circulating antibodies to bovine albumin in ulcerative colitis and Crohn's disease. *Gastroenterology* 70:5, 1976.
28. Kovacs A: Titre of Anti-E. coli antibodies in ulcerative colitis. *Digestion* 10:205, 1974.
29. Mitchell DN, Rees RT: Agent transmissable from Crohn's disease tissue. *Lancet* 2:168, 1970.
30. Cooper HS, Raffensperger EC, Jonas L, *et al*: Cytomegalovirus inclusions in patients with ulcerative colitis and toxic dilatation requiring colonic resection. *Gastroenterology* 72:1253, 1977.
31. Parent K, Mitchell PD: Bacterial variants: Etiologic agent in Crohn's disease? *Gastroenterology* 71:365, 1976.

32. Burnham WR, Jones JE, Stanford JL, et al: Mycobacteria as a possible cause of inflammatory bowel disease. *Lancet* 1:693, 1978.
33. Whorwell PJ, Davidson IW, Beeken WL, et al: Search by immunofluorescence for antigens of *Rotavirus, Pseudomonas maltophilia,* and *Mycobacterium kansaii* in Crohn's disease. *Lancet* 1:697, 1978.
34. Whorwell PJ, Beeken WL, Phillips CA, et al: Isolation of reovirus-like agents from patients with Crohn's disease. *Lancet* 1:1169, 1977.
35. Fawaz KA, Glotzer DJ, Goldman H, et al: Ulcerative colitis and Crohn's disease of the colon: A comparison of the long term postoperative courses. *Gastroenterology* 71:372, 1976.
37. Wilson FA, Dietschy JM: Differential diagnostic approach to clinical problems of malabsorption. *Gastroenterology* 61:911, 1971.
38. Gerson CD, Cohen N, Janowitz HD: Small intestinal absorptive function in regional enteritis. *Gastroenterology* 64:907, 1973.
39. Harris OD, Philip HM, Cooke WT, et al: 47 Ca studies in adult coeliac disease and other gastrointestinal conditions with particular reference to osteomalacia. *Scan J Gastroenterol* 5:169, 1970.
40. Krawitt EL, Beeken WL, Janney CD: Calcium absorption in Crohn's disease. *Gastroenterology* 71:251, 1976.
41. Beeken WL, Busch HJ, Sylwester DL: Intestinal protein loss in Crohn's disease. *Gastroenterology* 62:207, 1972.
42. Flynn A, Pories WJ, Strain WH, et al: Rapid serum-zinc depletion associated with corticosteroid therapy. *Lancet* 1:1169, 1971.
43. Sturniolo G, Molokhia M, Sandle L, et al: Mechanisms for zinc deficiency in Crohn's disease. *Gut* 19:443, 1978.
44. Solomons NW, Rosenberg IH, Sandstead HH: Zinc nutrition in celiac sprue. *Am J Clin Nutr* 29:371, 1976.
45. Grand RJ, Homer DH: Approaches to inflammatory bowel disease in childhood and adolescence. *Pediatr Clin North Am* 22:835, 1975.
46. Whittington PF, Barnes U, Bayless TM: Medical management of Crohn's disease in adolescence. *Gastroenterology* 72:1338, 1977.
47. Kelts DG, Grand RJ, Shen G, et al: Nutritional basis of growth failure in children and adolescents with Crohn's disease. *Gastroenterology* 76:720, 1979.
48. Stauffer JW, Humphreys MH, Weir GJ: Acquired hyperoxaluria with regional enteritis after ileal resection. *Ann Intern Med* 79:383, 1973.
49. Dobbins JW, Binder HJ: Effect of bile salts and fatty acids on the colonic absorption of oxalate. *Gastroenterology* 70:1096, 1976.
50. Strobel CT, Byrne WJ, Advent ME: Home parenteral nutrition in children with Crohn's disease: An effective management alternative. *Gastroenterology* 77:272, 1979.
51. Kramer P, Kearney MM, Ingelfinger FJ: The effect of specific foods and water loading on the ileal encreta of ileostomized human subjects. *Gastroenterology* 42:535, 1962.
52. Kramer P: The effect of varying sodium loads on the ileal excreta of human ileostomized subjects. *J Clin Invest* 45:1710, 1966.
53. Lennenberg E, Rowbotham JL: *The Ileostomy Patient.* Springfield: C C Thomas, 1970.
54. Hill GL: *Ileostomy, Surgery, Physiology and Management.* New York: Grune & Stratton, 1976.
55. Editorial: Nutritional management of enterocutaneous fistulas. *Lancet* 2:507, 1979.
56. Calam J, Crooks PE, Walker RJ: Elemental diets in the management of Crohn's perianal fistulae. *J Parent Ent Nutr* 4:4, 1980.

The Irritable Bowel Syndrome

16.1. DEFINITION

The irritable bowel syndrome (IBS) can be defined as unexplained lower abdominal pain, constipation, or diarrhea. In children, recurrent abdominal pain (RAP) is a similar syndrome. Patients suffering from this disease have no organic pathology, therefore the word "unexplained" in the definition. Once organic pathology is demonstrated, such as diverticula of the colon or gastric ulcer, it is better not to ascribe the pain, constipation, or diarrhea as IBS.

First descriptions of this syndrome began to appear in the clinical literature early in the 19th century[1]; however, it was not until 1944 that Peters and Bargen[2] used the phrase "irritable bowel syndrome." Since that time, observation of patients with the syndrome has led most workers in the field of gastroenterology to call it a motility disturbance. But the cause remains unknown, and consequently "irritable bowel syndrome" still best fits the clinical picture.

16.2. INCIDENCE

IBS patients make up the single largest group presenting to a gastroenterologist. It is estimated that 70% of gastroenterologic patients have IBS. Within this group of patients lie many missed diagnoses. For example, only in the past decade have we realized that lactose intolerance was the cause of mild abdominal pain and diarrhea following eating. These patients formerly were included in IBS. Similarly, it is only in the past three decades have we been able to diagnose gluten enteropathy, and these patients were frequently included in IBS. As discussed in Chapter 17, our present understanding of food allergies and the gut response to foods is in its infancy. This author is sure that it is only a matter of time before we will be able to demonstrate numerous food intolerances presently being included in IBS. The syndrome probably includes numerous other unidentified diseases. As knowledge in the field of gut motility evolves, it will probably yield a syndrome of a true motility disturbance that will wind up as the classic IBS.

16.3. CLINICAL PRESENTATION

The incidence is approximately twice as high in females than in males. Female patients more often suffer from constipation, and male patients present more often with diarrhea. It is relatively rare for a male to present as IBS with constipation, whereas females present with both symptoms. Children most often present with recurrent abdominal pain (RAP).

The age distribution of IBS has a range of 10 yr to the geriatric population. The majority of patients present between the ages of 20 and 40; the mean age is 33.

Abdominal pain is the classic symptom. It is usually described as persistent, mild, and located either in the right or left lower quadrant. Many patients describe it as radiating across the abdomen. If constipation is present, it is often associated with increased laxative intake. When diarrhea is present, the pain may be associated with increased bowel movements and, at times, relief after the movement.[3] Often, these patients complain of loose bowel movements following a meal and experience the movement after either the breakfast, lunch, or dinner meals. What is most peculiar is that the hyperactive gastrocolic reflex often occurs only after one of the meals and not all three. Some patients will complain of alternating diarrhea and constipation. Others complain of abdominal distention. Other clinical features, such as weight gain, mild nausea, intolerance to a variety of foods, cancer phobia, or associated urinary symptoms, are noted in a moderate percentage. It is also common for these patients to have minor to major emotional disturbances.

Often, the clinician can diagnose IBS on the basis of the clinical picture; however, the diagnosis can only be certain after a careful screening workup. The workup must include a careful history for such food intolerances as lactose. The physical examination must include a sigmoidoscopic evaluation, and the radiographic evaluation must include barium studies of the upper gastrointestinal tract, small bowel, and colon. If there is any history of weight loss, then there must be an evaluation for malabsorption.

Twenty percent of the patients who present with IBS do have weight loss and these patients must have malabsorption workups to rule out gluten enteropathy and other rare disorders. Once the workup reveals no organic disease, then the diagnosis of IBS can be established, but it behooves the physician to follow these patients carefully and reevaluate if necessary.

If this is the first presentation of the disease, then a reevaluation workup may be in order after 1 yr if therapy has not been successful. Often, these patients are extremely resistant to therapy and a repeat evaluation is necessary to ensure that occult disease has not been missed by the initial screening for organic disease.

16.4. CAUSE

The cause of IBS is unknown.[3a] As knowledge in the field of gastroenterology increases, specific disease entities are identified that result in patients being deleted from the category of IBS. At the present state of the art, there are three areas of investigation into the cause of this disorder: (1) Motility disturbances; (2) psychiatric disorders; and (3) dietary deficiencies or intolerance.

16.4.1. Motility Disturbances

Basic normal motility can be summarized as follows. The basal electrical rhythm, which can be measured by placing an electrode in the wall of the intestine, originates in the duodenal bulb at the rate of 17 to 18/min and progresses down the small bowel at approximately 19 to 20 cm/sec. Spike potentials, which signal a muscular contraction, recur intermittently. Eighty percent of contractions travel less than 3 cm. The basal electrical rhythm decreases as they descend the small bowel, so that there are approximately 12 to 13/min in the distal ileum. Segmentation is the main form of motility in the small bowel. It results as an interaction of the basal electrical rhythm spontaneously generated by cells of the longitudinal muscle layer. Short propulsive movements move chyme along the small intestine. Segmental movements appear to increase after feeding as propulsive movements decrease. Peristalsis is a progressive wave of contraction of successive rings of circular muscle. Distension of the small intestine usually causes a peristaltic reflex. First, the longitudinal muscle contracts, which is followed by a circular muscle contraction. The ileocecal sphincter is usually closed and represents a zone of elevated pressure approximately 4 cm long. Increased activity of the ileum usually occurs with increased gastric secretion. Gastric emptying may result in opening of the sphincter. This relationship is usually called gastroileal reflex resulting in colonic distention and the so-called gastrocolic reflex.[4]

The basic myoelectric activity in the colon is less well understood than in the small intestine. Snape and associates[5] demonstrated two types of basal electrical rhythm by means of bipolar electrodes attached to the mucosa of the rectosigmoid and rectal areas. There is a major component of approximately 6 cycles/min and a minor component of 3 cycles/min. They found the 3 cycle/min activity present in 44% of patients with IBS compared to only 10% of normal subjects, and an increased colonic responsiveness to stimuli during the slow-wave activity. The myoelectric activity results in two types of motor activity, segmental contractions and mass movements.[6] Colonic segmental contractions are normally present approximately 50% of the time and appear to be uncoordinated, nonprogressive, and nonpropulsive local contractions, which are increased by cholinergic stimulation and eating. They are usually reduced by anticholinergic activity. It is presently thought that segmental contractions slow down the passage of feces through the colon. Peristaltic or mass movements move feces along the colon and are stimulated by eating and the so-called gastrocolic reflex.

Ritchie[7] has studied movement and pain threshold through the bowel. He describes multihaustral propulsion, which increases after a meal, but is greatly increased by the administration of carbachol,[7] as seen in cineradiography. Balloon distension of the pelvic colon clearly reveals that patients with IBS have a lower threshold for pain.[7] IBS patients feel pain at less distension and more frequently than other subjects.[7]

Earlier studies employing telemetry capsules revealed that the motility of the intact human colon was intermittent rather than continuous and that the resting colon, both in ulcerative colitis and IBS, had decreased general activity with wave forms of lesser amplitude and shorter duration than controls.[8] They were unable to demonstrate any clear relationship of motility to pain.[8] However, Connell and associates[9] studied contractions via three fine polyethylene tubes placed in the sigmoid colon and found a syndrome of abdominal pain associated with increased colonic motility after meals. They also found these patients had pain with a paradoxical slowing of movement of feces and referred to the syndrome of spastic colon, which is now synonymous with IBS.[9]

The entry of food into the small intestine is one of the most important factors in initiating the colonic response to food and does not require the presence of the stomach, acid, gastrin, or vagal intervention.[10] When pelvic colonic pressures are recorded before, during, and after meals, in patients with either nonspecific diarrhea or constipation, the patients with diarrhea have far less colonic activity than constipated patients.[11] When pressure recordings are made during a meal, colonic activity markedly increases in diarrhea subjects, but not in subjects with constipation.[11] When patients with IBS were fed a meal consisting of a roast beef sandwich and an ice cream milk shake, both colonic myoelectric and motor activity were increased compared to normal subjects.[12] Although anticholinergics do not effect the basal state, they reduce the prolonged postprandial colonic spike and motor activity in patients with IBS,[12] while cholinergic agents increase it.[13]

The above studies suggest that the IBS patient has an increased susceptibility to both segmental and mass movements as well as an increased sensitivity to colonic distension, meals, and cholinergic agents. These findings may explain the symptoms of IBS. The questions that arise are whether they are secondary motility phenomena or primary in their origin. On the basis of the present information, Harvey[14] has classified the clinical patterns into the following categories:

Clinical type	Motor abnormality
Constipation	Increased motor activity
Painless diarrhea	Decreased motor activity in fasting state, increased activity after meals
Abdominal pain unrelated to food	Increased motor activity
Abdominal pain related to food	Increased motor activity at time of symptoms

Increasing knowledge in the field of gastrointestinal hormones has been of help in better understanding the relationship of food to motor activity. Postprandial motor activity is increased with the release of cholecystokinin or gastrin. The former is much stronger in its effect and the exact physiologic role of gastrin is still uncertain.[15,16] When substances that release CCK are given, there is a rapid increase in colonic segmental motor activity. When patients with IBS are given magnesium sulfate, which causes a significant release of CCK, they may develop an attack of classical abdominal pain associated with a marked increase in motility.[17] When either CCK or pentagastrin are given to subjects with IBS while myoelectrical activity is measured, there is a similar marked increase in 3-cycle activity that is more prominent in IBS patients,[5,18] thereby implying CCK may be the mediator of the increased or abnormal motor activity noted in IBS subjects.[17,18]

16.4.2. Psychiatric Disorders

Whenever the cause of a syndrome is not known and the symptoms are vague and not debilitating, then a psychiatric cause of the illness is suspect. The psychodynamics and psychophysiology of gastrointestinal symptomatology are extremely complex.[19] The question that always arises in the clinician's mind is whether the gastrointestinal symptom precedes or follows emotional stress. All too often, the two cannot be separated and, in reality, are related. When IBS patients are standardized on the basis of psychologic tests for neurologic symptoms and compared to the general population samples as well as to patients with definite neuroses, they fall between the normal and the definite neurotic.[20] In a careful study performed by Young and associates[21] of 29 outpatients with IBS, 72% had psychiatric illness, whereas only 18% of 33 controls had any manifestation of psychiatric illness.[21] It is difficult to draw conclusions from so few studies, but it does appear that patients with IBS have an increased number of neurotic symptoms that may be related to the cause or a result of the syndrome.

16.4.3. Diet

The third factor that may be important in the pathogenesis is the diet. IBS may represent a syndrome that has resulted from a deficiency of fiber in the diet. Other fiber-deficiency diseases have been discussed elsewhere. However, the recent increase in awareness of fiber deficiency has led physicians to recommend it for a variety of unexplained symptoms. A British postal survey revealed that 90% of British gastroenterologists were using bran or a high-fiber diet for the treatment of IBS.[22] Manning and associates[22] fed both high- and low-wheat-fiber diets for 6-week periods to 26 patients with IBS and found both significant improvement in symptoms as well as an objective change in colonic motor activity while the subjects were on a high-fiber diet. They found subjective improvement both in patients with diarrhea as well as those with constipation. This work confirms the initial observation of Painter[23] on the beneficial effect of bran on IBS as well as on diverticular disease. Piepmeyer[24] conducted a 4-month trial in IBS using 8 to 10 rounded teaspoons of bran per day. Twenty-three of his thirty patients reported definite improvement, and 17 of the 23 continued to use bran on a daily basis after the trial. Although most reports confirm the benefit of a high-fiber diet for IBS, there are some that question the effectiveness of bran.[25] Soltoft and associates[25] performed a double-blind trial employing three biscuits that either contained placebo or 10 g of wheat bran for a 6-week trial period. He found that there was a greater improvement in symptoms in the 23 patients in the placebo group versus the 29 patients in the bran group. He did not notice the well-documented laxative effect of bran and, therefore there is some question as to whether enough bran was eaten to have an effect in this study.[25] From available data, it appears dietary fiber may play an important role in certain patients with IBS.

16.5. MANAGEMENT OF IBS

It is accepted that the management of IBS should first include a careful explanation to the patient of the physical basis of the symptoms.[26] It is essential for short- and long-term treatment that the patient understand the possible three causes, simple motility, and the effect of eating in order to control the symptomatology. I often take anywhere from 30 to 60 min to carefully explain the possible causes of the irritable bowel syndrome, trying to make the patient understand what intestinal motility is, how it can be modified either by drugs or diet, and above all, making every effort to elicit any psychologic factors that may be controlled. The prognosis varies tremendously. Some patients can be helped by merely reassuring them, whereas others require repeated trials of therapy.

The following is an outline of management following patient education.[27]

1a. If pain is the predominant symptom, then a trial of anticholinergics may be of help.[12]

1b. In conjunction with, or in replacement of, anticholinergics, the high-residue diet should be tried.[22-24] The diets employed may vary (see Chapter 20, diet 27).

1c. In lieu of a broad high-fiber diet, one can attempt to increase fiber immediately by the use of simple bran, as outlined in Refs. 22–24, or by psyllium seed. Bran intake should be started slowly. A rapid increase of bran intake can cause abdominal distention and pain. Patients should begin with 1 teaspoon/day and work up to between 8 and 10 teaspoons/day. Some subjects cannot tolerate bran and would therefore have to substitute the high-fiber diet or another high-fiber substance that takes on water, such as psyllium seed. Anywhere from 2 to 6 teaspoons/day can have the same effect as bran. I have found that subjects with either constipation or diarrhea may benefit from the high-fiber or high-bran intake.

2. If constipation is the predominant symptom alone, then high-fiber diets and high bran are indicated. A trial of these is in order first. However, it may be necessary to include an anticholinergic. The use of laxatives should be a last resort.

3. If diarrhea is a predominant symptom, then high-fiber diets should be tried; if they fail, a more bland, soft diet is used. In these patients, it is often necessary to administer an anticholinergic or an antidiarrheal agent concomi-

tantly. Very often, these subjects have a hyperactive gastrocolic reflex and it is necessary to use the anticholinergic before eating or upon arising.

If there is a strong implication of neurotic symptomatology,[21] then it may be necessary to seek psychiatric aid if reassurance is not helpful. In such subjects, it is necessary to prescribe an antidepressant agent if depression is obvious, or to use an antianxiety drug if there appears to be overt anxiety that must be controlled before reassurance and modification of diet can take effect.

Treatment of the irritable bowel syndrome can be frustrating, but when the physician makes the effort to educate the patient, reassure the patient, and then work with the patient employing dietary management with or without anticholinergics or psychiatric assistance, the treatment can be most rewarding and the patient most appreciative.[26,27]

REFERENCES

1. Fielding JF: The irritable bowel syndrome. I. Clinical spectrum. *Clin Gastroenterol* 6:607, 1977.
2. Peters GA, Bargen JA: The irritable bowel syndrome.*Gastroenterology* 3:399, 1944.
3. Drossman DA: Diagnosis of the irritable bowel syndrome.*Ann Intern Med* 90:431, 1979.
3a. Almy T: The irritable bowel syndrome. Back to square one. *Dig Dis Sci* 25:401, 1980.
4. Davenport HW: *Physiology of the Digestive Tract*, ed 4. Chicago: Year Book Medical Publishers, 1977.
5. Snape NJ, Carlson GM, Cohen S: Colonic myoelectric activity in the irritable bowel syndrome. *Gastroenterology* 70:326, 1976.
6. Misiewicz JJ: Motility of the gastrointestinal tract, in Dietschy JM (ed): *Disorders of the Gastrointestinal Tract*. New York: Grune & Stratton, 1976.
7. Ritchie J: The irritable bowel syndrome. II. Manometric and cineradiographic studies. *Clin Gastroenterol* 6:622, 1977.
8. Bloom AA, Loresti P, Farrar JT: Motility of the intact human colon.*Gastroenterology* 54:232, 1968.
9. Connell AM, Jones FA, Rowlands EN: Motility of the pelvic colon. III. Abdominal pain associated with colonic hypermobility after meals.*Gut* 6:105, 1965.
10. Holdstock DJ, Misiewicz JJ: Factors controlling colonic motility: Colonic pressures and transit after meals in patients with total gastrectomy, pernicious anemia, or duodenal ulcer. *Gut* 11:100, 1970.
11. Waller SL, Misiewicz JJ, Kiley N: Effect of eating on motility of the pelvic colon in constipation or diarrhea.*Gut* 13:805, 1972.
12. Sullivan MA, Cohen S, Snape WJ: Colonic myoelectrical activity in irritable bowel syndrome. *N Engl J Med* 298:878, 1978.
13. Chaudhary NA, Truelove SC: Human colonic motility: A comparative study of normal subjects, patients with ulcerative colitis, patients with the irritable colon syndrome. The effect of prostigmine.*Gastroenterology* 40:18, 1961.
14. Harvey RF: The irritable bowel syndrome. III. Hormonal influences.*Clin Gastroenterol* 6:631, 1977.
15. Dinoso VP, Meshkinpour H, Lorber SH: The response of the sigmoid colon and rectum to exogenous cholecystokinin and secretin.*Gastroenterology* 62:844, 1972.
16. Misiewicz JJ, Waller SL, Holdstock DJ: Gastrointestinal motility and gastric secretion during intravenous infusions of gastrin II.*Gut* 10:723, 1969.
17. Harvey RF, Read AE: Effects of oral magnesium sulphate on colonic motility in patients with the irritable bowel syndrome.*Gut* 14:983, 1973.
18. Snape WJ, Carlson GM, Matarazzo SA, *et al:* Evidence that abnormal myoelectrical activity

produces colonic motor dysfunction in the irritable bowel syndrome. *Gastroenterology* 72:383, 1977.

19. Glaser JP, Engel JF: Psychodynamics, psychophysiology and gastrointestinal symptomatology. *Clin Gastroenterol* 6:507, 1977.

20. Palmer RL, Stonehill E, Crisp AH, *et al:* Psychological characteristics of patients with the irritable bowel syndrome. *Postgrad Med J* 50:416, 1974.

21. Young SJ, Alpers DH, Norland CC, *et al:* Psychiatric illness and the irritable bowel syndrome. *Gastroenterology* 70:162, 1976.

22. Manning AP, Heaton KW, Harvey RF, *et al:* Wheat fibre and irritable bowel syndrome. *Lancet* 1:8035, 1977.

23. Painter NS: Irritable or irritated bowel. *Br Med J* 2:46, 1972.

24. Piepmeyer JL: Use of unprocessed bran in treatment of irritable bowel syndrome. *Am J Clin Nutr* 27:106, 1974.

25. Soltoft J, Krag B, Gudmand-Hoyer E, *et al:* A double-blind trial of the effect of wheat bran on symptoms of irritable bowel syndrome. *Lancet* 1:270, 1976.

26. Editorial: Management of the irritable bowel. *Lancet* 2:557, 1978.

27. Goulston K: Diagnosis and treatment of the irritable bowel syndrome. *Drugs* 6:237, 1973.

Gastrointestinal Allergy and Food Hypersensitivity

17.1. DEFINITIONS

Allergy has been simply defined as altered reactivity, but to many it implies an adverse reaction that occurs due to the interaction of an antigen with antibody or lymphoid cells.[1,1a] Although there is tremendous interest, the clinical significance of gastrointestinal allergy and food hypersensitivity is uncertain, as the science of the field is in its infancy. The allergists have been inundated by patients who claim all sorts of symptoms from food intake and the art of allergy has responded with intermittent success, but the science of proving food hypersensitivity has been limited. The only clinical situations that have been established as disease entities due to food hypersensitivity are cow's milk intolerance and gluten enteropathy. Although eosinophilic gastroenteritis is suspected of being related to food and hypersensitivity, the association has not been proven. There is no question humans must have numerous intolerances and hypersensitivities to foods, but much research has to evolve before the subject is understood in more detail. The following facts appear certain and may be of help in management of suspected gastrointestinal allergy or food hypersensitivity. It is important to make the distinction that *gastrointestinal allergy* refers to allergic responses that are manifest within the gastrointestinal tract, whereas *food hypersensitivity* may be a hypersensitivity state caused by the ingestion of food with symptoms occurring in the gastrointestinal tract, or at other target organs.

17.2. PHYSIOLOGIC MECHANISMS

Allergy implies either one of three responses: immunity, tolerance, or hypersensitivity. The classic gastrointestinal hypersensitivity reactions are of four types.[2]

Type 1 is the immediate or reaginic reaction, which may be of the anaphylactoid type and characterized by shock, vomiting, diarrhea, colic, urticaria,

severe bronchospasm, pulmonary edema, and possibly eczema. All or any one
of these may occur. The Prausnitz–Kustner reaction is a classic example of an
anaphylactic reaction in food allergy. IgE antibodies are identified closely with
reaginic activity.[3] IgE has been identified by radioimmunologic techniques to
be specifically bound to constituents of foods.[4,5] IgE antibodies have been
demonstrated against castor bean[6] and β-lactoglobulin of milk.[7] Large numbers
of IgE-containing lymphoid cells are present in the lamina propriae of the gut
mucosa, and the apical portion of epithelial cells may also contain IgE.[8]

Type 2 are cytotoxic reactions and little is known of their occurrence in the
gastrointestinal tract.[2]

Type 3 reactions produce the typical Arthus phenomenon. They result in
immune complexes that fix complement and result in a fall in serum complement. They are reported in collagen vascular disease and have been noted in
milk allergy. Clinically, the response is delayed by approximately 8 to 12 hr
after exposure and requires repeated stimuli to develop. These type reactions
may be characterized by fever, leucocytosis, edema, hemorrhage, and necrosis
following the administration of an antigen to a subject who already has the
antibody. Several studies have indicated that this type of reaction may take
place in the small intestine after gluten challenge in the celiac patient and after
milk challenge in patients with hypersensitivity to cow's milk.[9]

Type 4 are classical cellular hypersensitivity reactions and are delayed by
24 to 72 hr after exposure to the antigen.[2]

The process of gut hypersensitivity begins with an antigen crossing the
mucosal barrier. Various large molecules and particles, mostly derived from
food and the intestinal flora, can cross the intestinal epithelium.[10-12] Using a
scanning electron microscope, Owen and Jones[13] have described an "N" cell
of the mucosa characterized as having luminal surface microfolds rather than
microvilli. These N cell processes form a latticework that allows lymphoid cells
to approach to within 0.3 μm of the intestinal lumen, while maintaining the
integrity of the intestinal epithelium. It is suspected that these cells are responsible for the transport function of antigenic material into lymphoid tissue or for
secretion of material into the lumen from lymphoid tissue.[8] They thus become
a gateway for the controlled presentation of antigens to the cells of the immune
system.[10] Antigenic absorption may also occur across the normal columnar
epithelium, as can immunoglobulins in passive reactions. Proteins may be absorbed whole, engulfed, or digested at the microvillus level (see Chapter 13).[13]
As much as 5% of ingested protein can be found in the bloodstream[14]; bovine
serum albumin can be absorbed directly into the blood. Antibodies to albumin
occur in normal subjects, but with greater frequency in patients with inflammatory bowel disease.[15, 16]

Once an antigen crosses the mucosal barrier, it enters the submucosa,
where abundant lamina propia plasma cells act as sources of IgA and IgE. The
antigen may also reach the lymph node circulation, where lymphocytes act as
sources of IgG and IgM. Antibodies produced by the plasma cells or lymphoid
population may then circulate to other parts of the body. Hypersensitivity
reactions then may occur locally at the site of antibody production, in the
epithelial cells of the mucosa, or at distal target organs.

It is important to note that the bacterial flora of the gut is related to the development of the lamina propria plasma and lymphoid cell population. Germ-free animals have poorly developed immune systems, whereas conventional animals with a gut microflora have typical development.[9] It is also important to note that most foods seem to have little effect on the morphologic development of the intestine. Synthetic diets, which contain no antigens when used as a sole source of energy, cause the same development of the small intestine as complex foods.[9]

Once antigens cross the mucosal barrier, there is a possible protective relationship of IgA antibodies. This relationship may occur at the mucosal surface, as absorption of antigens is reduced in the presence of IgA.[10] It is postulated that an IgA–antigen complex is degraded by pancreatic enzymes, and that in clinical situations where IgA deficiency or pancreatic insufficiency exists, there is a greater susceptibility to food sensitivity because of a defect in this protective mechanism.[10]

In summary, antigens and proteins are now known to easily cross the gut mucosal barrier; once they enter, they are either inactivated by a defense mechanism involving the antibodies produced by the gut mucosa, or they stimulate the production of IgA, IgE, IgG, or IgM antibodies, which may result in a hypersensitivity reaction.

17.3.   CLINICAL MANIFESTATIONS

Hypersensitivity reaction in the gastrointestinal tract may cause almost any gastrointestinal symptom—nausea, vomiting, diarrhea, pain, bleeding, anorexia, or weight loss. There are no specific signs or symptoms. When food hypersensitivity causes systemic symptoms, it may be shocklike, as in an anaphylaxis, or involve, more commonly, the skin, respiratory system, and, less commonly, other systems.[2,17] Neurologic symptoms such as fatigue, hyperreactivity, altered behavior, and headache are often blamed on allergy. However, there is little good scientific evidence to substantiate these claims and much more work is needed in order to be certain food hypersensitivity causes neurologic symptoms.

17.4.   CRITERIA FOR IDENTIFICATION OF FOOD HYPERSENSITIVITY

May and Bock[17] include the following criteria to establish food as a cause of hypersensitivity.

1. In symptomatic hypersensitivity, (a) any symptoms must be evoked in a double-blind food challenge, (b) sensitization must be demonstrated by either specific antibodies or skin testing reactions, and (c) other causes of adverse reactions to foods should be eliminated.

2. In asymptomatic hypersensitivity, (a) there are no symptoms from a challenge, and (b) sensitization is demonstrable.

17.5. DIAGNOSIS

17.5.1. Skin Tests

No single test is available to make a diagnosis, but clinical sensitivity to foods and an etiologic relationship between reagents can be suspected when skin tests are strongly positive.[18] Commercially available extracts of foods are reportedly reliable for specific identification of hypersensitive persons by skin reactions.[17-19] Some workers feel specific identification of hypersensitive persons can be accomplished when correct dilutions of the extract are used.[17] However, many clinicians feel there are numerous false positives with skin testing which cannot be relied upon as the only method for determining clinically significant food hypersensitivities.[18]

17.5.2. RAST (Radioallergosorpent Test)

The RAST method can be used to detect reagents in serum. The test approximates the level of specific reagent antibody and it can be confirmatory for the same substances tested by skin reaction.[19]

17.5.3. Food Challenge

Proof should be obtained by a classic food challenge whenever the history or skin test implicates a food as the cause of a hypersensitive reaction. The suspected food is eliminated from the diet for at least 2 weeks prior to the challenge and then administered in a double-blind study fashion. It is difficult to mix the food with other foods and still mask it; therefore, experts prefer to place the food in No. 1 gelatin capsules, which can hold anywhere from 200 to 500 mg of dried food. The capsules are administered with a meal. Depending on the degree of the hypersensitivity, anywhere from 20 to 1000 mg of the food is administered in a single dose and symptoms can be expected in anywhere from 2 to 24 hr. If there is no symptom, then the dose can be doubled or increased as much as ten times, depending on the food substance. A failure to respond to 8000 mg of dry food appears to rule out an immediate-type reaction to the food.[17,20]

Intermediate or delayed reactions are much more difficult to prove. However, some authors feel that a reaction can occur as long as 7 days after administration and the patient should be observed on the restricted diet for this time. The double-blind test method ensures that there is no psychologic component to the reaction. Although the response may be dramatic, it is wise to double-blind test on the restricted diet with capsules that do not contain the suspected food so that one can be absolutely certain psychologic components to reactions are eliminated. Some authors feel the double-blind challenge should be attempted three times before a true food hypersensitivity can be accepted.[17] However, all immunologic responses vary in intensity and time. Therefore, a symptomatic response can occur on one occasion that will be less or greater on another occasion. The clinician must assess the total picture. This may be extremely difficult and caution and expertise are clearly needed in this field. [17-20]

17.5.4. Elimination Diets

Elimination test diets are extremely difficult to use to identify suspected foods and require accepting a great deal of subjective evidence. As in all fields, when something is difficult to do, there are many varied ways to perform it.

Rowe's[21] method of using a cereal-free elimination diet for 7 days, followed by a fruit-free, cereal-free elimination diet for 7 days, is one acceptable method. If symptoms do not cease after the cereal-free diet, then the combined fruit-free, cereal-free diet is employed. If symptoms do not relent on either diet, then it should be maintained for 2 weeks. If the gross elimination diet is successful, then further breakdown of categories of foods is needed (see Chapter 20 for Rowe elimination diets).

Before beginning a challenge regimen, it is wise to employ a hypoallergenic-type diet that excludes most common offending foods.[17] Such a diet is listed in Chapter 20. A patient can be begun on this type of elimination diet and then progressed to a more vigorous regimen, such as suggested by Rowe. There are numerous other types of elimination diets: Those that eliminate one group of foods at a time and then evaluate objective symptomatic changes, and others that eliminate several groups of foods at a time for anywhere from several days to a week in order to evaluate groups of foods.

The following is an example of a simple, basic exclusion diet that permits some groups of foods and excludes others (modified from Ref. 13, pp 170–171).

Permitted	Excluded
Fresh fruit and vegetables	Meat and poultry
Fresh fruit juice	Eggs
Tea, coffee, sugar	Milk and dairy products
Rice	Tinned, frozen, and preserved food
Gluten-free bread	Bread, biscuits, cake, flour
Olive oil	
Barley sugar	
Tomor margarine	

An example of a rotation elimination regimen is the following abbreviated list to illustrate a 4-week rotation diet in which different food categories are eliminated (modified from Ref. 13, pp 170–171).

	Weeks			
Food groups	1	2	3	4
Cereals	Rice	Ryvita	Potatoes	—
Meat	Lamb	Chicken	Beef	—
Vegetables	Lettuce	Cabbage	Tomato	Carrots
Fruit	Oranges	Apricots	Grapefruit	—
Fats	Olive oil	Margarine	Margarine	—
Dairy products	—	—	Milk	Cheese Milk
Miscellaneous	Salt	Syrup	Jam	Sugar

Table 17-1. Food Family List[a]

Plant families

Apple—apple, pear, quince
Buckwheat—buckwheat, rhubarb
Cashew—cashew nut, pistachio, mango
Citrus—orange, lemon, lime, grapefruit, tangerine
Cola nut—chocolate (cocoa), cola
Fungi—yeast, mushroom
Ginger—ginger, tumeric, cardamom
Goosefoot—beet, spinach, Swiss chard
Gourd—watermelon, cucumber, cantaloupe, pumpkin, squash
Grass (grains)—wheat, corn, rice, oats, barley, rye, wild rice, brown cane sugar, molasses, bamboo
    shoots, millet, sorghum
Heath—blueberry, cranberry
Laurel—cinnamon, bay leaf, avacado, sassafras
Lily—onion, garlic, asparagus, chives, sarsaparilla, leek
Mallow—okra, cottonseed
Mint—peppermint, sage, thyme, spearmint, oregano, basil, balm, bergamot, hoarhound, mar-
    joram, savory, rosemary
Mustard—mustard, turnip, radish, horseradish, cabbage (and kraut), cauliflower, broccoli, brus-
    sells sprout, Chinese cabbage, collards, kale, watercress, rutabaga, kohlrabi
Myrtle—allspice, clove, guava
Nightshade—tomato, potato, eggplant, tobacco, red pepper, bell pepper, cayenne, paprika, pi-
    miento, chili pepper
Palm—coconut, date
Parsley—carrot, celery, parsnip, anise, celery seed, cumin (comino), coriander, angelica, caraway
    seed, fennel, lovage, samphire, sweet cicily
Pea (Legume)—pea, black-eyed pea, peanut, dry beans, green beans, soybean, lentils, licorice,
    tragacanth, acacia
Plum—almond, plum (prune), peach, apricot, cherry, nectarine
Rose—strawberry, blackberry, raspberry, other bramble berries
Sunflower—lettuce, chicory, endive, artichoke, dandelion, salsify, sunflower seed, tarragon (rag-
    weed and pyrethrum are related inhalants)
Walnut—walnut, pecan, hickory nut, butternut

Plant foods without relatives

These foods are *not* related to each other or to any other foods: banana, black (and white) pepper,
Brazil nut, coffee, fig, gooseberry and currant, grape and raisin, hazelnut, honey, nutmeg and
mace, olive, pineapple, sweet potato, tapioca, tea, vanilla, arrowroot, chestnut, chicle, elderberry,
juniper, flaxseed, karaya gum, macadamia nut, maple sugar, New Zealand spinach, papaya, per-
simmon, poppyseed, sesame seed, wintergreen.

Animal foods

Mollusks—oyster, clam, abalone
Crayfish—shrimp, crab, lobster, prawns
Fish—all true fish such as tuna, salmon, catfish, perch, etc.
Birds—egg, chicken, turkey, duck, goose, pheasant, quail
Mammals—cow's milk, beef, lamb, goat's milk, pork, rabbit, squirrel, venison

Chemicals and drugs

Food color, sweeteners, fruit acids, flavors, aspirin, sulfa drugs, antibiotics, barbiturates, tranquil-
izers, other drugs.

---

[a] Model list used by Dr. Robert Biondi. Material extracted from Speer F, Dockhorn RJ: *Allergy and Immunology
in Children.* Springfield, Ill: Charles C Thomas, 1973.

The type of elimination diet depends on the clinician and the allergist treating the patient. Often, a clinician will vary the elimination diet depending on the history. Because much of the effectiveness depends on subjective symptoms and a clinical evaluation, the type of elimination diet employed depends greatly on the individual clinician.

Another approach to the elimination diet is the use of the elemental diet.[24] Galant and associates[24] have shown that an elemental diet of synthetic nutrients (Vivonex) can be tolerated as an elimination diet. In animal experiments the synthetic diet is not immunogenic. Of 14 patients followed carefully on Vivonex, no significant subjective improvement in allergic symptomatology was noted, although they did not get worse. The authors concluded that Vivonex can be used on an outpatient basis as an elimination diet, but more rigorous elimination regimens are needed in patients who require hospitalization because of suspected food sensitivity.[24]

Table 17-1 lists foods according to their common families.[22,23] People allergic to one food are often allergic to related foods. The relationships of foods in plant families often govern their response. For example, someone allergic to peas might well be allergic to beans, which are in the pea family, but not to beets, which are in the goosefoot family. Table 17-1 categorizes foods roughly and the interested clinician is referred to other food texts for more detail.[23]

Food additives have become a way of life since the development of rapid food processing and long-term storage of food. In Table 17-1 food additives are grouped under chemicals. Cartrazine is one of the most common additives and may be present in great amounts. Sensitivity to it may be cross-reactive with aspirin, or may occur alone.[25]

It is important to remember that cooking may denature certain proteins so that people sensitive to uncooked food may not experience a sensitivity when the same food is cooked.

In summary, a very careful history is important in first establishing the presence of food hypersensitivity. This must be followed by some form of skin testing, with or without serum studies, such as the RAST, and then confirmed by the double-blind food challenge, which appears to be the most accurate, or some form of an elimination diet regimen, which is more subjective and difficult for therapeutic results.

17.6. TREATMENT

As May and Bock state, "If one is depressed by the seeming complexity of precise diagnosis of hypersensitivity to food, he may be cheered by the apparent simplicity of the treatment—avoid the food."[19,20] The patient truly allergic and suffering is greatly rewarded by the elimination of the offending food. A patient who is not suffering and cherishes the food more than relief of the symptom discomfort will often not comply with recommendations. Success is dependent on the motivation of the individual patient. Elimination of a single offending food is easy. However, when multiple food allergies are present, it is difficult for the patient to follow a rigid, restrictive regimen and breaks in the therapeutic recommendation often follow, with resultant therapeutic failure. In such patients the prognosis is poor and they often turn to drugs for therapy.

Hypersensitivity to food, in some persons, is occasionally gradually lost and they move from a symptomatic state to an asymptomatic hypersensitivity. This is seen in classic gluten enteropathy, where patients may have severe malabsorption for years and after extended therapy may begin to tolerate small amounts of gluten (see Chapter 13).

The loss of hypersensitivity in some subjects suggests that desensitization for food allergies may be successful. However, this has proved so variable that most allergists do not suggest its use on a regular basis.[17] Desensitization is far more successful for inhaled allergies than it is for food hypersensitivity.

At the present state of the art, the offending food should be identified and then eliminated. Whenever there is an anaphylactic type of reaction, it must be permanently eliminated and food challenges contraindicated.

Corticosteroids may be employed for the more seriously ill patient when drug therapy is needed. Antihistamines are usually of little help. In the anaphylactoid reaction, adrenalin must be used, followed by corticosteroid management.

Sodium cromoglycate has had mild success in certain regimens but remains controversial for use in food hypersensitivity.[26-30] Initial studies revealed little success in preventing reactions,[26] but one study showed administration of 50 mg before meals was able to prevent adverse reactions in 14 of 20 patients.[27] Others have shown immune complexes of food allergens and IgE could be decreased and attacks prevented by use of oral cromoglycate.[28,29] A study of children allergic to milk plus one or more other foods revealed 50 to 70 mg of cromolyn given at meals afforded statistically significant protection against reactions.[30] Conversely, in chronic reactions, a double-blind long-term study of subjects suffering from food-induced urticaria showed no difference between oral cromoglycate and placebo.[12] Certainly, it is worth a trial in selected subjects, but it should not replace elimination of an antigen if possible.

When food sensitivity is so severe that it hampers growth and development, or may cause malnutrition, then hospitalization may be necessary. In these cases, the use of intravenous feeding and, if necessary, TPN (see Chapter 19) may be needed. This is a very rare, but occasionally required, step in rehabilitating the severely allergic patient.[2]

17.7. HYPERSENSITIVITY TO COW'S MILK

Western societies use cow's milk as a major source of food. Throughout the world its use is varied. The variation depends on social, economic, and environmental factors as well as intolerance to substances in the milk. Lactose is the sugar found in cow's milk. A lactase deficiency results in lactose intolerance. The large majority of black and Asian peoples are lactase deficient and therefore tend to avoid any significant amounts of cow's milk intake. Mediterranean peoples are less deficient and Nordics appear to have higher lactase levels and therefore tolerate large amounts of cow's milk (see Chapter 13 for lactose intolerance).

The high cholesterol and saturated fat content of cow's milk may also have significance. Avoidance of milk for this reason is still controversial but may be of long-term benefit. Cow's milk can also pose a problem through food hypersensitivity. As discussed above, large protein antigens can easily cross the intestinal mucosa. Both IgE and IgA antibodies can be found in exclusively bottle-fed infants. Although most adults do not have circulating antibodies to milk protein, when systemic immunoglobulins are present, they represent true reactions that may or may not have any pathologic significance.[31] Serum of 50% of those persons extremely sensitive and 30% of those moderately sensitive to milk has a heat-labile reaginic antibody that can confer passive cutaneous sensitization in animals. Most of these people have IgE antibodies that bind radioisotope-labeled β-lactoglobulin, but it is not yet clear whether the allergy might result from IgE or IgG antibodies. However, it is clear that macromolecules can cross the intestinal barrier and can produce an immunologic responsive food hypersensitivity to cow's milk. Eastham and Walker[31] point out that the normal systemic response to cow's milk protein usually progresses from one of a responsiveness on an immunological basis in infancy to one of unresponsiveness or tolerance in adult life.[31] The mechanism is extremely complex, but its importance is further emphasized by the demonstration that infant colic can be cured by removing cow's milk from the diet of breast-feeding mothers.[33]

The estimates of the incidence of cow's milk hypersensitivity range from 0.1 to 8% of the population. When rigid criteria are employed, the incidence is felt to be approximately 0.5%. The criteria for establishing sensitivity are difficult because there may be a delayed response and some of the sensitivity is subjective.

Seventy-five percent of reported sensitivities occur in the first 2 months of life. In one series, 41% developed symptoms within 7 days of exposure. I have observed an apparent milk-induced colitis in a premature infant, and ulcerative colitis has been reported due to milk protein when rigid criteria and challenges have been followed.[34]

A severe malabsorption syndrome with jejunal mucosal damage has been described.[35] Although there are reports of normal mucosa in association with cow's milk intolerance, Kuitinen and colleagues[35] report 54 infants with a malabsorption syndrome they felt due to cow's milk intolerance. All had diarrhea, failed to thrive, and laboratory investigations revealed malabsorption, increased serum IgA, and precipitants to cow's milk. In approximately 50%, the jejunal mucosa revealed a loss of villi. Their patients did well on human milk but reacted clinically to cow's milk challenge. As observed in other studies, the cow's milk intolerance disappeared at about the age of 1 yr. Approximately 80% of the children had no evidence of malabsorption at that time, but only 30% had a normal mucosal biopsy. Many of their patients developed intolerances to other foods and several were followed on a subsequent gluten-free diet. On their final evaluation, 12% had subtotal villus atrophy and they concluded that those were celiac patients and not simple milk intolerance. They concluded that malabsorption due to milk intolerance is a definite entity, but it must be differentiated from gluten enteropathy.[35] The syndrome of milk malabsorption appears to be transient, whereas most gluten enteropathy subjects appear to have the disease for longer periods of time and are not greatly improved until gluten is removed from the diet. Prior to Kuitinen and colleagues' publication, an abnormal mucosa due to milk intolerance in children was established.[36,37] Confirmatory observations are needed to definitely establish cow's milk as a cause of malabsorption.

Because there is a wide overlap between normal controls and subjects suspected of cow's milk intolerance, the diagnosis is often difficult to make.[31] When the intolerance affects a child's growth and development, then cow's milk should be eliminated and a substitute obtained. In selected cases, there may be no substitute for human milk. Much more work is needed in this field that may well yield information in our understanding of all food intolerances and human immunologic reactivity to foods.

17.8. GLUTEN ENTEROPATHY

From the present state of knowledge, the malabsorption syndrome that results from gluten sensitivity appears to be a food hypersensitivity reaction. The exact cause, whether it be a mucosal enzyme deficiency or a hyperreactivity, is not clearly understood. However, there is no question that removal of gluten from the diet does cure these patients. Details of the disease and its pathophysiology are discussed in Chapter 13, but it is important to point out here that refractory sprue may be due to other protein reactions.[38] When the celiac does not respond easily, then other allergies should be investigated.

17.9. EOSINOPHILIC GASTROENTERITIS

There are two clinical entities frequently referred to as eosinophilic gastroenteritis, but they are clearly different. *Eosinophilic granuloma* forms a mass lesion either in the pylorus or upper small intestine and can present acutely as a surgical abdomen,[39] whereas *eosinophilic gastroenteritis* is a diffuse gastrointestinal lesion and can be diagnosed easily by peroral small-bowel or gastric biopsy.[40-42] In 1970, Leinbach and Rubin[43] reported a 24-year-old man with a peroral biopsy diagnosed as eosinophilic gastroenteritis whom they studied for a 2.5-yr period while performing extensive biopsies of the small bowel. Their patient had the classic symptoms of nausea, vomiting, and abdominal pain after eating, and laboratory evaluation revealed peripheral eosinophilia and, what is less commonly noted, steatorrhea and hypoalbuminemia. They also were able to document a protein-losing enteropathy. Biopsy studies revealed a patchy infiltrate throughout the small bowel. The infiltration varied from area to area. When the patient was treated with a diet free of animal protein, he had a dramatic clinical improvement, but unfortunately the peripheral eosinophilia persisted and symptoms returned, although he supposedly was maintained on an elimination diet. They challenged their patient repeatedly with different foods and their final conclusion was that eosinophilic gastroenteritis was not a simple reversible allergic reaction to specific foods but a self-perpetuating process that is aggravated by different foods.[43]

Caldwell and associates[44] studied another patient with the histologic picture of eosinophilic gastroenteritis and found increased IgE, atopy, severe food sensitivity, protein-losing enteropathy, iron deficiency anemia, and growth retardation.[44] They compared their patient to controls with no atopy, but eosinophilic enteritis, and concluded that there are two forms of the disease, their patient having the form that exhibits an intestinal reaginic mechanism.

Thus, the cause of eosinophilic gastroenteritis is still uncertain, although it is often accompanied by systemic allergic manifestations. The classic clinical picture is one of abdominal pain with nausea and vomiting. The peripheral eosinophilia is a hint of the diagnosis, which is confirmed by simple peroral mucosal biopsy. Most of these patients do extremely well. In the present state of the art, elimination diets, food challenge, and skin testing have not delineated any definite food hypersensitivity relationship. However, most clinicians following these patients do feel that response is due to some allergic phenomenon, although a clear demonstration of the allergen cannot be made, as evidenced in the case followed so closely by Leinbach and Rubin.[43]

Once the diagnosis of eosinophilic gastroenteritis is made, the patient will often respond to corticosteroid therapy. The response is often dramatic and the course of therapy can be short. If symptoms recur, then corticosteroids can be reinstituted. When there is associated gastrointestinal bleeding or protein-losing enteropathy, there may well be involvement of the stomach.[42] These patients have to be followed more closely and perhaps given a more prolonged course of corticosteroids.

REFERENCES

1. Editorial: Food allergy. *Lancet* 1:249, 1979.
1a. Gleich GJ: IgE, allergy, and the gut. *Dig Dis Sci* 25:321, 1980.
2. Freir S: Paediatric gastrointestinal allergy. *Clin Allergy* 3:597, 1973.
3. Ishizaka K, Ishizaka T: Identification of gamma E antibodies as a carrier of reaginic activity. *J Immunol* 99:1187, 1967.
4. Ishizaka K, Ishizaka T, Hornbrook MM: Physiochemical properties of reaginic antibody. V. Correlation of reaginic activity with E-globulin antibody. *J Immunol* 97:840, 1966.
5. Heiner DC, Rose B: Elevated levels of gamma E (IgE) in conditions other than classical allergy. *J Allergy* 45:31, 1970.
6. Coombs RRA, Hunter A, Jones WE, et al: Detection of IgE (IgND) specific antibody (probably ragin) to castor bean allergen by the red-cell-linked antigen–antiglobulin reaction. *Lancet* i:1115, 1968.
7. Kletter B, Gery I, Freier S, et al: Immunoglobulin E antibodies to milk proteins. *Clin Allergy* 1:249, 1971.
8. Tada T, Ishizaka K: Distribution of IgE forming cells in lymphoid tissues of the human and monkey. *J Immunol* 104:377, 1970.
9. Ferguson A: Models of intestinal hypersensitivity. *Clin Gastroenterol* 5:271, 1976.

10. Editorial: Antigen absorption by the gut. *Lancet* 2:715, 1978.
11. Walker WA, Isselbacher KJ: Uptake and transport of macromolecules by the intestine. *Gastroenterology* 67:531, 1974.
12. Hemmings WA: *Antigen Absorption by the Gut.* Lancaster: MTP Press, 1978, pp 170, 181.
13. Owen RL, Jones AL: Epithelial cell specialization within human Peyer's patches. An ultrastructural study of intestinal lymphoid follicles. *Gastroenterology* 66:189, 1974.
14. Hemmings WA, Williams EW: Transport of large breakdown products of dietary protein through the gut wall. *Gut* 19:715, 1978.
15. Warshan AL, Walker WA, Isselbacher KJ: Protein uptake by the intestine: Evidence for absorption of intact macromolecules. *Gastroenterology* 66:987, 1974.
16. Falchuk KR, Isselbacher KJ: Circulating antibodies to bovine albumin in ulcerative colitis and Crohn's disease. *Gastroenterology* 70:5, 1976.
17. May CD, Bock SA: Adverse reactions to food due to hypersensitivity, in Middleton E, Reed CE, Ellis CF (eds): *Allergy, Principles and Practice.* St Louis: CV Mosby, 1978, p 1159.
18. Goldstein GB, Heimer DC: Clinical and immunological perspectives in food sensitivity: A review. *J Allergy* 46:270, 1970.
19. Bock SA, Buckley J, Holst A, *et al:* Proper use of skin tests with food extracts in diagnosis of hypersensitivity to food in children. *Clin Allergy* 7:375, 1975.
20. May CD: Objective clinical and laboratory studies of immediate hypersensitivity reactions to food in children. *J Allergy Clin Immunol* 58:500, 1976.
21. Rowe AH: *Food Allergy.* Springfield, Ill: Charles C Thomas, 1972.
22. Biondi R: Personal communication.
23. Speen F: *Food Allergy.* Littleton, Mass: PSG Publishing, 1978.
24. Galant SP, Franz ML, Walker P, *et al:* A potential diagnostic method for food allergy: Clinical application and immunogenicity evaluation of an elemental diet. *Am J Clin Nutr* 30:512, 1977.
25. Zlotlow MJ, Settipane GA: Allergic potential of food additives: A report of a case of tartrazine sensitivity without aspirin. *Am J Clin Nutr* 30:1023, 1977.
26. Freier S, Berger H: Disodium cromoglycate in gastrointestinal protein intolerance. *Lancet* 2:916, 1973.
27. Vaz GA, Tan L, Gerrard JW: Oral cromoglycate in treatment of adverse reactions to foods. *Lancet* 1:1066, 1978.
28. Brostoff J, Carini C, Wraith DG, *et al:* Production of IgE complexes by allergen challenge in atopic patients and the effect of sodium cromoglycate. *Lancet* 1:1268, 1979.
29. Paganelli R, Levinsky RJ, Brostoff J, *et al:* Immune complexes containing food proteins in normal and atopic subjects after oral challenge and effect of sodium cromoglycate on antigen absorption. *Lancet* 1:1270, 1979.
30. Kocoshis S, Gryboski JD: Use of cromolyn in combined gastrointestinal allergy. *JAMA* 242:1169, 1979.
31. Eastham EJ, Walker WA: Adverse effects of milk formula ingestion on the gastrointestinal tract. *Gastroenterology* 76:365, 1979.
32. Parish WE: Detection of reaginic and short-term sensitizing anaphylactic or anaphylactoid antibodies to milk in sera of allergic and normal persons. *Clin Allergy* 1:369, 1971.
33. Jakobsson I, Lindberg T: Cow's milk as a cause of infantile colic in breast-fed infants. *Lancet* 2:438, 1978.
34. Grybowski JD: Gastrointestinal milk allergy in infants. *Pediatrics* 40:354, 1967.
35. Kuitinen P, Visakorpi JK, Savilahti E, *et al:* Malabsorption syndrome with cow's milk intolerance. *Arch Dis Child* 50:351, 1975.
36. Freier S, Kletter B, Gery I, *et al:* Intolerance to milk protein. *Pediatrics* 75:623, 1969.
37. Lubos MC, Gerrard JW, Buchan DJ: Disaccharidase activities in milk-sensitive and celiac patients. *J Pediatr* 70:325, 1967.
38. Baker AL, Rosenberg IH: Refractory sprue: Recovery after removal of nongluten dietary proteins. *Ann Intern Med* 89:505, 1978.
39. Ureles AL, Alscibaja T, Lodico D, *et al:* Idiopathic eosinophilic infiltration of the gastrointestinal tract: diffuse and circumscribed. *Am J Med* 30:899, 1961.
40. Klein N, Hargove R, Sleisenger AM, *et al:* Eosinophilic gastroenteritis. *Medicine* 49:299, 1970.

41. Pitchumoni CS, Dearani AC, Burke AF, *et al*: Eosinophilic granuloma of the gastrointestinal tract. *JAMA* 211:1180, 1970.
42. Katz AJ, Goldman H, Grand RJ: Gastric mucosal biopsy in eosinophilic (allergic) gastroenteritis. *Gastroenterology* 73:705, 1977.
43. Leinbach GE, Rubin CE: Eosinophilic gastroenteritis: A sample reaction to food allergens? *Gastroenterology* 59:874, 1970.
44. Caldwell JH, Sharma HM, Hurtubise PE, *et al*: Eosinophilic gastroenteritis in extreme allergy. *Gastroenterology* 77:560, 1979.

Methods of Nutritional Treatment

18

Enteral Alimentation

18.1. METHODS OF NUTRITIONAL SUPPORT THERAPY

Natural foods are the best method of supplying energy and basic requirements to humans. In disease states, however, patients may not be able to ingest, assimilate, or absorb enough nutrients from foods to maintain body weight or vital functions. Then supplementation of nutrients is necessary. Our understanding of diet support for maintenance or therapy has advanced greatly during the past decade.

Diet supplementation may be administered through several routes:

1. Orally or via tube into the gut—*enteral alimentation*
2. Intravenous
   a. Peripheral vein
   b. Central venous—*total parenteral nutrition (TPN)*

Several questions should be answered before a patient is placed on dietary supplementation. Can the patient tolerate oral feedings? How long will dietary supplementation be necessary? From what point in nutritional status does the patient begin his illness? Does the patient have a large energy store? Will the bowel have to be rested? Can it tolerate bulk? And will long-term high-energy requirements by necessary? Whenever a patient can eat, or foods can be instilled into the stomach, it is preferable to let natural gastrointestinal mechanisms do their job to maintain nitrogen balance. When this is not possible, intravenous or central venous therapy is needed. The determination must be made for each individual patient, depending on the disease state, energy requirements, and the time they will be needed.

18.2. ENTERAL ALIMENTATION

With the development of modern blenders, it is possible to liquefy almost all foods so that they can be ingested in a liquid form. With the physician's attempt to control the patient's intake better and simultaneously simplify the absorption of foods, a whole host of more defined liquid diets have been

327

developed. When the liquid diet contains protein in the form of prepared amino acids, in addition to other easily digestible mono-, di-, or oligosaccharides, minerals, and vitamins, the formula is referred to as an elemental, or chemically defined, liquid diet.[1] The major difference between elemental diets and liquid foods is the elimination of fiber and fats from the elemental diet. As the science of supplemented diet therapy advanced, fats were added to selected elemental diets. There has been no significant recent work to compare and define the fiber and fat substances in elemental diets. Although liquefied foods have been used extensively in the past, the recent interest in elemental diets has arisen because of our ability to define the contents of these diets more accurately.[2,3]

Credit is usually given to Rose,[4] who in 1949 demonstrated humans could be maintained in a positive nitrogen balance on simple defined liquid nutrient materials.[5] In the 1960s, through the efforts of Greenstein and colleagues,[6-8] Winitz and colleagues,[9,10] and Bury and colleagues,[11] chemically defined diets were introduced, used in experiments in space laboratories, and finally aggressively produced for clinical application. The diets were water soluble and contained simple amino acids. They were shown to support normal growth in animals and man.[1]

Progress in the use of elemental diets has led to their having the following properties. (1) They are liquid; (2) they are free of any fiber and indigestible long-chain polysaccharides; (3) the protein content is in the form of synthetic amino acids, protein hydrolysates, or pure protein food; (4) they are usually high in oligosaccharides and may or may not contain simple sugars; (5) they usually contain no fats, but may have medium-chain triglycerides added; (6) they may or may not contain small amounts of essential fatty acids to prevent fatty acid deficiency; (7) they may or may not contain vitamin and mineral supplements; and (8) they are usually acidic and hypertonic.

The less hypertonic, or conversely the more isotonic, supplements usually contain more fats and are protein–electrolyte balanced to be more isotonic.

The main characteristic of elemental diets is easy absorption from the proximal gut. Thus, the energy-needy patient is supplied a high-energy nutritious source. Because of the lack of fiber in these diets, the stools will become scanty and less than half that of the usual constipated patient. This in itself is a major advantage in treating certain conditions, but in others, where it will be important to maintain bowel function, it can be a disadvantage; in such cases, liquefied fiber-containing food should be considered rather than an elemental diet. Of course, one can maintain complete control by adding fiber to the elemental diet and that can be accomplished by adding bran, psyllium seed, or maize. Naturally, the other properties of these fiber substances have to be considered while the patient is being treated.

Studies of the effects on gastrointestinal physiology of chemical diets reveal there is increased gastric secretion[12] and probably decreased enzyme secretion by both the pancreas and the intestinal mucosa,[13] and some possible alterations in the intestinal microflora.[14-16] There is no question that the flora must be affected by elimination of bulk from the diet, but the significance of these changes and the exact effect on host metabolism has yet to be demonstrated. Although systemic changes such as decreased blood pressure, a fall in serum lipids, and hypoallergic effects for the diets are claimed, these have been little studied and need more clinical evaluation.

18.2.1. Indications

The following outline of Shils and colleagues[17] lists the clinical problems for which they use oral and tube feedings.

> Inborn errors of metabolism
>
> Gastrointestinal disease
>   Ulcerative colitis
>   Granulomatous bowel disease
>   Chronic partial obstruction
>   Malabsorption syndromes
>   Short-bowel syndrome
>   Infantile diarrhea
>   Fistulas
>   Pancreatitis
>
> Renal failure
>
> Hypermetabolic states
>   Severe trauma
>   Major burns
>
> Incidental uses
>   Preoperative bowel preparation
>   Nonallergic food source
>   Food supplement
>   Toilet management problems
>   Protection of bowel mucosa against damaging agents
>   Feeding of premature infants
>   Feeding of anorectic patients

There is general agreement that using liquefied foods and elemental diets may be helpful, but they are no panacea, and very careful consideration should be given to their long-term use. I find that most clinicians tend to use the diets short term with an inadequate trial, and then from frustration go on to another form of therapy. The patient must be carefully selected, the proper course outlined, and the diet meticulously introduced; then, and only then, has the therapy a chance of succeeding.

18.2.2. Enteral Alimentation Formulas

It is almost impossible to list all of the formulas on the market today. Tables 18-1 through 18-4, modified from Shils *et al.*,[17] list those commonly used.

Table 18-1. Hydrolyzed Protein—Lactose Fiber-Free Formulas[a,b]

	Flexical[c] (Mead Johnson)	Vivonex[c] (Eaton)	Vivonex HN[c] (Eaton)	Nutramigen[d] (Mead Johnson)	Pregestimil[d] (Mead Johnson)
Protein (g)	22.4	20.4	45.6	32.5	32.5
Source	Hydrolyzed casein Amino acids	Crystalline amino acids	Crystalline amino acids	Hydrolyzed casein	Hydrolyzed casein
Fat (g)	34.0	1.4	0.9	39.0	41.0
Source	Soy oil 27.4 MCT 6.6	Safflower oil	Safflower oil	Corn oil	MCT 36.6 Corn oil 5.2
Carbohydrate (g)	154.0	226.3	202.4	130.0	130.0
Source	Sugar 100.9 Dextrin 48.4 Citrate 4.7	Glucose oligo saccharides	Glucose oligo saccharides	Sucrose 93.6 Tapioca starch 36.4	Glucose 91.0 Tapioca starch 39.0
Volume to give 1000 kcal	1000	1000	1000	1500	1500
Minerals					
Calcium (mg)	600.0	443.3	266.6	945.0	945.0
Phosphorus (mg)	500.0	443.3	266.6	705.0	705.9
Magnesium (mg)	200.0	194.3	116.7	111.0	111.0
Iron (mg)	9.0	5.6	3.3	18.8	18.8
Iodine (μg)	75.0	80.0	48.0	72.0	72.0
Copper (mg)	1.0	1.1	0.6	0.9	0.9
Manganese (mg)	2.5	1.6	0.9	1.7	1.7
Zinc (mg)	10.0	6.9	3.7	6.3	6.3
Sodium (meq)	15.2	37.4	33.5	21.0	21.0
Potasium (meq)	32.0	30.0	18.0	25.5	25.5
Chloride (meq)	35.2	50.8	52.4	19.5	19.5
mOsm/kg	723	500	850	443	590
Volume needed to meet 100% RDA including vitamins (ml)	2000	1800	3000		

[a]From Shils et al.[17]
[b]List and contents subject to change at the discretion of the manufacturer.
[c]Solution yields approximately 1 cal/ml.
[d]Infant formulas.

Table 18-1 includes diets in which hydrolyzed protein material is used. These are also lactose free but usually of rather high osmolality, and they may cause nausea or diarrhea. They are also referred to as monomeric and have the uniqueness of being virtually fiber free.[18]

Table 18-2 lists the elemental diets whose protein content is supplied by an intact or whole protein. Most of these are also relatively fiber and lactose free. Because of their low osmolality (350 mOsm/kg), I have found these better tolerated. Ensure has excellent palatability, and patients tolerate Isocal well.

Table 18-3 lists diets that may have some residue and the protein supply is from foods. These diets usually have a moderate amount of lactose. Because these contain liquefied foods, they may not be as effective in "resting the gut" and some may actually contain unreported amounts of fiber.

Table 18-4 lists nutrient supplements and their therapeutic uses. Those such as medium-chain triglycerides have been useful; many others are marketed but have not proven helpful.

Medium chain triglycerides can be absorbed in the absence of bile acids and pancreatic lipase, transported across the intestinal epithelium, and passed rapidly through the portal circulation. They are rapidly oxidized in the liver and, consequently, can become an important source of energy for patients who have difficulty in absorbing long-chain triglycerides. Patients who have difficulty in absorbing fats may benefit from taking medium-chain triglycerides. It may also be of great help in subjects who have defective ehylomicron formation, i.e., β-lipoproteinemia.

The product comes as an oil and can be added to salads and foods and used in cooking (see Chapter 20). We have found this helpful at times for subjects who had difficulty in maintaining adequate caloric intake because of poor fat absorption. It can be used for extended periods, but patients often tire of using it. While used, it will help maintain nutrition and often helps a patient through a difficult period of malabsorption.[19-21]

18.2.2.1.   Choice of Formula

A large number of claims are made for each product by each of the manufacturers, and all appear to have publications by various authors to substantiate some of their claims. Needless to say, these are most confusing and there is no clear pattern in the literature. At the present time, consideration must be given to the disease situation to determine the type of supplement needed.

The most natural product tolerated is the best to use. Formulas listed in Table 18-3 use more food extracts than the chemical formulas, but they contain high amounts of lactose, and patients intolerant to lactose will usually develop some diarrhea.[22] When this becomes a difficult problem, a formula low in lactose listed in Tables 18-1 or 18-2 can be used. Lactose intolerance can be a major problem, especially when one has a disease involving the small intestine. In all instances, careful consideration should be given to the individual patient and disease process. Certainly, whenever diarrhea follows the institution of one of the diets that does contain lactose, then it should be changed in a clinical trial.

Heymsfield and associates[18] have roughly classified the formulas into (1) monomeric—thin-low residue (Table 18-1), (2) polymeric—contains lactose or fiber and requires more active digestion (Table 18-2), and (3) "high-density"—more concentrated mixing. On the basis of this functional classification and their clinical experience, they recommend the following selections.

This author's experience correlates well with those recommended in Table 18-5, but there are great variations in individual taste and gut function. When a trial with one formula fails, another may succeed.

Table 18-2. Whole Protein—Lactose and Fiber, Low or Free[a,b,c]

	Ensure (Ross)	Ensure Plus (Ross)	Isocal (Mead Johnson)	Precision HN (Doyle)	Precision Isotonic (Doyle)	Precision LR (Doyle)	Precision Mod N (Doyle)
Protein (g)	**35.0**	**36.6**	**32.5**	**41.7**	**30.0**	**23.7**	**32.5**
Source	Na+Ca caseinate 30.6 Soy protein 4.4	Na+Ca caseinate Soy protein	Na caseinate Soy protein	Egg white solids	Egg white solids	Egg white solids	Egg white solids
Fat (g)	**35.0**	**35.5**	**42.0**	**0.5**	**31.3**	**0.7**	**31.0**
Source	Corn oil	Corn oil	Soy oil 33.6 MCT 8.4	Vegetable oil Monodiglycerides	Vegetable oil Monodiglycerides	Vegetable oil Monodiglycerides	Vegetable oil Monodiglycerides
Carbohydrate (g)	**136.7**	**133.2**	**125.0**	**206.7**	**150.0**	**224.7**	**150.0**
Source	Corn syrup solids 98.3 Sucrose 38.4	Corn syrup solids Sucrose	Corn syrup solids	Maltodextrin 193.9 Sugar 12.8	Maltodextrin Sugar	Maltodextrin 209.4 Sugar 15.3	Maltodextrin 114.2 Sugar 35.8
Lactose (g)	0	0	0	0	0	0	0
Volume to give 1000 kcal	943	676	960	950	1042	900	825
Minerals							
Calcium (mg)	471.5	422.5	600.0	333.3	667.0	526.3	500.0
Phosphorus (mg)	471.5	422.5	500.0	333.3	667.0	526.3	500.0
Magnesium (mg)	188.6	211.3	200.0	133.3	266.0	210.5	200.0
Iron (mg)	9.4	9.5	9.0	6.0	12.0	9.5	9.0
Iodine (µg)	75.4	70.4	75.0	50.0	100.0	78.9	75.0
Copper (mg)	0.9	1.1	1.0	0.7	1.3	1.1	1.0
Manganese (mg)	1.9	1.4	2.5	1.3	2.7	2.1	2.0
Zinc (mg)	18.9	15.9	10.0	5.0	10.0	7.9	7.5
Sodium (meq)	28.7	30.6	21.7	40.6	34.8	27.5	37.0
Potassium (meq)	24.2	32.5	32.1	22.2	25.6	20.2	19.2
Chloride (meq)	26.6	30.2	28.2	32.0	30.1	28.2	25.4
mOsm/kg	450	600	350	557	300	500–545	395
Volume needed to meet 100% RDA including vitamins (ml)	1920	1920	1920	2950	1560	1710	1650

Table 18-2 (Continued)

	Portagen (Mead Johnson)	ProSobee[d] (Mead Johnson)	Mull-Soy[d] (Syntex)	Neo Mull-Soy[d] (Syntex)	Isomil[d] (Ross)	Lolactene (Doyle)	Sustacal Liquid (Mead Johnson)
Protein	**35.0**	**37.5**	**48.0**	**27.0**	**30.0**	**66.2**	
Source	Na caseinate	Soy protein isolate L-Methionine	Soy flour	Soy protein isolate L-Methionine	Soy protein isolate L-Methionine	Nonfat dry milk Na caseinate	Na+Ca caseinate Soy protein
Fat (g)	**47.7**	**50.0**	**55.5**	**52.5**	**54.0**	**23.5**	**23.0**
Source	MCT 41.0 / Corn oil 5.5 / Lecithin 1.3	Soy oil	Soy oil	Soy oil	Soy oil Coconut oil Corn oil	Vegetable oil Monodiglycerides	Soy oil
Carbohydrate (g)	**115.0**	**100.0**	**79.5**	**96.0**	**102.0**	**132.4**	**137.8**
Source	Maltodextrin 83.5 / Sucrose 28.8 / Other 2.5	Sucrose 62.0 / Corn syrup solids 38.0	Sucrose / Invert sucrose	Sucrose	Corn syrup / Sucrose	Corn syrup solids 27.3 / Sucrose 10.4 / Glucose 46.7 / Galactose 46.7	Sucrose 97.2 / Corn syrup solids 25.4
Lactose (g)	<0.3	0	0	0	0	<4.0	0
Volume to give 1000 kcal	**1000**	**1500**	**1500**	**1500**	**1500**	**1250**	**1000**
Minerals							
Calcium (mg)	937.8	1185.0	1875.0	1250.0	1050.0	2353.0	1000.0
Phosphorus (mg)	604.4	795.0	1250.0	950.0	750.0	2058.8	916.7
Magnesium (mg)	208.4	111.0	117.2	120.0	75.0	441.2	375.0
Iron (mg)	18.8	18.8	15.6	15.6	18.0	19.9	16.7
Iodine (µg)	72.9	72.0	234.4	234.4	225.0	166.2	138.9
Copper (mg)	1.6	0.9	1.6	0.6	0.8	2.2	1.9
Manganese (mg)	3.1	1.7	2.2	4.0	—	4.4	2.8
Zinc (mg)	9.4	8.0	12.5	4.8	7.5	16.5	13.9
Sodium (meq)	20.4	27.0	28.1	26.2	19.6	47.9	40.2
Potassium (meq)	32.1	28.5	66.1	41.2	27.3	79.0	52.7
Chloride (meq)	24.2	18.0		14.1	22.4	57.2	(43.8)
mOsm/kg	357	258	252	275	—	670	625
Volume needed to meet 100% RDA including vitamins (ml)	**960**					**1150**	**1080**

[a] From Shils et al.[17]
[b] List and contents subject to change at the discretion of the manufacturer.
[c] Solution yields approximately 1 cal/ml.
[d] Infant formulas.

Table 18-3. Whole Protein—Full Lactose and Varying Fiber Formulas[a,b,c]

	Compleat B (Doyle)	Formula 2 (Cutter)	Meatbase Formula 142 (Hosp. Diet Prod.)	C.I.B.[d] (Carnation)	Meritene + Milk[d] (Doyle)
Protein (g)	**40.0**	**37.5**	**33.0**	**55.2**	**60.0**
Source	Beef Nonfat milk	Nonfat milk Beef	Beef Soy protein isolate Nonfat dry milk	Nonfat milk Soy protein Na caseinate	Concd skim milk Na caseinate
Fat (g)	**40.0**	**40.0**	**48.2**	**27.6**	**33.3**
Source	Corn oil	Corn oil Egg Yolks	Corn oil	Milk fat	Vegetable oil Monodiglycerides
Carbohydrate (g)	**120.0**	**122.5**	**107.6**	**124.1**	**115.0**
Source (Fiber foods)	Sucrose 23.0 Maltodextrin Vegetables Fruits 73.4 Orange juice	Sucrose Vegetables Orange juice Farina Dextrose	Dextrose Fruit Vegetables	Sucrose Corn syrup solids Lactose	Corn syrup solids Sucrose
Lactose (g)	24.4	37.5		84.0	56.7
Volume to give 1000 kcal	**1000**	**1000**	**667**	**880**	**1000**
Minerals					
Calcium (mg)	625.0	720.0	646.8	1206.9	1250.0
Phosphorous (mg)	1687.5	560.0	710.2	972.4	1250.0
Magnesium (mg)	250.0	100.0	264.0	403.5	333.3
Iron (mg)	11.3	12.6	11.9	15.9	15.0
Iodine (mg)	93.8	75.0	100.0	129.3	125.0
Copper (mg)	1.3	1.0	1.7	2.0	1.7
Manganese (mg)	2.5	0.2	—		3.3
Zinc (mg)	9.4	7.5	9.9	13.7	12.5
Sodium (meq)	67.9	26.1	31.5	37.0	39.8
Potassium (meq)	33.7	45.1	27.1	63.3	42.7
Chloride (meq)	22.9	53.5	—		47.0
mOsm/kg (meq)	490	435–510	725		700–750
Volume needed to meet 100% RDA including vitamins (ml)	**1600**	**2000**	**1000**	**1373**	**1200**

Table 18-3 (Continued)

	Meritene+Milk[d] (Doyle)	Nutri-1000 (Syntex)	Sustacal+Milk[d] (Mead Johnson)	Sustagen+Water (Mead Johnson)
Protein (g)	**65.1**	**38.0**	**60.3**	**60.0**
Source	Nonfat milk Whole milk	Skim milk	Nonfat milk Whole milk	Nonfat milk Whole milk Ca caseinate
Fat (g)	**32.5**	**52.0**	**24.4**	**8.6**
Source	Milk fat	Corn oil	Milk fat	Milk fat
Carbohydrate (g)	**112.0**	**95.6**	**134.4**	**171.4**
Source	Corn syrup solids 14.4	Sucrose 30.7 Corn syrup solids 14.8	Sucrose 36.2 Corn syrup solids 11.8	Corn syrup solids 104.7 Glucose 9.3
Lactose (g)	97.5	50.1	85.8	57.3
Volume to give 1000 kcal	**995**	**960**	**600**	**600**
Minerals				
Calcium (mg)	2168.6	1150.0	1611.2	1828.6
Phosphorus (mg)	1807.2	900.0	1333.4	1371.4
Magnesium (mg)	361.4	200.0	375.0	228.6
Iron (mg)	16.3	9.0	16.7	10.3
Iodine (µg)	136.1	75.0	138.9	85.7
Copper (mg)	1.8	1.0	1.9	1.1
Manganese (mg)	3.6	1.3	2.8	2.9
Zinc (mg)	13.5	7.5	13.9	11.4
Sodium (meq)	39.3	21.7	40.2	29.8
Potasium (meq)	71.1	35.9	64.8	51.3
Chloride (meq)	64.6	31.0	37.6	
mOsm/kg	690	500	756	1334
Volume needed to meet 100% RDA including vitamins (ml)	**1095**	**1920**	**756**	**1050**

[a] From Shils et al. [17]
[b] List and contents subject to change at the discretion of the manufacturer.
[c] Solution yields approximately 1 cal/ml.
[d] Whole milk added.

Table 18-4. Liquid

	Lipomul-Oral (Upjohn)	Liprotein (Upjohn)	Lofenalac (Mead Johnson)	Lonalac (Mead Johnson)
Protein (g)	0.1	82.7	32.5	53.0
Fat (g)	111.1	64.3	39.0	54.7
Carbohydrate (g)	1.1	22.1	130.0	74.2
Sodium (meq)	2.90	8.00	21.00	1.70
Potassium (meq)	0.09	42.46	25.50	48.08
Amount needed to give 1000 kcal	166.7 ml liquid	183.7 g dry wt	1500 ml standard dilution	1500 ml standard dilution g dry wt or 196 gm dry wt
	Fat source	Calorie source	Low phenyl-alanine formula	Low-sodium, high-protein source

	Amin-Aid (McGraw)	Cal-Power (General Mills)	Casec (Mead Johnson)	Cho-Free (Syntex)
Protein (g)	9.4	0.6	237.6	46.9
Fat (g)	31.7	0	5.4	91.3
Carbohydrate (g)	168.8	272.0	0	0.5
Sodium (meq)	2.88	2.39	6.00	41.34
Potassium (meq)	2.88	0.70	2.10	59.21
Amount needed to give 1000 kcal	213 g dry wt or 489 ml standard dilution	550 g liquid	270.0 g dry wt	1304 ml undiluted
Use	Formula for renal failure therapy	Carbohydrate source	Protein source	Low-carbo hydrate formula

[a]From Shils et al.[17]
[b]List and contents are subject to change at the discretion of the manufacturer.
[c]All values based per 1000 kilocalories.

Supplements[a,b,c]

Lytren (Mead Johnson)	MCT OIL (Mead Johnson)	Pedlayte (Ross)	Polycose (Ross)	Probana (Mead Johnson)	Sumacal (Hosp. Diet Prod.)
0	0	0	0	60.0	0
0	120.5	0	0	32.0	0
253.0	0	250.0	235.0	118.0	250.0
100.00	0	150.00	12.00	40.50	10.40
83.30	0	100.00	0.30	46.50	1.15
268 g dry wt or 3333 ml standard dilution	120.5 g liquid	5000 ml liquid	250.0 g dry wt	1500 ml standard dilution	360 ml liquid
Calorie & electrolyte source	Medium-chain triglycerides	Calorie & electrolyte source	Oligo-saccharides	High-protein banana powder formula for celiac condition & diarrhea	Concentrated carbohydrate source & low electrolytes

Citrotein (Doyle)	Controlyte (Doyle)	dp High p.e.r. Protein (General Mills)	EMF (Control Drugs)	Gevral (Lederle)	Hy-Cal (Beecham-Massengill)
60.5	Trace	206.0	250.0	170.9	0.1
2.6	48.0	10.0	0	5.7	0.1
184.2	143.0	21.0	0	66.8	244.1
45.80	1.30	22.40	53.00	18.57	2.41
26.80	0.20	9.90	5.10	3.65	0.07
263.4 g dry wt	198.0 g dry wt	258 g dry wt	500 ml liquid	284.8 g dry wt	407 ml liquid
Calorie supplement	Low-protein, low-electrolyte calorie source	Protein supple-ment & low electrolytes	Protein source	Protein–calorie supplement	Carbohydrate source

Table 18-5

Clinical status	Recommended enteral solution [a]
Normal gastrointestinal tract	Isocal
Normal metabolic rate	Ensure
Inability to eat (cerebrovascular accident, anorexia, oropharyngeal or esophageal surgery)	Precision Istonic (lower cost, nearly isotonic)
Increased metabolic rate	Large quantities of Ensure
Normal gastrointestinal tract	Add Polycose to Isocal
	Ensure Plus (greater concentration of kilocalories)
	Precision HN (increased protein content)
Abnormal gastorintestinal tract (pancreatic insufficiency, short-bowel syndrome, enterocutaneous fistual, inflammatory bowel disease)	Vivonex
	Vivonex HN
	Flexical

[a]See Tables 18-1 and 18-2.

18.2.2.2. Method of Administration

Liquid formula diets are administered via three routes: oral, tube, and ostomy (gastric or jejunal).

The oral route is preferred as long as a patient can swallow and follow instructions, there is no obstruction of the upper gastrointestinal tract, and a compatible formula is available. It is relatively easy to establish the first three criteria by clinical observation, but it is often difficult to find a formula a patient will drink 2 to 3 liters of daily. Failure in the use of the oral route occurs most frequently because of the sheer boredom of having to drink the same fluid repeatedly, or because of a taste incompatibility. When a patient can tolerate the oral route, it is still most efficient, but frequently one of the tube techniques must be used if enteral hyperalimentation is to succeed over a long period of time.

Tube feeding can be accomplished by positioning a thin nasogastric tube in the stomach or duodenum. Several successful methods are reported.[23-31] All employ the same principle. A fine catheter-type tube is passed through the nose into the nasopharynx. Because it is soft and thin, it must be made more firm by attaching it to a larger Levine tube and wedging the ends into a dissolvable gelatin capsule so that the Levine tube can be removed,[23] or by using a more expensive, mercury-weighted special tube.[18,24,25] (These are available from Bioresearch Medical Products, Inc., Raritan, N.J., or Keofeed, Hedeco Corp., Palo Alto, Calif. The latter is a silicone tube.) Others have used a stylet, and others merely a No. 5 or No. 8 French cold stiffened pediatric feeding tube.[29-31] I have found the silicone tubes most compatible when they can be used. Once the tube is in the nasopharynx, it is then passed by having the patient swallow a few sips of water. At times, it is necessary to use pressure. Once past the epiglottis, the tube drops into the stomach. This position can be determined by listening to an air injection over the stomach. The tube then can be localized in the duodenum, if desired, by X-ray. Once the tube tip is in place, it is taped to the side of the face or forehead for comfort. Catheter and tube lengths range from 24 to 36 in.

Vivonex, Isocal, and Flexical[23-31] flow freely through the thinnest tubes. The rate of flow can be controlled by an infusion pump, or more simply by hanging the bag container at 2 to 3 ft above the level of the head. If the tube is thin, the flow rate is naturally safely limited. The catheter can be disconnected from the container at any point, but it may be necessary to run it through the night in order to administer the required amount. It is best to keep a constant flow rate than distend the stomach. Thicker formulas may plug or cling to the tubes; therefore, it is wisest to use one of the thinner formulas if the tube is difficult to change. Frequency of tube change will vary greatly depending on the cleanliness and compatibility of the individual.

Gastrostomy and jejunostomy permit the feeding of more complex liquid diets. If the patient has a normal, functioning intestine, then blended foods may be optimal. Once a feeding ostomy is functioning, the only diet limitations are the presence of disease caudad. If blended foods are too difficult to prepare, formulas in Tables 18-3 and 18-4 can be employed.

18.2.2.3.   Complications

Elemental diets are relatively free of complications when administered orally. The problems that arise are nausea, vomiting, or diarrhea. The diarrhea may often be due to a lactose intolerance; consequently, a switch to a nonlactose diet is important. However, diarrhea may be due to the hyperosmolar effect of the solution or too rapid intake. The nausea and vomiting are often also due to the hyperosmolar solution. In these cases, the patient rapidly expresses dissatisfaction with the diet. At times, nausea can be related to the boredom that occurs with such a diet or the unpleasant taste experienced by certain patients. In these instances, the patient should be encouraged to sip the drink instead of gulping it down. At no time should he be forced to take more than a glass of the liquid. It often helps for it to be placed on ice rather than to be drunk at room temperature. At times, diarrhea may be corrected by careful, slow infusion-pump administration.

Elemental hyperalimentation via tube has few complications that cannot be managed by manipulation of the formula. Heymsfield and colleagues[18] have estimated the frequency of complications. Table 18-6 lists suggested alternatives to correct problems. Only hyperosmolar coma or an erosive ulcer requires discontinuing therapy.

Weight gain is the goal, but *edema* must be guarded against. Small amounts are no problem except to the cardiac patient. Edema occurs transiently as proteins are repleted. Elemental diets appear to produce less fluid retention than TPN.[32,33] It is well to point out that amino acid diets induce more fluid

Table 18-6.   Frequency and Management of Complications of Enteral Hyperalimentation[a]

Type of complication	Frequency (%)	Therapy
Mechanical		
Tube lumen clogged by solution	Infrequent (10)	Flush with water; replace tube if unsuccessful
Pulmonary aspiration of stomach contents	Rare (1)	Unlikely with head of bed elevated; discontinue if aspiration occurs
Esophageal erosion	Rare (1)	Discontinue tube
Gastrointestinal symptoms		
Vomiting and bloating	10–15	Reduce flow rate and add peripheral hyperalimentation if needed
Diarrhea and cramping	10–20	Reduce flow, dilute solution, consider different type solution, add antidiarrheal drug, use infusion pump
Metabolic, fluid, and electrolyte abnormalities		
Hyperglycemia and glucosuria	10–15	Reduce flow, administer insulin
Hypersomolar coma	Rare (1)	Discontinue therapy
Edema	20–25	Usually none; may reduce Na content or slow hyperalimentation rate; rarely use diuretics
Congestive heart failure	1–5	Slow hyperalimentation, administer diuretics and digoxin as indicated
Hypernatremia, hypercalcemia	5	Adjust electrolyte content of hyperalimentation
Essential fatty acid deficiency	Common[b]	Add linoleic acid supplement

[a] Modified from Heymsfield *et al.*[18]
[b] If enteral feeding mixture lacks linoleic acid.

secretion into the gut than peptide diets.[34] It is not clear how this relates to edema, but it may be helpful in the management of diarrhea.[34]

The best way to prevent complications is to observe the patient carefully. Pulse, blood pressure, signs of edema, and urine for glucose and electrolytes should all be checked. The mechanical complications are obvious if checked and simple laboratory tests aid in evaluating the metabolic problems.

18.2.3.   Treatment of Fistulas of the Gastrointestinal Tract

This is an area of success for elemental diet therapy. The patient with a fistula gets complete lower-bowel rest while adequate nutrition is taking place. It is the same physiologic approach as bypass surgery for the healing of a fistula.

One of the goals of elemental diet therapy is to decrease the number of bowel movements in subjects with fistulas or inflammatory bowel disease. A comparison of three formulas (Vivonex, Flexical, and Precision LR—Tables 18-1 and 18-2) reveals all effectively decrease stool weight and number in volunteers, but Vivonex and Flexical are more effective than Precision LR.[41]

(See discussion on types of formulas for more detail on choice of diet supplements.)

Fistulas that involve the upper small bowel do poorly with this therapy. The elemental diet increases secretion in the duodenum and requires the jejunum for absorption. Fistulas of the lower intestine do best. Analysis of some reports from several institutions[35-39] indicates that there may be an overall spontaneous fistula closure rate of as high as 70% of patients treated with elemental diets. The mortality in this group may be as high as 50%, with deaths frequently occurring in disease outside of the gastrointestinal tract. Others have had less success and conservative opinion feels the benefit of elemental diets is still open to question.[40] In a cooperative patient who can tolerate the diet well, it should be considered as one of the first forms of treatment. As it has been one of the most successful areas of diet therapy, it certainly should be tried in the patient who is a poor risk for surgery.

18.2.4. Treatment of Inflammatory Bowel Disease

Successful use of fiber-free elemental diets has now been reported in ulcerative colitis, Crohn's disease involving the small and large intestines, and diverticular disease.[42-45] These reports have been sporadic and there is no study on a large number of patients. In most studies, the supplemental diets have been effective and helpful in managing complications of inflammatory bowel disease, such as intractable diarrhea, poor nutrition, and fistula. Total parenteral nutrition also has been successful and used more often in seriously ill patients (see Chapter 19). The elemental oral diets are reserved for those less ill, and as a consequence, a question always arises whether these patients would have done well on less rigorous restrictions. In any event, the diet is indicated to restrict bulk and consequently rest the bowel. This is described as a so-called "medical ostomy."[1,46] The patient who can tolerate elemental diets has a good source of nutrition during an acute exacerbation of the disease. If the oral method fails, then one can turn to TPN. During diet therapy, other accepted forms of therapy for the disease, such as sulfasalizine and steroids, naturally are indicated.

18.2.5. Treatment of Malnutrition and Maldigestion

Because the elemental diets require little digestion, they may be helpful in patients with a disturbance in their digestive processes; specifically, in aiding absorption in the patient with chronic pancreatic insufficiency, maintaining the patient with a short-bowel syndrome, and in sustaining the debilitated cancer patient. Patients receiving chemotherapy and/or radiotherapy that effects the gut may be maintained in positive nitrogen balance if they tolerate these diets.[17,47-49]

Uses of oral elemental diets in nondigestive disease have been reviewed by others.[1,3,17] There appears to be a definite benefit in selected cases of phenylketonuria and in selected patients with severe burns and trauma. In addition, it may be beneficial in selected infantile diarrheas and anorectic children.[1,17] The significance of preoperative use and helping maintain the patient undergoing successive diagnostic tests is not yet clear, but certainly elemental diet supplementation does no harm in such patients and may be of benefit.[1,50]

In summary, the use of oral elemental diets is helpful but difficult. Many patients cannot tolerate long-term feeding because of its marked hyperosmolal effect or, if they get used to it, the boredom that sets in with prolonged use. As a consequence, such problems can be a relative contraindication. A trial of other elemental formulas may help, but in the intolerant patient tube feeding or parenteral nutrition may be needed immediately. This remains a clinical bedside decision. In patients who cannot take these feedings orally, one should rapidly consider a tube feeding.[51] Caution must always be observed concerning the possibility of aspiration. The patient should be observed carefully and feedings given slowly through small tubes and in the upright position or semirecumbent. At times, it will be necessary to insert a gastric tube via surgical procedure or perform a gastrostomy. These conditions are growing rare today, but they can be lifesaving when instituted. It is essential that each case be analyzed carefully for the correct route of nutritional support in order to guarantee its success.

REFERENCES

1. Russell RI: Progress report elemental diets. *Gut* 16:68, 1975.
2. Robinson CH: Liquid diets. *J Clin Nutr* 1:476, 1953.
3. Kark RM: Liquid formula and chemically defined diets. *J Am Diet Assoc* 64:476, 1974.
4. Rose WC: Amino acid requirements of man. *Fed Proc Fed Am Soc Exp Biol* 8:546, 1949.
5. Rose WC, Wixom RL: The amino acid requirements of man. XVI. The role of nitrogen intake. *J Biol Chem* 217:997, 1955.
6. Greenstein JP, Birnbaum SM, Winitz M, *et al:* Quantitative nutritional studies with water-soluble, chemically defined diets. I. Growth, reproduction and lactation in rats. *Arch Biochem* 72:396, 1957.
7. Birnbaum SM, Winitz M, Greenstein JP: Quantitative nutritional studies with water-soluble, chemically defined diets. III. Individual amino acids as sources of 'non-essential' nitrogen. *Arch Biochem* 72:428, 1957.
8. Greenstein JP, Otey MC, Birnbaum SM, *et al:* Quantitative nutritional studies with water-soluble, chemically defined diets. X. Formulation of a nutritionally complete liquid diet. *J Natl Cancer Inst* 24:211, 1960.
9. Winitz M, Graff J, Gallagher N, *et al:* Evaluation of chemical diets as nutrition for man-in-space. *Nature (London)* 205:741, 1965.
10. Winitz M, Seedman DA, Graff J: Studies in metabolic nutrition employing chemically defined diets. I. Extended feeding of normal human adult males. *Am J Clin Nutr* 23:525, 1970.
11. Bury KD, Stephens RV, Randall HT: Use of a chemically defined, liquid, elemental diet for nutritional management of fistulas of the alimentary tract. *Am J Surg* 121:174, 1971.
12. Rivilis J, McArdle H, Wlodek GK, *et al:* The effect of an elemental diet on gastric secretions. *Ann R Coll Surg Engl* Canada 5:57, 1972.
13. Perrault J, Devroede G, Bounous G: Effects of an elemental diet in healthy volunteers. *Gastroenterology* 64:569, 1973.
14. Winitz M, Adams RF, Seedman DA, *et al:* Studies in metabolic nutrition employing chemically defined diets. II. Effects on gut microflora populations. *Am J Clin Nutr* 23:546, 1970.

15. Bounous G, Devroede GJ: Effects of an elemental diet on human faecal flora. *Gastroenterology* 66:210, 1974.
16. Crowther JS, Drasar BS, Goddard P, *et al:* The effect of a chemically defined diet on the faecal flora and faecal steroid concentration. *Gut* 14:790, 1973.
17. Shils ME, Bloch AS, Chernoff R: Liquid formulas for oral and tube feeding. *Clin Bull* 6:151, 1976.
18. Heymsfield SB, Bethel RA, Ansley JD, *et al:* Enteral hyperalimentation: An alternative to central venous hyperalimentation. *Ann Intern Med* 90:63, 1979.
19. Senior JR: *Medium Chain Triglycerides*. Philadelphia: University of Pennsylvania Press, 1968.
20. Hashim SA, Arteaga A, Van Itallie TB: Effect of a saturated medium-chain triglyceride on serum-lipids in man. *Lancet* 1:1105, 1960.
21. Gracey M, Burke V, Anderson CM: Medium chain triglycerides in paediatric practice. *Arch Dis Child* 45:445, 1970.
22. Walike BC, Walike JW: Relative lactose intolerance. *JAMA* 238:958, 1977.
23. Kaminski MV: Enteral hyperalimentation. *Surg Gynecol Obstet* 143:12, 1976.
24. Dobbie RP, Hoffmeister JA: Continuous pump-tube enteric hyperalimentation. *Surg Gynecol Obstet* 143:273, 1976.
25. Hoffmeister JA, Dobbie RP: Continuous control pump-tube feeding of the malnourished patient with Isocal. *Am Surg* 43:6, 1977.
26. Page CP, Ryan JA, Haff RC: Continual catheter administration of an elemental diet. *Surg Gynecol Obstet* 142:184, 1976.
27. Metz G, Dilawari J, Kellock TD: Simple technique for nasoenteric feeding. *Lancet* 2:454, 1978.
28. Bethel RA, Jansen RD, Heymsfield SB, *et al:* Nasogastric hyperalimentation through a polyethelene catheter: An alternative to central venous hyperalimentation. *Am J Clin Nutr* 32:1112, 1979.
29. Rivard J-Y, Lapoint R: Clinical experience in using elemental diet in the management of various surgical nutritional problems. *Can J Surg* 18:90, 1975.
30. Freeman JB, Egan MC, Millis BJ: The elemental diet. *Surg Gynecol Obstet* 142:925, 1976.
31. Randall HT: Enteric feeding, in Ballinger WF, Collins JA, Drucker WR, *et al* (eds): *Manual of Surgical Nutrition*. Philadelphia: WB Saunders, 1975, pp 267-284.
32. Yeung CK, Smith RC, Hill GL: Effect of an elemental diet on body composition. A comparison with intravenous nutrition. *Gastroenterology* 77:652, 1979.
33. Jeejeebhoy KN: Is more better? Is weight water? The significance of weight gain during parenteral nutrition with amino acids and dextrose. *Gastroenterology* 77:799, 1979.
34. Silk DBA, Chung YC, Berger KL, *et al:* Comparison of oral feeding of peptide and amino acid meals to normal human subjects. *Gut* 20:291, 1979.
35. Voitk A, Echave V, Brown R, *et al:* Elemental diet in the treatment of fistulas of the alimentary tract. *Surg Gynecol Obstet* 137:68, 1973.
36. Rocchio MA, Chung-Ja C, Haas K, *et al:* Use of a chemically defined diet in the management of patients with high output gastrocutaneous fistulas. *Am J Surg* 127:148, 1974.
37. Bode HH: Healing of faecal fistula initiated by synthetic low residue diet. *Lancet* 2:954, 1970.
38. Chapman R, Foran R, Dunphy JE: Management of intestinal fistulas. *Am J Surg* 108:157, 1964.
39. Lorenzo GA, Beal JM: Management of external small bowel fistulas. *Arch Surg* 99:394, 1969.
40. Editorial: Nutritional management of enterocutaneous fistulas. *Lancet* 2:507, 1979.
41. McCamman S, Beyer PL, Rhodes JB: A comparison of three defined formula diets in normal volunteers. *Am J Clin Nutr* 30:1655, 1977.
42. Voitk AJ, Echave V, Feller JH, *et al:* Experience with elemental diet in the treatment of inflammatory bowel disease. Is this primary therapy? *Arch Surg* 107:329, 1973.
43. Rocchio MA, Cha CJM, Haas KF, *et al:* Use of chemically defined diets in the management of patients with inflammatory bowel disease. *Am J Surg* 127:469, 1974.
44. Bury KD, Turnier E, Randall HT: Nutritional management of granulomatous colitis with perianal ulceration. *Can J Surg* 15:108, 1972.
45. Giorgini GL, Stephens RV, Thayer WR Jr: Use of 'medical by-pass' in the therapy of Crohn's disease. *Am J Dig Dis* 18:153, 1973.
46. Moss G: Physiological colostomy. *Arch Surg* 106:741, 1973.

47. Voitk A, Brown RA, Echave V, *et al:* Use of an elemental diet in the treatment of complicated pancreatitis. *Am J Surg* 125:223, 1973.

48. Thompson WR, Stephens RV, Randall HT, *et al:* Use of 'space diet' in the management of a patient with extreme short bowel syndrome. *Am J Surg* 117:449, 1969.

49. Bourous G, Gentile JM, Hugon JS: Elemental diet in the intestinal lesion produced by 5-fluoruracil in man. *Can J Surg* 14:312, 1971.

50. Johnson WC: Oral elemental diet. *Arch Surg* 108:32, 1974.

51. Voitk AJ: The place of elemental diet in clinical nutrition. *Br J Clin Pract* 29:55, 1975.

Peripheral and Total Parenteral Nutrition

19.1. PERIPHERAL INTRAVENOUS THERAPY

It is amazing how much dextrose in water is administered by clinicians in hospital settings while the patient loses weight and resistance. The administration of glucose and water, or glucose and electrolytes, does little but supply a small carbohydrate energy source. Patients who receive merely glucose and electrolyte solution go into rapid negative nitrogen balance. These patients lose weight. They are energy starved. Furthermore, patients receiving protein supplementation have decreased morbidity. It certainly behooves every clinician to think critically when prescribing simple glucose and electrolyte solutions. If it is for a short-term illness and the diagnosis is certain, then merely maintaining fluid balance is important and some carbohydrate assists in that function. However, if the patient is seriously ill, will be ill for more than several days, and does not have any evidence of obesity, then the physician should consider more than glucose for intravenous therapy.

If protein is added peripherally, glucose does tend to facilitate whole-body protein synthesis. Sim and associates,[1] using [^{15}N]glycine in healthy fasting volunteers, showed glucose plus amino acid infusion increased protein synthesis compared to amino acid alone.[1] Hence, glucose is an important intravenous energy source for promoting body building in combination with other nutrients.

19.1.1. Protein

The administration of amino acid solutions into peripheral veins became therapeutically feasible in 1972 when Blackburn[2,3] and Freeman[4,5] demonstrated that approximately 75 to 100 g of amino acid infusion daily maintains positive nitrogen balance. This occurs in the absence of fulfilling basal metabolic energy requirements. The explanation offered by Blackburn and his colleagues[2,3] is that insulin plays a major role in regulating energy metabolism by increasing the rate of glucose utilization and by controlling the rate of free fatty acid release from adipose tissue. They referred to it as "protein-sparing." Normally, protein can be converted to glucose, whereas free fatty acids are converted into ketone bodies. If a starving patient receives only glucose, or glucose in conjunction with protein, then insulin activity increases so that its antilipolytic activity is increased, and consequently there are less fatty acids and ketone bodies available. This further causes protein breakdown so that gluconeogenesis can occur. They also point out that the brain functions well on ketone bodies and that glucose is not necessary for cerebral function. Blackburn and associates have shown that peripheral protein administration results in maintaining nitrogen balance and, consequently, maintaining protein structures, while adipose tissue is utilized for energy.

The indications for this form of therapy seem to be best during acute trauma and acute burn situations. Some physicians feel that it has use in the preoperative and postoperative states. This observer is anxious to see whether the technique of simple protein amino acid infusion gains acceptance in other institutions.

In one study, patients receiving parenteral isotonic amino acid infusion during the postoperative period were compared to those receiving protein plus glucose, protein plus a soybean oil emulsion, and only simple glucose.[6] The authors confirmed the protein-sparing effect of amino acid infusion compared to simple intravenous glucose. However, they found that the addition of glucose did not decrease the effect of the amino acid. Actually, Sim and associates[1] have shown that glucose can aid whole-body protein synthesis.

The theory of the beneficial effect of isotonic amino acid infusion during the postoperative period and in other catabolic states has been reviewed carefully by Felig.[7] He stresses five points.

1. During the postoperative period, patients receiving only saline reveal peak urinary nitrogen loss that coincides with fivefold to tenfold increments in blood ketones (approximately the second to fourth days).

2. Administration of simple glucose (100 g/day) will reduce nitrogen loss from 30 to 60% regardless of suppression of ketosis and stimulation of insulin secretion.

3. The nitrogen-sparing effect of ketone infusions is only seen after long fasting, where nitrogen losses are usually much less than seen in the postoperative state.

4. The work of Greenburg and colleagues[6] reveals that the addition of glucose to isotonic amino acid infusions does not always impair nitrogen-sparing.

5. Increased fat utilization by intravenous lipid emulsions does not improve nitrogen-sparing above that of the amino acid solutions alone.

Felig concludes that the nitrogen-sparing effect of intravenous amino acids does not require either ketosis or hyperinsulinemia, but that the sparing effect does occur and is a marked improvement in intravenous therapy.

The clinical role of protein sparing was tested in a controlled study at the General Infirmary in Leeds.[8] They found that postoperative amino acid infusion prevented nitrogen and potassium loss but only patients on hyperalimentation had significantly fewer postoperative complications. More evaluations such as this one are needed.

The administration of an isotonic 3% amino acid solution that provides between 1 and 1.5 g of protein per kg of body weight per day is preferred. It is recommended that patients receive no more than 1 or 2 liters of amino acid infusion during the first few days of therapy and that adequate electrolytes, vitamins, and minerals be added to the solutions. Metabolic acidosis has been noted with isotonic amino acid infusion in some patients. Blackburn recommends that in order to counteract this trend, half of the daily sodium requirement may be given as sodium acetate and half of the potassium requirement as potassium phosphate. It is recommended that the solutions contain 4.5 to 9 meq of calcium glutonate, 8 meq of magnesium sulfate, and oral vitamins each day.

Blood glucose and BUN should be measured daily during the first few days and then at least twice weekly if therapy is to be maintained. It is expected that ketonuria and ketosis will develop, but if the BUN begins to rise more than 5 to 10 mg/100 ml per day for three consecutive days, then administration should be discontinued.

Caution should be observed in the patients with hepatic failure, renal failure, or insulin-dependent diabetes. In these patients, it is recommended that TPN be considered.

The indications for the use of this therapy await further evaluation. However, from initial evaluations,[1-8] appears that amino acid peripheral and vein administration is helpful in preserving the body protein mass. *In conditions where one can safely estimate that the patient will not require intravenous feeding for more than a few weeks, and that there is an adequate fat or adipose reserve, then this therapy may well become the therapy of choice. It obviates the need for central venous catheter while still helping the body maintain normal protein stores.* When the patient has to be maintained for more than a few weeks, severe weight loss will ensue and unless that is wanted the patient must be considered for TPN.

The major complications of amino acid infusion is occasional phlebitis or thrombosis of the peripheral veins involved. A small dose of cortisol (5 mg/liter iv) may help prevent this complication.[9] However, in early studies, there appears to be a very low incidence of this phenomenon. The major complication that concerns most clinicians is the development of metabolic hyperchloremic acidosis. This is counteracted by administration of sodium acetate and potassium phosphate, or cessation of therapy with appropriate electrolyte balance treatment.

In summary, peripheral vein amino acid therapy results in maintaining nitrogen balance, preserving the so-called lean body mass while blood glucose levels are maintained and the body develops ketosis with utilization of adipose tissue. This appears to be relatively contraindicated in patients with chronic renal and hepatic disease and most indicated in patients with acute traumatic injuries where insulin resistance appears to be a significant phenomenon and where protein sparing is helpful over a 1- to 2-week period.[1-8]

19.1.2. Intravenous Lipid Therapy

In the initial development of total parenteral nutrition in the United States, total caloric requirement was supplied with glucose–protein solutions because of government restriction on the use of intravenous fats. As a consequence, essential fatty acid deficiency was clinically recorded that led to the obvious need for intravenous fat to be included in TPN. As a consequence, fat emulsions became acceptable for intravenous administration in the United States.

Fat emulsions employed successfully contain 10% soybean oil, 1.2% egg yolk phospholipids, and 2.25% glycerine in a 10% fat emulsion now referred to as Intralipid. The soybean oil is composed of predominantly unsaturated fatty acids containing approximately 54% linoleic, 26% oleic, 9% palmitic, and 8% linolenic. It is safe to give this solution through a peripheral vein, and it acts as a high-energy source when given alone.

Several protocols have been described to use fat emulsion for total parenteral nutrition.[10-13] The system described by Silberman and colleagues[14] from the University of Southern California Medical Center, Los Angeles County, was used successfully. They treated 77 patients over a 9-month period and had relatively few complications, one mild hypotension in a septic patient, and two suppurative phlebidites in severe burn patients. They have seen mild phlebitis in several patients, but no other infections. Mild hyponatremia and anemia develop in a few. The system they employed is a two-component system. One bottle consists of 500 ml of 10% fat emulsion and a second bottle of 500 ml of 5.95% crystalline amino acid–10% glucose with 45 meq of sodium acetate, 40 meq of potassium chloride, 8 meq of magnesium sulfate, 5 meq of calcium gluconate, and 1000 units of heparin sodium with 5 ml of multiple vitamins. This liter unit is given together with the fat emulsion directly administered into the vein so that the glucose–amino acid solution is piggybacked into the tubing. The glucose–amino acid solution is hyperosmolar at 1500 mOsm/liter. However, when dripped into the emulsion solution, the concentration of the solution going into the vein is approximately 900 mOsm/liter. (Isaacs' group has also shown that a 900 mOsm/liter solution can be tolerated peripherally.[9]) It provides 4.4 g of nitrogen per liter, the protein equivalent of 27.3 g/liter, with 250 carbohydrate cal and 500 lipid cal. It is recommended that one start with the 500 ml of 10% fat emulsion over a 4- to 8-hr period, and if there are no reactions, such as fever, chills, shivering, or an occasional side effect of chest and back pain, then a 1-liter unit may be given the first day and each additional day another 1-liter unit added, so that 3 to 4 liters can be given per day as the optimal administration. The authors have used this protocol for as long as 63 days at 5 liters/day.

This form of therapy has not gained wide acceptance throughout the United States, but other modified systems are employed more readily in Canada[10,15] and Europe because lipid solutions have been approved in other countries for a greater period of time. Because they can be given intravenously and do not require central venous therapy, there are less potentially serious complications and greater ease in administration. These peripheral TPN systems have great promise for maintaining positive nitrogen balance with a high-energy source and should gain more acceptance in the future.

The most important complications are immediate histaminelike reactions, hypotension, and fever.[14] As indicated by Silberman and colleagues,[14] they occur rarely. Anemia was noted by Silberman and colleagues in a few patients, and was recorded in conjunction with thrombocytopenia in patients treated with Lipomul. That cottonseed oil fat emulsion is no longer available. The cause of the anemia and blood abnormalities was never truly understood. Complications that have been observed with Lipomul have not been recorded as yet with Intralipid. For some reason, soybean oil has fewer side effects than cottonseed oil emulsions.

Caution should be observed in the use of Intralipid in patients with abnormal fat transport, diabetes mellitis, severe liver disease, and a hypercoagulable state. As greater experience is gained with Intralipid, the significance of its use in these clinical situations may become more apparent.

In summary, peripheral venous administration of lipid emulsions holds great promise for dietary supplementation therapy. It is a useful energy source, and added to glucose–amino acid infusion it may come to replace central venous total parenteral nutrition, as it appears to have less complications from early studies and may well be more physiologic if given in the formulas described by Silberman and colleagues[14] or Jeejeebhoy and colleagues.[15]

19.2. CENTRAL INTRAVENOUS THERAPY—TOTAL PARENTERAL NUTRITION (TPN)

Total parenteral nutrition is a procedure capable of maintaining patients in positive nitrogen energy balance for prolonged periods of time while a primary disease is being treated or the gastrointestinal tract is not capable of function. It consists of the deep-vein administration of hyperosmolar glucose solutions in conjunction with synthetic amino acid or protein hydrolysate solutions.

In 1968, Dudrick and colleagues[16,17] introduced the concept and demonstrated that TPN was possible. Credit for popularizing TPN or hyperalimentation must go to Dudrick. Since that time, central venous administration of total nutritional requirements has gained wide acceptance.

Guidelines for TPN have been published by the Council on Food and Nutrition of the American Medical Association.[18] These initially stressed that the *indication* for TPN was when oral or tube feeding is contraindicated or inadequate or when conventional parenteral support is insufficient for the needs of the patient. At the time of publication of this book, peripheral venous support with lipid emulsion solutions are being tested, but the most widely used system for TPN is central venous hyperosmolar glucose–amino acid administration with lipid added once or twice weekly to prevent fatty acid deficiency. The indications for the use of TPN can be listed in a general fashion. However, they are similar to those listed in Table 18-1 as clinical problems for which oral elemental diets may be employed. When those indications exist and the patient either no longer can tolerate oral feedings, or tube feeding is impossible, or the condition is rapidly deteriorating, then TPN is indicated. TPN has been most effective in the patient suffering from hypermetabolic states, such as severe trauma, burns, and pancreatitis. This is a text primarily on gastrointestinal disorders; consequently, the benefits seen in inflammatory bowel disease, ulcerative colitis, and granulomatous disease as well as in malabsorption syndromes and short-bowel syndrome will be discussed in greater detail under those particular conditions.

Of note is the fact that TPN can be of extreme help in infants and children. Feliciano and Telander[19] have reviewed their experience at the Mayo Clinic, which substantiates that TPN is an effective form of maintaining normal growth and development. For use in the pediatric population, one is referred to one of the most recent texts on the subject.[20]

The technique of administration, delivery of the fluid, and the types of solutions chosen have now gained wide acceptance so that most pharmacists are familiar with the products and systems available. Before beginning TPN, the patient should be properly instructed and educated on the need and the course of the therapy.

It is essential that an adequate central venous portal be obtained. This is accomplished by subclavian vein catheterization. The catheter is used for an extended period of time. Very strict aseptic technique should be employed with necessary recurrent infection control. Insertion of the catheter should be done by an expert under sterile conditions. The protocol may include changing of the dressing every 2 to 3 days. It is best to suture the catheter in place. Appropriate nursing education is essential. The greatest hazard associated with indwelling catheters is infection, and consequently aseptic techniques are essential. Techniques for this are described in detail by Fischer.[21] For prolonged long-term use, peripheral access fistulas can be created for use in the ambulatory patient.[22] However, Broviac and co-workers[23,24] have described a silicone rubber (Silastic) atrial catheter that is used successfully in ambulatory patients undergoing prolonged TPN at home. These catheters have remained *in situ* for 6 to 19 months before needing replacement.[25]

The delivery system may be employed merely by a drip technique. However, infusion pumps are readily available today and more accurately deliver regulated amounts of fluids. They are recommended for ease of administration and monitoring. Certainly for the ambulatory patient receiving 12-hr therapy at home during sleep, it is best to have a pump regulator than to depend on a gravity drip.

19.2.1. Recommended Solutions

The basic formula should be prepared fresh in the pharmacy daily. At the Massachusetts General Hospital they maintain 16 basic solutions. The variable factors are the protein, which is either of mixed amino acids or a protein hydrolysate, the sodium content, potassium content, and the insulin content. Solutions free of some substances or additives are maintained. In hospitals where there is not a great demand for TPN, the solutions can be prepared daily and factors added or subtracted, depending on the patient's condition. The basic components are:

Protein equivalent	Phosphate
Dextrose	Chloride
Potassium	Acetate
Sodium	Insulin
Magnesium	Vitamins
Calcium	

At the onset of ordering TPN, the basic formula to be used for an individual patient should be discussed with the pharmacist and then modifications made on a daily basis in accordance with the patient's response. Recommended formulas are published in the text by Fischer.[21] A basic formula for TPN for adults is listed in Table 19-1.

Table 19-1. Basic Crystalline Amino Acid Solutions (8.5%)[a]

L-Amino acids	FreAmine II	FreAmine III	Travasol
Essential (g/liter)	8.7[c]	8.7	4.92[d]
L-Lysine	1.3	1.3	1.52
L-Tryptophan	4.8	4.8	5.26
L-Phenylalanine	4.5	0[b]	4.92[e]
DL-Methionine	0	4.5[b]	0
L-Methionine	7.7	7.7	5.26
L-Leucine	5.9	5.9	4.06
L-Isoleucine	5.6	5.6	3.9
L-Valine	3.4	3.4	3.56
L-Threonine			
Nonessential (g/liter)			
L-Arginine	3.1	8.1[b]	8.8
L-Histidine	2.4	2.4	3.72
L-Glycine	17.0	11.9[b]	17.6
L-Cysteine • HCl•H$_2$O	<0.2	<0.2	—
L-Proline	9.5	9.5	3.56
L-Alanine	6.0	6.0	17.6
L-Serine	5.0	5.0	—
L-Glutamate	—	—	—
L-Tyrosine	—	—	0.34
L-Aspartate	—	—	—
Electrolytes (meq/liter)			
Sodium	10		70
Chloride			70
Acetate			135
Phosphate	20		60
Magnesium	—		10
Potassium	—		60

[a] Available in U.S.A. For most recent details, consult hospital pharmacist and Ref. 21.
[b] FreAmine III changes from II.
[c] As acetate (free base 6.2).
[d] Added as the hydrochloride salt.
[e] L form.

A major controversy concerning these solutions has been the relative merit of protein hydrolysates versus crystalline amino acids. Hydrolysates contain peptides plus varying degrees of free amino acids. Their use has been questioned by the U.S. Food and Drug Administration. The mixed amino acids appear to supply more utilizable nitrogen, but they are suspect of causing a higher incidence of metabolic hyperchloremic acidosis. Some report this is more common with the synthetic amino acids, whereas others report it occurring with both types of protein preparations.[27-30] Heird and colleagues[27] feel the explanation is metabolism of cationic amino acids, which results in a net excess of hydrogen ions causing the acidosis, whereas Chan and colleagues[28,29] feel that the titratable acidity of the amino acid mixtures is at fault in conjunction with a simple net increase in hydrogen ion secretion due to either type of the protein solutions. Felig[31] points out that the exact cause of this phenomenon is still uncertain and that it is still not clear whether anionic, cationic, or a mixture of amino acids are better. Of further interest is the work done by Olney and colleagues[32] that shows that casein and fibrin hydrolysates used in human parenteral alimentation can produce acute degeneration of neurons in the developing hypothalamus. They caution against the use of these hydrolysates and recommend a low combined concentration of acidic amino acids as a criteria for safe parenteral alimentation.[32] Anderson and co-workers[33] have attempted to design their amino acid mixture on the basis of blood aminograms in patients with gastrointestinal disease. This may become the most effective method in the future, but at present it is not readily adaptable in most institutions. Until there is more long-range experience, it is evident that mixed amino acid solutions supply the most available nitrogen and hence the best nitrogen balance and appear to have the least serious side effects.[34]

Administration of TPN requires careful observation. At the onset, a team approach is most often successful: the physician, pharmacist, and nurse or physician extender collaborate to give the necessary advice and repeated daily observation, the physician checking the need for changes in the protocol, the pharmacist carefully preparing all solutions, and the nurse caring for the administration and site. Best short- and long-term results have been obtained by a well-trained team approach.[21,35] In St. Mark's Hospital, London, a team approach results in no catheter changes for a mean of 6 week's administration in 80% of TPN patients.[35]

When the patient is cooperative and ambulatory and TPN still needed, home delivery can be safely accomplished.[25,26,35,36] The home systems may be conventional or actually permit enough freedom to work. The system devised by Dudrick et al.[37] uses a vest on which bag containers are held and a pump (Cormed) to infuse the solution through a Broviac catheter. It has been successful in 25 patients. There were only 9 catheter complications, 2 patients who died of their cancers, 8 who recovered from their illnesses, and 15 still on long-term alimentation. Hospital or home TPN is successful so long as there is expert and concerned care.

In monitoring the care of a TPN patient it is of help to use standardized flow sheets and checklists. Figure 19-1 is a standard solution order form, and Table 19-2 is a daily and weekly program checklist for complications.

Directions:
1. Routine TPN orders should be written daily to cover a 24-hour period.
2. Maintain a continuous sequence of bottle numbers for each course of therapy.
3. Use the special instruction section of a new order form to discontinue bottles. Do not re-use a discontinued bottle number.

Ingredients	Standard amounts	Maintain bottle number sequence			
Base solution		Bottle #	Bottle #	Bottle #	Bottle #
Amino acid solution (FreAmine II) 8.5%	500 ml	ml	ml	ml	ml
Dextrose injection 50%	500 ml	ml	ml	ml	ml

Multiple vitamin concentrate will be added to 1 liter per day.

Standard electrolyte additive*
　*Check box(es) above for each bottle required
The final concentration of electrolytes per liter of standard base solution and standard electrolyte additive is:

　　Sodium 40 meq　　　Magnesium 5 meq　　　Chloride 35 meq　　　Acetate 29.5 meq
　　Potassium 20 meq　　Calcium 4.5 meq　　　Phosphate 10 meq

To Add electrolytes to the standard electrolyte additive, check the box(es) above and indicate the additions below *for each bottle.*

To reduce or omit electrolytes from the standard electrolyte additive, *do not check* the box(es) above and order *all* the required electrolytes *for each bottle.*

Calcium gluconate		meq	meq	meq	meq
Insulin		U	U	U	U
Magnesium sulfate		meq	meq	meq	meq
Potassium acetate		meq	meq	meq	meq
Potassium chloride		meq	meq	meq	meq
Potassium phosphate		meq	meq	meq	meq
Sodium acetate		meq	meq	meq	meq
Sodium chloride		meq	meq	meq	meq
Sodium phosphate		meq	meq	meq	meq
Infusion rate					
Special instructions:					

Ordered by:　　　　　　　　　　　M.D.　Time:　　　　　Date:

Figure 19-1.　　Standard TPN order form used at Norwalk Hospital.

Table 19-2. Clinical Observation Program for TPN

A. Information obtained at onset

1. Height	6. Serum Magnesium
2. Weight	7. CBC + diff.
3. SMA—Electrolytes	8. PT + PTT
4. SMA—Chemistries	9. TIBC
5. Phosphorus [if not in (4)]	10. 24-hr urine for urea and creatinine
	11. Anthropometric measurements (as desired)

B. Daily clinical checklist
   1. Observe patient's appearance, color, mood, orientation, and physical findings.
   2. Check TPN flow rate.
   3. Examine label on TPN bottle.
   4. Check weight.
   5. Check bedside chart for record of vital signs.
   6. Check record of fluid intake and output.
   7. Check results of fractional urine tests for glucose and acetone.
   8. Examine dressing covering site of infusion.
   9. Make appropriate notes in chart.

C. Blood chemical tests monitored daily at onset until stable and then twice weekly

1. Glucose	4. K
2. BUN	5. Cl
3. Na	6. CO_2

D. Tests monitored weekly

1. CBC and diff.	9. Creatinine
2. PT + PTT	10. Calcium
3. Albumin/globulin–total protein	11. Phosphorus
4. Bilirubin	12. Magnesium
5. Alkaline phosphatase	13. Uric acid
6. SGOT	14. Cholesterol
7. Creatinine phosphokinase	15. 24-hr urine urea and creatinine, TIBC, and
8. Lactic dehydrogenase	anthropometrics as desired

19.2.2. Complications

The complications as a result of TPN are as follows:

1. Due to catheter insertion: Incidence varies with experience and the institution.
2. Sepsis: Overall incidence in the literature is 7%.
3. Metabolic:
   a. Glucose intolerance.
   b. Hyperosmolar, hyperglycemic, nonketonic coma.
   c. Hyperchloremic metabolic acidosis.
   d. Selective mineral deficiencies.

Complications of catheter insertion are managed at the time of insertion. The final position of the catheter tip is checked by X-ray. Pneumothorax is feared but a rarity in the expert's hands. It is treated conventionally should it occur.[21]

Infection and sepsis are the most common problems. When a TPN patient begins to run a fever, all possible sources of infection, urine, lungs, etc., are checked. Should blood cultures be negative and fever persist for 48 to 72 hr with no obvious cause, then the catheter must be removed and the tip cultured.

Of interest is the fact that Candidemia occurs with an inordinately high frequency in TPN septic patients and that phagocyte dysfunction has been observed in patients with hypophosphatemia while on TPN.[40] The constant use of infection control is important.[38,39] However, the treatment of the sepsis usually responds rapidly to appropriate antimicrobial therapy and appropriate removal of the infecting catheter.

Glucose intolerance may be mild. A low renal threshold should be checked against the fasting blood sugar. If glycosuria of 3^+ to 4^+ persists and the blood level exceeds 300 mg/100 ml, then insulin is probably needed. In Fischer's[21] experience, 15% of patients have at least one glucose level greater than 400 mg/100 ml, but he had no cases of hyperosmolar nonketonic coma. When the sugar rises and is associated with the classic findings of hyperosmolar coma, the TPN solutions must be stopped and hypoosmolar fluid given.[21] Ordinarily, TPN should not be stopped abruptly, but over 48 to 72 hr so the body can readjust its glucose tolerance.

Hyperchloremic metabolic acidosis is reported and discussed under amino acid infusion.[27-31] The acidosis is usually prevented by Na and K acetate additives, but it can also be treated by those salts.[21]

Hypophosphatemia may occur, due either to inadequate additive or excessive carbohydrate metabolism. If the serum phosphorus levels fall, then the hyperosmolar solution can be decreased and phosphate added. When serum levels rise, TPN can be resumed.

Rare complications such as supraventricular arrhythmias associated with parenteral hyperalimentation[41] and selective copper deficiency occurring on hyperalimentation are reported intermittently.[42] Although copper requirements are often met, and less understood, the need for zinc in normal metabolism is more apparent. Zinc deficiency and its metabolism in TPN gastrointestinal disease patients has been studied at the University of Toronto.[43] Increased urinary zinc excretion was associated with a negative nitrogen loss, and a positive zinc balance was associated with improved nitrogen retention. It is postulated that zinc deficiency may impair insulin response and the utilization of glucose and amino acids.[43] Obviously, all essential minerals must be replaced when long-term TPN is used, including iron, zinc, etc.

Vitamin additives do not contain K. In long-term TPN use the prothrombin time should be checked and aqueous vitamin K given. Some vitamin additives are incomplete, and others may be in high doses.[44] Care should be given to balanced vitamin administration when long-term TPN is used.

Essential fatty acid deficiency can occur when only carbohydrate–protein solutions are given long term. Present solutions do not include essential fatty acids, especially linoleic acid. Riella and associates[45] carefully documented three patients who developed classic skin lesions, had accumulation of 5,8,11-eicosatrienoic acid, and high plasma triene to tertaene ratio (the most sensitive index of essential fatty acid deficiency). The lesions and abnormal fatty acid ratios were corrected when parenteral fat (Intralipid) was administered. In TPN of long-standing duration, if both exogenous linoleic acid and that from adipose tissue is cut off as a peripheral circulating supply, the symptoms of fatty acid deficiency will occur. Essential fatty acid deficiency can develop in most patients on TPN. In a recent study on volunteer subjects, biochemical studies revealed deficiency could begin as early as 10 days after therapy was begun.[46,47] Certainly, it behooves the clinician who is going to keep a patient on long-term TPN that *some form of essential fatty acid replacement is essential.* Intralipid is needed once or twice a week.

Little is known about the physiologic effects of TPN on the gastrointestinal tract. Intravenous amino acid administration does increase gastric HCl acid secretion,[48] but TPN decreases hepatobiliary, pancreatic, and intestinal secretions, which are reversible even after long-term use.[49] The decrease in gut secretion may be a reflection of decreased secretory cell turnover in a resting gut.[1] A retrospective review of patients receiving carbohydrate–protein for at least 2 weeks reveals mild elevations of SGOT levels in 68% of patients, alkaline phosphatase 54%, and bilirubin 21%.[50] Enough of the patients were not studied in detail for any clinical conclusions, but of the four with liver biopsies, three had steatosis and one cholestosis. These abnormalities may be related to starvation, the primary disease, or TPN; therefore, no conclusions can be drawn.

Also of interest is the observation that patients on TPN appear to gain more weight, but no more body or plasma protein or fat than those on comparable elemental diets.[51] This raises the question of whether TPN patients are really water overloaded.[52] Further studies on both gut and body metabolism should reveal interesting information that may have clinical significance.

All of the complications of TPN are readily manageable. Those occurring due to insertion of a catheter are most disturbing but are manageable by early detection and treatment, if necessary. Those that result in infection and metabolic disturbance are managed by appropriate removal of any infecting source, antibiotic treatment, and appropriate manipulation of the solutions. *The key to success in TPN is careful monitoring of the patient.*

REFERENCES

1. Sim AJW, Wolfe BM, Young VR, et al: Glucose promotes whole-body protein synthesis from infused amino acids in fasting men. *Lancet* 1:68, 1979.
2. Blackburn GL: Peripheral intravenous feeding with isotonic amino acid solutions. *Am J Surg* 125:447, 1973.
3. Blackburn GL: Protein-sparing therapy during periods of starvation with sepsis or trauma. *Ann Surg* 177:588, 1973.
4. Freeman JB: Evaluation of amino acid infusions as protein-sparing agents in normal adult subjects. *Am J Clin Nutr* 28:477, 1975.
5. Freeman JB: Metabolic effects of amino acid vs. dextrose infusion in surgical patients. *Arch Surg* 110:916, 1975.

6. Greenberg GR, Marliss EB, Anderson GH, et al: Protein-sparing therapy in postoperative patients. N Engl J Med 294:1411, 1976.

7. Felig P: Intravenous nutrition: Fact and fancy. N Engl J Med 294:1455, 1976.

8. Collins JP, Oxby CB, Hill GL: Intravenous amino acids and intravenous hyperalimentation as protein-sparing therapy after major surgery. A controlled clinical trial. Lancet 1:789, 1978.

9. Isaacs JW, Millikan WJ, Stackhouse J, et al: Parenteral nutrition of adults with a 900 milliosmolar solution via peripheral veins. Am J Clin Nutr 30:552, 1977.

10. Deitel M, Kaminsky V: Total nutrition by peripheral vein: The lipid system. Can Med Assoc J 111:152, 1974.

11. Hansen LM, Hardle WR, Hidalgo J: Fat emulsion for intravenous administration: Clinical experience with Intralipid 10%. Ann Surg 184:80, 1976.

12. Yeo MT, Cazzaniga AB, Bartlett RH, et al: Total intravenous nutrition: Experience with fat emulsions and hypertonic glucose. Arch Surg 106:792, 1973.

13. Zohrab WJ, McHattie JD, Jeejeebhoy KN: Total parenteral alimentation with lipid. Gastroenterology 64:582, 1973.

14. Silberman H, Freehauf M, Fong G, et al: Parenteral nutrition with lipids. JAMA 238L;380, 1977.

15. Jeejeebhoy KN, Anderson GH, Nakhooda AF, et al: Metabolic studies in total parenteral nutrition with lipid in man: Comparison with glucose. J Clin Invest 57:125, 1976.

16. Dudrick SJ, Wilmore DW, Vars HM, et al: Long-term total parenteral nutrition with growth, development and positive nitrogen balance. Surgery 64L:34, 1968.

17. Wilmore DW, Dudrick SJ: Safe long-term venous catheterization. Arch Surg 98:256, 1969.

18. Shils ME: Guidelines for total parenteral nutrition. JAMA 220:1721, 1972.

19. Feliciano DV, Telander RL: Total parenteral nutrition in infants and children. Mayo Clin Proc 51:647, 1976.

20. Gryboski J: Gastrointestinal Problems in the Infant. Philadelphia: WB Saunders, 1975.

21. Fischer JE: Total Parenteral Nutrition. Boston: Little, Brown, 1976.

22. Buselmeier TJ, Kjellstrand CM, Sutherland DER, et al: Peripheral blood access for hyperalimentation. Use of expanding polytetrafluorethylene arteriovenous conduit. JAMA 238:2399, 1977.

23. Broviac JW, Cole JJ, Scribner BH: A silicone rubber Atrial catheter for prolonged parenteral alimentation. Surg Gynecol Obstet 136:602, 1973.

24. Broviac JW, Scribner BH: Prolonged parenteral nutrition in the home. Surg Gynecol Obstet 139:24, 1974.

25. Grundfest S, Steiger E: Experience with the Broviac catheter for prolonged parenteral alimentation. J Parenteral Nutr 3:45, 1979.

26. Scribner BH, Cole JJ: Evolution of the technique of home parenteral nutrition. J Parenteral Nutr 3:58, 1979.

27. Heird WC, Dell RB, Driscoll JM, et al: Metabolic acidosis resulting from intervenous alimentation mixtures containing synthetic amino acids. N Engl J Med 287:944, 1972.

28. Chan JCM, Malekzadeh M, Hurley J: pH and titratable acidity of amino acid mixtures used in hyperalimentation. JAMA 220:1119, 1972.

29. Chan JCM, Asch MJ, Lin S, et al: Hyperalimentation with amino acids and casein hydrolysate solutions. JAMA 220:1700, 1972.

30. Fraley DS, Adler S, Bruns F, et al: Metabolic acidosis after hyperalimentation with casein hydrolysate. Occurrence in a starved patient. Ann Intern Med 88:352, 1978.

31. Felig P: Parenteral nutrition: Substrate and secreagogue. N Engl J Med 287:982, 1972.

32. Olney JW, Ho OL, Rhee V: Brain-damaging potential of protein hydrolysates. N Engl J Med 289:391, 1973.

33. Anderson GH, Patel DG, Jeejeebhoy KN: Design and evaluation by nitrogen balance and blood aminograms of an acid mixture for total parenteral nutrition of adults with gastrointestinal disease. J Clin Invest 53:904, 1974.

34. Long CL, Zikria BA, Kinney JM, et al: Comparison of fibrin hydrolysates and crystalline amino acid solutions in parenteral nutrition. Am J Clin Nutr 27:163, 1974.

35. Powell-Tuck J, Nielsen T, Farwell JA, et al: Team approach to long-term intravenous feeding in patients with gastrointestinal disorders. Lancet 2:825, 1978.

36. Strobel CT, Byrne WJ, Ament ME: Home parenteral nutrition in children with Crohn's disease: An effective management alternative. *Gastroenterology* 77:272, 1979.

37. Dudrick SJ, Englert DM, Van Buren CT, *et al:* New concepts of ambulatory home hyperalimentation. *J Parenteral Nutr* 3:72, 1979.

38. Goldmann DA, Maki DG: Infection control in total parenteral nutrition. *JAMA* 223:1360, 1973.

39. Dillon JD, Schalfner W, VanWay CW, *et al:* Septicemia and total parenteral nutrition. *JAMA* 223:1341, 1973.

40. Craddock PR, Yawata Y, VanSanten L, *et al:* Acquired phagocyte dysfunction. *N Engl J Med* 290:1403, 1974.

41. Schulze VE: Supraventricular arrhythmias caused by parenteral hyperalimentation. *JAMA* 228:341, 1974.

42. Vilter RW, Bozian RC, Hess EV, *et al:* Manifestations of copper deficiency in a patient with systemic sclerosis on intravenous hyperalimentation. *N Engl J Med* 291:188, 1974.

43. Wolman SL, Anderson GH, Marliss EB, *et al:* Zinc in total parenteral nutrition: Requirements and metabolic effects. *Gastroenterology* 76:458, 1979.

44. Kishi H, Nishii S, Ono T, *et al:* Thiamin and pyridoxine requirements during intravenous hyperalimentation. *Am J Clin Nutr* 32:332, 1979.

45. Riella MC, Broviac JW, Wells M: Essential fatty acid deficiency in human adults during total parenteral nutrition. *Ann Intern Med* 83:786, 1975.

46. Editorial: Intravenous fat. *Lancet* 1:1059, 1976.

47. Connor WE: Pathogenesis and frequency of essential fatty acid deficiency during total parenteral nutrition. *Ann Intern Med* 83:895, 1975.

48. Law DH: Current concepts in nutrition: Total parenteral nutrition. *N Engl J Med* 297:1104, 1977.

49. Kotler DP, Levine GM: Reversible gastric and pancreatic hypersecretion after long-term total parenteral nutrition. *N Engl J Med* 300:241, 1979.

50. Lindor KD, Fleming CR, Abrams A, *et al:* Liver function values in adults receiving total parenteral nutrition. *JAMA* 241:2398, 1979.

51. Yeung CK, Smith RC, Hill GL: Effect of an elemental diet on body composition. A comparison with intravenous nutrition. *Gastroenterology* 77:652, 1979.

52. Jeejeebhoy KN: Is more better? Is weight water? The significance of weight gain during parenteral nutrition with amino acids and dextrose. *Gastroenterology* 77:799, 1979.

Diets and Food Lists

Diet manuals are available at all hospitals. Common references employed to create these manuals are listed.[1-4] The purpose of this chapter is not to create but to correlate new diet information and to recommend and reproduce diets and food lists reported to be successful and those found useful by the author at Norwalk Hospital and the Yale University School of Medicine in the treatment of gastrointestinal disease. Local groups should create their own diet manuals from this information.

Many of the following diets are modified from the listed references.[1-4] Each is introduced by a general statement and cross-referenced to the text for clinical use. Where the diet is modified from a specific reference, the diet plan is included. Where it is a recommended plan of this author, the format of (1) *daily food exchanges* from (2) *the food exchange lists* (Tables 20-1 through 20-7) is used in conjunction with a reproducible list of foods categorized as "allowed" or "avoid."

It should be noted that most exchange lists are only guidelines and are not accurate for all nutrient compositions.[4] In order to design an individual, accurate food list, the nutrient components should be calculated from the U.S. Agriculture,[3] Church and Church,[1] and Paul and Southgate[2] references.

Table 20-1. Protein Exchange List[a]

One meat exchange is approximately equivalent to 1 oz of meat or 30 g and contains:	7 g protein 5 g fat 73 cal	
	Portion	Grams
Meat (medium fat		
Beef, ham, lamb, pork, or veal	1 oz	30
Sausage, pork (omit 2 fat exchanges)	2 links	40
Beef, dried, chipped (add 1 fat exchange)	2 thin slices	20
Cold cuts:		
Bologna, luncheon meat, minced ham, liverwurst	1 slice	45
Salami (omit 1 fat exchange)	1 slice	30
Frankfurters or wieners (omit 1 fat exchange)	1 (8–9/lb)	50
Fowl		
Chicken, duck, goose, or turkey	1 oz	30
Egg	1	50
Fish		
Salmon or tuna, canned	$^1/_4$ cup	30
Sardines	3 medium	35
Shellfish:		
Clams (add 1 fat exchange)	5 small	50
Lobster (add 1 fat exchange)	1 small tail	40
Oysters (omit $^1/_2$ bread exchange and add 1 fat exchange)	5 small	70
Scallops (add 1 fat exchange)	1 large (12/lb)	50
Shrimp (add 1 fat exchange)	5 small	30
Cheese		
American, brick, cheddar, Roquefort, Swiss, and processed cheeses (omit 1 fat exchange)	1 slice ($3^1/_2 \times 3^1/_2 \times ^1/_8$)	30
Cheese foods, American	1 slice	30
Cheese spreads, American	2 tbsp	30
Cottage cheese, creamed	$^1/_4$ cup	50

[a]Modified from *Mayo Clinic Diet Manual.*[4]

Table 20-2. Fat Exchange List[a]

One fat exchange is equivalent to approximately 1 tsp or 5 g of fat and contains:	4 g fat 36 cal	
	Portion	Grams
Avocado	$^{1}/_{8}$ (4-in. diam)	30
Bacon, crisp	1 strip	5
Butter or margarine[b]	1 tsp	5
Cooking fats[b]	1 tsp	5
Cream		
Half and half	2 tbsp	30
Sour	2 tbsp	30
Whipped	1 tbsp	15
Cream cheese	1 tbsp	15
Mayonnaise[b]	1 tsp	5
Nuts		
Almonds, slivered	5 (2 tsp)	6
Pecans, shelled	4 halves	5
Walnuts, shelled	5 halves	10
Oil, salad[b]	1 tsp	5
Olives, green	3 medium	30

[a]Modified from *Mayo Clinic Diet Manual.*[4]
[b]Polyunsaturated fats are available in these commercial products and natural vegetable fat products.

Table 20-3. Bread, Cereal, and Starchy Vegetable Exchange List[a,b]

One bread exchange is equivalent to approximately 1 slice of bread or 25 g and contains:	2 g protein 1 g fat 13 g carbohydrate 69 cal	
	Portion	Grams
Breads		
Bread	1 slice	25
Biscuit (omit 1 fat exchange)	1 (2-in. diam)	35
Muffin (omit 1 fat exchange)	1 (2-in. diam)	35
Cornbread	1 (1$\frac{1}{2}$-in. cube)	35
Roll	1 (2-in. diam)	25
Bun, hamburger or frankfurter	1-in.	30
Pancake	1 (4-in. diam)	45
Waffle (omit 1 fat exchange)	$\frac{1}{2}$ square	35
Cereals		
Cooked	$\frac{2}{3}$ cup	140
Dry, flaked	$\frac{2}{3}$ cup	20
Dry, puffed	1$\frac{1}{2}$ cups	20
Shredded wheat	1 biscuit	20
Crackers		
Graham	3 (2$\frac{1}{2}$-in. square)	20
Melba toast	4 (3$\frac{3}{4}\times$2-in.)	20
Oyster	20 ($\frac{1}{2}$ cup)	20
Ritz, plain or cheese	6	20
Rye Krisp	3	30
Saltines	6 (2-in. square)	20
Soda	3 (2-in. square)	20
Desserts		
Commercial flavored gelatin	$\frac{1}{2}$ cup	100
Ice cream (omit 2 fat exchanges)	$\frac{1}{2}$ cup	75
Sherbert	$\frac{1}{3}$ cup	50
Sponge or angel food cake	1$\frac{1}{2}\times$1$\frac{1}{2}$	25
Vanilla wafers	5	15
Flour products		
Cornstarch	2 tbsp	15
Flour	2 tbsp	15
Macaroni		
Noodles		
Rice } Cooked	$\frac{1}{4}$ cup	50
Spaghetti		
Tapioca, dry	2 tbsp	15

(continued)

Table 20-3. (Continued)

	Portion	Grams
Vegetables[b]		
Beans		
Baked (no pork) ⎫		
Kidney ⎪		
Lima ⎪		
⎬ Cooked	$^{1}/_{2}$ cup	90
Navy ⎪		
Pinto ⎪		
White marrow ⎭		
Corn		
Canned or frozen	$^{1}/_{3}$ cup	80
Fresh on cob	$^{1}/_{2}$ ear (4 in. long)	50
Hominy	$^{1}/_{2}$ cup	100
Parsnips	$^{2}/_{3}$ cup	100
Peas		
Canned, fresh, or frozen	$^{1}/_{2}$ cup	100
Dry, split, cooked		30
Popcorn (without butter)	1 cup	15
Potatoes		
Potato chips (omit 2 fat exchanges)	10 or 1-oz bag	30
White, baked or boiled	1 (2-in. diam)	100
White, mashed	$^{1}/_{2}$ cup	100
Sweet or yams	$^{1}/_{4}$ cup	50
Soup, canned, undiluted	$^{1}/_{2}$ cup	100

[a]Modified from *Mayo Clinic Diet Manual.*[4]
[b]For additional information on carbohydrate and fiber content, see Table 20-7.

Table 20-4. Milk Exchange List[a]

One milk exchange is approximately equivalent to an 8-oz cup or 240 g and contains:		8 g protein 8 g fat 11 g carbohydrate 156 cal
	Portion	Grams
Whole milk	1 cup	240
Evaporated whole milk	1/2 cup	120
Powdered whole milk, dry	1/4 cup	30
Buttermilk, fat free (add 2 fat exchanges)	1 cup	240
Skim milk, fat free (add 2 fat exchanges)	1 cup	240
Powdered skim milk, dry (add 2 fat exchanges)	1/4 cup	30
Cream, half and half (4 g protein, 14 gm fat, 5.5 g carbohydrate)	1/2 cup	120
Ice cream, chocolate (5 g protein, 33 g fat, 16 mg carbohydrate)	1/2 cup	120

[a]The Mayo Clinic[4] exchange uses 10 g of fat and 12 g of carbohydrate per 240 g, but most milks sold today in groceries are as listed.

Table 20-5. Fruit Exchange List[a,b]

One fruit exchange is approximately equivalent to the amount or weight listed and contains:		12 g carbohydrate 48 cal
	Portion	Grams
Apple		
Fresh	1 (2-in. diam)	80
Sauce	1/2 cup	120
Juice	1/2 cup	120
Apricots		
Canned	1/2 cup	120
Dried	4 halves	20
Fresh	3 small	120
Nectar	1/3 cup	80
Banana (EP)		
Whole	1/2 small	60

(continued)

Table 20-5. (Continued)

	Portion	Grams
Berries		
Fresh		
Blackberries	³/₄ cup	100
Blueberries	¹/₂ cup	80
Boysenberries	1 cup	120
Gooseberries	³/₄ cup	120
Loganberries	³/₄ cup	100
Raspberries	³/₄ cup	100
Strawberries, whole	1 cup	150
Cherries		
Canned	¹/₂ cup	120
Fresh	15 small	80
Dates		
Pitted	2	15
Figs		
Canned	¹/₂ cup	120
Dried	1 small	15
Fresh	1 large	60
Fruit cocktail		
Canned	¹/₂ cup	120
Grapes		
Canned	¹/₃ cup	80
Fresh	15 (¹/₂ cup)	80
Juice		
Bottled	¹/₄ cup	60
Frozen	¹/₃ cup	80
Grapefruit		
Fresh	¹/₂ medium (3¹/₂ in. diam)	120
Juice	¹/₂ cup	120
Sections	³/₄ cup	150
Mandarin orange		
Canned	³/₄ cup	200
Mango		
Fresh	¹/₂ small	70
Melon		
Fresh		
Cantaloupe	¹/₂ small	200
Honeydew	¹/₄ medium	200
Watermelon	¹/₂ slice (10×³/₄ in.)	200

(continued)

Table 20-5. (Continued)

	Portion	Grams
Nectarine		
Fresh	1 medium	80
Orange		
Fresh, whole	1 medium	100
Juice	$1/2$ cup	120
Sections		
Fresh or canned	$1/2$ cup	100
Papaya		
Fresh	$1/2$ cup	120
Peach		
Canned	$1/2$ cup	120
Dried	2 halves	20
Fresh		
Whole	1 medium	120
Sliced	$1/2$ cup	120
Nectar	$1/2$ cup	120
Pear		
Canned	$1/2$ cup	120
Dried	2 halves	20
Fresh, whole	1 small	80
Nectar	$1/3$ cup	80
Pineapple		
Canned	$1/2$ cup	120
Fresh	$1/2$ cup	80
Juice	$1/3$ cup	80
Plums		
Canned	$1/2$ cup	120
Fresh	2 medium	80
Prunes	2 medium	15
Prune juice	$1/4$ cup	60
Raisins	1 Tbs.	15
Tangerine		
Fresh, whole	2 small	100
Juice	$1/2$ cup	120
Sections	$1/2$ cup	100

[a]Modified from *Mayo Clinic Diet Manual.*[4]
[b]For information on carbohydrate and fiber, see Table 20-7. Carbohydrates in fruits are primarily mono- and disaccharides.

Table 20-6. Vegetable Exchange List

A. Uncooked low-calorie vegetables

These exchanges are approximately equivalent to less than 30 cal/100 g.

Asparagus	Kale
Artichokes	Kohlrabi
Bean sprouts	Leeks
Beets	Lettuce
Beet Greens	Mushrooms
Broccoli	Mustard greens
Brussels sprouts	Okra
Cabbage	Onions
Carrots	Peppers
Cauliflower	Pumpkin
Celery	Radish
Chard	Rutabagas
Collards	Sauerkraut
Cress	Spinach
Cucumber	Squash, summer
Dandelion greens	Tomatoes
Eggplant	Turnip greens
	Turnips

B. Caloric vegetables

There is tremendous variation in nutrient and caloric content in cooked and uncooked starchy vegetables. If a variety of vegetables are used, it is common sense to substitute freely for it matters little over long periods of time. For specific short-term diets, the starch list (Table 20-3), and the carbohydrate and fiber list (Table 20-7) should be used.

Table 20-7. Nutrient, Fiber, and Water Content of Foods (g/100 g)[a]

Food	kcal	Water	Carbohydrate			Fiber		Protein	Fat
			Total	Sugar[b]	Starch	Dietary	Crude		
Cereals and breads									
Arrowroot	355	12.2	94.0	Tr	94.0	—		0.4	0.1
Barley (pearl), raw	360	10.6	83.6	Tr	83.6	6.5	0.5	7.9	1.7
Barley, boiled	120	69.6	27.6	Tr	27.6	2.2	—	2.7	0.6
Bemax	347	6.0	44.7	16.0	28.7	—	2.5	26.5	8.1
Bran (wheat)	206	8.3	26.8	3.8	23.0	44.0	9.1	14.1	5.5
Corn flour	354	12.5	92.0	Tr	92.0	—	0.7	0.6	0.7
Custard powder	354	12.5	92.0	Tr	92.0	—		0.6	0.7
Flour (whole meal 100%)	318	14.0	65.8	2.3	63.5	9.6	2.3	13.2	2.0
Flour, brown (85%)	327	14.0	68.8	1.9	66.9	7.5		12.8	2.0
Flour, white (72%)	337	14.5	74.8	1.5	73.3	3.0	1.9	11.3	1.2
Flour, household, plain	350	13.0	80.1	1.7	78.4	3.4	0.2	9.8	1.2
Flour, self-rising	339	13.0	77.5	1.4	76.1	3.7	0.4	9.3	1.2
Patent (40%)	347	14.1	78.0	1.4	76.6	—	0.3	10.8	1.3
Macaroni, raw	370	10.4	79.2	Tr	79.2	—	0.3	13.7	2.0
Macaroni, boiled	117	71.5	25.2	Tr	25.2	—	0.1	4.3	0.6
Oatmeal, raw	401	8.9	72.8	Tr	7.28	7.0	1.2	12.4	8.7
Porridge	44	89.1	8.2	Tr	8.2	0.8		1.4	0.9
Rice, polished, raw	361	11.7	86.8	Tr	86.8	2.4	0.3	6.5	1.0
Rice, boiled	123	69.9	29.6	Tr	29.6	0.8	0.1	2.2	0.3
Rye flour (100%)	335	15.0	75.9	Tr	15.9	—	2.0	8.2	2.0
Sago, raw	355	12.6	94.0	Tr	94.0	—		0.2	0.2
Semolina, raw	350	14.0	77.5	Tr	77.5	—		10.7	1.8
Soya flour, full fat	447	7.0	23.5	11.2	12.3	11.9	2.4	36.8	23.5
Soya flour, low fat	352	7.0	28.2	13.4	14.8	14.3	2.5	45.3	7.2
Spaghetti, raw	378	10.5	84.0	2.7	81.3	—		13.6	1.0
Spaghetti, boiled	117	71.7	26.0	0.8	25.2	—	0.3	4.2	0.3
Spaghetti, canned, in tomato sauce	59	83.1	12.2	3.4	8.8	—	0.1	1.7	.07
Tapioca, raw	359	12.2	95.0	Tr	95.0	—	0.1	0.4	0.1
Bread									
Whole meal	216	40.0	41.8	2.1	39.7	8.5	1.6	8.8	2.7
Brown	223	39.0	44.7	1.8	42.9	5.1		8.9	2.2
Hovis	228	40.0	45.1	2.4	42.7	4.6		9.7	2.2
White	233	39.0	49.7	1.8	47.9	2.7	0.2	7.8	1.7
White, fried	558	4.0	51.3	1.7	49.6	(2.2)[c]	0.2	7.6	37.2[a]
Toasted	297	24.0	64.9	2.1	62.8	(2.8)	0.3	9.6	1.7
Dried crumbs	354	4.7	77.5	2.6	74.9	(3.4)		11.6	1.9
Currant	250	37.7	51.8	13.0	38.8	(1.7)		6.4	3.4
Malt	248	39.0	49.4	18.6	30.8	—		8.3	3.3
Soda	264	34.2	56.3	3.0	53.3	2.3	—	8.0	2.3
Rolls, brown, crusty	289	28.6	57.2	2.1	55.1	(5.9)	1.6	11.5	3.2
Rolls, brown, soft	282	31.0	47.9	1.9	46.0	(5.4)		11.7	6.4
Rolls, white, crusty	290	28.8	57.2	2.1	55.1	(3.1)	0.2	11.6	3.2
Rolls, white, soft	305	28.8	53.6	1.9	51.7	(2.9)	0.2	9.8	7.3
Rolls, starch reduced	384	8.5	45.7	1.6	44.1	(2.0)		44.0	4.1
Chapatis with fat	336	28.5	50.2	1.8	46.5	3.7		8.1	12.8
Chapatis without fat	202	45.8	43.7	1.6	42.1	(3.4)		7.3	1.0

(continued)

Table 20-7 (Continued)

Food	kcal	Water	Carbohydrate			Fiber		Protein	Fat
			Total	Sugar[b]	Starch	Dietary	Crude		
Breakfast cereals									
All-Bran	273	2.3	43.0	15.4	27.6	26.7	8.2	15.1	5.7
Corn Flakes	368	3.0	85.1	7.4	77.7	11.0	0.7	8.6	1.6
Grape Nuts	355	3.9	75.9	9.5	66.4	7.0		10.8	3.0
Muesli	368	5.8	66.2	26.2	40.0	7.4		12.9	7.5
Puffed Wheat	325	2.5	68.5	1.5	67.0	15.4	2.0	14.2	1.3
Ready Brek	390	6.3	69.9	2.2	67.7	7.6		12.4	8.7
Rice Krispies	372	3.8	88.1	9.0	79.1	4.5		5.9	2.0
Shredded wheat	324	7.6	67.9	0.4	67.5	12.3	2.3	10.6	3.0
Special K	388	2.7	78.2	9.6	68.6	5.5		18.0[a]	
Sugar Puffs	348	1.8	84.5	56.5	28.0	6.1		5.9	0.8
Weeta Bix	340	3.8	70.3	6.1	66.5	12.7		11.4	3.4
Biscuits									
Chocolate, full coated	524	2.2	67.4	43.4	24.0	3.1	0.8	5.7	27.6
Cream crackers	440	4.3	68.3	Tr	68.3	(3.0)		9.5	16.3
Crisp bread, rye	321	6.4	70.6	3.2	67.4	11.7	2.2	9.4	2.1
Crisp wheat, starch reduced	388	4.9	36.9	7.4	29.5	4.9	45.3	7.6	
Digestive, plain	471	4.5	66.0	16.4	49.6	(5.5)		9.8	20.5
Digestive, chocolate	493	2.5	66.5	28.5	38.0	3.5		6.8	24.1
Ginger nuts	456	3.4	79.1	35.8	43.3	2.0	0.1	5.6	15.2
Homemade	469	8.4	65.5	26.8	38.7	1.7	0.1	6.4	22.0
Matzo	384	6.7	86.6	4.2	82.4	3.9		10.5	1.9
Oatcakes	441	5.5	63.0	3.1	59.9	4.0		10.0	18.3
Sandwich	513	2.6	69.2	30.2	39.0	1.2	0.1	5.0	25.9
Semisweet	457	2.5	74.8	22.3	52.5	2.3		6.7	16.6
Short-sweet	469	2.6	62.2	24.1	38.1	1.7		6.2	23.4
Shortbread	504	5.0	65.5	17.2	48.3	2.1	0.2	6.2	26.0
Wafers, filled	535	2.3	66.0	44.7	21.3	1.6	0.1	4.7	29.9
Water biscuits	440	4.5	75.8	2.3	73.5	(3.2)		10.8	12.5
Fruits									
Apples, just flesh	46	84.3	11.9	11.8	0.1	2.0[a]	0.6	0.3	Tr
Apples, flesh, skin, core	35	64.9	9.2	9.1	0.1	1.5	1.0	0.2	Tr
Apples, cooking, raw	37	85.6	9.6	9.2	0.4	2.4		0.3	Tr
Apples, stewed, no sugar	32	87.7	8.2	7.9	0.3	2.1		0.3	Tr
Apples, stewed, with sugar	66	79.3	17.3	17.0	0.3	1.9		0.3	Tr
Apricots, fresh, raw	28	86.6	6.7	6.7	0	2.1	0.6	0.6	Tr
Apricots, stewed, no sugar	23	87.9	5.7	5.6	0	1.7		0.4	Tr
Apricots, stewed, with sugar	60	79.4	15.6	15.6	0	1.6		0.4	Tr
Apricots, dried, raw	182	14.7	43.4	43.4	0	24.0	3.0	4.8	Tr
Apricots, dried, stewed, without sugar	66	68.4	16.1	16.1	0	8.9		1.8	Tr
Apricots, dried, stewed, with sugar	81	76.4	19.9	19.9	0	8.5		1.7	Tr
Apricots, canned	106	67.8	27.7	27.7	0	1.3	0.6	0.5	Tr
Avocados	223	68.7[a]	1.8	1.8	Tr	2.0	1.6	4.2	22.2[a]
Bananas, raw	79	70.7	19.2	16.2	3.0	3.4	0.5	1.1	0.3
Blackberries, raw	29	82.0	6.4	6.4	0	7.3	4.1	1.3	Tr

(continued)

Table 20-7 (Continued)

Food	kcal	Water	Carbohydrate			Fiber		Protein	Fat
			Total	Sugar[b]	Starch	Dietary	Crude		
Fruits (Continued)									
Blackberries, stewed, no sugar	25	84.6	5.5	5.5	0	6.3		1.1	Tr
Blackberries, stewed, with sugar	60	76.5	14.8	14.8	0	5.7		1.0	Tr
Cherries, eating, raw	47	81.5	11.9	11.9	0	1.7	0.2	0.6	Tr
Cherries, cooking, raw	46	79.8	11.6	11.6	0	1.7		0.6	Tr
Cherries, stewed, no sugar	39	83.4	9.8	9.7	0	1.4		0.5	Tr
Cherries, stewed, with sugar	77	72.8	20.1	19.7	0	1.2		0.4	Tr
Cranberries, raw	15	87.0	3.5	3.5	0	4.2	1.4	0.4	Tr
Currants, black, raw	28	77.4	6.6	6.6	0	8.7	2.4	0.9	Tr
Currants, stewed, no sugar	24	80.7	5.6	5.6	0	7.4		0.8	Tr
Currants, stewed, with sugar	59	72.4	15.0	15.0	0	6.8		0.8	Tr
Currants, red, raw	21	82.8	4.4	4.4	0	8.2	3.4	1.1	Tr
Currants, stewed, no sugar	18	85.3	3.8	3.8	0	7.0		0.9	Tr
Currants, stewed, with sugar	53	88.2	13.3	13.3	0	6.4		0.9	Tr
Currants, white, raw	26	83.3	5.6	5.6	0	6.8	3.4	1.3	Tr
Currants, stewed, no sugar	22	85.7	4.8	4.8	0	5.8		1.1	Tr
Currants, stewed, with sugar	57	77.5	14.2	14.2	0	5.3		1.0	Tr
Currants, dried	243	22.0	63.1	63.1	0	6.5		1.7	Tr
Dates, dried	248	14.6	63.9	63.9	0	8.7	2.3	2.0	Tr
Dates, dried, with pits	213	12.6	54.9	54.9	0	7.5		1.7	Tr
Figs, green, raw	41	84.6	9.5	9.5	0	2.5	1.2	1.3	Tr
Figs, dried, raw	213	16.8	52.9	52.9	0	18.5	5.6	3.6	Tr
Figs, stewed, no sugar	118	53.8	29.4	29.4	0	10.3		2.0	Tr
Figs, stewed, with sugar	136	50.7	34.3	34.3	0	9.7		1.9	Tr
Fruit pie filling, canned	95	72.6	25.1	23.2	1.9	(1.8)		0.3	Tr
Fruit salad, canned	95	71.1	25.0	25.0	0	1.1	0.4	0.3	Tr
Gooseberries, green, raw	17	89.9	3.4	3.4	0	3.2	1.9	1.1	Tr
Gooseberries, stewed, no sugar	14	91.4	2.9	2.9	0	2.7		0.9	Tr
Gooseberries, stewed, with sugar	50	82.7	12.5	12.5	0	2.5		0.9	Tr
Gooseberries, ripe, raw	37	83.7	9.2	9.2	0	3.5		0.6	Tr
Grapes, black, raw	61	80.7	15.5	15.5	0	0.4	0.6	0.6	Tr
Grapes, white, raw	63	79.3	16.1	16.1	0	0.9	0.5	0.6	Tr
Grapefruit, raw	22	90.7	5.3	5.3	0	0.6	0.2	0.6	Tr
Grapefruit, canned	60	81.8	15.5	15.5	0	0.4	0.2	0.5	Tr
Green gages	47	78.2	11.8	11.8	0	2.6	0.4	0.8	Tr
Green gages, stewed, no sugar	40	81.4	10.0	10.0	0	2.2	0.2	0.6	Tr
Green gages, stewed, with sugar	75	72.8	19.4	19.2	0	2.1		0.6	Tr
Guavas, canned	60	77.6	15.7	15.7	Tr	3.6		0.4	Tr
Lemons, whole	15	85.2	3.2	3.2	0	5.2	0.4	0.8	Tr
Lemon juice, fresh	7	91.3	1.6	1.6	0	0	Tr	0.3	Tr
Loganberries, raw	17	85.0	3.4	3.4	0	6.2	3.0	1.1	Tr
Loganberries, stewed, no sugar	16	86.1	3.1	3.1	0	5.7		1.0	Tr
Loganberries, stewed, with sugar	54	77.3	13.4	13.4	0	5.2		0.9	Tr
Loganberries, canned	101	66.3	26.2	26.2	0	3.3	2.0	0.6	Tr
Lychees, raw	64	82.0	16.0	16.0	0	(0.5)	0.3	0.9	Tr
Lychees, canned	68	79.3	17.7	17.7	0	0.4		0.4	Tr
Mandarin oranges, canned	56	84.3	14.2	14.2	0	0.3	(0.1)	0.6	Tr
Mangoes, raw	59	83.0	15.3	15.3	Tr	(1.5)	0.9	0.5	Tr
Mangoes, canned	77	74.8	20.3	20.2	0.1	1.0		0.3	Tr

(continued)

Table 20-7 (Continued)

Food	kcal	Water	Carbohydrate			Fiber		Protein	Fat
			Total	Sugar[b]	Starch	Dietary	Crude		
Fruits *(Continued)*									
Melons:									
Cantaloupe, raw	24	93.6	5.3	5.3	0	1.0	0.3	1.0	Tr
Yellow honeydew, raw	21	94.2	5.0	5.0	0	0.9	0.6	0.6	Tr
Watermelon, raw	21	94.0	5.3	5.3	0	—	0.3	0.4	Tr
Mulberries, raw	36	85.0	8.1	8.1	0	1.7		1.3	Tr
Nectarines, raw	50	80.2	12.4	12.4	0	2.4	0.4	0.9	Tr
Olives, in brine	103	76.5	Tr	Tr	0	4.4	1.3	0.9	11.0
Oranges, raw	35	86.1	8.5	8.5	0	2.0	0.5	0.8	Tr
Orange juice, fresh	38	87.7	9.4	9.4	0	0	0.1	0.6	Tr
Passion fruit, raw	37	73.3	6.2	6.2	0	15.9	—	2.8	Tr
Pawpaw, canned	65	80.4	17.0	17.0	0	0.5		0.2	Tr
Peaches, fresh, raw	37	86.2	9.1	9.1	0	1.4	0.6	0.6	Tr
Peaches, dried, raw	212	15.5	53.0	53.0	0	14.3	3.1	3.4	Tr
Peaches, stewed, no sugar	79	68.7	19.6	19.6	0	5.3		1.3	Tr
Peaches, stewed, with sugar	93	66.0	23.3	23.3	0	5.1		1.3	Tr
Peaches, canned	87	74.3	22.9	22.9	0	1.0	0.4	0.4	Tr
Pears, eating	41	83.2	10.6	10.6	0	2.3	1.4	0.3	Tr
Pears, cooking, raw	36	83.0	9.3	9.3	Tr	2.9		0.3	Tr
Pears, stewed, no sugar	30	85.5	7.9	7.9	Tr	2.5		0.2	Tr
Pears, stewed, with sugar	65	77.3	17.1	17.1	Tr	2.3		0.2	Tr
Pears, canned	77	76.2	20.0	20.0	0	1.7	0.8	0.4	Tr
Pineapple, fresh	46	84.3	11.6	11.6	0	1.2	0.4	0.5	Tr
Pineapple, canned	77	77.1	20.2	20.2	0	0.9	0.3	0.3	Tr
Plums, Victoria dessert, raw	38	84.1	9.6	9.6	0	2.1	0.4	0.6	Tr
Plums, cooking, raw	26	85.1	6.2	6.2	0	2.5		0.6	Tr
Plums, stewed, no sugar	22	86.3	5.2	5.2	0	2.2		0.5	Tr
Plums, stewed, with sugar	59	77.7	15.3	15.1	0	1.9		0.4	Tr
Pomagranate juice	44	85.4	11.6	11.6	0	0		0.2	Tr
Prunes, dried, raw	161	23.3	40.3	40.3	0	16.1	1.6	2.4	Tr
Prunes, stewed, no sugar	82	60.5	20.4	20.4	0	8.1	0.8	1.3	Tr
Prunes, stewed, with sugar	104	57.1	26.5	26.5	0	7.7	0.6	1.2	Tr
Raisins, dried	246	21.5	64.4	64.4	0	6.8	0.9	1.1	Tr
Raspberries, raw	25	83.2	5.6	5.6	0	7.4	3.0	0.9	Tr
Raspberries, stewed, no sugar	26	82.2	5.9	5.9	0	7.8		0.9	Tr
Raspberries, stewed, with sugar	68	72.6	17.3	17.3	0	7.0		0.8	Tr
Raspberries, canned	87	74.0	22.5	22.5	0	(5.0)	2.6	0.6	Tr
Rhubarb, raw	6	94.2	1.0	1.0	0	2.6	0.7	0.6	Tr
Rhubarb, stewed, no sugar	6	94.6	0.9	0.9	0	2.4		0.6	Tr
Rhubarb, stewed, with sugar	45	85.0	11.4	11.4	0	2.2	0.6	0.5	Tr
Strawberries, raw	26	88.9	6.2	6.2	0	2.2	1.3	0.6	Tr
Strawberries, canned	81	79.4	21.1	21.1	0	1.0	0.6	0.4	Tr
Sultanas, dried	250	18.3	64.7	64.7	0	7.0	0.9	1.8	Tr
Tangerines, raw	34	86.7	8.0	8.0	0	1.9	0.5	0.9	Tr
Nuts									
Almonds	565	4.7	4.3	4.3	0	14.3	2.6	16.9	53.5
Barcelona nuts	639	5.7	5.2	3.4	1.8	10.3		10.9	64.0
Brazil nuts	619	8.5	4.1	1.7	2.4	9.0	3.1	12.0	61.5

(continued)

Table 20-7 (Continued)

Food	kcal	Water	Carbohydrate			Fiber		Protein	Fat
			Total	Sugar[b]	Starch	Dietary	Crude		
Nuts (Continued)									
Chestnuts	170	51.7	36.6	7.0	29.6	6.8	1.1	2.0	2.7
Cob or hazelnuts	380	41.1	6.8	4.7	2.1	6.1	3.0	7.6	36.0
Coconut, fresh	351	42.0	3.7	3.7	0	13.6	4.0	3.2	36.0
Coconut, milk	21	92.2	4.9	4.9	0	(Tr)	—	0.3	(0.2)
Coconut, desiccated	604	2.3	6.4	6.4	0	23.5	3.9	5.6	62.0
Peanuts, fresh	570	4.5	8.6	3.1	5.5	8.1	2.4	24.3	49.0
Peanuts, roasted, salted	570	4.5	8.6	3.1	5.5	8.1	2.4	24.3	49.0
Peanut butter, smooth	623	1.1	13.1	6.7	6.4	7.6	1.5	22.6	53.7
Walnuts	525	23.5	5.0	3.2	1.8	5.2	14.8	10.6	51.5
Vegetables									
Artichokes, globe, boiled	15	84.4	2.7[a]	—	0	—	2.4	1.1	Tr
Asparagus, boiled	18	92.4	1.1	1.1	0	1.5	0.7	3.4	Tr
Aubergine, raw	14	93.4	3.1	2.9	0.2	2.5		0.7	Tr
Beans, French, boiled	7	95.5	1.1	0.8	0.3	3.2	1.1	0.8	Tr
Beans, runner, raw	26	89.0	3.9	2.8	1.1	2.9		2.3	0.2
Beans, broad, boiled	48	83.7	7.1	0.6	6.5	4.2		4.1	0.2
Beans, red kidney, raw	272	11.0	45.0	(3.0)	(42.0)	(25.0)	4.2	22.1	1.7
Bean sprouts, canned	9	95.4	0.8	0.4	0.4	3.0	0.7	1.6	Tr
Broccoli, tops, raw	23	89.0	2.5	2.5	Tr	3.6	1.5	3.3	Tr
Broccoli, boiled	18	89.9	1.6	1.5	0.1	4.1	1.5	3.1	Tr
Brussels sprouts, raw	26	88.1	2.7	2.6	0.1	4.2	1.6	4.0	Tr
Brussels sprouts, boiled	18	91.5	1.7	1.6	0.1	2.9	1.6	2.8	Tr
Cabbage, red, raw	20	89.7	3.5	3.5	Tr	3.4	1.0	1.7	Tr
Cabbage, white, raw	22	90.3	3.5	3.7	0.1	2.7	0.8	1.9	Tr
Carrots, old, raw	23	89.9	5.4	5.4	0	2.9	1.0	0.7	Tr
Carrots, boiled	19	91.5	4.3	4.2	0.1	3.1	1.0	0.6	Tr
Carrots, young, boiled	20	91.1	4.5	4.4	0.1	3.0	1.0	0.9	Tr
Carrots, canned	19	91.2	4.4	4.4	Tr	3.7	0.8	0.7	Tr
Cauliflower raw	13	92.7	1.5	1.5	Tr	2.1	1.0	1.9	Tr
Cauliflower, boiled	9	94.5	0.8	0.8	Tr	1.8	1.0	1.6	Tr
Celery, raw	8	93.5	1.3	1.2	0.1	1.8	0.6	0.9	Tr
Celery, boiled	5	95.7	0.7	0.7	0	2.2	0.6	0.6	Tr
Chicory, raw	9	96.2	1.5[a]	—	0	—	0.6	0.8	Tr
Cucumber, raw	10	96.4	1.8	1.8	0	0.4	0.3	0.6	0.1
Endive, raw	11	93.7	1.0	1.0	0	2.2	0.9	1.8	Tr
Horseradish, raw	59	74.7	11.0	7.3	3.7	8.3	2.4	4.5	Tr
Leeks, raw	31	86.0	6.0	6.0	0	3.1	1.3	1.9	Tr
Leeks, boiled	24	90.8	4.6	4.6	0	3.9		1.8	Tr
Lentils, raw	304	12.2	53.2	2.4	50.8	11.7	3.9	23.8	1.0
Lentils, split, boiled	99	72.1	17.0	0.8	16.2	3.7	1.2	7.6	0.5
Lettuce, raw	12	95.9	1.2	1.2	Tr	1.5	0.5	1.0	0.4
Mushrooms, raw	13	91.5	0	0	0	2.5	0.8	1.8[a]	0.6
Mustard and cress, raw	10	92.5	0.9	0.9	0	3.7	1.1	1.6	Tr
Okra, raw	17	90.0	2.3	2.3	Tr	(3.2)	1.0	2.0	Tr
Onions, raw	23	92.8	5.2	5.2	0	1.3	0.6	0.9	Tr
Onions, boiled	13	96.6	2.7	2.7	0	1.3	0.6	0.6	Tr
Parsley, raw	21	78.7	Tr	Tr	0	9.1	1.5	5.2	Tr

(continued)

Table 20-7 (Continued)

Food	kcal	Water	Carbohydrate			Fiber		Protein	Fat
			Total	Sugar[b]	Starch	Dietary	Crude		
Vegetables *(Continued)*									
Parsnips, raw	49	82.5	11.3	8.8	2.5	4.0	2.0	1.7	Tr
Parsnips, boiled	56	83.2	13.5	2.7	10.8	2.5	2.0	1.6	Tr[6]
Peas, fresh, raw	67	78.5	10.6	4.0	6.6	5.2	2.0	5.8	0.4
Peas, fresh, boiled	52	80.0	7.7	1.8	5.9	5.2	2.0	5.0	0.4
Peas, frozen, raw	53	79.1	7.2	4.1	3.4	7.8	1.9	5.7	0.4
Peas, frozen, boiled	41	80.7	4.3	1.0	3.3	12.0	1.9	5.4	0.4
Peas, canned, garden	47	81.6	7.0	3.6	3.4	6.3	2.3	4.6	0.3
Peas, processed	80	71.5	13.7	1.3	12.4	7.9	2.2	6.2	0.4
Peas, dried, raw	286	13.3	50.0	2.4	47.6	16.7	4.9	21.6	1.3
Peas, dried, boiled	103	70.3	19.1	0.9	18.2	4.8	0.4	6.9	0.4
Peas, split, dried, raw	310	12.1	56.6	1.9	54.7	11.9	1.2	22.1	1.0
Peas, split, dried, boiled	118	67.3	21.9	0.9	21.0	5.1	0.4	8.3	0.3
Peas, chick Bengal gram, raw	320	9.9	50.0	(10.0)	(40.0)	(15.0)	5.0	20.2	5.7
Peas, red pidgeon, raw	301	10.0	54.0	(9.0)	(45.0)	(15.0)	7.0	20.0	2.0
Peppers, green, raw	15	93.5	2.2	2.2	Tr	0.9	1.4	0.9	0.4
Peppers, green, boiled	14	93.7	1.8	1.7	0.1	0.9	1.4	0.9	0.4
Plantain, green, raw	112	67.0	28.3	0.8	27.5	(5.8)	0.4	1.0	0.2
Plantain, green, boiled	122	63.9	31.1	0.9	30.2	6.4		1.0	0.1
Potatoes, old, raw	87	75.8	20.8	0.5	20.3	2.1	0.5	2.1	0.1
Potatoes, boiled	80	80.5	19.7	0.4	19.3	1.0	0.5	1.4	0.1
Potatoes, mashed, with margarine and milk	119	76.9	18.0	0.6	17.4	0.9	0.4	1.5	5.0
Potatoes, baked	105	71.0	25.0	0.6	24.4	2.5	0.6	2.6	0.1
Potatoes, new, boiled	76	78.8	18.3	0.7	17.6	2.0	0.5	1.6	0.1
Potatoes, new, canned	53	84.2	12.6	0.4	12.2	2.5	0.2	1.2	0.1
Potatoes, instant powder	318	7.2	73.2	2.2	71.0	16.5	(1.6)	9.1	0.8
Potatoes, instant powder, made up	70	79.4	16.1	0.5	15.6	3.6	0.3	2.0	0.2
Potato crisps	533	2.7	49.3	0.7	48.6	11.9	(1.6)	6.3	35.9
Pumpkin, raw	15	94.7	3.4	2.7	0.7	0.5	1.1	0.6	Tr
Radishes, raw	15	93.3	2.8	2.8	0	1.0	0.8	1.0	Tr
Spinach, boiled	30	85.1	1.4	1.2	0.2	6.3	0.6	5.1	0.5
Spring greens, boiled	10	93.6	0.9	0.9	0	3.8		1.7	Tr
Sweet corn, on-the-cob, raw	127	65.2	23.7	1.7	22.0	3.7	0.7	4.1	2.4
Sweet corn, on-the-cob, boiled	123	65.1	22.8	1.7	21.1	4.7	0.7	4.1	2.3
Sweet corn, canned, kernels	76	73.4	16.1	8.9	7.2	5.7	0.8	2.9	(0.5)
Sweet potatoes, raw	91	70.0	21.5	(9.7)	(11.8)	(2.5)	0.7	1.2	0.6
Sweet potatoes, boiled	85	72.0	20.1	9.1	11.0	2.3	0.7	1.1	0.6
Tomatoes, raw	14	93.4	2.8	2.8	Tr	1.5	0.5	0.9	Tr
Tomatoes, canned	12	94.0	2.0	2.0	Tr	0.9	0.4	1.1	Tr
Turnips, raw	20	93.3	3.8	3.8	0	2.8	0.9	0.8	0.3
Turnips, boiled	14	94.5	2.8	2.3	0	2.2	0.9	0.7	0.3
Turnip tops, boiled	11	92.8	0.1	0	0.1	3.9	0.8	2.7	Tr
Watercress, raw	14	91.1	0.7	0.6	0.1	3.3	0.7	2.9	Tr
Yam, raw	131	73.0	32.4	1.0	31.4	(4.1)	0.9	2.0	0.2
Yam, boiled	119	65.8	29.8	0.2	29.6	3.9		1.6	0.1
Squash, all varieties									
Summer, raw	19	94.0					0.6	1.1	0.1
Summer, boiled	14	95.5					0.6	0.9	0.1

(continued)

Table 20-7 (Continued)

Food	kcal	Water	Carbohydrate		Fiber		Protein	Fat
			Total	Sugar[b] Starch	Dietary	Crude		
Vegetables (Continued)								
Squash (Continued)								
Zucchini, raw	17	94.6			0.6	1.2	0.1
Zucchini, boiled	12	96.0			0.6	1.0	0.1
Winter, all varieties								
Raw	50	85.1			1.4	1.4	0.3
Cooked: Baked	63	81.4			1.8	1.8	0.4
Boiled, mashed	38	88.8			1.4	1.1	0.3

[a] All data are taken from Paul and Southgate[2] except those under Crude Fiber and Squash, which are taken from the U.S. Agriculture Department.[1,3]

[b] Sugar includes all free monosaccharides and disaccharides.

[c] Values in parentheses are taken from the literature.

20.1. LIQUID BLAND DIET

This diet is needed when the mouth, esophagus, or stomach cannot tolerate solid or potentially irritating foods. Ulceration of the mucosa—either peptic, traumatic, or iatrogenic—is the most common reason. The diet is used temporarily and usually is the first step in a progression to a soft bland diet and then a full bland diet. It is recommended in Chapters 7, 8, and 9 for the early treatment of acute esophagitis, gastritis, intractable ulcers, early recovery from surgery, etc.

Obviously, this diet is deficient in many vitamins and minerals. If used long term, it must be supplemented. The hospitalized patient will often be on concurrent or crossover intravenous therapy, but at no time should the diet act as the sole source of energy and nutrient supply for longer than 3 to 7 days, depending on clinical conditions and fat stores.

Approximate Composition[a]	
Calories	1800
Carbohydrate	245 g
Protein	50 g
Fat	65 g
[a]Modified from Ref. 4.	

Suggested Eating Plan	
Breakfast	*Number of servings*
Fruit juice	1
Cereal drink	1
Milk	1
Sugar or candy	2 tsp added
Midmorning	
Milk or fruit juice	1
Lunch	
Soup or vegetable drink	1
Milk	1
Dessert	1
Sugar or candy	2 tsp
Midafternoon	
Milk drink	1
Dinner	
Soup or vegetable drink	1
Dessert	1
Milk	1
Sugar or candy	2 tsp
Evening	
Milk or fruit drink	1

Food Lists		
	Allowed	Avoid
Milk	Whole, or with egg powder or egg added	—
Cereal	Strained or enriched whole-grain cereal	—
Vegetable	Juice or pureed	Tomato
Fruit	Juice or strained	Orange, grapefruit, pineapple, any citrus
Soup	Strained cream soups	All spiced, broth, bouillon
Dessert	Custard, gelatin, nonspiced ice cream, sherbert, fruit ice	Spiced, citrus flavorings
Candy	Sugar cubes, plain sugar candy	Spiced, e.g., caramel, etc.

20.2. SOFT BLAND DIET

This diet is intended as a progression from the liquid bland diet or as a direct prescription if clinically indicated. It is intended primarily for the damaged upper gastrointestinal mucosa, as described in Chapters 7, 8, and 9, for esophagitis, gastritis, or ulcer disease. It should also be used in recovery periods for any illness where eating is mechanically difficult for the patient. When ready, the patient usually progresses to the full bland diet.

Approximate Composition	
Calories	2000
Carbohydrate	230 g
Protein	70 g
Fat	95 g

The following food exchanges are recommended in any combination that is most satisfactory for the patient, based on Tables 20-1 to 20-6 for exchanges, and the allowed food lists.

Suggested Eating Plan				
Exchange	Breakfast	Lunch	Dinner	Snacks
Protein[a] (Table 20-1)	1	2	2	—
Fat (Table 20-2)	3	3	3	—
Cereal (Table 20-3)	1	—	—	—
Milk or cream (Table 20-4)	1	1	1	1
Fruit or soup (Table 20-5)	1	1	1	1
Vegetables[a] (Table 20-6)	—	1	1	—
Dessert	—	1	1	—
Candy	—	1	1	1

[a]All foods are ground, pureed, or mashed by hand or in any standard blender.

Food Lists		
	Allowed	Avoid
Protein foods	Pureed, ground, or blended veal, beef, fish, or fowl; egg; cottage or cream cheese	*All seasoned foods*, meats, cheeses
Fat	Cream cheese, butter, margarine, oils	All others
Milk	All simple products	—
Cereal	Refined or whole, but soft and cooked	All spiced or with seasoning additives
Vegetable	Pureed or mashed	Tomato
Fruit	Pureed or juice	All citrus, orange, grapefruit, pineapple
Soup	Purees or creamed	Broth, bouillon, canned, spiced
Dessert	Jello, custard, ice cream, sherbet, angel food cake	All spiced, highly seasoned, or chocolate
Sweets	Sugar, simple sugar, hard candies	All spiced, e.g., caramels, chocolates, etc.
Miscellaneous	All beverages such as coffee (ground or decaffeinated), carbonated drinks, and those with added spices are not recommended. Juices, milk drinks, and water are recommended. Alcohol is a definite irritant and should be avoided in all foods.	

20.3.   FULL BLAND DIET

This diet is recommended as a progression from a soft bland diet or for all patients with nonspecific gastrointestinal complaints. It is particularly useful for subjects with peptic disease or intolerance to spices and is recommended in Chapters 7, 8, 9, 15, and 16 for inflammatory gastrointestinal diseases without fiber restriction (also see diet 29).

Along with advice on the quality of foods, the patient should be instructed to eat in moderation and not overeat at any one meal or with any one food.

And, of course, alcohol is to be avoided.

Approximate Composition	
Calories	2200
Carbohydrate	220 g
Protein	90 g
Fat	111 g

The following food exchanges are recommended in any combination that is most satisfactory for the patient based on Tables 20-1 to 20-6 and the allowed list.

Suggested Eating Plan[a]				
Food exchange	Breakfast	Lunch	Dinner	Snacks
Protein (Table 20-1)	1	2	2	—
Fat (Table 20-2)	3	1	1	—
Cereal and bread (Table 20-3)	2	1	1	—
Milk (Table 20-4)	—	1	1	—
Fruit (Table 20-5)	1	1	1	—
Vegetable (Table 20-6)	—	3	2	—
Dessert	—	—	1	1
Candy or sugar	—	1	—	1
[a]Modified from Ref. 4.				

Food Lists	
Allowed	Avoid
All others	Alcohol, caffeinated beverages and limited carbonated beverages, all citrus fruits and juices, orange, grapefruit, pineapple, tomato, all highly seasoned foods including pepper, chili powder, seed spices, cloves, garlic, spiced pickles, and gravies

20.4. ALTERNATING BLAND DIET AND ANTACID REGIMEN

This regimen is used when there is severe ulceration of the esophageal, gastric, and/or duodenal mucosa. The causes of such inflammations are delineated in Chapters 7 through 9. The purpose of this diet regimen is to decrease the contact of such irritants as HCl with the mucosal surface and to produce a coating effect for the relief of pain, if that is possible.

Some clinicians also use this type of regimen in intractable ulcer pain, but usually in association with bed rest. Another associated therapy is the use of the H_2 antagonist cimetidine to decrease acid production.

This regimen is intended for short-term acute therapy and frequently requires intravenous supplementation.

Approximate Composition[a]	
Calories	1400
Carbohydrate	105 g
Protein	75 g
Fat	75 g
[a]Based on nine feedings.	

Suggested Eating Plan[a]	
On awakening (approx)	8 AM: 1 cup or glass of milk[b]
One hour later	9 AM: 1 oz of antacid[c]
One hour later	10 AM: milk or substitute
One hour later	11 AM: antacid
Continue cycle hourly *until asleep*	

[a]The regimen can be extended to use soft bland foods as listed in Diet 2.
[b]I prefer the use of milk; some use cream in lesser quantities. Ice cream, skim milk–corn syrup mixtures, etc., may be substituted. Wherever lactose intolerance is a problem, this regimen should be avoided or milk prepared with prior lactose digestion (Lactaid) used.
[c]The choice of the antacid is discussed in Chapters 7 and 8 and is largely guided by patient tolerance, salt content, and whether constipation or diarrhea is a problem. The aluminum antacids help binding and the magnesium ones increase fluid stools.

Some clinicians formerly awakened patients to keep this regimen perpetuated, but with the use of cimetidine there is sufficient acid suppression to permit the great benefit of uninterrupted sleep.

20.5. HIGH-CALCIUM FOODS

Some diseases require diets low in milk, e.g., lactose intolerance, milk allergy, etc. When this occurs, the therapeutic diet is often low in calcium. The RDA for calcium (Chapter 2) is between 0.8 and 1.4 g/day after infancy. When milk is eliminated, this requirement is often not reached. Calcium tablet supplements can be used, but food supplements are always preferred. The following list includes food portions with significant amounts of calcium (greater than 25 mg/100 g).

High-Calcium Food Portions[a]			
Food group	Portion	Weight (g)	Calcium (mg)
Protein-rich			
Egg	2	100	54
Fish			
Pike	3¹/₂ oz	100	140
Salmon (with bones)	3¹/₂ oz	100	154
Sardines	3¹/₂ oz	100	303
Clams	3¹/₂ oz	100	90
Oysters	3¹/₂ oz	100	90
Shrimp	3¹/₂ oz	100	105
Smelt	3¹/₂ oz	100	272
Cheese			
Cheddar	1 oz	30	218
Cheese foods	1 oz	30	160
Cheese spread	1 oz	30	158
Cottage cheese	¹/₄ cup	50	53
Meat (cooked)			
Lamb quarters	3¹/₂ oz	100	258
(other meats are low)			
Bread, starches, flour			
Bread			
Biscuit	2-in. diam	35	42
Muffin	2-in. diam	35	36
Cornbread	1¹/₂-in. cube	35	36
Pancake	4-in. diam	45	45
Waffle	¹/₂ square	35	39
Beans, dry (canned or cooked)	¹/₂ cup	90	45
Lima beans	¹/₂ cup	100	42
Parsnips	²/₃ cup	100	45
Soybean flour	²/₃ cup	100	199
Bran, crude wheat	²/₃ cup	100	119
Milk products			
Whole	1 cup	240	288
Evaporated whole milk	¹/₂ cup	120	302
Powdered whole milk	¹/₂ cup	30	252

(continued)

High-Calcium Food Portions[a] (Continued)

Food group	Portion	Weight (g)	Calcium (mg)
Milk products *(Continued)*			
Buttermilk	1 cup	240	296
Skim milk	1 cup	240	298
Powdered skim milk, dry	1/4 cup	30	367
Cream, half and half	2 tbsp	30	32
Cream, sour	2 tbsp	30	31
Breakfast foods			
Oat cereals with additives	3 1/2 oz	100	150+
Fruit			
Blackberries	3/4 cup	100	32
Orange	1 medium	100	41
Raspberries	3/4 cup	100	30
Rhubarb	1 cup	100	96
Tangerine	2 small	100	40
Vegetables (cooked)			
Artichokes	1/2 cup	100	51
Beans, green, wax, white, red	1/2 cup	100	50
Beet greens	1/2 cup	100	99
Broccoli (raw)	1/2 cup	100	88 (103)
Brussels sprouts	1/2 cup	100	32
Cabbage	1/2 cup	100	49
Cabbage, Chinese	1/2 cup	100	43
Carrots	1/2 cup	100	33
Celery	1/2 cup	100	39
Chickpeas	1/2 cup	100	150
Chard	1/2 cup	100	73
Collards	1/2 cup	100	188
Cress	1/2 cup	100	81
Dandelion greens	1/2 cup	100	140
Kale	1/2 cup	100	187
Kohlrabi	1/2 cup	100	33
Leeks, raw	3–4	100	52
Mustard greens	1/2 cup	100	138
Okra	1/2 cup	100	92
Pumpkin	1/2 cup	100	25
Rutabagas	1/2 cup	100	59
Sauerkraut	1/2 cup	100	36
Spinach	1/2 cup	100	93
Squash, summer	1/2 cup	100	25
Squash, winter	1/2 cup	100	28
Turnip greens	1/2 cup	100	184
Turnips	1/2 cup	100	35
Desserts			
Cake, white	1 piece	50	32
Custard, baked	1/3 cup	100	112
Ice cream	1/2 cup	75	110
Ice milk	1/2 cup	75	118
Pie, cream	1/6 of 9-in. pie	160	120
Pudding	1/2 cup	100	117
Sherbet	1/3 cup	50	25

(continued)

High-Calcium Food Portions[a] (Continued)			
Food group	Portion	Weight (g)	Calcium (mg)
Nuts and seeds			
Almonds, dried	2/3 cup	100	234
Butternuts	2/3 cup	100	1238
Hazelnuts	2/3 cup	100	209
Pistachio nuts	2/3 cup	100	131
Sunflower seeds	3 1/2 oz	100	120
Walnuts, English	3 1/2 oz	100	99

[a]Modified from Ref. 4.

20.6. LOW-CALCIUM DIET

A low-calcium diet is infrequently used. It is recommended only for iatro-genic overdose, such as the milk-alkali syndrome (see Chapters 8 and 9), cal-cium stone formers, and severe hypercalcemia.

The RDA for adults is 800 to 1400 mg. The following diet has an approxi-mate intake of 350 mg of calcium per day and is adequate in all other nutrients except riboflavin. When more vigorous control of calcium intake is desired, all calcium foods can be eliminated, but this type of therapy should probably be done in a hospital setting since it is used only for acute hypercalcemia that is life threatening and usually serious enough to warrant concurrent drug therapy.

Approximate Composition	
Calories	2050
Carbohydrate	220 g
Protein	90 g
Fat	90 g
Calcium	0.350 g

Suggested Daily Eating Plan

8 oz meat or substitute only
3–4 eggs per week only
1 serving refined cereal (no bran)
4 servings bread or substitute
1 serving potato or substitute
2 servings vegetables, 1 should be raw
3 servings fruit, 1 should be citrus
3 tbsp fat

Food Lists		
	Allowed	Avoid[a]
Beverage	Coffee, tea, carbonated beverages	Milk, milk beverages, chocolate-flavored drinks, cereal beverages, e.g., Postum, Sanka, Nescafe
Bread	White and light rye bread or crackers	Whole-grain or soybean bread, biscuits, muffins, crackers, rolls
Cereal	All refined cereals, e.g., Cream of Wheat, farina	Chocolate-flavored cereal, oatmeal, cold cereals, bran, whole-grain cereals, e.g., Maltex, Ralston, Wheatena
Meat, fish, poultry, cheese	Only 8 oz daily of any meat, fish, or poultry except those listed under "Avoid"	Clams, oysters, mackerel, shrimp, sardines, scallops, salmon lobster, cheese
Egg	Not more than 3 to 4 per week, including those used in cooking	More than 3 to 4 per week
Potato or substitute	Potato, hominy, refined rice, macaroni, spaghetti, noodles	Sweet potato, whole-grain rice, more than one serving of potato daily
Vegetable	At least 2 servings, one of which should be raw, of any canned, cooked, or fresh vegetables except those listed under "Avoid"	Beet greens, broccoli, chard, green cabbage, celery, collards, endive, dandelion greens, leaf lettuce, kale, okra, parsley, scallions, parsnips, rutabaga, watercress, spinach, dried beans and peas, lentils
Fruit	At least 3 servings of any canned, cooked, or fresh fruit or juice, except only 4 oz of citrus fruit per day	Dried apricots; currants, dates, figs, prunes, raisins, rhubarb, strawberries, raspberries, honeydew melon
Fat	Fortified margarine, French dressing, salad oil, shortening, 2 tbsp cream daily	Cream, except for amount allowed; mayonnaise
Dessert	Cake, cookies, gelatin desserts, pastries, puddings, sherbert (all made without chocolate, milk, nuts, and dried fruits). If egg yolk is used, it must be from egg allowance	Ice cream and all others
Soup	Broth, vegetable soup made from vegetables allowed	Bean or pea soup, cream or milk soups
Sweets	Candy without chocolate, molasses, milk, or nuts; honey, jam, jelly, sugar, syrups except molasses and maple syrup	Candy made from chocolate, molasses, milk, or nuts; molasses and maple syrup
Miscellaneous	Vinegar, spices, salt, herbs, pickles, popcorn, relishes, catsup, chili sauce, prepared mustard	Nuts, chocolate, cocoa, gravies, cream sauces, olives, brewer's yeast, peanut butter, potato chips, soy sauce, Worcestershire sauce

[a]See Section 20.5 for high-calcium foods.

20.7. MEDIUM CHAIN TRIGLYCERIDE SUPPLEMENTS

Medium chain triglycerides (MCT) are a mixture of triglycerides containing 6 to 12 chain carbon fatty acids. Their use and physiology are described in Chapters 1 and 13. Essentially, they are absorbed more easily than long-chain triglycerides because they do not require chylomicron formation, lymphatic involvement, and go directly to the portal circulation. They are useful as a high-calorie fat supplement in malabsorptive disorders (Chapter 13), pancreatic deficiency (Chapter 12), and generally where an easy source of energy is needed (hence, some have added them to enteral alimentation formulas; see Chapter 18).

The product is available as an oil and contains approximately 230 cal/oz. The percentages of fatty acids in the oil are listed below.

Composition of Medium-Chain Triglyceride Oil[a]		
	Fatty acid	MCT oil (% of total fat)
$C_{6:0}$	Caproic	1
$C_{8:0}$	Caprylic, octanoic	75
$C_{10:0}$	Capric, decanoic	23
$C_{12:0}$	Lauric	1
$C_{14:0}$	Myristic	—
$C_{16:0}$	Palmitic	—
$C_{18:0}$	Stearic	Tr
$C_{18:1}$	Oleic	—
$C_{18:2}$	Linoleic	—
[a]From Ref. 4.		

The oil can be used on salads or in cooking, or a basic MCT formula can be made by mixing the following ingredients[5]:

Basic Formula for Suggested Diet Plan		
Sodium or potassium caseinate	20.9 g	18.8 g protein
		0.3 g fat
		77.9 kcal
MCT oil	22.2 g	22.2 g fat
		186.5 kcal
Dextrose (anhydrous)	53.3 g	53.3 g carbohydrate
		213.2 kcal
Water, mixed in and beaten to make a total of 350 g of formula containing approximately 15% of its total caloric value as protein, 40% as fat, and 45% as carbohydrate.		

This formula can be used in the following diet plan.[5]

Approximate Composition

Calories	2754
Carbohydrate	46%
Protein	20%
Fat	34% (60% MCT)

Suggested Eating Plan

Breakfast

8 oz orange juice or fruit
2 eggs, poached or boiled
2 pieces of toast
2 tsp butter
2 tsp jelly

Midmorning

8 oz MCT formula

Lunch

8 oz tomato or fruit juice
4 oz very lean meat, fish, or poultry
1 serving of potato
2 slices of bread
2 servings of vegetables
1 salad plus 1 tbsp MCT oil
1 serving fruit
8 oz skim milk
1 tsp jelly

Midafternoon

8 oz MCT formula

Dinner

4 oz fruit juice or 1 serving fruit cocktail
4 oz very lean meat, fish, or poultry
2 slices of bread
1 serving of potato
1 tsp butter
2 servings of vegetables
1 salad plus 1 tbsp MCT oil
1 serving fruit
1 tsp jelly
8 oz skim milk

Bedtime

8 oz MCT formula

Liberal substitutions for the food groups can be used depending on the clinical condition.

MCT oil can be used in cooking. The following recipes are successful in incorporating the oil into palatable food and have been taken from the excellent reference by Senior.[5]

MCT Breakfast Foods

MCT Jelly Muffins

1 tbsp jelly (per muffin)	3 tbsp skim milk powder
2 cups sifted flour	2 tbsp MCT oil
$^1/_2$ tsp salt	1 cup water
$^1/_4$ cup sugar	

Method: Sift together flour, salt, sugar, skim milk powder, and baking powder. Combine oil and water. Add to sifted ingredients, stirring only until flour is moistened. Fill one-third of each oiled (MCT) muffin tin. Place 1 tbsp jelly in center; add another third of mixture over jelly. Bake 25 min in 400°F oven.
Yield: 12 muffins
1 muffin: $^1/_2$ tsp MCT oil

MCT French Toast

1 egg white	3 drops butter flavoring
2 tbsp skim milk	2 slices bread
Pinch of salt	2 tbsp MCT oil

Method: Beat the egg white together with the milk and salt. Dip the bread into the egg white mixture. Brown in the MCT oil.
1 serving: 1 tbsp MCT oil

Griddle Cakes

2 cups sifted flour	1 tsp salt
$1^1/_2$ tbsp sugar	$1^1/_2$ cups skim milk
$2^1/_2$ tsp baking powder	2 tbsp MCT oil

Method: Sift together flour, sugar, baking powder, and salt. Combine milk and MCT oil. Add skim milk and MCT oil to dry ingredients, stirring only enough to moisten dry ingredients. Cook in hot, lightly greased (with MCT oil) pan until golden brown. Turn only once and continue cooking until second side is brown.
Yield: 12 cakes, $4^1/_2$-in. diam

MCT Breads

Popovers

1 cup flour	1 tbsp MCT oil
$1^1/_2$ tbsp sugar	3 egg whites
1 cup skim milk	

Method: Sift dry ingredients. Add milk gradually. Beat egg whites and oil together. Add to milk and flour mixture. Beat 2 min. Pour into tins $^3/_4$ full. Bake in 425°F oven for 40 to 45 min.
Yield: 6 popovers
One popover: $^1/_2$ tsp MCT oil

(continued)

MCT Breads (Continued)

Baking Powder Biscuits

2 cups sifted flour	1/3 cup MCT oil
1/2 tsp salt	2/3 cup skim milk
3 tsp baking powder	

Method: Mix and sift dry ingredients together. Combine oil and milk. Pour all at once over entire surface of flour mixture. Mix with fork to make a soft dough. Shape lightly with hand to make a round ball. Place on waxed paper and knead lightly ten times or until smooth. Pat out to 1/2-in. thickness or roll between two pieces of waxed paper. Remove top sheet; cut biscuits with unfloured medium-sized cookie cutter. Place on ungreased baking sheet. Bake in hot oven (425° to 450°F) 12 to 15 min.

Yield: 12 medium-sized muffins
1 biscuit:1 1/4 tsp MCT oil

MCT Salad Dressings

MCT French Dressing

2/3 cup MCT oil	1/3 cup vinegar or lemon juice

Seasoning to taste: 1 tsp salt, 1/2 tsp pepper, 1/2 tsp sugar, 1/4 tsp paprika

Method: Combine ingredients in jar and shake well. Vary seasonings to taste with dry mustard, minced onion, or garlic.
Yield: 1 cup dressing
1 tbsp = 2 tsp MCT oil

Oil Spread

1 tbsp cornstarch	2/3 cup water
2/3 cup nonfat dry milk	2 cups MCT oil
1 tsp salt	1 tbsp lemon juice
Few drops yellow food coloring	

Method: Sift cornstarch, nonfat dry milk, and salt together into top of double boiler. Combine lemon juice and water; gradually add to starch mixture mixing until smooth. Cook over boiling water, stirring constantly until mixture thickens—about 4 min. Remove from heat. Add MCT oil, 1/4 cup at a time, beating with rotary beater after each addition. Add coloring. Do not use electric blender.
Yield: 1 lb 5 oz (2 2/3 cups)
1 tsp oil spread = 3.3 g MCT oil

MCT Entrees

Tomato and Macaroni Skillet

1/3 cup MCT oil	1 1/2 cups water
1/2 cup chopped onion	1 tbsp salt
1 medium green pepper, chopped	1 tsp dry mustard
3 1/2 cups canned tomatoes	8 oz elbow macaroni

(continued)

MCT Entrees (Continued)

Tomato and Macaroni Skillet (Continued)

Method: Heat MCT oil in heavy saucepan. Add onion and green pepper and cook slowly until tender. Add remaining ingredients except macaroni. When mixture returns to boil, add macaroni, reduce heat, cover, and cook, stirring occasionally, about 30 min or until macaroni is tender.

Yield: 6 servings

1 serving: 1 tbsp MCT oil.

Pasta with Cottage Cheese

1 pkg large manicotti, noodles, lasagna, or elbow macaroni
1 lb skim-milk cottage cheese
2 small cans of Hunt's tomato sauce (fat free)
4 tbsp MCT oil
2 egg whites
Salt, pepper, oregano to taste

Method: Mix the cheese, egg whites, and seasoning well. Add 2 tbsp of MCT oil. Beat well. Oil the bottom of a rectangular pan with the remaining oil. Fill the manicotti noodles with the cheese mixture. Place noodles in pan, cover with tomato sauce. Bake covered with aluminum foil in 350°F oven for about 1 hr. Check to make sure that sauce is not evaporating too fast. If needed, add more sauce. If lasagna noodles are used, put the cheese mixture between two layers of noodles. Cover with suace and bake as above. If elbow macaroni is used, boil first. Drain. Mix with all the rest of the ingredients. Bake for a short time (15 min) at 350°F.

MCT Creole Sauce

3 tbsp MCT oil	$1/2$ tsp salt (or to taste)
2 tbsp chopped onion	$1/2$ tsp basil
2 tbsp chopped green pepper	Dash pepper
$1/4$ cup sliced mushrooms	Few drops tabasco sauce
2 cups stewed tomatoes	

Method: Cook onion, green pepper, and sliced mushrooms in the MCT oil over low heat for about 5 min. Add tomatoes and seasonings and simmer until sauce is thick, about 30 min. For shrimp Creole: Add 12 oz drained shrimp to sauce.

Yield: 6 servings

1 serving: $1^1/2$ tsp MCT oil

MCT Desserts

Sugar Cookies

2 cups all-purpose flour, sifted	3 egg whites
2 tsp baking powder	$2/3$ cup MCT oil
$1/2$ tsp salt	1 tsp vanilla
$1/4$ tsp nutmeg	$3/4$ cup sugar

Method: Preheat oven to 400°F. Sift flour, baking powder, salt, and nutmeg. In large bowl, beat the egg whites lightly. Stir in the oil and vanilla. Blend in sugar. Stir flour mixture into egg mixture. Drop by teaspoonfuls 2 in. apart on an ungreased cookie sheet. Flatten each cookie with an oiled glass dipped in sugar. Bake 8 to 10 min. until a delicate brown. Remove immediately from baking sheet.

Yield: About 60 cookies

2 cookies: 1 tsp MCT oil

(continued)

MCT Desserts (Continued)

Butterscotch Brownies

^{2}/$_3$ cup sifted cake flour or ^{1}/$_2$ cup ^{1}/$_4$ cup MCT oil
plus 1^1/$_3$ tbsp all-purpose flour   2 egg whites
1 tsp baking powder ^{1}/$_2$ cup raisins
^{1}/$_2$ tsp salt   1 tsp vanilla extract
1 cup brown sugar

Method: Sift together flour, baking powder, and salt. Combine sugar and oil. Add egg whites and beat well. Stir raisins and vanilla. Mix in dry ingredients. Turn into greased (using MCT oil) 8-in.-square pan. Bake in moderate oven (350°F) 25 to 30 min. Cut into bars while warm.
Yield: 12 (2^2/$_3$ × 2 × 1 in.) bars
1 piece: 1 tsp MCT oil.

MCT Frozen Dessert

1 tbsp gelatin   2 cups MCT oil cream*
^{1}/$_4$ cup cold water ^{1}/$_4$ tsp salt
1^3/$_4$ cup skim milk ^{1}/$_2$ frozen strawberries
^{1}/$_2$ cup sugar

*MCT oil cream:

1 cup cold water   4 tbsp MCT oil
^{1}/$_3$ cup nonfat dry milk   2 to 3 drops butter flavoring

Method: Soften gelatin in ^{1}/$_4$ cup cold water. Scald the milk in a double boiler. Dissolve the softened gelatin in milk. Add sugar to milk mixture and stir until dissolved. Add milk mixture to MCT oil cream. Add salt and strawberries. Freeze in regular ice cream freezer or in freezing section of refrigerator.
Note: Different flavors may be used. If fresh bananas are used, it is recommended that yellow food coloring be added.

20.8. LOW–CARBOHYDRATE, HIGH–PROTEIN, HIGH–FAT DIET

This diet can be used for the dumping syndrome (Chapter 9) and hypoglycemia. It is the regimen recommended by the American College of Surgeons.[6] The basic principles for treating the dumping syndrome are elimination of mono- and disaccharides and liquids taken 30 to 45 min before or after solid foods. When the diet is used for hypoglycemia, liquids can be taken with meals. This diet stresses meats and is high in fat to compensate for low sugar. If it is needed long term, then vegetable protein and polyunsaturated fats should be used.

It has proven successful to surgeons. It does encourage vegetables. Because calcium is limited, added calcium supplements (tablets or Diet 5 should be encouraged) are needed for males and females 11 to 18 years of age and pregnant or lactating women.

Approximate Composition	
Calories	2200
Carbohydrate	105 g
Protein	130 g
Fat	140 g

Suggested Eating Plan and Menu

Total food for the day:

Milk or substitute	¹/₂ cup
Meat or substitute	16 servings or more
Starchy foods	4 servings
Vegetables	Any amount (low calorie exchange)
Fruit	2 servings
Fats	3 tablespoons or more
Miscellaneous	Any amount

	Meal plan	Amount	Sample menu
Breakfast	Fruit	0	
	Meat	2 servings	1 egg, 1 oz ham
	Starchy foods	¹/₂ serving	¹/₂ slice toast
	Fats	Any amount	Margarine
30–45 min after meal[a]	Milk or substitute	¹/₄ cup	2 oz milk
	Miscellaneous	Any amount	Coffee, sugar substitute
Midmorning	Meat	2 servings	2 oz cheese
	Starchy foods	¹/₂ serving	3 saltines

(continued)

Suggested Eating Plan and Menu (Continued)			
	Meal plan	Amount	Sample menu
30–45 min after meal[a]	Miscellaneous	Any amount	Water
Lunch	Meat	4 servings	4 oz broiled hamburger
	Starchy foods	1 serving	1 small potato
	Vegetables—raw, cooked	Any amount	Carrots (drained), lettuce
	Fats	Any amount	Mayonnaise, margarine
	Fruit	1 serving	1/2 cup fresh fruit salad (drained)
30–45 min after meal[a]	Miscellaneous	Any amount	Dietetic gelatin
Midafternoon	Meat	2 servings	2 oz tuna fish
	Starchy foods	1/2 serving	1/2 slice bread
	Vegetables—raw, cooked	Any amount	Celery sticks
	Fats	Any amount	Mayonnaise, margarine
30–45 min after meal[a]	Miscellaneous	1/2 cup	4 oz tomato juice
Supper	Meat	4 servings	4 oz chicken breast
	Starchy foods	1 serving	1/2 cup rice
	Vegetables—raw, cooked	Any amount	Green beans (drained), tossed green salad
	Fats	Any amount	Margarine, mayonnaise
	Fruit	1 serving	1/2 grapefruit
30–45 min after meal[a]	Milk or substitute	1/4 cup	2 oz milk
	Miscellaneous	Any amount	Coffee, sugar substitute
Evening	Meat	2 servings	2 slices bologna with mustard
	Starchy foods	1/2 serving	1/2 slice bread
	Fats	Any amount	Margarine
30–45 min after meal[a]	Miscellaneous	12 oz limit	Dietetic ginger ale

[a]Liquids should be taken 30 to 45 min before or after the time solid foods are eaten when treating the dumping syndrome, but not for hypoglycemia.

Food Selection List

Milk*a* or substitute
Limit milk (whole or skim) to $1/2$ cup (4 oz) daily, if tolerated.

Meat or substitute	1 serving
Meat, fish, poultry	1 oz
Egg	1
Luncheon meat	1 slice
Cheese	1 oz (daily limit 2 oz)
Cottage cheese	$1/4$ cup (daily limit)

Starchy foods

Bread or substitute	1 serving
Bread	1 slice
Biscuit, roll, muffin	1 small
English muffin	$1/2$
Plain doughnut	1
Corn bread	2-in. square
Pancake, waffle	1 (3-in. diam)
Cereal	
Cooked	$1/2$ cup
Dry (flakes, puffed)	$3/4$ cup
Shredded wheat	1 biscuit
Crackers	
Saltines	5
Uneedas	3
Graham	3
Starchy vegetables or substitute	
Potato, corn, lima beans, baked beans	$1/2$ cup
Pasta, rice	$1/2$ cup
French fried potatoes	8 pieces
Potato chips	15 pieces
Occasional treats	
Ice cream	$1/2$ cup
Popcorn	$1^1/2$ cups

Vegetables
All except those listed under "Starchy foods"

Fruit
1 serving is $1/2$ cup fresh, frozen, or canned fruit (drained) without added sugar

Fats
3 tbsp or more daily—butter, margarine, oil, mayonnaise, bacon, olives
2 tbsp (1 oz) daily limit—cream*a* or nondairy cream*a*

Miscellaneous*a*

Beverages (sugar free)	Flavoring aids
Coffee, decaffeinated coffee	Sugar substitutes
Tea	Lemon
Lemonade, limeade	Vinegar
Kool-Aid	Spices, seasonings

(continued)

Food Selection List (Continued)

Miscellaneous[a] *(Continued)*
- Carbonated beverage (daily limit 12 oz)
- Tomato juice (daily limit $\frac{1}{2}$ cup)
- Cranberry juice

Soups
- Clear soups
- Consomme
- Broth

Sugarless chewing gum

Flavoring aids *(Continued)*
- Unsweetened pickles
- Low-calorie coffee syrup
- Low-calorie jelly
- Low-calorie maple syrup

Desserts (sugar free)
- Unflavored gelatin
- Gelatin dessert mixes
- Pudding mixes
- Junket tablets
- Cranberries, rhubarb

[a]Liquids should be taken 30 to 45 min before or after the time solid foods are eaten when treating the dumping syndrome, but not for hypoglycemia.

20.9. PECTIN SUPPLEMENTS

Jenkins and colleagues have shown that pectin decreases the attacks and symptoms of the dumping syndrome (Chapter 9, Ref. 13). If 5-g sachets of pectin are taken twice daily in juice at the time of the main meals, the symptoms of dumping that correlate with hypoglycemia are reported to be corrected.

Pectin powder rapidly takes on water and can make the patient feel as though he were eating cotton. It should be mixed thoroughly with any diluent.

The following foods contain significant amounts of pectin. Diets employing these foods in large amounts have not received clinical trials. It should also be kept in mind that pectin is added in varying quantities to jellies and pies.

Pectin Content of Food[a,b]	
Food	g/100 g
Apples	0.71–0.84
Apricots	0.71–1.32
Asparagus	Tr
Bananas	0.59–1.28
Beans	0.27–1.11
Blackberries	0.68–1.19
Carrots	1.17–2.92
Cherries	0.24–0.54
Cucumbers	0.10–0.50
Dewberries	0.51–1.00
Grapes	0.09–0.28
Grapefruits	3.30–4.50
Lemons	2.80–2.99
Loganberries	0.59
Oranges	2.34–2.38
Raisins	0.82–1.04
Raspberries	0.97
Squash	1.00–2.00
Sweet potatoes	0.78

[a]Modified from Campbell LA, Palmer GH. Pectin, in Spiller GA (ed): *Topics in Dietary Fiber Research*. New York: Plenum Press, 1978, pp 105–115.
[b]Includes peel and/or skin.

20.10. LOW-PROTEIN DIETS

The RDA for protein is approximately 0.7 g/kg body wt. These diets are necessarily deficient in protein when they fall below that figure. As pointed out in Chapter 10, anywhere between 40 and 60 g will enable albumin synthesis. These diets are employed primarily in hepatic encephalopathy and chronic renal disease. The goal should be to give as much protein as the patient can tolerate.

During hepatic coma, or in trying to bring encephalopathy under control, a zero protein diet may be attempted for a short period of time. Here all protein foods are eliminated. (Diet 13 deals with the newer theory of branched-chain amino acids.)

The following is a 30-g basic protein diet.

To *decrease* the protein, an ounce of meat (approximately 7 g of protein) may be eliminated to make it a 23-g protein diet. To *increase*, an ounce of meat (7 g) or egg (6.5 g) can be added, and so on, in so many units to get the desired amount. For example, if 2 eggs (13 g protein) and 1 oz of meat (7 g protein) were added to the basic 30-g protein diet, it would then be a 50-g protein diet. The modified Mayo exchange lists in Tables 20-1 to 20-4 can be used for varying the food content of these diets. Keep in mind fruits have little protein, but vegetables may have a lot.

Calories can be added by carbohydrate and fat pure foods as needed.

Approximate Composition	
Calories	2300
Carbohydrate	340 g
Protein	30 g
Fat	90 g

Suggested Daily 30-g Protein Food Plan[a]			
	Breakfast	Lunch	Dinner
Protein exchange (meat, etc.)	1	—	2
Cereal	1	—	—
Bread[a]	1	2	2
Fat[b]	2 tsp	4	3
Milk	1/2 cup	—	—
Fruit	2	2	—
Dessert (sugared fruit)	—	—	2
Vegetable	—	1	1
Salad	—	—	1
Sugar/jelly	2 tbsp	1	1
Beverages	Coffee, tea, sodas as tolerated for Na and fluid balance		

Food list: All foods are allowed in accordance with protein restriction. Therefore, nuts, candies, ice cream, high-protein breads *are all essentially eliminated.* Patients must be cautioned against these high-protein snacks.

[a]See Tables 20-1 to 20-6.
[b]Low-protein breads.
[c]Butter and margarines and oils may be used liberally.

20.11. HIGH-PROTEIN DIET

The RDA for protein is 0.7 g/kg body wt. Therefore, these diets are in excess. There is no scientific evidence of any definite benefit from a high-protein diet. To derive 120–150 g of protein from food, other excesses result such as high-cholesterol intake.

This type of diet is used in severe nutritional depletion due to poor intake and should be a short-term diet recommendation.

Approximate Composition	
Calories	2700
Protein	150 g
Carbohydrate	300 g
Fats	100 g

Suggested Basic Eating Plan and Menu[a,b]

Breakfast

Fruits	$1/2$ cup orange juice
Meat, eggs, cheese	1 medium poached egg
Fats	2 strips bacon
Breads and cereals	1 cup bran flakes
Milk beverage	1 cup nonfat milk
Sweets	1 tsp sugar
Beverage	Coffee or tea

Midmorning

Milk beverage	1 cup milk

Noon

Meat, eggs, cheese	2 oz cheese
Meat, eggs, cheese	3 oz grilled hamburger patty (lean)
Breads and cereals	1 hamburger bun
Vegetable, fruit	3 slices tomato on a lettuce leaf
Fats	2 tsp mayonnaise
Vegetable, fruit	French fries (10)
Beverage	Coffee or tea

Midafternoon

Milk or ice cream	1 cup ice cream or sherbet

(continued)

Suggested Basic Eating Plan and Menu[a,b] *(Continued)*

Dinner

Vegetable salad, fruit	Fresh salad made with $1/2$ cup lettuce, $1/2$ cup spinach
Fats	1 tbsp Italian dressing
Meat, eggs cheese	5 oz broiled fish
Starchy vegetable	$1/2$ cup steamed rice
Vegetable, fruit	$1/2$ cup broccoli
Fats	2 tsp margarine
Breads, cereals	1 hard roll
Fats	1 tsp margarine
Dessert	$1/2$ cup sweetened rhubarb
Beverage	Coffee or tea

Bedtime

Milk or ice cream	1 cup milk, sherbet, or ice cream

[a]Modified from Carnation Co. High Protein Diet Plan.
[b]Exchanges can be used liberally. There are no foods eliminated.

20.12. LOW-SODIUM DIETS

Low-sodium diets are used when the body retains sodium, causing edema. By far, cardiac failure is the most common condition requiring this restriction, but the edema and ascites associated with cirrhosis are similarly treated (see Chapter 10).

The 20-meq (approx 500 mg) Na diet represents the high-restriction diet. More restrictive sodium diets require obtaining foods in which Na has been removed. As the daily requirement is well above 500 mg, this should be a depleting diet if followed closely and should be effective. Often the reason for failure is lack of patient compliance. The 40-meq ($\sim$1000 mg) diet is merely an extension of the 20-meq diet.

The 85-meq ($\sim$2000 mg) Na diet permits cooking with salt, a drastic change from the other two diets. It is essentially a free diet except that no Na is added. It is not restrictive in the same sense as the others and hence will not be as effective. It is not a depleting diet, but is used to avoid Na overload.

Basic 20-meq Na Diet, Approximate Composition	
Calories	2250
Carbohydrate	310 g
Protein	100 g
Fat	60 g
Sodium	500 mg

All labels must be read carefully to avoid Na- or salt-containing products. Preservatives such as sodium benzoate, seasonings such as monosodium glutamate, and neutralizing agents such as sodium bicarbonate must be avoided. Read all labels for sodium (Na) content.

Suggested Eating Plan[a]		Sodium	
		(mg)[b]	(meq)[b]
Breakfast			
1 serving fruit or juice	½ grapefruit	1	0.04
1 serving cereal	1 cup puffed rice cereal (enriched)	Tr	—
1 serving nonfat milk	1 cup nonfat milk	116	5.0
1 serving fruit	1 small banana	1	0.04
1 serving salt-free bread	1 small salt-free whole-wheat toast	5	0.2
1 serving fat	1 tsp unsalted margarine	—	—
1 serving sweets	1 tsp jam	—	—
Beverage	Coffee or tea	2	0.1

(continued)

		Sodium	
		(mg)[b]	(meq)[b]
Lunch			
2 oz cooked fresh meat	2 oz unsalted roasted chicken (light meat)	36	1.6
2 servings salt-free bread	2 slices salt-free white bread	5	0.2
1 serving fat	1 tsp salt-free mayonnaise	—	—
1 serving vegetable	lettuce (3 small leaves)	1	0.04
2 servings fruit	1 box (1½ oz) raisins	12	0.5
	½ medium apple	1	0.04
1 serving nonfat milk	1 cup nonfat milk	116	5.0
Dinner			
2 servings vegetables	1 cup chopped fresh spinach	39	1.7
	½ medium tomato	2	0.1
2 servings fat	2 tbsp oil and vinegar dressing	1	0.04
1 serving salt-free bread	1 slice salt-free whole-wheat bread	5	0.2
1 serving fat	1 tsp unsalted margarine	—	—
1 serving wine (optional)	7 oz wine	10	0.4
4 oz cooked fresh meat	4 oz broiled lean steak	68	3.0
1 serving vegetable	1 baked potato	6	0.3
2 servings fat	2 tbsp sour cream	12	0.5
2 servings vegetable	6 asparagus spears	1	0.04
	½ cup cooked rhubarb with sugar	2	0.1
1 serving dessert	½ cup ice cream	42	1.8
Beverage	Coffee or tea	2	0.1
Snack			
1 serving fruit	1 medium orange	1	—
	Total	490 mg[b]	21.0 meq[b]

Suggested Eating Plan[a] **(Continued)**

[a]This basic-type plan is adopted from Carnation Co., Sodium-Restricted Diet Plan.
[b]Approximate values.

Note: This plan could be reduced to 250 mg Na if the 2 cups of nonfat milk were eliminated or exchanged for fruit juice. In designing another plan Na foods could be substituted from lists that follow.

The user, or food preparer for the user, should make frequent estimates of the daily intake similar to this plan. A rough guide of allowed and not allowed foods for the user is included. The more detailed analysis of use should be done from the sodium-content list. The *intelligent* patient on a long-term low-sodium diet must memorize the list. A less cooperative patient must be educated. Failure results when a patient is either not educated or refuses to follow the lists.

Rough Guide for Low-Na Diet

Allowed	Avoid
Beverages	
Coffee, tea. Use only carbonated drinks free of sodium preservatives and sodium bicarbonate; natural waters. Check sodium content.	Check all carbonated drinks.
Meat	
Unsalted meat, fowl, liver, heart, egg, cheese, peanut butter; fish, fresh clams, oysters, shrimp	Salted meat, fish, fowl, cheese, peanut butter; other organ meats; commercially frozen fish; other shellfish
Fat	
Unsalted; cream, $1/3$ cup or less per day	Salted butter, margarine, salad dressings; bacon; olives; salted nuts
Milk	
Must be calculated in diet	Buttermilk, commercial chocolate milk, beverage mixes; condensed milk
Bread, Cereals, and Starches	
Salt-free bread, quick breads or rolls made with the following leavening agents: cream of tartar, potassium bicarbonate, sodium-free baking powder, yeast; unsalted crackers	Any made with salt, baking powder, or baking soda; commercially prepared mixes; self-rising flour
Unsalted cooked cereal; unsalted ready-to-eat cereals	Instant, quick-cooking, or ready-to-eat cereals with salt or sodium compound added
Unsalted potatos and potato substitutes	Commercially prepared mixes; potato chips; commercially frozen lima beans or peas; dried legumes; hominy; salted popcorn
Vegetables	
Unsalted canned, cooked, fresh, or frozen except "Avoid"; no more than *one* serving daily of the following:	Any prepared with salt or sodium compounds
Fruit	
Any canned, dried, fresh, or frozen	
Soup	
Unsalted soup or broth; cream soup made with milk allowance	Any other

Vegetables list:

artichokes	collards
beet greens	dandelion greens
beets	kale
carrots	mustard greens
celery	spinach
chard	white turnips

(continued)

Rough Guide for Low-Na Diet (Continued)	
Allowed	Avoid

Dessert

Unsalted cake and cookies made with the following leavening agents: cream of tartar, potassium bicarbonate, sodium-free baking powder; unsalted pie; sodium-free gelatin; pudding or frozen dairy dessert when used as part of milk allowance	Desserts made with baking powder, baking soda, or salt; commercial dessert mixes

Sweets

Sugar; honey, jam, jelly, marmalade without a sodium preservative; plain sugar candy	Chocolate and cream candies

Miscellaneous

Pepper, spices, and herbs except "Avoid"; flavoring extracts; bitter or sweet chocolate, cocoa powder; vinegar; unsalted white sauce; unsalted meat sauces	Salt or seasoning salts, mixed spices, dried celery products; bottled meat sauces, meat tenderizer, monosodium glutamate; pickles; gravy; regular baking powder, baking soda, salt substitutes, artificial sweeteners, saccharin

Sodium Content of Foods[4]

Food	Portion	Weight (g)	Sodium (meq)
Meat			
Meat (cooked)			
Beef	1 oz	30	0.8
Ham	1 oz	30	14.3
Lamb	1 oz	30	0.9
Pork	1 oz	30	0.9
Veal	1 oz	30	1.0
Liver	1 oz	30	2.4
Sausage, port	2 links	40	16.5
Beef, dried	2 slices	20	37.0
Cold cuts	1 slice	45	25.0
Frankfurters	1	50	24.0
Fowl			
Chicken	1 oz	30	1.0
Goose	1 oz	30	1.6
Duck	1 oz	30	1.0
Turkey	1 oz	30	1.2
Egg	1	50	2.7
Fish	1 oz	30	1.0
Salmon			
Fresh	1/4 cup	30	0.6
Canned	1/4 cup	30	4.6

(continued)

Sodium Content of Foods[4] (Continued)

Food	Portion	Weight (g)	Sodium (meq)
Meat *(Continued)*			
Fish *(Continued)*			
Tuna			
Fresh	1/4 cup	30	0.5
Canned	1/4 cup	30	10.4
Sardines	3 medium	35	12.5
Shellfish			
Clams	5 small	50	2.6
Lobster	1 small tail	40	3.7
Oysters	5 small	70	2.1
Scallops	1 large	50	5.7
Shrimp	5 small	30	1.8
Cheese			
Cheese, American or cheddar	1 slice	30	9.1
Cheese foods	1 slice	30	15.0
Cheese spreads	2 tbsp	30	15.0
Cottage cheese	1/4 cup	50	5.0
Peanut butter	2 tbsp	30	7.8
Peanuts, salted	25	25	—
Fats			
Avocado	1/8	30	—
Bacon	1 slice	5	2.2
Butter or margarine	1 tsp	5	2.2
Cooking fat	1 tsp	5	—
Cream			
Half and half	2 tbsp	30	0.6
Sour	2 tbsp	30	0.4
Whipped	1 tbsp	15	0.3
Cream cheese	1 tbsp	15	1.7
Mayonnaise	1 tsp	5	1.3
Nuts			
Almonds, slivered	5 (2 tsp)	6	—
Pecans	4 halves	5	—
Walnuts	5 halves	10	—
Oil, salad	1 tsp	5	—
Olives, green	3 medium	30	31.3
Bread			
Bread	1 slice	25	5.5
Biscuit	1 (2-in. diam.)	35	9.6
Muffin	1 (2-in. diam.)	35	7.3
Cornbread	1 (1 1/2-in. cube)	35	11.3
Roll	1 (2-in. diam.)	25	5.5
Bun	1	30	6.6
Pancake	1 (4-in diam.)	45	8.8
Waffle	1/2 square	35	8.5
Cereals			
Cooked	2/3 cup	140	8.7
Dry, flaked	2/3 cup	20	8.7
Dry, puffed	1 1/2 cups	20	—
Shredded wheat	1 biscuit	20	—

(continued)

Sodium Content of Foods[4] (Continued)			
Food	Portion	Weight (g)	Sodium (meq)
Crackers			
Graham	3	20	5.8
Melba Toast	4	20	5.5
Oyster	20	20	9.6
Ritz	6	20	9.5
Rye-Krisp	3	30	11.5
Saltines	6	20	9.6
Soda	3	20	9.6
Flour products[a]			
Cornstarch	2 tbsp	15	—
Macaroni	1/4 cup	50	—
Noodles	1/4 cup	50	—
Rice	1/4 cup	50	—
Spaghetti	1/4 cup	50	—
Tapioca	2 tbsp	15	—
Dessert			
Commercial gelatin	1/2	100	2.2
Ice cream	1/2 cup	75	2.0
Sherbet	1/3 cup	50	—
Angel food cake	$1^1/_2 \times 1^1/_2$ in.	25	3.0
Sponge cake	$1^1/_2 \times 1^1/_2$ in.	25	1.8
Vanilla wafers	5	15	1.7
Vegetable[a]			
Asparagus			
Cooked	1/2 cup	100	—
Canned[b]	1/2 cup	100	10.0
Frozen	1/2 cup	100	—
Artichokes	1 large bud	100	1.3
Beans, dried (cooked)	1/2 cup	90	—
Beans, lima	1/2 cup	90	—
Beans, green or wax			
Fresh or frozen	1/2 cup	100	—
Canned[b]	1/2 cup	100	10.0
Bean sprouts	1/2 cup	100	—
Beet greens	1/2 cup	100	3.0
Beets	1/2 cup	100	1.8
Broccoli	1/2 cup	100	—
Brussels sprouts	2/3 cup	100	—
Cabbage			
Cooked	1/2 cup	100	0.6
Raw	1 cup	100	0.9
Carrots			
Cooked	1/2 cup	100	1.4
Raw	1 large	100	2.0
Cauliflower, cooked	1 cup	100	0.4
Celery, raw	1 cup	100	5.4
Chard, Swiss	3/5 cup	100	3.7
Collards	1/2 cup	100	0.8

(continued)

Sodium Content of Foods [4] (Continued)

Food	Portion	Weight (g)	Sodium (meq)
Vegetable *(Continued)*			
Corn			
Canned[b]	1/3 cup	80	8.0
Fresh	1/2 ear	100	—
Frozen	1/3 cup	80	—
Cress, garden (cooked	1/2 cup	100	0.5
Cucumber	1 medium	100	0.3
Dandelion greens	1/2 cup	100	2.0
Eggplant	1/2 cup	100	—
Hominy (dry)	1/4 cup	36	4.1
Kale, cooked	3/4 cup	100	2.0
Kohlrabi	2/3 cup	100	—
Leeks, raw	3–4	100	—
Lettuce	Varies	100	0.4
Mushrooms, raw	4 large	100	0.7
Mustard greens	1/2 cup	100	0.8
Okra	1/2 cup	100	—
Onions, cooked	1/2 cup	100	—
Parsnips	2/3 cup	100	0.3
Peas			
Canned[b]	1/2 cup	100	10.5
Dried	1/2 cup	90	1.5
Fresh	1/2 cup	100	—
Frozen	1/2 cup	100	2.5
Pepper, green or red			
Cooked	1/2 cup	100	—
Raw	1	100	0.5
Popcorn	1 cup	15	—
Potato			
Potato chips	1 oz	30	13.0
White, baked	1/2 cup	100	—
White, boiled	1/2 cup	100	—
Sweet, baked	1/4 cup	50	0.4
Pumpkin	1/2 cup	100	—
Rutabagas	1/2 cup	100	—
Squash, winter			
Baked	1/2 cup	100	—
Boiled	1/2 cup	100	—
Milk			
Whole milk	1 cup	240	5.2
Evaporated whole milk	1/2 cup	120	6.0
Powdered whole milk	1/4 cup	30	5.2
Buttermilk	1 cup	240	13.6
Skim milk	1 cup	240	5.2
Powdered skim milk	1/4 cup	30	6.9
Fruits			

All fruits are essentially free of sodium. All labels of canned and frozen products should be read, but most are also free of sodium.

[a]Estimated for products without added salt.
[b]Estimated on basis of added salt.

20.13. BRANCHED-CHAIN AMINO ACID THERAPY

Valine, leucine, and *iso*-leucine, the branched-chain amino acids, are decreased in the serum of hepatic encephalopathics, but not aromatic amino acids (see Chapter 10, Refs. 27 to 29).

Trials of amino acid formulas enriched with branched-chain amino acids given orally or by infusion have been reported to be successful in treating hepatic encephalopathy (see Chapter 10, Ref. 28).

At the present time, it is difficult to arrange any diet that stresses branched-chain amino acids without using supplements. Because clinical trials are not available, the use of branched-chain amino acid supplementation can be only recommended on the basis of clinical judgment.

Hepatic-Aid (McGaw) Composition		
Calories	560	
Carbohydrate	97.7 g (maltdextrose + sucrose)	
Fat	12.3 g (hydrogenated soybean oil, lecithin, mono- and diglycerides)	
Protein	14.5 g *amino acid* (g)	
	L-leucine	2.008
	L-isoleucine	1.643
	glycine	1.643
	L-lysine (acetate)	1.575
	L-lysine (free base)	1.116
	L-valine	1.533
	L-proline	1.460
	L-alanine	1.397
	L-arginine	1.095
	L-serine	0.912
	L-threonine	0.820
	L-histidine	0.438
	L-methionine	0.182
	L-phenylalanine	0.180
	L-tryptophan	0.120

When the powder is mixed to make 1 liter, it yields an osmolality of 900. This may be difficult for some patients to tolerate, but it is necessary in order to increase the branched-chain amino acids.

Hepatic-Aid may be used alone or in conjunction with a protein-free *or* low-protein diet (see Diet 10). Each packet of branched-chain amino acids contains 14.5 g of protein; therefore, if three or four are given daily, the total intake will be adequate for protein as well as calories. *Vitamins and minerals must be added.* (The insert circulated with the product recommends additional protein be used with Hepatic-Aid. It is still not clear as to whether the mixture supplies adequate essential amino acids. More studies are needed on the ability of subjects on this mixture to synthesize proteins. As four packets per day supply 60 g protein, caution must still be used in giving additional protein during encephalopathy.)

20.14.   HIGH-NH_4^+-PRODUCING FOODS

Hepatic encephalopathy is related to serum NH_4^+ levels. Rudman and associates have shown that foods containing high NH_4^+ can increase blood levels and potentially aggravate encephalopathy.

The following foods are sufficiently high in NH_4^+ as related to N content to raise the serum above 150 µg/100 ml in 25% of cirrhotic patients when eaten in standard portions. Although only two cheeses are listed, it is safest to avoid all cheese while treating encephalopathy.

Domestic blue cheese	Frozen fish
Cheddar cheese	Gelatin
Salami	Large amounts
	of onions
	Peanut butter

The following table (from Ref. 30, Chapter 10) lists the ammonia content of foods, *but* serving size and body weight must be calculated from the data. The above servings of foods produce more than 0.04 g NH_3/50 kg body wt.

NH_3 Content of Foods (Uncooked)[a]					
Food item	N g/100 g wet wt	N of NH_3 as % of total N	Food item	N g/100 g wet wt	N of NH_3 as % of total N
American cheese	3.712	2.19	Domestic blue cheese	3.440	4.00
Apples	0.032	3.36	Egg white	0.744	0.05
Bacon	4.864	0.33	Egg yolk	2.560	0.16
Banana	0.176	0.00	French dressing	0.096	14.00
Beer	0.048	1.62	Grapefruit	0.80	2.07
Beer cheese	3.760	2.44	Grapes	0.208	4.20
Bread	0.320	0.94	Grape wine	0.016	11.20
Breakfast cereal (Rice Krispies)	0.944	0.00	Gelatin	13.696	0.25
Brewer's yeast	6.208	0.35	Green peas	1.008	0.60
Broccoli	0.496	1.26	Grits (corn)	0.192	0.00
Brussels sprouts	0.512	2.15	Ground beef (hamburger)	3.872	0.26
Buttermilk	0.576	2.75	Half milk/ half cream	0.512	2.27
Cabbage	0.208	0.82			
Carrots	0.176	0.81	Ham	2.704	0.58
Catsup	0.320	11.00	Hoap cheese	2.000	0.32
Cauliflower	0.302	1.41	Hot dog	2.000	0.32
Celery	0.144	0.00	Idaho potatoes	0.416	2.33
Cheddar cheese	4.000	2.76	Lemon juice (frozen)	0.080	2.90
Chicken	3.808	0.45			
Corn	0.560	0.25	Lettuce	0.144	0.55
Cucumbers	0.144	3.27			

(continued)

	NH$_3$ Content of Foods (Uncooked)[a] (Continued)				
Food item	N g/100 g wet wt	N of NH$_3$ as % of total N	Food item	N g/100 g wet wt	N of NH$_3$ as % of total N
Lima beans	1.344	0.21	Pickle relish	0.080	10.84
Margarine	0.096	21.96	Potato chips	0.848	2.83
Mayonnaise	0.176	23.33	Radishes	0.160	2.77
Milk	0.560	0.35	Rice	0.320	0.04
Mushrooms	0.304	2.18	Salami	2.800	3.97
Mustard	0.352	1.00	Spanish olives	0.224	4.16
Onions	0.240	11.20	Spinach	0.480	0.20
Orange juice (frozen)	0.112	3.16	Squash	0.192	4.28
			String beans	0.160	0.46
Peaches	0.064	3.78	Sweet potatoes	0.288	0.61
Peanut butter	4.448	1.10	Tilsit cheese	4.000	1.38
Pears	0.112	2.67	Tomatoes	0.176	2.09
Pecans	1.472	0.48	Turnip greens	0.400	0.73

[a]From Ref. 30, Chapter 10.

20.15.   FLUID-RESTRICTION DIET (1000 ml)

Fluid retention is associated with cirrhosis. It becomes a problem when it is intractable to simple salt restriction. Ascites associated with cirrhosis also can be intractable. When this occurs, concurrent fluid restriction can be helpful (see Chapter 10). Often thirst becomes a problem and fluid restriction is difficult. However, it can be accomplished when patients are cooperative.

When one is using a fluid-restriction diet, sodium and protein intake should also be calculated depending on the clinical findings—see Diets 20.10 and 20.12.

Suggested Eating Plan[a]	
Breakfast	
¹/₂ cup orange juice	120 ml
1 cup cold cereal	
1 soft-cooked egg	
1 slice toast	
1 cup milk	240 ml
¹/₂ cup coffee	120 ml
creamer	30 ml
	510 ml
Lunch	
¹/₂ cup consomme	120 ml
3 oz roast beef	
1 hard roll	
1 cup mixed greens	
salad dressing	30 ml
¹/₂ cup drained peaches	
¹/₂ cup milk	120 ml
	270 ml
Dinner	
3 oz baked chicken	
¹/₂ cup mashed potatoes	
¹/₂ cup green beans	
1 cup mixed salad	
salad dressing	30 ml
1 apple	
¹/₂ cup milk	120 ml
	150 ml
Total	930 ml

[a] All beverages must be limited. If thirst becomes a problem, hard candies can be helpful. Liberal substitutions can be made, but all fluid intake must be calculated.

20.16.   GLUTEN-FREE DIET

This diet is used in the treatment of gluten enteropathy (see Chapter 13).

The patient requiring vigorous gluten-free therapy must be motivated and educated to eat absolutely gluten-free foods. After the disease goes into remission (months to years), most patients begin to eat small amounts of gluten without effect. However, any recommendation to use gluten can only be made by the physician on a clinical basis.

The following table lists all foods that must be avoided. *All labels must be checked.* Foods prepared in restaurants, e.g., soups and gravies, will often use a flour that contains gluten.

Besides this guideline list, I give my patients two texts (Refs. 7 and 8) that offer elaborate gluten-free recipes and recommendations.

Corn and rice cereals and flours are permitted—all others are not. Commercial foods use others as "jellies"—be careful!

Food Lists		
	Allowed	Avoid
Milk	All without additives	Postum, Ovaltine, malted milk or commercial chocolate milk
Vegetable	All	Any vegetables seasoned with glutamate (canned) or prepared with commercial cream sauce or breading (frozen)
Fruit	All processed fruits that are processed naturally and without added ingredients (except honey)	None
Bread	Only those made from rice, corn, potato, soybean, gluten-free flour, arrowroot starch, cornstarch	All bread, rolls, crackers, cake, and cookies made from wheat or rye, including Rye-Krisp, muffins, biscuits, doughnuts, waffles, pancake flour and prepared mixes, rusk, zwieback, pretzels; products containing oatmeal, barley, or buckwheat; breaded foods, bread crumbs
Cereal and cereal products	Cream of Rice, corn meal, hominy, rice, Rice Krispies, Puffed Rice, Rice Honey, Sugar Corn Pops, Oriental cellophane noodles	All wheat and rye cereals, including wheat germ, barley, oatmeal, buckwheat, noodles, macaroni, spaghetti, ravioli, dumplings, corn flakes containing malt flavoring; any stuffing
Starchy vegetables	Sweet potatoes or yams, white potatoes, dried peas, beans or lentils, lima beans or soybeans, hominy	Commercial baked beans
Egg, cheese, meat, fish, and poultry	All varieties (*not* sauces)	Meat patties or meat loafs made with bread or bread crumbs; croquettes; breaded meat, fish, or chicken; bread stuffing; chili con carne and other canned meats; cold cuts unless guaranteed pure meat; cream cheese dips that have soup mix added; processed cheeses

(continued)

Food Lists (Continued)

	Allowed	Avoid
Fat and oil	Butter, oleo, and other fats as desired; yogurt; sweet, sour, or whipped cream; homemade salad dressing made of allowed foods; pure mayonnaise (read all labels carefully); all oil except wheat germ oil, peanut butter (read label for possible wheat additives)	"Whipped cream" substitutes (whipped topping); commercial salad dressings (except pure mayonnaise); wheat germ oil
Sweets and dessert	Candies made of allowed foods; sugar, honey, molasses, chocolate, coconut; homemade puddings with cornstarch; gelatine desserts; custards; cakes made with allowed foods; homemade ice cream; junket, tapioca, sago, fruits of all kinds; fruit ice; sherbets; rice pudding	Package-mix puddings; commercial ice cream; commercial cakes, cookies, or pastry; commercial candies containing cereal products
Miscellaneous	Homemade broths, and cream soups thickened with cornstarch or potato flour only; popcorn, plain potato chips, plain corn chips, salt, spices, herbs, olives, vinegar; pickles; sauces and gravies prepared with flour or starches "allowed"; nuts; meat tenderizer made from papaya (Adolph's is one)	All commercial canned soups; read all labels of instant coffee (may be stabilized with cereal foods); commercial crackers in general; special seasonings on corn and potato chips (e.g., barbecued); dry roasted nuts (these contain cereal grains); soysauce, Worcestershire sauce, monosodium glutamate (uncertain)
Beverage	Coffee (if labeled 100% pure coffee), pure cocoa, tea, fruit juices, carbonated beverages (perhaps), wine, vermouth, brandy, rum, most liqueurs, tequila (made from the century plant), cognac	Check instant coffee, check cocoa. Root beer is considered a carbonated beverage but can be troublesome; beer and ale, gin, scotch, vodka, rye, bourbon

Gluten-Free Commercial Foods[a,b]

Fruit-flavored drinks
  General Foods Bird's Eye Awake Frozen
    Concentrate for Imitation Orange Juice
  General Foods Kool-Aid Regular or Pre-
    sweetened Soft Drink Mix
  General Foods Kool Pops Pop Bar
  General Foods Start Instant Breakfast Drink
  General Foods Tang Instant Breakfast
    Drink
  General Foods Twist Imitation Lemon,
    Grape, Orange, and Punch Mixes

Fruit-flavored drinks (Continued)
  General Foods Bird's Eye Orange Plus
    Frozen Concentrate for Orange Juice Drink
  General Foods Great Shakes Shake Mixes
  Carnation Instant Breakfast
Breads, cereals, crackers, flour
  Cellu Gluten Free Bread Mix
  Cellu Low Protein Bread Mix
  General Mills Corn Muffin Mix
  Beechnut Rice Cereal
  General Foods Post Rice Krinkles Cereal

(continued)

Gluten-Free Commercial Foods[a,b] (Continued)

Breads, cereals, crackers, flour (Continued)

Gerber's Rice Cereal, Rice Cereal with
   Applesauce and Banana
Heinz Rice Cereal
Kellogg's Puffed Rice, Sugar Pops
Quaker Puffed Rice, Diet Frosted Rice Puffs
Cellu Whole Rice Wafers
Devonsheer Brown Rice Wafers
Frito's Corn Chips
General Mills Bugles, Bows, Onyums, Pizza
   Spins
Nabisco Corndiggers
Wise Cheese Pixies
Wise Onion Flavored Rings
Borden's Cracker Jacks
Cellu Arrowroot Flour
Cellu Corn Flour
Cellu Grainless Mix (potato–soybeans)
Cellu Potato Starch Flour
Cellu Rice Flour
Cellu Soybean Flour
Cellu Tapioca Flour
Cellu Wheatless Mix
Cellu Cereal-Free Baking Powder
General Foods Calumet Baking Powder
Argo Cornstarch
Duryea's Cornstarch
Kingsford Cornstarch
General Mills Au Gratin Potatoes
General Mills Potato Buds
General Mills Scalloped Potatoes
General Foods Bird's Eye French Fried
   Potatoes
French's Mashed Potatoes
Mrs. Paul's Candied Sweet Potatoes
Penobscot Stuffed Potatoes with Cheese
   Flavor
Pillsbury Hungry Jack Mashed Potatoes
Tato Mix Potato Pancakes
Cellu Spanish Rice, unsalted
General Foods Minute Rice
General Mills, Rice Provence

Desserts and sweets

Cellu Grainless Cookies
Cellu Rice Cookies
Cellu Wheat-Free Cake
Cellu Butterscotch Pudding
Featherweight Chocolate Pudding
Featherweight Vanilla Pudding

Desserts and sweets (Continued)

Cherry, lemon, lime, orange, raspberry,
   strawberry gelatins
General Mills Frosting Mixes
General Mills Ready-to-Serve Puddings
General Mills Coconut Macaroon Mix
My-T-Fine Lemon Pudding
My-T-Fine Tapioca Pudding
Ice cream
   Brock Hall
   Howard Johnson's
   McDonald's
   Sealtest
Hershey's Milk Chocolate with Almonds
Hershey's Semi-Sweet Chocolate
Nestle's Crunch Milk Chocolate
Nestle's Fruit and Nut Bar
Nestle's Crackle Bar
Mars Marsettes Caramel Chocolate
Mars Three Musketeers
DeMet's Turtles
Hollywood's Butternut
Kraft Caramels
Pearson's Mint Patty
Brach's Chocolate Mint
Schutter's Bit O'Honey
Halloway's Milk Duds
Necco Canada Mints
Welch's Junior Mints
Harde's Jujy Fruits
Chuckles
Tootsie Roll
Heath Bar
Lifesavers
Reeds (all flavors)
General Foods
   Baker's Chocolate products
   Baker's Cocoa
   Baker's Coconut (all varieties)
   Bird's Eye Cool 'n Creamy Puddings
   Bird's Eye Cool Whip Non-Dairy Whipped
      Topping
   Certo Fruit Pectin
   Dream Whip Whipped Topping Mix
   D'zerta Low Calorie Gelatin Desserts
   D'zerta Low Calorie Pudding and Pie
      Fillings
   D'zerta Low Calorie Whipped Topping Mix
   Jell-O Gelatin Desserts
   Jell-O Golden Egg Custard Mix

(continued)

Gluten-Free Commercial Foods[a,b] (Continued)

Desserts and sweets (Continued)
  Jell-O Lemon Chiffon Pie Filling
  Jell-O Puddings and Pie Fillings
  Jell-O Tapioca Puddings
  Jell-O Whip 'n Chill Dessert Mixes (except
    chocolate flavor)
  Minute Tapioca
  Sure Jell Fruit Pectin

Fruits
  General Foods Bird's Eye Fruits
  General Foods Bird's Eye Concentrated Fruit
    Juices

Salad dressings
  Caine's
    Mayonnaise
    Sandwich Spread
  Cellu
    French or French Style Dressing
    Fancy Fruit Dressing
    Italian Dressing
    Neu Blue Dressing
    Russian Dressing
    Soyamaise Dressing
    Thousand Island Dressing
    Whipped Salad Dressing
  General Foods
    Good Season's Salad Dressing Mix
    Good Season's Open Pit Barbeque Sauce
  Best Foods
    Hellmann's Family French Dressing
    Old Homestead Garlic French Dressing
    True Italian
    Real Mayonnaise
    Tartar Sauce
  Kraft
    Casino French Dressing
    French Dressing
    Green Onion Dressing
    Miracle Whip Dressing
    Oil and Vinegar Dressing
    Roquefort Dressing
  Seven Seas
    Cheddar 'n' Wine Dressing
    Creole French Dressing
    Italian
    Italian Bleu Dressing
    Tomato 'n' Spice Dressing
  Sweet Life Pure Mayonnaise

Salad dressings (Continued)
  Wishbone
    Creamy Onion Dressing
    Italian Rose Dressing
    Russian Dressing
    Thousand Island Dressing

Margarines, regular and "diet"
  Bluebonnet
  Imperial
  Mazola
  Mrs. Filbert's
  Nucoa
  Parkay

Meats
  DAK Danish Cooked Ham
  Kosher Best All-Beef Cocktail Franks
  Roessler's All-Beef Frankfurters
  Swanson's
    Boned Chicken
    Chunks o'Chicken
    Chicken Spread
    Boned Turkey
    Chunks o'Turkey
  Swift's Premium Brown'n' Serve Sausage

Soups
  Campbell's
    Bean with Bacon
    Beanbroth
    Black Bean
    Chicken Gumbo
    Chicken with Rice
    Chunky Turkey
    Chunky Vegetable
    Consomme
    Frozen Fish Chowder
    Frozen Green Pea with Ham
    Frozen Old-Fashioned Vegetable with Beef
    Frozen Oyster Stew
    Onion
    Vegetable Bean
  Heinz
    Beef Liver Soup
    Chicken Soup
  Lipton's Onion
  Red Kettle
    Onion Soup Mix
    Potato Soup Mix
  Swanson Chicken Broth

(continued)

Gluten-Free Commercial Foods[a,b] (Continued)

Vegetables
  General Foods Bird's Eye Vegetables
    (without sauces)
  Campbell's
    Franks and Beans
    Old-Fashioned Beans
    Pork and Beans

Miscellaneous
  Carnation Coffee-Mate Cream Substitute
  French's Prepared Mustard
  Heinz Catsup
  Karo Syrups
  Skippy Peanut Butter, creamy or chunk style
  General Mills Bacos
  General Foods Log Cabin Syrup (all varieties)

[a]Modified from Ref. 9.
[b]Commercial food products may change their ingredients without notice. Therefore, all labels should be checked periodically before use. No claim is made that these products will remain gluten free.

20.17. LACTOSE–INTOLERANCE DIETS

Lactose intolerance (discussed in Chapters 13 through 17) occurs both as a primary disease and secondary to damage of the small intestine. Subjects deficient in lactase vary in their clinical sensitivity. Lactose causes no known physical harm to the lactase-deficient patient, but there is the discomfort of symptoms. Therefore, limitation of lactose intake depends on the degree of discomfort. Some will need complete abstinence and others will be able to tolerate milk and ice cream.

20.17.1. Lactose–Free Diet

All milk, milk products, foods prepared from milk, or with milk added must be eliminated. Because milk is one of the main sources of calcium in some societies, the patient on a lactose-free diet must supplement the diet with calcium foods (see Diet 5) or take calcium tablets.

All pills, tablets, and commercial foods must be checked to make sure that they are lactose free.

Lactose Content of Some Commercial Products [a,b]

More than 30 mg/30 g (oz)	More than 30 mg/30 g (oz) (Continued)
Sweetine Low Calorie Sweetener	Pine Tree Strawberry Yogurt
Sugar Twin granulated sugar substitute	Parkay Soft Margarine
Gerber Strained Creamed Spinach Baby Food	Blue Bonnet Margarine
Gerber Macaroni & Tomato Baby Food	Land O'Lakes Salted Butter
Cheddar Cheese Crouton	Fleischman's Margarine
Monk's Whole Wheat Bread	Evan's Chocolate Fudge Dessert Topping
Kellogg's Eggo Waffles	M & M's
Chock Full O'Nuts Pound Cake	Hershey's Mr. Goodbar
Hostess Twinkees	Evan's Butterscotch Dessert Topping
Kaufman's Cinnamon Swirl Coffee Cake	Rich's Orange Sherbet
Monk's White Bread	Seven Seas Green Goddess Salad Dressing
Betty Crocker Pie Crust Mix	Potassium phenoxyonethyl penicillin
Thomas' English Muffins	Gantrisin
Home Pride White Bread	Tetracycline
Gaylord Waffles	Erythromycin
Wonder White Bread	Aspirin
Strohman's White Bread	
Peppridge Farm Whole Wheat Bread	*Less than 30 mg/30 g (oz)*
Kraft Grated American Cheese	
Gaylord Chocolate Marble Ice Cream	Heinz Strained Beef Baby Food
Sealtest Chocolate Ice Cream	Wegman White Bread
Sealtest Vanilla Ice Cream	Nabisco Vanilla Creme Cookies
Breyer's Coffee Ice Cream	Arnold Bread Crumbs
Star's Vanilla Ice Cream	Kraft Swiss Cheese
Kraft American	Wegman's Extra Sharp Cheddar Cheese
Bison Brand Diet Cottage Cheese	Nestle's Butterscotch Bits
Borden's Strawberry–Vanilla Swirl Ice Cream	Campbell's Tomato Bisque Soup
Breyer's Chocolate Ice Cream	Zweigle's Hot Dogs

(continued)

Lactose Content of Some Commercial Products[a,b] (Continued)

Not detectable
Heinz Lamb Baby Food
Beechnut Pork Baby Food
Gerber Chicken Baby Food
Gerber Strained Veal Baby Food
Gerber Mixed Vegetable Baby Food
Nabisco Oreo Cookies
Keebler Cheese & Peanut Butter Crackers
Abel's Frozen Bagels
Kaufman's Pumpernickel Bread
Arnold Rye Bread
Arnold Soft Roll
DiPaolo Italian Bread
Borden's Cremora
Bird's Eye Cool Whip
Jell-O Orange Gelatin
Hershey's Special Dark Chocolate

Not detectable (Continued)
Fanta Orange Soda
Nestle's Quick
Franco American Spaghetti
Franco American Spaghettio's
Staff Spaghetti Rings
Maplecrest Lebanon Bologna
Tylenol
Calcium gluconate
APC Tablets
Maalox
Temaril
Ferrous gluconate
Calcium lactate
Ampicillin
Dimetap Extentabs

[a]Modified from Ref. 113, Chapter 13.
[b]Commercial food products may change their ingredients without notice. Therefore, all labels should be checked periodically before purchase. No claim is made that these products will remain lactose free.

Lactose-Free Commercial Products[a,b]

Fruit-flavored drinks
  General Foods:
    Bird's Eye Awake
    Bird's Eye Orange Plus
    Instant Postum
    Kool-Aid, Regular or Pre-sweetened
    Kool Pops
    Start Instant Breakfast Drink
    Tang Instant Breakfast Drink
    Twist Imitation Lemon, Grape, Orange,
      Punch Mixes
Breads, cereals, crackers, flour
  Breads
    Epicile's Italian Bread
    Rositani French Bread
    Kosher bagels
    Most French and Italian breads
  Cereals
    Post Brand Cereals except Fortified Oat
      Flakes
    General Mills Cereals except Clackers
    Kellogg's Cornflakes, Frosted

Cereals *(Continued)*
    Flakes, Product-19, 40% Bran
    Flakes, Raisin Bran, Rice Krispies
  Crackers
    Premium
    Ritz
    Zesta
    General Mills Bows, Bugles
  Flour
    General Mills Gold Medal Flour (Regular,
      Softasilk Cake Flour, Wondra)
    General Foods Swans Down Cake Flour
      and Swans Down Self-Rising Cake Flour
  Others
    General Foods Bird's Eye potato products
    General Foods Minute Rice
    General Foods Minute Rice Mixes
      (Drumstick and Rib Roast only)
    General Foods Calumet Baking Powder

Desserts
    Duncan Hines Angel Food Cake Mix

(continued)

Lactose-Free Commercial Products [a,b] (Continued)

Desserts *(Continued)*

General Foods

Baker's Chocolate (unsweetened, semi-sweet, German sweet)

Baker's Semi-Sweet Chocolate Chips (not glazed chips)

Baker's Redi-Blend chocolate products for baking

Baker's Cocoa

Baker's Coconut (all varieties)

Certo Fruit Pectin

D'zerta Gelatin Desserts

D'zerta Pudding (chocolate flavor only)

Jell-O Gelatin Desserts

Jell-O Lemon Chiffon Pie Filling

Jell-O Pudding and Pie Fillings (except milk chocolate flavor)

Jell-O Tapioca Pudding

Minute Tapioca

Sure-Jell Fruit Pectin

Swans Down Angel Food Cake Mix

General Mills

Angel Food Cake Mixes

All Chiffon Cake Mixes

All Fluffy Frosting Mixes

Chocolate Chip Fudge Brownie Mix

Chocolate Fudge Brownie Mix

Graham Cracker Pie Crust

Ready-to-Serve Lemon Pudding

Walnut Brownie Mix

Gelatins

Royal Pudding and Pie Fillings (except milk chocolate flavor)

Fruits

General Foods Bird's Eye Frozen Fruits

General Foods Bird's Eye Frozen Concentrated Fruit Juices

Margarines

Diet Bluebonnet

Diet Fleischman's

Diet Imperial

Mazola

Mother's Brand

Meat products

Roessler's All-Beef Frankfurters

Armour Frankfurters

Kosher Best All-Beef Cocktail Franks

Nepco Frankfurters

Oscar Mayer Frankfurters

Swift's Premium Frankfurters

Swift's Premium Brown 'n' Serve Sausage (fully cooked)

Salad dressings

General Foods Good Season's Dressing Mixes (Creamy French, Garlic, Italian, Low Calorie Italian, Old-Fashioned French, Onion)

General Foods Good Season's Open Pit Barbeque Sauces (Original, Hickory Smoke, and Mild Garlic Flavors)

General Foods Good Season's Thick 'n Creamy Salad Dressing Mixes (Coleslaw, French, Thousand Island)

Soups

Campbell's

New England Clam Chowder

Chili Beef

Noodles and Ground Beef

Turkey Vegetable

Chicken Gumbo

Manhattan Clam Chowder

Chicken Noodle

Vegetarian Vegetable

Old-Fashioned Vegetable

Chicken with Rice

Chunky Turkey

Habitant

Chicken Noodle Soup

Chicken Rice Soup

Lipton's

Country Vegetable Soup

Noodle Mixed with Chicken Broth

Onion Soup

Pepperidge Farm Petite Marmite Beef and Vegetable Soup

Sweet Life Vegetable Beef

Vegetables

General Foods "Plain" Bird's Eye Vegetables (without sauces or butter)

Miscellaneous

General Mills Bacos

General Foods Log Cabin Syrup (except Log Cabin Buttered Syrup)

Cream substitutes:

Borden Cremora

Carnation Coffee-mate

Peanut butter:

Jiffy

Peter Pan

Planter's

Sweet Life

---

[a] Modified from Ref. 113, Chapter 13.

[b] Commercial food products may change their ingredients without notice. Therefore, all labels should be checked periodically before purchase. No claim is made that these products will remain lactose free.

20.17.2. Lactose–Restricted Diet

Most patients with lactose intolerance can tolerate small amounts of lactose without discomfort (Ref. 109, Chapter 13). These patients should be carefully instructed to try small amounts of lactose not greater than 2–4 g in any one feeding, depending on their tolerance. An 8-oz glass of milk has approximately 10 g of lactose. If milk is limited, then calcium-rich foods or tablets must be added (see Diet 5).

The following list gives the exact lactose content found in 10 g of common foods and medicines. The values should be extrapolated to portions: e.g., Wonder white bread has 30 mg/10 g, therefore, approximately 90 mg/30-g slice —less than 0.1 g (100 mg) per slice of bread. More than 80% of lactose-intolerant patients should easily tolerate the white bread.

Grams of Lactose in 10-g (ca. 1/3 oz) Food Samples[a,b]	
Product	Lactose (g)
Artificial sweeteners	
Sweetine Low Calorie Sweetener	11.08
Sugar Twin granulated sugar substitute	7.52
Baby foods	
Gerber Strained Creamed Spinach	0.12
Gerber Macaroni & Tomato	0.02
Heinz Strained Beef	Trace
Heinz Lamb	None
Beechnut Port	None
Gerber Chicken	None
Gerber Strained Veal	None
Gerber Mixed Vegetable	None
Bread	
Cheddar Cheese Crouton	0.23
Monk's Whole Wheat Bread	0.17
Kellogg's Eggo Waffles	0.16
Chock Full O'Nuts Pound Cake	0.12
Hostess Twinkees	0.12
Kaufman's Cinnamon Swirl Coffee Cake	0.09
Monk's White Bread	0.08
Betty Crocker Pie Crust Mix	0.06
Thomas' English Muffins	0.05
Home Pride White Bread	0.04
Gaylord Waffles	0.03
Wonder White Bread	0.03
Strohman's White Bread	0.03
Peppridge Farm Whole Wheat Bread	0.02
Wegman White Bread	0.01
Nabisco Vanilla Creme Cookies	0.0004

(continued)

Grams of Lactose in 10-g (ca. 1/3 oz) Food Samples[a,b] (Continued)	
Product	Lactose (g)
Bread *(Continued)*	
Arnold Bread Crumbs	Trace
Nabisco Oreo Cookies	?
Keebler Cheese & Peanut Butter Crackers	?
Abel's Frozen Bagels	None
Kaufman's Pumpernickel Bread	None
Arnold Rye Bread	None
Arnold Soft Roll	None
DiPaolo Italian Bread	None
Dairy	
Kraft Grated American Cheese	1.42
Kraft Velveeta Cheese	0.93
Gaylord Chocolate Marble Ice Cream	0.84
Sealtest Chocolate Ice Cream	0.82
Sealtest Vanilla Ice Cream	0.81
Breyer's Coffee Ice Cream	0.69
Star's Vanilla Ice Cream	0.68
Kraft American Cheese	0.52
Bison Brand Diet Cottage Cheese	0.34
Borden's Strawberry–Vanilla Swirl Ice Cream	0.32
Breyer's Chocolate Ice Cream	0.31
Pine Tree Strawberry Yogurt	0.19
Parkay Soft Margarine	0.10
Blue Bonnet Margarine	0.08
Land O'Lakes Salted Butter	0.08
Fleischman's Margarine	0.07
Land O'Lakes Margarine	0.05
Kraft Swiss Cheese	Trace
Wegman's Extra Sharp Cheddar Cheese	Trace
Borden's Cremora	None
Bird's Eye Cool Whip	None
Desserts	
Evan's Chocolate Fudge Dessert Topping	0.29
M & M's	0.23
Hershey's Mr. Goodbar	0.12
Evan's Butterscotch Dessert Topping	0.12
Rich's Orange Sherbet	0.06
Nestle's Butterscotch Bits	?
Jell-O Orange Gelatin	None
Hershey's Special Dark Chocolate	None
Drinks	
Fanta Orange Soda	None
Nestle's Quick	None
Medicines	
Potassium phenoxyonethyl penicillin	2.16
Gantrisin	1.07

(continued)

Grams of Lactose in 10-g (ca. 1/3 oz) Food Samples[a],[b] (Continued)	
Product	Lactose (g)
Medicines *(Continued)*	
Tetracycline	0.69
Erythromycin	Yes
Aspirin	Yes
Tylenol	None
Calcium gluconate	None
APC Tablets	None
Maalox	None
Temaril	None
Ferrous gluconate	None
Calcium lactate	None
Ampicillin	None
Dimetap Extentabs	None
Miscellaneous	
Seven Seas Green Goddess Salad Dressing	0.04
Campbell's Tomato Bisque Soup	0.001
Zweigle's Hot Dogs	Trace
Franco American Spaghetti	None
Franco American Spaghettio's	None
Staff Spaghetti Rings	None
Maplecrest Lebanon Bologna	None

[a] Adapted from Ref. 113, Chapter 13.
[b] Commercial food products may change their ingredients without notice. Therefore, all labels should be checked periodically before purchase. No claim is made that these products will remain at the level reported.

20.18.   FRUCTOSE-FREE, SUCROSE-FREE DIET

Patients who cannot digest sucrose or are intolerant to fructose must eliminate both the disaccharide and the hexose from this diet.

Foods to Eliminate[a]	
Food group	Eliminate
Milk	Infant formulas containing fructose or sucrose, sweetened condensed milk, commercial chocolate milk, milk drinks with added sugar, ice cream, sherbet
Meat	Meats processed in sugar brine, e.g., ham, bacon, luncheon meat
Cereal	Sugar-coated cereals, defatted wheat germ rice, bran
Dessert	Cookies, cakes, and other desserts made with sugar, sweet or chocolate milk syrup or molasses
Potato	Sweet potatoes (regular white cooking potatoes may be a significant source of fructose, depending on harvesting, storage, cooking techniques)
Vegetable	Broccoli, cucumber, peas, rhubarb, beets, carrots parsnips, pumpkins, rutabagas, winter squash, turnips, corn, hominy
Fruit	ALL FRUITS AND FRUIT JUICES
Beverage	Any containing fructose or sucrose; read all labels
Miscellaneous	Granulated, powdered, and brown sugars, milk and sweet chocolate, honey, jelly, syrup, molasses, sorghum, peanuts and other nuts
[a]Modified from Ref. 10.	

20.19.   SUCROSE-FREE, MALTOSE-FREE DIET

This intolerance is a rare condition, especially in adults. It requires the elimination of all sucrose and maltose from the diet. Therefore, all starches and fruits cannot be eaten.

Meats, fish, eggs, poultry, vegetable and animal fats, sugar-free carbonated drinks, artifical sweeteners, unprocessed cheeses, and milk are permitted.

Foods to Eliminate[a]	
Food group	Eliminate
Milk	Milk substitutes that may contain starch or soy milks; infant formulas containing fructose or sucrose; sweetened condensed milk, commercial chocolate, milk drinks with added sugar; ice cream, sherbet
Meat	Sausages, processed meats containing starch; meats processed in sugar brine, e.g., ham, bacon, and luncheon meat
Bread, Cereal	Bread, cakes, biscuits, pastries, wheat, oats, rye, barley, rice flours, tapioca, cornstarch, spaghetti, macaroni, breakfast cereals; sugar-coated cereals, defatted wheat germ, rice, bran
Fruit	*All fruits and fruit juices*
Vegetable	Beans, peas, all potatoes, parsnips, corn, broccoli, cucumber, peas, rhubarb, beets, carrots, pumpkins, rutabagas, winter squash, turnips
Dessert	Ice cream (unless homemade), cookies, cakes, and other desserts made with sugar, sweet or chocolate milk, syrup or molasses
Beverage	Malted milks, cocoa, chocolate; all drinks made with fructose, sucrose, maltose, malt
Miscellaneous	Nuts, thickened soups, pickles, sauces; granulated, powdered, and brown sugars; milk and sweet chocolate; honey, jelly, syrup, molasses, sorghum

[a]Prepared from Ref. 10.

20.20.   LOW-FAT DIET

Low-fat diets are recommended when it is desirable to decrease biliary or pancreatic function. This occurs in gallbladder and pancreatic diseases (see Chapters 11 and 12). Decreased fat intake *may* also be desired in severe steatorrhea when a high intestinal lumenal fat content is aggravating function (see Chapter 13). The amount of fat that can be tolerated by patients with steatorrhea varies; therefore, the amount given to them is titrated clinically.

The RDA for fat is not clear. It is recommended in prudent diets that approximately 30% of the total calories be from fat. A 2250-cal diet should therefore contain approximately 75 g of fat. The compensation in calories comes from increased protein in this diet but may be in carbohydrates.

The following is a basic 40-g fat diet (approximately 15% fat). Fat content can be decreased or increased by varying the use of eggs, fat substitutes, or protein exchanges high in fat (meats, etc.).

Other low-fat diets are Diet 20.25, for the management of hyperlipoproteinemia, and Diet 20.26, for the management of glucose intolerance, constipation, and serum lipids. These low-fat diets are now popular and are recommended by this author for general good health. (Diet 20.11 is a 100-g fat diet, Diet 20.8 a 140-g fat diet.)

Approximate Composition[a]	
Calories	2250
Carbohydrate	365 g
Protein	130 g
Fat (primarily polyunsaturates)	40 g

[a]Eliminating fats from eating plan reduces fat content to less than 30 g.

Suggested Eating Plan				
	Breakfast	Lunch	Dinner	Snacks
Protein exchange	1	3	3	
Cereal, Bread	3	3	1	
Vegetable	—	1	2	
Fruit	1	1	1	
Soup	—	1	—	
Salad	—	—	1	
Fat	1	1	1	
Sweets	3	—	—	1
Nonfat milk	1	1	—	2
Beverage	With no cream, as desired			

[a]Use Tables 20-1 to 20-7.

Basic Meal Plan (Food Cooked in No Fat)[a]

Breakfast

Fruit	1 small banana
Bread, cereal	1 cup bran flakes
Sweets	1 tsp sugar
Egg	1 medium poached egg
Bread, cereal	2 slices whole-wheat toast
Fat	1 tsp margarine
Sweets	2 tsp jelly
Milk	1 cup nonfat milk
Beverage	Coffee or tea without cream

Midmorning

Milk	1 cup nonfat milk

Lunch

Soup	1 cup vegetable soup, prepared with water
Bread, cereal	4 saltine crackers
Vegetable	3 celery sticks
Meat	3 oz roast turkey, white meat
Bread cereal	2 slices white bread
Fat	1 tsp mayonnaise
Fruit	1 medium apple
Milk	1 cup nonfat milk
Beverage	Coffee or tea without cream

Dinner

Salad	1 cup fresh spinach with 1/2 medium tomato and 1 tbsp diet salad dressing
Meat	3 oz broiled halibut with lemon wedge
Starchy vegetable	1/2 cup steamed rice
Vegetable	1/2 cup green beans
Fats	1 tsp margarine
Breads or cereals	1 hard roll
Fruits	1/2 cup sweetened rhubarb
Beverages	Coffee or tea without cream

Bedtime

Milk	1 cup milk

[a]Modified from Carnation Co. Low Fat Diet Plan.

Food Lists to Limit Fat[a]		
	Allowed	Avoid
Milk	Nonfat milk, buttermilk	Whole milk, low-fat milk, whipped cream, ice cream, ice milk, half and half, chocolate milk, evaporated milk, yogurt, nondairy creamer
Egg	1 daily, cooked without fat, egg white permitted	More than 1 daily, eggs fried in fat
Cheese	Cottage cheese, whey cheeses; especially made low-fat yellow cheeses clearly labeled as such	Any other
Meat, fish	All lean meats, fish, fowl, vegetable meat substitutes (if fowl is breaded and fried, remove skin and breading before eating), waterpacked canned fish; choose poultry and fish often	Fried meats, fish, and poultry; fish canned in oil, fatty cuts of meat such as brisket, short ribs, pork chops, bacon, sausage, luncheon meats, frankfurters, duck, goose; peanut butter
Vegetable	All	Creamed vegetables, vegetables in cheese sauce, hollandaise sauce or other rich sauces, fried vegetables, buttered vegetables unless buttered with fat allowance
Fruit	All fruits and fruit juices	Avocado, coconut
Soup	Broths, bouillon, canned broth-based soups prepared with water or nonfat milk	Creamed or canned soups prepared with whole milk
Bread, cereal	Bread, cereals, pastas, crackers, popcorn (no butter)	Biscuits, waffles, rich rolls, pancakes, sweet rolls, doughnuts, popovers, other rich breads
Dessert	Gelatin, puddings prepared with nonfat milk, angel food cake, sherbert, ices, meringues, vanilla wafers	Custards and puddings prepared with whole milk or egg yolks, cakes (except angel), cookies, ice cream, pie crusts, pastries, any dessert made with chocolate or nuts
Sweets	Jelly, jam, sugar, syrup, honey, molasses, plain hard candy, gumdrops, cocoa	Chocolate, any candy containing chocolate, cream, or fat
Fat	Diet salad dressing, *limit of 1 tsp* oil, mayonnaise, salad dressing, or margarine per fat exchange	Butter, shortening, lard; oil, mayonnaise, salad dressing, or margarine in amounts *greater* than 1 *tsp per meal*
Beverage	Coffee, tea, coffee substitute, carbonated beverages	None
Miscellaneous	Salt, spices, herbs, condiments, cocoa, vinegar, pickles, catsup, mustard, steak sauce, etc.	Olives, nuts, gravy, rich sauces such as cream sauce

[a]Modified from Carnation Co. Low Fat Diet Plan.

20.21. ZERO-FAT DIET

Zero-fat diets are used only for acute inflammatory gallbladder or pancreatic disease (Chapters 11 and 12), the object being to rest the organ. Some controversy exists as to the actual benefit of complete fat withdrawal, but it is an accepted diet approach.

This diet therapy is only prescribed for short periods of time during acute disease, when patients are too ill to maintain a full-calorie diet and need intravenous supplementation. If they can tolerate full feedings, they probably should be *advanced to a low-fat diet.*

The fat-free foods may be taken in quantities tolerated.

Food List		
Food group	Allowed	Avoid
Milk	None	All
Protein food exchanges (meat, etc.)	Gelatin, egg white (only a trace of fat)	All others
Vegetable, fruit	*None* if zero fat desired, but some fruits can be used if minute amounts are permitted	All have some amounts (<0.1%); coconut, avocado are high
Fruit juice	Apple, blueberry, lemon, loganberry, papaya, peach (only traces)	All others have small amounts (<0.1%)
Soup	Campbell beef broth or consomme	All others
Bread, cereal	None	All have small amounts
Sweets	Sugar, jellies, jelly beans, plain marshmallows, gumdrops, corn syrup, molasses	All others
Fats	None	All
Beverages	Coffee, tea, all carbonated sugar drinks, ginger ale, Coca-Cola, Pepsi, lemonade	All others
Desserts	Jello, some artificially flavored ices	All others

20.22.   DIET SUPPLEMENT FOR CYSTIC FIBROSIS

The primary treatment is enzyme replacement. However, several authors have reported that elemental supplements improve growth (see Refs. 39 and 40, Chapter 12). They are easily absorbed and compensate for pancreatic enzyme deficiency.

Berry et al.[40] described a regimen of diet supplementation used in children aged 1 to 18 consisting of a beef serum protein hydrolysate, a glucose polymer, and a medium-chain triglyceride (MCT) oil supplemented with 2.5% safflower oil. The supplement was used to supply 100% of the RDA for protein and approximately 50% of the caloric requirement for the age. The formula contains 55 to 70 percent of calories as carbohydrate, 14 to 16% as amino acids, and 20 to 30% as fat. It was mixed with flavored liquids and taken in small divided doses with ad lib diet, vitamin, and mineral substitution.

Allan et al.[39] similarly used a beef serum hydrolysate, glucose polymer, and MCT successfully.

The Mead-Johnson Company has developed a product, Pregestimil, that can be used as a supplement in children. One quart contains:

Calories	20 kcal/fl oz
Carbohydrate	86.4 g
Fat	25.7 g
Protein	18.0 g

Casein hydrolysate, corn syrup, starch, and MCT are used with vitamins and minerals added. The recommended formulations for each age should be followed.

To add easily absorbed MCT, see Diet 20.7.

A wide variety of other liquid supplements are listed in Chapter 18. These can be tried if palatability is a problem. Tables 18-1 to 18-4 list the protein, carbohydrate, and fat contents. Several contain MCT.

20.23.   HIGH-FIBER AND HIGH-CARBOHYDRATE/HIGH-FIBER DIABETIC DIETS

Several authors report the benefit of high-carbohydrate, high-fiber diets for glucose tolerance in diabetic subjects (Refs. 50 to 53, Chapter 12). Evidence substantiates that high fiber promotes stabilization of glucagon and insulin metabolism. Further reports are awaited.

I. High-Fiber Miranda Diet: Approximate Composition[a,b]	
Calories	1750
Carbohydrate	185 g (43%)
Protein	90 g (21%)
Fat	70 g (36%)
Fiber (crude)	20 g

[a]Diet modified from Ref. 51, Chapter 12.
[b]This is not a high-carbohydrate diet.

This diet can be varied by using protein exchanges for the meat, vegetable, and fruit exchanges, etc. (Tables 20-1 thru 20-7).

It is suggested that no further fat be added. If more calories are needed, they can be added in the form of vegetables and starches from bread, cereal, and vegetable lists.

Sample Miranda Menu			
	Weight (g)	Carbohydrate (g)	Fiber (g)
Breakfast			
Frozen fresh blueberries	100	13.6	1.50
Scrambled egg, 1	30	0.3	—
Bacon, 2 strips	15	0.5	—
High-fiber wheat bread, 2 slices	60	20.6	4.50
Butter, 1 tsp	5	—	—
Whole milk, 1/2 cup	120	5.8	—
Coffee, 1 cup	240	—	—
Lunch			
Beef patty, 2 oz	60	—	—
American cheese, 1 small slice	20	0.4	—
High-fiber wheat bread, 1 slice	30	10.3	2.25
Mustard, 2 tsp	10	0.6	0.10
Sliced tomato, 1 medium	100	4.7	0.50
Lettuce, 1 small wedge	50	1.5	0.25
Oil for dressing, 1 tsp	5	—	—

(continued)

Sample Miranda Menu (Continued)

	Weight (g)	Carbohydrate (g)	Fiber (g)
Lunch (Continued)			
Vinegar for dressing, 1 tsp	5	0.3	—
Green peas, 1 serving	100	7.1	1.90
Butter, 1 tsp	5	—	—
Fresh apple, 1 medium	150	21.6	1.50
Apple juice, 2/3 cup	160	19.0	0.16
Dinner			
Broccoli, 1 serving	70	3.3	0.77
Broiled chicken, 1 serving	100	—	—
Whole kernel corn, 1 serving	120	16.3	0.84
Butter for vegetables, 2 tsp	10	—	—
Sliced tomato, 1 medium	100	4.7	0.50
Oil for dressing, 1 tsp	5	—	—
Vinegar for dressing, 1 tsp	5	0.3	—
Fresh pear, 1 medium	200	30.6	2.80
Tea, 1 cup	240	—	—
Snack			
Roast beef, 1 oz	30	—	—
High-fiber wheat bread, 1 slice	30	10.3	2.25
Mayonnaise, 1 tsp	5	0.1	—
Whole milk, 3/4 cup	180	8.6	—
	Approximate total:	185 g	20 g

II. Albrink High-Carbohydrate, High Fiber Diet: Approximate Composition[a]

Calories (adjust to keep weight stable)	2600
Carbohydrate	455 g (70%)
Protein	95 g (15%)
Fat	45 g (15%)
Fiber (crude)	20 g

[a]Modified from Ref. 52, Chapter 12.

Daily Intake

Food	Amount/day (g)	Crude fiber (g)
All Bran	56	4.6
Whole milk	244	—
Whole-wheat bread	115	2.0
Cooked dried beans	396	6.0
Cooked brown rice	900	2.4

(continued)

Daily Intake (Continued)		
Food	Amount/day (g)	Crude fiber (g)
Oranges (4)	400	2.0
Butter	20	—
Cottage cheese (skim)	50	—
Lettuce	100	0.5
Celery	75	0.5
	Total	18.0 g

This was reported as an experimental diet and would be virtually impossible to maintain over a long period of time. If one should attempt to use it, then fruits, vegetables, and cereals with high fiber can be substituted for the beans, rice, and oranges as suggested in exchanges in Diet 20.26.

The following diet has been used successfully at the Lexington Veterans Administration Hospital. Successful results were first published by Kiehm *et al.* (Ref. 50, Chapter 12).

III. Anderson 70% High-Carbohydrate, High-Fiber, Low-Fat Diet: Approximate Composition [a]	
Calories	1800
Carbohydrate (total)	314 g
(simple)	78 g
Protein	81 g
Fat	24 g
Fiber[b] (soluble)	21 g
(total plant)	65 g

[a] Diet modified from Ref. 11.
[b] Calculated according to Ref. 12.

Suggested Eating Plan			
Breakfast		**Lunch**	
Skim milk	180 g	Navy beans	200 g
Whole oats (dry)	20 g	Ham flavoring	10 g
with eggs	4 g	Cornbread	100 g
Fresh orange	100 g	Broccoli (cooked)	100 g
Whole-wheat bread	75 g	Cabbage (cooked)	100 g
Margarine	6 g	Banana	63 g
1 Multivitamin		Butter	4 g

(continued)

Suggested Eating Plan (Continued)			
Dinner		*Evening snack*	
White bread	70 g	Skim milk	240 g
Potatoes (raw wt)	185 g	Corn flakes	30 g
Peas (cooked)	100 g	Graham crackers	30 g
Corn (cooked)	200 g		
Fresh apple	100 g		
Salad			
Celery	20 g		
Lettuce	50 g		
Onions	20 g		
Carrots	20 g		

Exchanges may be calculated from Tables 20-1 through 20-7 and from Refs. 11 and 12.

20.24.   LOW-CARBOHYDRATE DIABETIC DIET

The goal of this diet therapy is to reduce the intake of carbohydrate so that blood glucose levels and urine glucose loss will be decreased. This is the conventional and traditional diet therapy for diabetics, but there is controversy (as reviewed in Chapter 12) concerning the role of diet and which diet approach is best (see Diet 23). There are also problems with patient compliance. *Concurrent control of obesity and total caloric intake is equally important* in increasing carbohydrate tolerance. The diet recommendations must limit total intake as well as control carbohydrate intake.

In the following diet (modified from Ref. 4 and the Norwalk Hospital Manual), the estimated desired caloric intake should be determined first and then the exact content of the diet established.

Approximate Composition[a]						
Calories	1000	1400	1800	2200	2600	3000
Carbohydrate (g)	90	120	170	210	230	285
Protein (g)	70	75	80	100	115	130
Fat (g)	40	70	90	110	135	150

[a]Modified from Ref. 4.
[b]Use Tables 20-1 to 20-7 for exchanges.

Suggested Daily Eating Plan, 1800-cal Diet[a,b]				
Exchanges	Breakfast	Lunch	Dinner	Snacks
Protein	1	3	3	—
Bread	2	2	2	2
Fruit	1	1	1	—
Vegetable	—	1	2	—
Fat	2	2	2	1
Milk	—	1	—	1
Sugar-free beverages, diet sodas, etc., as desired and tolerated				

[a]Modified from Ref. 4.
[b]Use Tables 20-1 to 20-7 for exchanges.

Daily Exchanges for Other Calorie Plans[a]

Calories	1000	1400	2200	2600	3000
Protein	7	7	8	9	10
Bread	2	4	9	10	13
Fruit	3	3	4	4	5
Vegetable	1	1	1	1	1
Fat	1	3	8	10	12
Milk	2	2	3	4	4

[a]Use Tables 20-1 to 20-7.

Sample Menu for 1800-cal Diabetic Diet (Norwalk Hospital)

Breakfast
1/2 cup orange juice or 1 fruit exchange
1 egg, cooked without extra fat
2 slices toasted bread
1 tsp butter or 1 fat exchange
1 cup whole milk
Coffee or tea, no cream or sugar

Lunch
Sandwich:
   2 slices bread
   3 oz lean meat or meat exchanges
   2 tsp mayonnaise or 1 fat exchange
   Lettuce and tomato or other vegetable
1 piece of fresh fruit
Coffee or tea, no cream or sugar
Broth as desired

Dinner
3 oz lean meat or 3 meat exchanges
1 potato or 1 bread exchange
1 serving cooked vegetable
Tossed green salad with no-calorie dressing
2 slices bread or 2 bread exchanges
1 fresh fruit cup or 1 fruit exchange
Coffee or tea, no cream or sugar

Snack
1 cup milk or exchange
1 tsp butter or 1 fat exchange
2 slices toast or 2 bread exchanges

20.25. DIET MANAGEMENT OF HYPERLIPOPROTEINEMIA

The following quote from the classic National Institute of Health[13] monograph best describes the problem.

> An abnormally increased plasma concentration of cholesterol or triglyceride is commonly encountered in clinical practice. Such hyperlipidemia may arise from one of many causes. Sometimes it is "secondary" to other diseases known to affect lipid metabolism. Typical examples are hyperlipidemia caused by hypothyroidism, insulin-dependent diabetes, obstructive liver disease or the nephrotic syndrome. Treatment is directed toward the causative disorder, not the hyperlipidemia. "Primary" hyperlipidemia is that which is not obviously caused by a well-known disease. It may be due to some disorder of lipid metabolism, to an abnormal diet (or abnormal response to a normal diet), to excess alcohol, or to other still obscure causes. Primary hyperlipidemia is sometimes inheritable. Treatment of primary hyperlipidemia is usually undertaken for one of three major reasons: Most commonly it is because hyperlipidemia of certain kinds is associated with a high risk of premature atherosclerosis; this is especially important in certain children and young adults. Another reason, less common but sometimes urgent, is to relieve the recurrent abdominal pain and the risk of acute pancreatitis that attend severe hyperglyceridemia. Finally, annoying and alarming skin lesions (xanthomas) may be due to hyperlipidemia and frequently can be made to disappear with proper treatment.

The hyperlipoproteinemias are roughly classified into five types.[13] The recommended diet management varies with each group and is summarized as follows (adopted from the Norwalk Hospital Diet Manual).

Summary of Diets for Types I–V Hyperlipoproteinemia

	Type I	Type IIa	Types IIb & III	Type IV	Type V
Diet prescription	Low fat (25–35 g)	Low cholesterol; polyunsaturated fat increased	Low cholesterol; approx 20% cal protein, 40% cal fat, 40% cal CHO	Controlled CHO, approx 45% of calories; moderately restricted cholesterol	Restricted fat, 30% of calories; controlled CHO, 50% of calories; moderately restricted cholesterol
Calories	Not restricted	Not restricted	Achieve and maintain "ideal" weight, i.e., reduction diet if necessary	Achieve and maintain "ideal" weight, i.e., reduction diet if necessary	Achieve and maintain "ideal" weight, i.e., reduction diet if necessary
Protein	Total protein intake is not limited	Total protein intake is not limited	High protein	Not limited other than control of patient's weight	High protein
Fat	Restricted to 25–35 g; kind of fat not important	Saturated fat intake limited; polyunsaturated fat intake increased	Controlled to 40% of calories (polyunsaturated fats recommended in preference to saturated fats)	Not limited other than control of patient's weight (polyunsaturated fats recommended in preference to saturated fats)	Restricted to 30% of calories (polyunsaturated fats recommended in preference to saturated fats)
Cholesterol	Not restricted	As low as possible; the only source of cholesterol is the meat in the diet	Less than 300 mg; the only source of cholesterol is the meat in the diet	Moderately restricted to 300–500 mg	Moderately restricted to 300–500 mg
Carbohydrate (CHO)	Not limited	Not limited	Controlled— concentrated sweets are restricted	Controlled— concentrated sweets are restricted	Controlled— concentrated sweets are restricted
Alcohol	Not recommended	May be used with discretion	Limited to 2 servings (substituted for CHO)	Limited to 2 servings (substituted for CHO)	Not recommended

Types II and IV are the most common disturbances. The following are the diets recommended by the NIH.[13]

20.25.1. Low-Cholesterol Modified-Fat (High Polyunsaturated Fat) Diet[13]

This diet is used to treat type II hyperlipoproteinemia. Cholesterol is limited to less than 300 mg/day, and polyunsaturated fats of plant origin are liberally substituted for animal fats.

Suggested Daily Eating Plan[a] (1700–2000 cal)

1 pint or more skim milk
Cooked poultry, fish, or lean trimmed meat
5 servings of vegetable and fruit; include:
    1 serving citrus fruit[b]
    1 serving dark-green or deep-yellow vegetable[c]
7 or more servings of whole-grained or enriched bread or cereal[d]
1 or more servings of potato, rice, etc.
Allowed fat
Allowed desserts and sweets

Breakfast

Citrus fruit or juice
Cereal
Toast
Allowed fat
Jelly and sugar
Skim milk
Coffee or tea if desired

Lunch

Poultry, fish, or lean meat
Potato or substitute
Vegetables
Bread
Allowed fat
Fruit or allowed dessert
Skim milk

Dinner

Poultry, fish, or lean meat
Potato or substitute
Vegetable
Bread
Allowed fat
Fruit or allowed dessert
Skim milk

Between-meal snack

Fruit
Skim milk

---

[a]From Ref. 13.
[b]One serving of citrus fruit is recommended daily to provide adequate vitamin C.
[c]One dark-green or deep-yellow vegetable is recommended daily to provide adequate vitamin A.
[d]Enriched cereal or bread should be included in the diet to provide adequate vitamin B complex and iron.

Food List		
Food group	Allowed	Avoid
Beverage (nondairy)	Coffee, tea, carbonated beverages, fruit and vegetable juice	None
Bread	Whole wheat, rye, and white bread; matzo, saltines, graham crackers; baked goods containing no whole milk or egg yolk and made with allowed fat	Biscuits, commercial muffins, sweet rolls, cornbread, pancakes, waffles, French toast, hot rolls, corn chips, potato chips, cheese crackers or other flavored crackers
Cereal	All cereals; also grain products such as rice, macaroni, noodles, spaghetti, flour	None
Dairy	Skim milk, nonfat buttermilk, dried nonfat milk, evaporated skim milk, dry (nofat) cottage cheese, yogurt made from skim milk, skim milk cheese such as Sapsago and specially prepared low-fat cheese and specially prepared cheese high in polyunsaturated fat	Whole milk, whole-milk drinks, dried whole milk, evaporated milk, condensed milk, cream (sweet or sour), ice cream, ice milk, sherbet, commercial whipped toppings, cream substitutes, cream cheese, all hard cheese except skim-milk cheese, creamed cottage cheese unless substituted for meat; one-fourth cup creamed cottage cheese equals 1 oz of meat
Meat, poultry, fish	*Limit as recommended:* Lean beef, lamb, veal, pork and ham, tongue, chicken, turkey, dried or chipped beef, fish except those excluded, egg white, peanut butter	Egg yolk, luncheon meat, cold cuts, hot dogs, sausages, bacon, goose, duck, poultry skin, shellfish including oysters, lobster, scallops, shrimp, clams, and crab; fish roe including caviar, all organ meats such as heart, liver, brains, and kidney, all fatty meats, regular fried meats and fried fish unless fried with allowed oils, corned beef, *regular* ground beef or hamburger, spareribs, pork and beans, meats canned or frozen in sauces or gravies, frozen or packaged dinners, frozen or packaged prepared products (convenience foods)
Soup	Bouillon, clear broth, fat-free vegetable soup, cream soup made with skim milk, packaged dehydrated soups	All others

(continued)

	Food List (Continued)	
Food group	Allowed	Avoid
Dessert	Fruit ices (water ices), angel food cake including angel food cake mix, puddings made with skim milk, "jello," frostings made with allowed fat, meringues; cakes, cookies, and pies made with allowed fats and skim milk (no egg yolks); fruit whips, junkets made with skim milk	Desserts that contain whole milk, saturated or hydrogenated fat, and egg yolks, including commercial pies, cakes, and cookies; all cake and cookie mixes (except angle food cake mixes)
Fat	Safflower oil, corn oil, commercial mayonnaise and soft safflower margarine (liquid safflower oil—not hardened, partially hardened, or hydrogenated—should be the first listed ingredient (the first listed item on the label indicates the predominant ingredient)	Butter, lard, hydrogenated margarine and shortening, coconut oil, and all other oils not listed, salt pork, suet, bacon and meat drippings, gravies unless made with allowed polyunsaturated fat, sauces, such as cream sauce, etc., unless made with allowed fat and skim milk
Fruit	Any fresh, canned, frozen, or dried fruit or juice; avocado may be used in small amounts	None
Sweets	Hard candies, jam, jelly, honey, sugar, syrup containing no fat	All other candies or chocolate
Vegetable	Any fresh, frozen, or canned, cooked without saturated fat	Buttered, creamed, or fried vegetables unless prepared with allowed fat
Miscellaneous	Olives, pickles, salt, spices, herbs, nuts, except those excluded, and cocoa	Coconut, cashew and macadamia nuts

In order to get enough polyunsaturated fats into the diet, either foods cooked in vegetable oils or salad dressings containing vegetable oils must be eaten.

The following list of cholesterol content may be helpful, but it must be used with cognizance of the saturated-fat content. Two foods may have the same cholesterol content, but one can be high in saturated fat. Both factors must be considered.

mg Cholesterol/100 g Food[a]

Cereal products

Cakes

Fancy ice cakes	—
Fruit cake, rich	50
Iced	40
Plain	—
Gingerbread	60
Madeira cake	—
Rock cakes	40
Sponge cake	
With fat	260
Without fat	130
Jam-filled	—

Buns and pastries

Currant buns	—
Doughnuts	—
Eclairs	90
Jam tarts	—[b]
Mince pies	—[b]
Pastry	
Choux, raw	110
Choux, cooked	170
Flaky, raw	—[b]
Flaky, cooked	—[b]
Shortcrust, raw	—[b]
Shortcrust, cooked	—[b]
Scones	(5)[c]
Scotch pancakes	50

Puddings

Apple crumble	—[b]
Bread and butter pudding	100
Cheesecake	95
Christmas pudding	60
Custard, egg	100
Made with powder	16
Custard tart	60
Dumpling	8
Fruit pie, individual, with pastry top and bottom	—
Fruit pie with pastry top	—[b]
Ice cream, dairy	21
Ice cream, nondairy	11
Jelly packet, cubes	0
Made with water	0
Made with milk	6
Lemon meringue pie	90
Meringues	0
Milk pudding	15
Canned, rice	—

Puddings (Continued)

Pancakes	65
Queen of puddings	100
Sponge pudding, steamed	80
Suet pudding, steamed	4
Treacle tart	—[b]
Trifle	50
Yorkshire pudding	70

Milk, milk products, and eggs

Milk

Cows'	
Fresh, whole	14
Sterilized	14
Fresh, skimmed	2
Condensed, whole, sweetened	34
Condensed, skim, sweetened	3
Evaporated, whole, unsweetened	34
Dried, whole	120
Dried, skim	18
Goats'	—
Human, mature	16
Transitional	—
Butter, salted	230
Cream single	66
Whipping	100

Cheese

Cambembert	72
Cheddar	70
Danish blue	88
Edam	72
Parmesan	90
Stilton	120
Cottage	13
Cream	94
Processed	88
Spread	71

Yogurt (low-fat)

Natural	7
Flavored	7
Fruit	6
Hazlenut	7

Eggs

Whole, raw	450
White, raw	0
Yolk, raw	1260
Dried	1780

(continued)

mg Cholesterol/100 g Food[a] (Continued)

Milk, milk products, and eggs (Continued)		Meat (Continued)	
Eggs (Continued)		*Lamb (Continued)*	
Boiled	450	Raw, lean only	69
Fried	—d	Cooked, lean and fat	110
Poached	480	Cooked, lean only	110
Omelette	410	Chicken	
Scrambled	410	Raw, light meat	69
Egg and cheese dishes		Raw, dark meat	110
Cauliflower cheese	17	Boiled, light meat	80
Cheese pudding	130	Broiled, dark meat	110
Cheese souffle	180	Roast, light meat	74
Macaroni and cheese	20	Roast, dark meat	120
Pizza, cheese, and tomato	20	Duck	
Quiche Lorraine	130	Raw, meat only	110
Scotch egg	220	Roast, meat only	160
Welsh rabbit	67	Turkey	
Products containing eggs		Raw, light meat	49
Lemon curd, homemade	150	Raw, dark meat	81
Marzipan, homemade	35	Roast, light meat	62
Mayonnaise	260	Roast, dark meat	100
		Rabbit, raw	71
Fats and oils		Offal	
Compound cooking fat	—e	Brain	
Drippings, beef	(60)c	Calf, raw	2200
Lard	(70)	Calf, boiled	3100
Low-fat spread	Tr	Lamb, boiled	2200
Margarine	—e	Heart	
Suet, block	(60)	Lamb, raw	140
Shredded	74	Sheep, roast	260
Vegetable oils	Tr	Ox, raw	140
		Ox, stewed	230
Meat		Kidney	
Bacon		Lamb, raw	400
Raw, lean and fat	57	Lamb, fried	610
Raw, lean only	51	Ox, raw	400
Fried, lean and fat	80	Ox, stewed	690
Fried, lean only	87	Pig, raw	410
Grilled, lean and fat	74	Pig, stewed	700
Grilled, lean only	77	Liver	
Beef		Calf, raw	370
Raw, lean and fat	65	Calf, fried	330
Raw, lean only	59	Chicken, raw	380
Cooked, lean and fat	82	Chicken, fried	350
Cooked, lean only	82	Lamb, raw	430
Lamb		Lamb, fried	400
Raw, lean and fat	78	Ox, raw	270
Raw, lean only	79	Ox, stewed	240
Cooked, lean and fat	110	Pig, raw	260
Cooked, lean only	110	Pig, stewed	290
Pork		Oxtail, raw	75
Raw, lean and fat	72	Oxtail, stewed	110

(continued)

mg Cholesterol/100 g Food[a] (Continued)

Meat (Continued)			Meat Products and dishes (Continued)	
Offal (Continued)				
Sweetbread			Meat and pastry products (Continued)	
Lamb, raw	260		Sausage roll, flaky pastry	20
Lamb, fried	380		Sausage roll, short pastry	30
Tongue			Steak and kidney pie, pastry top only	125
Lamb, raw	180			
Sheep, stewed	(270)		Cooked dishes	
Ox, pickled, raw	78		Beef steak pudding	30
Ox, pickled, boiled	(100)		Beef stew	30
Tripe, dressed	95		Bolognese sauce	25
Tripe, stewed	160		Curried meat	25
			Hot pot	25
Meat products and dishes			Irish stew	35
Canned meats			Moussaka	40
Beef, corned	85		Shepherd's pie	25
Ham	33			
Ham and pork, chopped	60		Fish	
Luncheon meat	53		White fish	
Stewed steak with gravy	44		Cod, fresh	
Tongue	110		raw	50
Veal, jellied	97		Baked	60
			Fried	—
Offal products			Grilled	60
Black pudding, fried	68		Poached	60
Faggots	79		Steamed	60
Haggis, boiled	91		Cod, smoked	
Liver sausage	120		Raw	(50)
			Poached	(60)
Sausages			Haddock, fresh	
Frankfurters	46		Raw	60
Polony	40		Fried	—
Salami	79		Steamed	75
Sausages			Haddock, smoked	
Beef, raw	40		Steamed	(75)
Beef, fried	42		Halibut	
Beef, grilled	42		Raw	50
Pork, raw	47		Steamed	60
Pork, fried	53		Lemon sole	
Pork, grilled	53		Raw	60
Saveloy	45		Fried	—
			Steamed	60
Other products			Plaice	
Beefburgers, frozen, raw	59		Raw	70
Beefburgers, frozen, fried	68		Fried in batter	—
Brawn	52		Fried in crumbs	—
Meat paste	68		Steamed	90
White pudding	22		Saithe	
			Raw	60
Meat and pastry products			Steamed	75
Cornish pastie	49		Whiting, steamed	110
Pork pie, individual	52			

(continued)

mg Cholesterol/100 g Food[a] (Continued)

Fish (Continued)		Fish (Continued)	
Fatty fish		Crustacea	
Eel		Crab	
Raw	—	Fresh	100
Stewed	—	Canned	100
Herring		Lobster	150
Raw	70	Prawns	200
Fried	(80)	Scampi	110
Grilled	80	Shrimp	200
Bloater, grilled	80		
Kipper, baked	80	Mollusks[f]	
Mackerel		Cockles	40
Raw	80	Mussels, raw	100
Fried	(90)	Oysters, raw	50
Pilchards, canned in		Scallops, raw	40
tomato sauce	(70)	Whelks	100
Salmon		Winkles	100
Raw	(70)		
Steamed	(80)	Fish products and dishes	
Canned	90	Fishs cakes	
Smoked	(70)	Frozen	—
Sardines		Fried	—
Canned in oil		Fish fingers	
Fish only	100	Frozen	(50)
Fish plus oil	80	Fried	(50)
Canned in tomato sauce	(100)	Fish paste	—
Sprats, fried	—	Fish pie	20
Trout, steamed	(80)	Kedgeree	120
Tuna, canned in oil	65	Roe, cod, hard	
		Raw	(500)
		Fried	(500)
Cartilaginous fish		Roe, herring, soft	
Dogfish	—	Raw	700
Skate	—	Fried	(700)

[a]Adapted from Ref. 2.
[b]Trace if made with vegetable fats.
[c]Values in parenthess from the literature.
[d]Cholesterol content depends on the fat used for frying.
[e]Cholesterol content depends on the blend of oils used; only a trace if vegetable oils used.
[f]Other sterols are present in these fish; cholesterol forms about 40% of the total sterols in cockles, mussels, oysters, and scallops. It forms about 90% of the total sterols in whelks and about 70% in winkles.

20.25.2.   Low-Carbohydrate, Low-Cholesterol, Modified-Fat (Increased Polyunsaturate) Diet[11]

This diet is used to treat type IV hyperlipoproteinemia. The carbohydrate intake is limited to approximately 40% of calories, while cholesterol intake is maintained between 300 and 500 mg, and polyunsaturates are encouraged. The diets are modified from the NIH[13] handbook.

Suggested Food Distribution at Various Caloric Levels							
Calories	1500	1800	2000	2200	2400	2600	2800
Carbohydrate (g)	135	180	195	210	225	255	285
Food groups							
Skim milk, servings	2	2	3	3	3	3	3
Breads, cereals, etc., servings	5	8	8	9	10	12	14
Fruit, servings[a]	3	3	3	3	3	3	3

These foods are not limited:
  Lean, well-trimmed meat, fish, and poultry[b]
  Vegetables except those grouped with breads and cereals
  Unsaturated fat

[a]Ten grams of carbohydrate has been calculated for 1/2 cup of any fresh or unsweetened fruit or juice.
[b]Eight grams of protein and three grams of fat have been calculated for one ounce of meat, fish, or poultry.

Sample 1800-cal Menu	
Breakfast:	1 servings citrus fruit or juice
	2 servings cereal or toast and allowed fat
	1 serving skim milk
Lunch:	Unlimited serving poultry, fish, or lean-trimmed meat
	1 serving potato or substitute vegetable
	2 servings bread and allowed fat
	1 serving fruit
Dinner:	Unlimited serving poultry, fish, or lean-trimmed meat
	1 serving potato or substitute vegetable
	2 servings bread and allowed fat
	1 serving fruit
	1 serving skim milk

Food Lists		
	Allowed	Avoid
Meat, fish, and poultry	Select lean, well-trimmed cuts Beef Lamb Pork or lean ham Veal Poultry Fish; if canned, drain oil Specially prepared low-fat cheese Creamed cottage cheese Egg whites	All fatty meats Bacon and sausage Canned meat products Corned beef Duck and goose Fish roe including caviar Fried meats and fried fish Frozen and packaged dinners Frozen and packaged prepared products (convenience foods) Luncheon meats, cold cuts, and hot dogs Meats canned or frozen in gravy or sauce Poultry skin Regular ground beef or hamburger Spareribs Egg yolk Cheese, Organ meats Shellfish Except as specified
Milk and eggs	You may use 3 egg yolks per week (this includes egg yolks used in cooking) Use low-fat or skim milk, buttermilk, yogurt made from skim milk	Whole milk Condensed milk Dried whole milk Evaporated whole milk Whole-milk drinks Commercial whipped toppings Cream (sweet or sour) Cream substitutes Ice cream and ice milk
Breads, cereals, and starch vegetables	Any one of the following is a serving: 1 slice of whole-grained or enriched bread ³/₄ oz bread sticks, rye wafers, or pretzels 2 graham crackers (2¹/₂-in. sq) 3 soda crackers (2¹/₂-in. sq) 20 oyster crackers (¹/₂ cup) ¹/₂ hamburger or hot dog roll 1 small or ¹/₂ large hard roll 1 piece matzo (6-in. sq) 4 pieces melba toast (3¹/₂ × 1¹/₂ × ¹/₈) 3 tbsp flour or cornmeal or dry grated bread crumbs ¹/₃ cup corn or × ¹/₂ large ear of corn on the cob	Biscuits Butter rolls Cheese bread and rolls Egg breads Flavored crackers French toast Muffins Pancakes Sweet rolls Waffles Corn chips Potato chips Pork and beans Soups made with milk All sweets Candies Chocolate

(continued)

Food Lists (Continued)

	Allowed	Avoid
Breads, cereals, and starch vegetables *(Cont.)*	½ cup cooked, dried peas, beans, lentils, chick peas, or lima beans 1 small white baked potato or ½ cup mashed ½ cup cooked spaghetti, rice, or noodles ½ cup (cooked) cereal ¾ cup ready-to-eat (dry) cereal 1½ cups popcorn, popped (no fat) 3½ tbsp tomato catsup 1½-in. cube angle food cake ⅓ cup "jello" ¼ cup sherbert or fruit ice ½ cup plain pudding, cornstarch, or tapioca (prepared with skim milk)	Cakes, including mixes, except angel food cake Jams and jellies Pies, including mixes Sweetened frozen or canned beverages Syrups, honey, sugar All other desserts, except those allowed
Fats and oils	Avocados Nuts Olives Peanut butter All vegetable oil except coconut oil Commercial salad dressings containing no sour cream "Specially prepared" margarines (made from one of the allowed oils)	Bacon and meat drippings Butter and products made with butter Coconut Coconut oil Gravies, except those containing no animal fat as specified elsewhere Solid vegetable shortenings (hydrogenated) Lard, salt pork, or suet Sauces, such as cream sauce, etc., unless made with allowed margarine or oil
Fruits	One serving of citris fruit is recommended daily to provide adequate vitamin C One serving is ½ cup of any fresh, unsweetened canned or frozen fruit, and unsweetened juice OR ½ cup cooked, unsweetened dried fruit	Sweetened fruit and juice
Vegetables	Any fresh, frozen, or canned without added fat or sauces One dark-green or deep-yellow vegetable is recommended daily to provide adequate vitamin A Most vegetables are not limited; however, it is expected that they be eaten in normal amounts, that is, medium-size servings should be selected Potatoes, corn, lima beans, dried peas, and beans are limited as designated in the breads and cereals group	Buttered, creamed, or fried vegetables, unless prepared with allowed fat

(continued)

Food Lists (Continued)	
Allowed	**Avoid**
Beverages Use skim milk or fruit or vegetable juices as allowed in diet; coffee, tea, and noncarbohydrate carbonated drinks	Alcohol, except on occasion in small amounts; all carbohydrate carbonated beverages; milk
Miscellaneous Bouillon (without fat) Clear broth Club soda Coffee Flavoring essence Gelatin, unflavored Herbs Mustard Rennet tablets Soy sauce Spices Tea Vinegar Worcestershire sauce	

20.26. HIGH-FIBER, HIGH-CARBOHYDRATE, LOW-CHOLESTEROL, LOW-FAT DIET

This diet regimen is reported to be of benefit in stabilizing glucose tolerance (Chapter 12), correcting constipation (Chapter 14), and altering serum lipids to a better antiatherosclerotic spectrum (Chapters 4, 11, and 14). It is also theorized to prevent other degenerative diseases of Western society (Chapter 4).

The diet below is from the Lexington Veterans Administration Hospital and uses oat bran, which may be difficult to obtain.[11] Anderson and Sieling[11] claim that oat bran is more effective in reducing cholesterol and elevating high-density lipoprotein.

Approximate Composition	
Calories	1800
Carbohydrate (simple)	63 g
(total)	249 g
Protein	91 g
Fat	50 g
Fiber[a] (soluble)	29 g
(total plant)	68 g
Cholesterol	35 mg

[a]Calculated from Ref. 12.

Sample Menu			
Breakfast		*Dinner*	
Oat muffins	102 g	Roast beef	45 g
Oat meal (dry)	40 g	Corn (cooked)	200 g
Skim milk	120 g	Broccoli (cooked)	200 g
Egg beaters	80 g	Oat muffin	100 g
Margarine	10 g	Margarine	5 g
1 Multivitamin			
		Evening snack	
Lunch		Skim milk	120 g
Potatoes, boiled	250 g	Corn flakes	40 g
with margarine	7 g		
Peas (cooked)	200 g		
Carrots sticks	50 g		
Orange	100 g		
Oat muffin	51 g		
Margarine	5 g		

Exchanges and increases in intake can be determined from Tables 20-1 to 20-7 and from Refs. 11 and 12.

20.27. HIGH-FIBER DIET

High-fiber diets maintain a larger stool volume than low-fiber diets. For this reason, they are effective in correcting constipation due to low fecal volume (Chapter 14). Claims have also been made for the diet's effectiveness in treating diverticular disease (Chapter 15) and the irritable bowel syndrome (Chapter 16). Some have found it helpful in treating the loose stool of inflammatory bowel disease (Chapter 15). Trials of its effectiveness as a preventative for other degenerative diseases will be difficult to obtain. The low-fat, high-fiber diet is another approach to prevention of degenerative disease (Diet 26) that awaits trial results.

All fiber substances vary in their physical properties. Increasing stool volume is best achieved by those that hold onto water and are poorly digested by bacteria (cellulose and some hemicelluloses, bran, maize, etc., see Chapter 4).

20.27.1. Whole-Bran Supplement

A significant increase in stool weight can be accomplished by taking 2 tbsp of whole bran daily. Patients must be cautioned to add it gradually to the diet. Suddenly taking 2 full or heaping tablespoons can cause bloating.

20.27.2. Cereal and Bread Supplement

The same effect can be accomplished by using 3 oz of commercial cereal (All-Bran, Kellogg's), equivalent to 5.4 g of crude fiber. It is difficult to eat so much of this cereal, but the desired effect can be accomplished by adding both cereals and breads made with high fiber. Approximately a total of 25 g of dietary fiber or 10 g of crude fiber (see food lists) should be eaten daily to significantly increase stool output. This could be reached by eating the following for breakfast.

Sample Breakfast		
Food	Amount (approx.)	Dietary fiber (g)
All-bran cereal	50 mg (1 $\frac{1}{2}$ oz)	13.4
Wholemeal bread	3 slices	6.3
Almonds	10 mg ($\frac{1}{3}$ oz)	1.4
Dried prunes	30 mg (1 oz)	4.8
		25.9

20.27.3. Fruit and Vegetable High-Fiber Gluten-Free Diet Exclusive of Cereals and Grains

Because this diet was designed for men, the quantities should be reduced for women and light men to achieve the caloric intake that maintains stable weight for that individual.

This diet was effective in a controlled clinical trial. It contains no cereals or bran. If desired or permitted, those could be added to replace any foods not tolerated. The diet is relatively high in fats and those could be reduced to save calories (e.g., bacon, butter, whole milk).

Approximate Composition [a]	
Calories	2800
Carbohydrate	343–348 (50%)
Sugar	266–270
Starch	73–82
Protein	95-96 (14%)
Fat	112–116 (36%)

[a]Diet modified from Ref. 23, Chapter 4.

Suggested Eating Plan

Breakfast

Grapefruit, 200 g
Puffed rice, 7.5 g
Dates, 80 g
Milk, 244 g
Bread, white, 25 g
Egg, 50 g
Butter, 43.2 g[a]
Half and half, 90 g[a]
Sugar, 60 g[a]

Lunch

Roast beef, 85 g
Bread, white, 25 g
Corn, 82.5 g
Pineapple tidbits, 123 g
Pudding, 130 g

Dinner

Ham, 85 g
Spinach, 102.5 g
Carrots, 55 g
Cabbage, 45 g
Dressing, 48 g[c]
Roll, white, 28 g
Angel cake, 53 g
Blackberries, 252 g
Milk, 122 g

[a]Amount for entire day.
[b]Cooked weight.
[c]Dressing for salad; used on macaroni with low-fiber diet.

An Alternative Plan

Breakfast	*Lunch*
Oranges, 200 g	Tuna fish, 78 g
Milk, 244 g	Mayonnaise, 28 g
Bread, white, 25 g	Bread, white, 50 g
Bacon, 15 g[b]	Peas, 85 g
Butter, 28.2 g[a]	Apple, 160 g
Half and half, 90 g[a]	Raisins, 72.5 g
Sugar, 72 g[a]	

Dinner

Ground beef, 113 g
Broccoli, 92.5 g
Squash, 120 g
Lettuce, 30 g
Tomato, 100 g
Dressing, 32 g[c]
Roll, white, 28 g
Blueberries, 284 g
Ice cream, 66 g
Milk, 122

[a]Amount for entire day.
[b]Cooked weight.
[c]Dressing for salad; used on macaroni with low-fiber diet.

To save calories the butter, milk, sugar, and bacon could be reduced.

Wholemeal bread or cereal could be substituted as desired to replace fruits and vegetables, if permitted; however, *not in gluten enteropathy.*

20.28.   FIBER-FREE LIQUID DIET

Fiber increases intestinal activity. A fiber-free diet will rest the lower intestine. This is used short-term during acute illness and after surgery. When used, it should be supplemented if the clinical condition indicates an immediate need for energy. If this diet is needed long term, then elemental diets are recommended (see Chapter 18) to supply the RDAs. The fiber-free diet implies no residue.

Fiber-Free Liquid Foods

Coffee
Tea
Carbonated beverages, except those made with fruit, milk, or chocolate
Fruit juices that are clear and free of any pulp— they must be clear extracts
Soups made as clear extracts such as broth, bouillon, clam juice, etc.
Sweets such as honey, sugar cubes, and hard clear candies that can be sucked
Desserts such as Jello without additives other than flavor; ices; sherbet made without milk

20.29. LOW-FIBER BLAND DIET (PREVIOUSLY LOW-RESIDUE)

These diets, previously called low-residue diets, are used when there is active disease in the intestinal tract or rectum and are recommended for inflammatory bowel diseases such as colitis, ileitis, and diverticulitis (see Chapter 15). Fiber increases both gut content and bowel and rectal activity. Eliminating fiber decreases intestinal activity and largely rests the rectum. This diet can be an extension of a liquid fiber-free diet. The amount of fiber in the diet can be varied; an increase will increase the amount of feces.

Two diets are recommended. The first has less than 1 g of crude fiber and the second less than 4 g. Vitamins and minerals should be added.

I. Low-Fiber (1 g Crude Fiber) Diet: Approximate Composition[a]	
Calories	2400
Carbohydrate	225 g
Protein	100 g
Fat	120 g

[a]Modified from Ref. 4.

Suggested Eating Plan			
	Breakfast	Lunch	Dinner
Protein exchange	2	4	5
Bread exchange[a]	2	2	3
Fat exchange	4 tsp	4	4
Potato	—	1	1
Dessert	—	1	1
Sweets, jellies	2 tbsp	1	1
Beverages as tolerated			

[a]Only breads made from refined flour should be used.

Food List		
	Allowed	Avoid
Beverage	Coffee, tea, carbonated beverage; avoid large amounts; avoid caffeinated coffee and colas	Alcohol
Meat	Meat, fish, fowl, egg, cheese	
Fat	All others	Avocado, nuts, olives, salad dressings
Milk[a]	A small amount used in cooking ($1/2$ cup or less)	Milk, milk drinks
Bread, cereals	Refined wheat or rye bread, quick breads, rolls, or crackers Refined cereals: cooked or ready to eat Potato and potato substitutes except "avoid"	Any with graham, *whole-grain* flour, *bran, seeds, nuts* Whole-grain or bran cereals Whole-grain rice, dried legumes, popcorn, corn, lima beans, peas
Vegetable	None	All
Fruit	None	All
Soup	Bouillon, broth, clear clam	All others
Dessert	All others; puddings and ice cream within milk allowance	Any with coconut, nuts, or *fruits*
Sweets	All others	Jam, marmalade; candy with coconut, nuts, or fruit
Miscellaneous	Salt	Garlic, pickles, cocoa, all spices and herbs

[a]Milk is usually discouraged because of the high incidence of lactose intolerance in inflammatory bowel disease, but when a patient is not lactose intolerant it can be used liberally.

II. *Low-Fiber (4 g Crude Fiber) Diet: Approximate Composition*[a]	
Calories	2200
Carbohydrate	340 g
Protein	110 g
Fat	55 g

[a]Modified from Carnation Co. Residue Controlled Diet Plan.

Suggested Eating Plan

Breakfast

Fruit juice	$1/2$ cup orange juice
Egg	1 poached egg
Bread	2 slices white toast
Sweets	4 tsp jelly
Sweet (optional)	1 tsp honey
Beverage	Coffee or tea (decaffeinated)

Snack

Milk beverage or substitute	1 cup
Fruit	1 small banana

Lunch

Lean meat or substitute	2 oz cheese, grilled with
Bread	2 slices white bread
Fat	1 tsp margarine
Vegetable	6 oz tomato juice
Fruit	$1/2$ cup canned pears

Snack

Milk beverage or substitute	1 cup nonfat milk
Sweets	2 sugar cookies

Dinner

Fruit or vegetable	1 cup fruit cocktail packed in water
Lean meat or substitute	4 oz roast turkey
Potato or substitute	$1/2$ cup white, enriched rice
Vegetable	$1/2$ cup pureed asparagus
Bread	1 hard roll
Fat	2 tsp margarine
Dessert	1 slice angel food cake topped with icing
Beverage	Coffee

Food Lists		
	Allowed	Avoid
Eggs	Prepared in any manner	
Meat and meat substitutes	Roasted, baked, or broiled beef, chicken, turkey, bacon, ham, lamb, veal, or fish; cottage, cream, and mild American cheese	Highly seasoned or highly salted meats and fish; fatty meats, luncheon meats, frankfurters (crisp bacon is allowed in moderate amounts); fried meats or shellfish with tough connective tissues; highly flavored cheese
Vegetables, fruits	Those low in crude fiber as listed in Table 20-7; those with less than 1 g fiber 100 g	Those high in fiber (see Table 20-7)
Soup	Bouillon, broth, strained cream soups made from allowed foods	All others
Fats	Butter, margarine, oils	Highly spiced salad dressings
Breads, cereals	Refined bread, toast, rolls, and crackers; those made with refined flour, cereals with less than 1 g fiber/ 100 g	Any with whole-grain or graham flour, bran, seeds, or nuts; whole-grain or bran cereals; whole-grain rice, dried legumes, popcorn
Desserts	Gelatin desserts, tapioca, angel food or sponge cake, plain custard, ice cream without fruit or nuts, cookies, pudding	All made with coconut, nuts, or fruits
Sweets	Sugar, honey, jelly, syrups, hard candy, milk chocolate	Jam, marmalade, candy with coconut, nuts, or whole fruits
Beverages	Milk as directed by physician or dietician; coffee, tea, carbonated beverages	Alcohol
Miscellaneous	Salt, plain gravy	Garlic, seed spices, pickles, olives, nuts, pepper, chili powder, popcorn, rich gravies, vinegar

20.30. ORAL GLUCOSE–ELECTROLYTE LIQUID DIET

Severe secretory diarrhea (Chapter 14) can be treated successfully with glucose–electrolyte solutions. They reverse the net secretion of fluid because glucose activates absorption, which decreases water secretion.

The following formulas have been used successfully.

World Health Organization[a]

1000 ml H_2O
3.5 g NaCl (salt)
2.5 g $NaHCO_3$ (baking soda)
1.5 g KCl (potassium chloride)
20.0 g glucose (2%)
Yields an electrolyte concentration of:

Na	90 mEq/liter
K	20 mEq/liter
Cl	80 mEq/liter
HCO_3	30 mEq/liter
Glucose	111mM

[a]From Ref. 38, Chapter 14.

GE-SOL[a]

1000 ml H_2O
1/2 tsp NaCl
1/2 tsp $NaHCO_3$
1/4 tsp KCl
2 tbsp glucose
Yields an electrolyte concentration of:

Na	81 mM
K	18 mM
Cl	71 mM
HCO_3	28 mM
Glucose	139 mM
	(within a range of 5%)

[a]From Ref. 39, Chapter 14.

Phillips Fluid[a,b]

1 tsp	(4 g)	NaCl (table salt)
1/2 tsp	(2 g)	NaHCO$_3$ (baking soda)
3 tsp	(12 g)	K bitartrate (cream of tartar)
2 tbsp	(30 mg)	glucose (Karo syrup)
1 quart	(1000 g)	H$_2$O

Yields an electrolyte concentration of:

Na	92 mM
K	18 mM
Cl	68 mM
HCO$_3$	24 mM
Glucose	125 mM

[a]*Am J Clin Nutr* 28:804, 1975.

[b]The authors stress that all of these ingredients are available commonly in a kitchen or local grocery. They use 4 tsp equal to 12 g and 1 tbsp equal to 30 g, which I have changed to 3 tsp and 2 tbsp to reflect more accurate measurements. The same general looseness exists in the formula with 1 quart published as equal to 1000 g, but this points out that the final dilutions need not be exact, can be made in a household kitchen, and can be used safely with variation in dilutions as long as they are within a general range.

The dosages for children and adults are determined by the clinicians. The goal is to combat small-bowel secretion of fluid by glucose absorption and thereby rehydrate and stop diarrhea. The amount and rate of oral administration depend on the degree of dehydration and diarrhea. A sick adult may consume as much as 750 to 1000 ml/h in extreme situations.

20.31.   ROWE ELIMINATION DIET (Modified from Rowe[14])

These diets are used to diagnose allergic disorders (see Chapter 17).

Cereal-Free Elimination Diet

Only Foods Allowed

Tapioca (pearl or minute)	Lettuce
White potatoes	Lima beans
Sweet potatoes	
Breads made with any	Grapefruit
combination of soy, lima,	Lemon
potato starch, and tapioca	Peach
flour	Pineapple
	Prunes
Soy milk: Mull-Soy and Neo-	Pear
Mull Soy[a]	
	Cane or beet sugar
Lamb	Salt
Chicken, fryers, rooster,	Sesame oil (not Chinese)
capon (no hens)	Soybean oil
Bacon	Margarine[b]
Liver (lamb, chicken)	Gelatin (Knox)
	White vinegar
Peas	Vanilla extract
Spinach	Lemon extract
Squash	Cornstarch-free baking powder
String beans	Baking soda
Tomatoes	Cream of tartar
Artichokes	
Asparagus	Maple syrup or syrup made with
Carrots	cane sugar flavored with maple

Fruit-Free, Cereal-Free Elimination Diet

Only Foods Allowed

Tapioca (pearl only)

White potatoes

Sweet potatoes

Breads made with any
   combination of soy, lima,
   potato starch, and tapioca
   flour

Soy milk: Mull-Soy and Neo-
   Mull-Soy[a]

Lamb

Chicken, fryers, rooster,
   capon (no hens)

Bacon

Liver (lamb, chicken)

Cooked carrots

Squash

Artichokes

Peas

Lima beans

String beans

Cane or beet sugar

Soybean oil

Margarine[b]

Gelatin (Knox)

Salt

Syrup made from cane sugar (no maple syrup)

Corn-free, tartaric-acid-free
   baking powder

[a]Free of corn syrup solids and corn oil (Syntex Laboratories, Palo Alto, Calif. 94304).

[b]Free of milk solids and corn oil.

20.32.   MAY ELIMINATION DIET

See Chapter 17 for use of this diet in the diagnosis of allergic disorders.

Food Lists

Foods allowed at mealtime

Rice
Puffed rice
Rice flakes
Rice Krispies
Pineapples ⎫
Apricots   ⎪
Cranberries ⎬ Also juice of
Peaches   ⎪   these
Pears   ⎭

Lamb
Chicken
Asparagus
Beets
Carrots
Lettuce
Sweet potato

Tapioca

White vinegar
Olive oil

Honey (2 oz/day
Cane sugar
Salt
Oleomargarine without milk
  (Mazola margarine)
Crisco, Spry

Bubble Up (a carbonated dye-free beverage)

Snacks

1 box rice cereal
  midmorning and
  midafternoon

Avoid

Any food and drink
  suspected to cause
  reactions or not on
  this list
Pepper and spices
Coffee
Tea
Chewing gum

REFERENCES

1. Church CF, Church HN: *Food Values of Portions Commonly Used*, ed 12. Philadelphia: Lippincott, 1975.
2. Paul AA, Southgate DAT: *McCance and Widdowson's The Composition of Foods*, ed 4. New York: Elsevier/North-Holland Biomedical Press, 1978.
3. Watt BK, Merrill AL: *Composition of Foods*. Agriculture Handbook No. 8. Washington, DC: Superintendent of Documents, U.S. Government Printing Office.
4. *Mayo Clinic Diet Manual*, ed 4. Philadelphia: Saunders, 1971.
5. Senior JR: *Medium Chain Triglycerides*. Philadelphia: University of Pennsylvania Press, 1968.
6. *Manual of Surgical Nutrition*. Philadelphia: Saunders, 1975.
7. Wood MN: *Gourmet Food on a Wheat-Free Diet*. Springfield: Charles C Thomas, 1967.
8. Sheedy CB, Keifetz N: *Cooking for Your Celiac Child*. New York: Dial Press, 1969.
9. Gryboski J: *Gastrointestinal Problems in the Infant*. Philadelphia: Saunders, 1975.
10. Francis DEM: *Diets for Sick Children*, ed 3. Oxford: Blackwell Scientific Publications, 1974.
11. Anderson JW, Sieling B: *HFC Diets: A Professional Guide to High-Carbohydrate, High-Fiber Diets*. Lexington: The University of Kentucky Diabetes Fund, 1979.
12. Anderson JW, Lin W–J, Ward K: Composition of foods commonly used in diets for persons with diabetes. *Diabetes Care* 1:293, 1978.
13. *Dietary Management of Hyperlipoproteinemia: A Handbook for Physicians*. Bethesda: National Heart and Lung Institute, 1970.
14. Rowe AH: *Food Allergy*. Springfield: Charles C Thomas, 1972.

Index